Low Temperature Materials and Mechanisms

Low Temperature Materials and Mechanisms

edited by
Yoseph Bar-Cohen

CRC Press
Taylor & Francis Group
Boca Raton London New York

CRC Press is an imprint of the
Taylor & Francis Group, an **informa** business

CRC Press
Taylor & Francis Group
6000 Broken Sound Parkway NW, Suite 300
Boca Raton, FL 33487-2742

First issued in paperback 2019

© 2016 by Taylor & Francis Group, LLC
CRC Press is an imprint of Taylor & Francis Group, an Informa business

No claim to original U.S. Government works

ISBN-13: 978-1-4987-0038-2 (hbk)
ISBN-13: 978-0-367-87134-5 (pbk)

Library of Congress Cataloging-in-Publication Data

Names: Bar-Cohen, Yoseph, editor.
Title: Low temperature materials and mechanisms / editor, Yosef Bar-Cohen.
Description: Boca Raton : Taylor & Francis, a CRC title, part of the Taylor & Francis imprint, a member of the Taylor & Francis Group, the academic division of T&F Informa, plc, [2016] | Includes bibliographical references and index.
Identifiers: LCCN 2016000772 | ISBN 9781498700382 (alk. paper)
Subjects: LCSH: Materials at low temperatures. | Machinery--Effect of environment on. | Low temperature engineering.
Classification: LCC TA418.28 .L69 2016 | DDC 621.5/6--dc23
LC record available at http://lccn.loc.gov/2016000772

Visit the Taylor & Francis Web site at
http://www.taylorandfrancis.com

and the CRC Press Web site at
http://www.crcpress.com

Contents

Preface

Operating at cold temperatures is essential to many science and engineering processes including refrigeration, space exploration, electronics, physics, chemistry, thermodynamics, and medicine. For thousands of years, humans have used temperatures near the freezing point of water and taken advantage of the resulting benefits. Over the last two centuries, it has become feasible to reach cryogenic temperatures as a result of the development of effective coolers. The production of cryogenic temperatures involves several challenges and disadvantages and, to be useful or marketable, applications need to have benefits that outweigh the disadvantages.

Generally, cryogenics is the science and engineering of creating and investigating low-temperature conditions. The word cryogenics is derived from the Greek words *kryos*, meaning "frost" or "cold," and *–genic*, meaning "to produce." Until around the mid-eighteenth century, it was believed that some gases could not be liquefied and were named permanent gases. These gases were carbon monoxide, hydrogen, methane, nitric oxide, nitrogen, and oxygen, with the most attention given to the primary constituents of air—oxygen and nitrogen. While cold temperatures can include temperatures at the level of water freezing (0°C), the term "cryogenic environment" refers to the temperature range below the point at which the permanent gases begin to liquefy. The term was applied initially to temperatures from approximately –100°C (–148°F) down to absolute zero Kelvin, and today the definition refers to temperatures below approximately –150°C (–238°F). In 1894, Kamerlingh Onnes, of the University of Leiden, Netherlands, was the first to coin the term "cryogenics" as the science of producing very low temperatures to liquefy the gases that were considered permanent. The English physicist Michael Faraday was one of the most successful in liquefying permanent gases; by 1845, he had managed to liquefy most of the permanent gases known at that time. He liquefied the gases by cooling them via immersion in a bath of ether and dry ice and then pressurizing until they reached the liquid state. The liquefaction of air was accomplished in 1877 by Louis Cailletet in France and by Raoul Pictet in Switzerland. In 1898, the liquefaction of hydrogen was successfully accomplished by the Scottish chemist James Dewar. In 1908, the last remaining gas element, helium, was successfully liquefied at 4.2 K by the Nobel laureate Dutch physicist Heike Kamerlingh Onnes.

Increasingly, there is a growing interest in technologies that are applicable at low temperatures for planetary exploration of bodies in the solar system that are extremely cold. These include potential NASA *in situ* exploration missions to Europa and Titan where the ambient temperatures are around –200°C. Elsewhere, as a method of slowing or halting chemical and biological processes, cooling is widely used to preserve food and chemicals, as well as biological tissues, organs, and embryos. Furthermore, cooling is used to increase electrical conductivity, leading to superconductors that enable such applications as levitation, highly efficient electromagnets, etc. The subject of low temperature materials and mechanisms is multidisciplinary and includes chemistry, material science, electrical engineering, mechanical engineering, metallurgy, and physics.

The temperature scale that is based on absolute zero with units in the same size as the Celsius degree is known as units Kelvin and is abbreviated as K. This symbol of the Kelvin unit was adopted in 1968. In the absolute, or Kelvin, scale, the lowest temperature is written as 0 K (without a degree sign). The Rankine scale with the symbol R represents the English absolute scale and has the same unit increments as the Fahrenheit scale.

The temperature of a material is a measure of the energy that it contains, which occurs in various forms of motion among the atoms or molecules that make up the material. A gas with a lower temperature has slower-moving atoms and molecules than gases at higher temperatures do. The possibility that a material can have a state at which all forms of motion are halted was predicted in 1848 by the English physicist William Thomson, later known as Lord Kelvin. The absence of all forms of motion would result in a complete absence of heat and temperature; this condition was defined by Thomson as absolute zero, which is equal to −273.15°C (−459.67°F). With the advent of quantum mechanics in the early twentieth century, we now know that at absolute zero some atomic motion and energy still exists in materials, but it is in the lowest possible energy state, known as the ground state. Much of low temperature physics involves the study of materials close to the ground state where unusual behavior can occur.

The measurement of cryogenic temperatures requires methods that are not commonly used in daily life, which are mostly based on the use of mercury or alcohol thermometers, since these two fluids freeze and become useless at such low temperatures. Platinum resistance thermometers are based on the relation between electrical resistance and temperature, and these provide relatively accurate measurements down to about 20 K. For temperatures down to 1 K and below, measuring the electrical resistance of certain semiconducting materials, such as doped germanium, also provides temperature gauging, but these thermometers require calibration over the range of temperatures at which the measurements are needed. The calibration of such secondary thermometers is done by calibration against the primary ones and they are based on measuring physical variable changes for which the response is determined theoretically.

Chapter 1 of this book provides an introduction to the physical parameters, methods of generating cryogenic temperatures, the measurement of temperature, and applications of low temperatures. The other chapters of this book cover some of the key aspects of the field including chemistry and thermodynamics (Chapter 2); materials science of solids and fluids (Chapter 3); characterization methods (Chapter 4) as well as nondestructive testing and health monitoring (Chapter 5); the methods of cooling to cryogenic temperatures (Chapter 6); actuation materials and mechanisms (Chapters 7 and 8); instruments for planetary exploration (Chapter 9); methods of drilling in ice (Chapter 10); applications to medicine and biology (Chapter 11); low temperature electronics (Chapter 12); and applications to the field of physics (Chapter 13), as well other applications and the challenges in the field (Chapter 14). Given the health hazards that are associated with working at cryogenic temperatures, a chapter has been dedicated to this topic as well (Chapter 15).

Acknowledgments

Some of the research reported in this book was conducted at the Jet Propulsion Laboratory (JPL), California Institute of Technology (Caltech), under a contract with the National Aeronautics and Space Administration (NASA).

Yoseph Bar-Cohen, PhD
Jet Propulsion Laboratory (JPL)/Caltech,
Pasadena, CA

Acknowledgment of the Reviewers

The Editor would like to express his appreciation of the many scientists and engineers who contributed to the reported technologies and also would like to express a special thanks to the reviewers of each of the chapters as follows:

Chapter 1
Paul E. Hintze, NASA Kennedy Space Center, FL
Michael Mischna, JPL/Caltech, Pasadena, CA
Colin H. Smith, JPL/Caltech, Pasadena, CA
Robert Youngquist, NASA Kennedy Space Center, Merritt Island, FL

Chapter 2
Will Grundy, Lowell Observatory, Flagstaff, AZ
Victoria (Maria) Iglesias, JPL/Caltech, Pasadena, CA
Gabriel Tobie, Université de Nantes, Nantes, France

Chapter 3
Jeffrey D. Hein, JPL/Caltech, Pasadena, CA
Robert P. Hodyss, JPL/Caltech, Pasadena, CA
Victoria Munoz-Iglesias, JPL/Caltech, Pasadena, CA

Chapter 4
Arvind Agarwal, Florida International University, Miami, FL
Donald Lewis, JPL/Caltech, Pasadena, CA

Chapter 5
Dave Arms, Level III NDT Inc., Santa Clara, CA
Tod Sloan, University of Colorado School of Medicine, and University of Texas at Dallas, School of Behavior and Brain Sciences
Russell (Buzz) Wincheski, NASA Langley Research Center, Hampton, VA

Chapter 6
Dean Johnson, JPL/Caltech, Pasadena, CA
Saul Miller, The Aerospace Corp., El Segundo, CA
Jeff Raab, Northrop Grumman Aerospace Systems—retired, Redondo Beach, CA
Chao Wang, Cryomech, Syracuse, NY

Chapter 7
Zhenxiang Cheng, Institute for Superconducting and Electronic Materials, University of
    Wollongong, Australia
Dragan Damjanovic, Ecolepolytechniquefédérale de Lausanne (EPFL), Lausanne,
    Switzerland
Laurent Lebrun, Laboratoire de Génie Electrique et Ferroélectricité (LGEF), Institut
    National Des Sciences Appliquees, Lyon, France

Chapter 8
Mircea Badescu, JPL/Caltech, Pasadena, CA
Ted Iskenderian, JPL/Caltech, Pasadena, CA

Chapter 9
James Cutts, JPL/Caltech, Pasadena, CA
Robert Hodyss, JPL/Caltech, Pasadena, CA
Tibor Kremic, NASA Glenn Research Center, Cleveland, OH

Chapter 10
George Cooper, University of California, Berkeley, CA
Chris P. McKay, NASA Ames Research Center, Division of Space Science, Moffett Field, CA
Lori Shiraishi, JPL/Caltech, Pasadena, CA

Chapter 11
Kenneth Diller, The University of Texas at Austin, Austin, TX
Brian Grout, University of Copenhagen, Denmark
David Pegg, University of York, York, UK

Chapter 12
Wayne Johnson, Tennessee Tech University, Cookeville, TN
Colin Johnston, Oxford University, Oxford, UK
Mike Hamilton, Auburn University, Auburn, AL

Chapter 13
Georg Raffelt, Max Planck Institute for Physics, Munich, Germany
William Wester, Fermilab, Batavia, IL

Chapter 14
Adam M. Swanger, Cryogenics Test Laboratory, NASA Kennedy Space Center, Merritt
    Island, FL
Ron Ross, JPL/Caltech, Pasadena, CA

Chapter 15
Michael C. Kumpf, University of California, Berkeley, CA
Philip F. Simon, University of California, Berkeley, CA
Ron Welch, JPL/Caltech, Pasadena, CA
Robert Boyle, NASA—Goddard Space Flight Center (GSFC), Greenbelt, MD

The photo in the circle on the front cover is showing the University of Copenhagen (UCPH) Drill in action at the NEEM Camp, Greenland. The photo is the courtesy of H. Thing, who photographed the drill as part of the NEEM ice core drilling project, http://www.photo.neem.dk/2008.

Editor

Dr. Yoseph Bar-Cohen is a Senior Research Scientist and a Group Supervisor at Jet Propulsion Lab (JPL), California Institute of Technology (Caltech), Pasadena, California (http://ndeaa.jpl.nasa.gov/). He received his PhD in Physics from the Hebrew University, Jerusalem, Israel, in 1979. His research is focused on electroactive mechanisms and biomimetics. He edited and coauthored nine books, coauthored over 380 publications, cochaired 47 conferences, has 28 registered patents, and coauthored 116 New Technology Reports. His notable initiatives include challenging engineers and scientists worldwide to develop a robotic arm driven by artificial muscles to wrestle with humans and win. For his contributions to the field of artificial muscles, in April 2003, *Business Week* named him one of the five technology gurus who are "Pushing Tech's Boundaries." His accomplishments have earned him two NASA Honor Award Medals, two SPIE's Lifetime Achievement Awards, and Fellow of two technical societies (ASNT and SPIE), as well as many other honors and awards.

Contributors

Mircea Badescu
Jet Propulsion Laboratory
California Institute of Technology
Pasadena, California

Xiaoqi Bao
Jet Propulsion Laboratory
California Institute of Technology
Pasadena, California

Yoseph Bar-Cohen
Jet Propulsion Laboratory
California Institute of Technology
Pasadena, California

Olivier Bollengier
University of Washington,
Seattle, Washington

Morgan L. Cable
Jet Propulsion Laboratory
California Institute of Technology
Pasadena, California

Gianpaolo Carosi
Physics & Life Sciences Directorate
Lawrence Livermore National Laboratory
Livermore, California

Julie C. Castillo
Jet Propulsion Laboratory
California Institute of Technology
Pasadena, California

Mathieu Choukroun
Jet Propulsion Laboratory
California Institute of Technology
Pasadena, California

Robert Cormia
Foothill College
Los Altos Hills
California

Joe Flynn
QM Power Inc
Lee's Summit, Missouri

Stephanie A. Getty
NASA—Goddard Space Flight Center
Greenbelt, Maryland

Murthy Gudipati
Jet Propulsion Laboratory
California Institute of Technology
Pasadena, California

Tom Guettinger
UniWest
Pasco, Washington

Wenbin Huang
Department of Mechanical and Aerospace
    Engineering
North Carolina State University
Raleigh, North Carolina

Corey Jamieson
SETI Institute
Mountain View, California

Xiaoning Jiang
Department of Mechanical and Aerospace
    Engineering
North Carolina State University
Raleigh, North Carolina

W. Kinzy Jones
Florida International University
Miami, Florida

Baptiste Journaux
Laboratoire de Glaciologie et Géophysique
    de l'Environnement
Université Joseph Fourier
Grenoble, France

Alasdair G. Kay
Keele University
Institute for Science and Technology in
    Medicine
Staffordshire, England

Matthew M. Kropf
Energy Institute
University of Pittsburgh at Bradford
Bradford, Pennsylvania

Lilia L. Kuleshova
Leibniz Universität Hannover
Hannover, Germany

Hyeong Jae Lee
Jet Propulsion Laboratory
California Institute of Technology
Pasadena, California

Thomas Loerting
Institute of Physical Chemistry
University of Innsbruck
Innsbruck, Austria

Patrick McCluskey
CALCE/Department of Mechanical
    Engineering
University of Maryland
College Park, Maryland

Gale Paulsen
Honeybee Robotics
Pasadena, California

Ray Radebaugh
Material Measurement Laboratory
National Institute of Standards and
    Technology
Boulder, Colorado

Terry D. Rolin
NASA/Marshall Space Flight Center
Huntsville, Alabama

Ronald G. Ross, Jr.
Jet Propulsion Laboratory
California Institute of Technology
Pasadena, California

Stewart Sherrit
Jet Propulsion Laboratory
California Institute of Technology
Pasadena, California

Victor Sloan
Victor Aviation Service, Inc.
Palo Alto, California

Josef Stern
Institute of Physical Chemistry
University of Innsbruck
Innsbruck, Austria

Pavel Talalay
National Mineral Resources University
St Petersburg, Russia

Lonnie Thompson
Byrd Polar Research Center
Ohio State University
Columbus, Ohio

Mimi Ton
Jet Propulsion Laboratory
California Institute of Technology
Pasadena, California

Melissa G. Trainer
NASA—Goddard Space Flight Center
Greenbelt, Maryland

Karl van Bibber
Department of Nuclear Engineering
University of California Berkeley
Berkeley, California

Steve Vance
Jet Propulsion Laboratory
California Institute of Technology
Pasadena, California

Nathan Valentine
CALCE/Department of Mechanical
    Engineering
University of Maryland
College Park, Maryland

Kris Zacny
Honeybee Robotics
Pasadena, California

Victor Zagorodnov
Byrd Polar Research Center
Ohio State University
Columbus, Ohio

Shujun Zhang
Materials Research Institute
The Pennsylvania State University
University Park, Pennsylvania

1

Introduction to Low Temperature Materials and Mechanisms

Yoseph Bar-Cohen and Ray Radebaugh

CONTENTS

1.1 Introduction

Operating at low temperatures is essential to many processes in numerous fields of science and engineering including refrigeration, space exploration, electronics, physics, chemistry, thermodynamics, and medicine [Barron, 1985; Gutierrez et al., 2000; Halperin, 1995; Kent, 1993; Pobell, 2007; Weisend, 1998]. There is a growing interest in technologies that are applicable at low temperatures for planetary exploration of bodies in the solar system that are extremely cold. These include potential NASA *in situ* exploration missions to Europa and Titan where ambient temperatures are around –200°C. Elsewhere, as a method of slowing or halting chemical and biological processes, cooling is widely used to preserve food and chemicals, as well as biological tissues and organs. Furthermore, cooling is used to increase electrical conductivity leading to superconductors that enable such applications as levitation, highly efficient electromagnets, and so on. The subject of low temperature materials and mechanisms is multidisciplinary including chemistry, materials science, electrical engineering, mechanical engineering, metallurgy, and physics. This book covers some of the key aspects of the field including the chemistry and thermodynamics (Chapter 2), materials science (Chapter 3), the methods of characterizations (Chapter 4) as well as nondestructive testing and health monitoring (Chapter 5), the methods of cooling to cryogenic temperatures (Chapter 6), actuation materials and mechanisms (Chapters 7 and 8), instruments for planetary exploration (Chapter 9), methods of drilling in ice (Chapter 10), applications to medicine and biology (Chapter 11), low temperature electronics (Chapter 12), applications to fields of physics (Chapter 13), as well as other applications and challenges to the field (Chapter 14). Given the health hazards that are associated with working at cryogenic temperatures, a chapter has been dedicated to this topic as well (Chapter 15).

1.2 Cryogenic Temperatures

Cryogenics is the science and engineering of creating and investigating low temperature conditions [Scurlock, 1993; Kittel, 1996; Flynn, 1997; Radebaugh, 2002; Callister and Rethwish, 2012]. The word cryogenics is derived from the Greek words *kryos*, meaning "frost" or "cold," and *–genic*, meaning "to produce." Until around the mid-eighteenth century, it was believed that some gases could not be liquefied—these were named permanent gases. These gases were carbon monoxide, hydrogen, methane, nitric oxide, nitrogen, and oxygen, with the maximum attention given to the primary constituents of air—oxygen and nitrogen.

The temperature scale that is based on absolute zero with units in the same size as the Celsius degree is known as units Kelvin and is abbreviated as K. This symbol of the Kelvin unit was adopted in 1968. In the absolute, or Kelvin, scale, the lowest temperature is written as 0 K (without a degree sign). The Rankine scale with the symbol R represents the English absolute scale and has the same unit increments as the Fahrenheit scale.

Although cold temperatures can include temperatures at the level of water freezing (0°C), the term "cryogenic environment" refers to the temperature range below the point at which the permanent gases begin to liquefy. The term was applied initially to temperatures from approximately –100°C (–148°F) down to absolute zero Kelvin [Shachtman, 1999], and today the definition is related to temperatures below approximately –150°C (–238°F). In 1894, Onnes, of the University of Leiden, the Netherlands, was the first to coin the term "cryogenics" as the science of producing very low temperatures to liquefy the gases that were considered permanent [Onnes, 1894]. The English physicist Michael Faraday was one of the most successful in liquefying permanent gases; by 1845, he had managed to liquefy most of the permanent gases known at that time [Ventura and Risegari, 2007]. He liquefied the gases by cooling them via immersion in a bath of ether and dry ice and then pressurizing until they reached the liquid state. The liquefaction of air was accomplished in 1877 by Louis Cailletet in France and by Raoul Pictet in Switzerland [Timmerhaus and Reed, 2007]. In 1898, the liquefaction of hydrogen was successfully accomplished by the Scottish chemist James Dewar. In 1908, successful attempts on the last remaining gas element to be liquefied, helium, were carried out by Onnes at 4.2 K [Soulen, 1996].

The measurement of cryogenic temperatures requires methods that are not commonly used in daily life. The commonly used methods are mostly based on the use of mercury or alcohol thermometers, and these two fluids freeze and become useless at such low temperatures. Platinum resistance thermometers are based on the relation between electrical resistance and temperature and these provide relatively accurate measurements down to about 20 K. For temperatures down to 1 K and below, measuring the electrical resistance of certain semiconducting materials, such as doped germanium, also provides temperature gauging, but these thermometers require calibration over the range of temperatures at which the measurements are needed. The calibration of such secondary thermometers is done against the primary ones and they are based on measuring physical variable changes for which the response is determined theoretically.

The temperature of a material is a measure of the energy that it contains, which occurs in various forms of motion among the atoms or molecules that make up the material [Enss and Hunklinger, 2005]. A gas at a lower temperature has slower-moving atoms and molecules than gases at higher temperatures do. The possibility that a material can have a state at which all forms of motion are halted was predicted in 1848 by the English physicist William Thomson, later known as Lord Kelvin. The absence of all forms of motion would result in a complete absence of heat and temperature; this condition was defined by Thomson as absolute zero, which is equal to –273.15°C (–459.67°F). With the advent of quantum mechanics in the early twentieth century, we now know that at absolute zero some atomic motion still exists in materials, but it is in the lowest possible energy state, known as the ground state. Much of low temperature physics involves the study of materials close to the ground state where unusual behavior can occur.

1.3 Materials at Low Temperatures

The mechanical and electrical properties of many materials significantly change at temperatures that are in the range below 100 K; for example, most plastics, rubber, and some metals become brittle [Russell, 1931; Teed, 1952; Chapters 2 and 3]. Moreover, many metals and ceramics exhibit the phenomenon of superconductivity, which is the loss of all resistance to the flow of electricity [Seeber, 1998; Tsuneto and Nakahara, 2005]. Another phenomenon that takes place at temperatures very close to absolute zero is the state of superfluidity. Specifically, at 2.17 K helium becomes superfluidic, allowing it to flow through significantly narrower passages without exhibiting friction than associated with being in this quantum mechanical ground state [Tsuneto and Nakahara, 2005]. At temperatures below 2.17 K, helium has zero viscosity and produces a film that can creep upward over the walls of an open container. Superfluid helium also has an extremely high thermal conductivity, which makes it useful in cooling superconducting magnets. Another material property of great interest is the boiling point of various gases (see Table 1.1). In Chapter 3, the subject of materials and their properties at low temperatures is covered in greater detail.

1.4 Cooling to Cryogenic Temperatures

The tools that allow materials to reach cryogenic temperatures are called cryocoolers. Generally, there are four basic methods of reaching cryogenic temperatures [de Waele, 2011; Kittel, 1996; Barron, 1999; Kakaç et al., 2003; Chapter 6]:

  a. Heat conduction—When two bodies are in contact, heat flows from the body at a higher temperature to the one at a lower temperature. Generally, conduction

TABLE 1.1

The Boiling Point of Various Gases

Gas Element	K	°C	°F
Argon	87	–186	–302
Helium	4.2	–269	–452
Hydrogen	20	–253	–423
Krypton	120	–153	–242
Neon	27	–246	–411
Nitrogen	77	–196	–320
Oxygen	90	–183	–297
Xenon	166	–107	–161

takes place among all forms of matter, that is, gas, liquid, or solid. Materials can be cooled to cryogenic temperatures by immersing them directly in a cryogenic liquid or by placing them in an atmosphere that is cooled by cryogenic refrigeration. In each of these techniques, the material is cooled by transferring heat (via conduction) to the colder material or the surrounding environment.

b. Evaporative cooling—This is also a commonly used method of cooling. Atoms and molecules move faster in the gas state than in the liquid state. When thermal energy is added to the particles in a liquid so they become gas, the remaining liquid is cooled. The gas is then pumped away and, as more heat is added, more liquid particles are converted into gas. The longer this process continues, the more heat is removed from the liquid, and the lower the temperature that is reached, as long as the pressure above the liquid is continually reduced. When the desired temperature is reached, the pumping is continued at a slower rate to allow maintenance at the specific low temperature. This method can be used to reduce the temperature of liquid nitrogen to its freezing point as well as to lower the temperature of liquid helium down to approximately 1 K.

c. Cooling by rapid expansion—This method uses the Joule–Thomson, or throttling, process and is the method of cooling that is most widely applied in household refrigerators and air conditioners, although it is also used in heat pumps and liquefiers. Cooling is achieved by expanding a gas or liquid through a valve or porous plug that is insulated from exchanging heat with the environment. In this process, the gas is first pumped into a container under high pressure. When a valve is opened, the gas escapes and expands quickly, and the temperature of the gas drops. The cyclic process involves continuous evaporation and condensation of a fixed amount of refrigerant in a closed system (see Figures 1.1 and 1.2). The evaporation takes place at a low temperature and low pressure, while the condensation takes place at a high temperature and high pressure. Thus, heat is transferred from an area of low temperature to an area of high temperature. This thermodynamic cycle involves compression of the vaporized refrigerant at constant entropy, which exits the compressor superheated. The superheated vapor travels through the condenser that cools it and removes vapor superheat, and then the vapor is condensed to a liquid by removing additional heat at constant pressure and temperature. The liquid refrigerant travels through the expansion valve, which rapidly reduces its pressure and causes evaporation of refrigerant to a mixture of liquid and vapor that travels though the evaporator coil until it is completely vaporized and cools the warm space by using a fan across the evaporator. The resulting refrigerant vapor returns to the compressor inlet, thus completing the thermodynamic cycle. In Figure 1.1, the aforementioned steps are taking place as follows: 1 to 2—the vapor is compressed; 2 to 3—superheat is removed from the vapor in the condenser; 3 to 4—the vapor is converted to liquid in the condenser; 4 to 5—the liquid is flashed into liquid and vapor across the expansion valve; and 5 to 1—the liquid and vapor are converted to a fully vapor state in the evaporator. Using this process with many stages, Onnes, in 1908, was able to liquefy gases such as helium down to 4.2 K.

d. Adiabatic demagnetization—This method provides a means for reaching temperatures much less than 1 K. The adiabatic demagnetization phenomenon uses paramagnetic salts that consist of a very large collection of magnetic particles with random polarity. The salt is placed in the magnetic field and it causes alignment of its magnetic particles. If the external magnet is removed and the paramagnetic

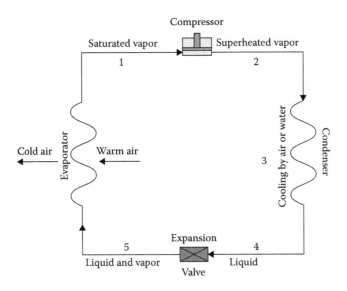

FIGURE 1.1
Typical vapor-compression refrigeration.

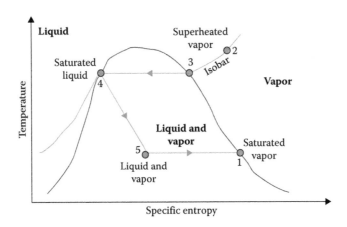

FIGURE 1.2
Temperature versus entropy diagram of the vapor-compression cycle.

salt is allowed to absorb heat, the particles' polarity is randomized again. However, this change requires input of energy, which is taken from the material that is being cooled, causing its temperature to drop. This method has been used to produce some of the coldest temperatures that have ever been achieved, effectively, within a few thousandths of a Kelvin from absolute zero. A related phenomenon is nuclear demagnetization, which involves the magnetization and demagnetization of atomic nuclei. This method has been successfully used to lower temperatures of systems to within a few millionths of a degree from absolute zero; however, temperatures within a few billionths of a degree above absolute zero have also been achieved, although these temperatures are only those of the nuclei, which are not in equilibrium with the warmer electrons or lattice.

1.5 Applications

The applications of low temperatures were conceived of, and applied by, physicists, chemists, materials scientists, and biologists who study the properties of metals, insulators, semiconductors, plastics, composites, and living tissue [Weisend, 1998; White and Meeson, 2002; Pobell, 2007]. Once it became possible to reach temperatures approaching absolute zero, many discoveries and related applications emerged. Generally, the use of cryogenic temperatures offers many benefits, which can be sorted into seven categories: (1) long-term preservation of biological material and food; (2) densification (liquefaction and separation of gases); (3) creation of macroscopic quantum phenomena (superconductivity and superfluidity); (4) reduced thermal noise; (5) low vapor pressures (cryopumping); (6) temporary or permanent property changes; and (7) tissue destruction (cryoablation). Applications of cryogenics usually make use of one or more of these benefits. To be useful, the benefit must be significant enough to warrant the extra effort to reach cryogenic temperatures.

Earlier, superconductivity and its application for levitation were mentioned [Seeber, 1998]. Another important cryogenic property is the fact that many materials become brittle at extremely low temperatures. This property is used by the recycling industry, where recyclables are immersed in liquid nitrogen, making them easy to pulverize and separate for reprocessing. Also, the fact that materials contract at low temperatures is used in the manufacture of automobile engines, where tight-fit applications are made by shrinking components and inserting them into the required cavity. Once expanded at room temperature, a part with a very tight fit is produced.

Cryogenic liquids, including rocket fuel and coolants, are widely used in the various programs of international space agencies. Typically, for rockets, a pair of tanks is used, where one is filled with liquid hydrogen as the fuel that is burned and the second one is filled with liquid oxygen used for the combustion. Elsewhere, liquid helium is used for space applications to cool orbiting infrared telescopes that are designed to detect objects that emit heat; they need to be cooled in order to prevent the instrumentation from being blinded to the infrared radiation from stars by its own emitted heat (Chapters 6 and 9). Since the temperature of superfluid liquid helium is 1.8 K, the telescope can easily pick up infrared stellar radiation with temperatures that are as high as about 3 K.

Cryogenic liquids such as oxygen, nitrogen, and argon are often used in industrial and medical applications (Chapter 11). An example of the use of liquid nitrogen for testing thermal insulation systems is shown in Figure 1.3 [Augustynowicz et al., 1999]. This application is used to provide energy conservation and involves establishing a cold boundary temperature as well as providing a direct measure of the heat flow. This includes fast freezing of some foods for prolonged preservation. The fact that the electrical resistance of most metals decreases as the temperature decreases is used in hospitals for the magnets in magnetic resonance imaging (MRI) systems. Electromagnetic coils with wires that are made of niobium alloys are cooled to 4.2 K and produce extremely high magnetic fields with no generation of heat and no consumption of electric power.

Other medical applications of cryogenic temperatures include the freezing of portions of the body to destroy tissues, a process known as cryosurgery. This process is used to treat cancers and abnormalities of the skin, cervix, uterus, prostate gland, and liver as well as surficial warts. Also, cryogenic temperatures are used for freezing and preserving biological materials including livestock semen as well as human blood, tissue, and embryos. Moreover, there is a practice of freezing the entire human body after death in the hope of restoring the life of the person at a later time. This practice, known as cryonics, is, however, not an accepted scientific application.

FIGURE 1.3
Liquid nitrogen is used in testing of thermal insulation systems for both establishing a cold boundary temperature and providing a direct measure of the heat flow. (Photo courtesy of James E. Fesmire/NASA-KSC.)

1.6 Summary

The range of temperatures that are called cryogenic is from as high as –150°C (–238°F) to absolute zero (–273°C or –460°F), where the latter is the theoretical level at which matter is at its lowest energy, or ground, state. In the absolute, or Kelvin, scale, this lowest temperature is written as 0 K (without a degree sign). Generally, these are extreme conditions at which the properties of materials such as strength, thermal conductivity, ductility, and electrical resistance are altered. Various methods are used to reach cryogenic temperatures and, as described in this chapter, these methods are based on heat conduction, evaporative cooling, cooling by rapid expansion, and adiabatic demagnetization. The tools that allow materials to reach cryogenic temperatures are called cryocoolers and are mostly based on heat exchange, as described in Section 1.4. Applications of cryogenic temperatures involve such fields as physics, chemistry, materials science, and biology. Further details about the applications and effects of such low temperatures are discussed in this book.

Acknowledgments

Some of the research reported in this chapter was conducted at the Jet Propulsion Laboratory (JPL), California Institute of Technology, under a contract with the National Aeronautics and Space Administration (NASA). The authors would like to thank Robert Youngquist, NASA Kennedy Space Center, Florida; Michael Mischna, JPL/Caltech, Pasadena, CA; Colin H. Smith, Oxford University, UK; and Paul E. Hintze, Auburn University, AL, for reviewing this chapter and providing valuable technical comments and suggestions.

References

Augustynowicz, S. D., J. E. Fesmire, and J. P. Wikstrom, Cryogenic insulation systems, *20th International Congress of Refrigeration*, IIR/IIF, Sydney, (1999), pp. 1–7. http://321energy.us/attachments/File/Public_Files/1999_ICR_Syndey_-_Cryogenic_Insulation_Systems.pdf

Barron, R. F., *Cryogenic Systems*, Oxford Press, Oxford, UK (1985).

Barron, R. F., *Cryogenic Heat Transfer*, Edwards Brothers, Ann Arbor, MI (1999).

Callister W, D. Rethwish, *Fundamentals of Materials Science and Engineering*, John Wiley & Sons; Hoboken, New Jersey, 4th Edition, ISBN-10: 111832269X; ISBN-13: 978-1118322697 (2012), 944 pages.

Davenport, J., *Animal Life at Low Temperature*, ISBN-10: 0412403501, ISBN-13: 978-0412403507, Springer, London (1991).

de Waele, A. T., Basic operation of cryocoolers and related thermal machines, *Journal of Low Temperature Physics*, Open Access, Volume 164, Issue 5–6, (September 2011), pp. 179–236.

Echlin, P., *Low-Temperature Microscopy and Analysis*, ISBN-10: 0306439840, ISBN-13: 978-0306439841, Springer, New York (1992).

Enss, C., and S. Hunklinger, *Low-Temperature Physics*, ISBN-10: 3540231641, ISBN-13: 978-3540231646, Springer, Berlin (2005).

Flynn, T., *Cryogenic Engineering*, Marcel Dekker, New York (1997).

Gutierrez-D, E. A., J. Deen, and C. Claeys, *Low Temperature Electronics: Physics, Devices, Circuits, and Applications*, ISBN-10: 0123106753, ISBN-13: 978-0123106759, Academic Press, London (2000).

Halperin, W. (Ed.), *Progress in Low Temperature Physics*, 1st Edition, Proceedings, ISBN 9780444822338, eBook ISBN 9780080539935, Elsevier, North Holland (20 Oct 1995).

Kakaç, S., M. R. Avelino, and H. F. Smirnov (Eds.), *Low Temperature and Cryogenic Refrigeration*, Proceedings of the NATO Advanced Study Institute, Altin Yunus-Çesme, Izmir, Turkey, June 23–July 5, 2002, Series: Nato Science Series II, SBN 978-1-4020-1274-7, Vol. 99 (2003), 485 pages.

Kent, A., *Experimental Low-Temperature Physics*, ISBN-10: 1563960303, ISBN-13: 978-1563960307, Macmillan Physical Science, American Institute of Physics, New York (1993).

Kittel, P., (Ed.), *Advances in Cryogenic Engineering*, Springer, New York, ISBN-10: 0306453002, ISBN-13: 978-0306453007, Vol. 41, Part B (1996), pp. 1–1014.

Onnes, H. K., On the cryogenic laboratory at Leyden and on the production of very low temperature, Communication of the Physics Laboratory, Leiden University, Netherlands. Vol. 14, 1894.

Pobell, F., *Matter and Methods at Low Temperatures*, ISBN-10: 3540463569, ISBN-13: 978-3540463566, 3rd revision, Springer, New York (2007).

Radebaugh, R., *Cryogenic Temperatures, The MacMillan Encyclopedia of Chemistry*, New York (2002). http://www.cryogenics.nist.gov/MPropsMAY/materialproperties.htm

Russell, H. W., Effect of low temperatures on metals and alloys, *Proceedings of the Symposium on Effect of Temperature on the Properties of Metals*, ASTM, Chicago, IL (1931).

Scurlock, R. G. (Ed.), *History and Origins of Cryogenics Monographs on Cryogenics*, Oxford University Press, Oxford, England, ISBN-13: 978-0198548140; ISBN-10: 0198548141 (1993), 680 pages.

Seeber, B. (Ed.), *Handbook of Applied Superconductivity*, Institute of Physics Publishing, Bristol, UK (1998).

Shachtman, T., *Absolute Zero and the Conquest of Cold*, Houghton Mifflin, Company Boston (1999).

Soulen, R., James Dewar, his flask and other achievements. *Physics Today*, doi:10.1063/1.881490, Vol. 49, No. 3 (March 1996), pp. 32–37.

Teed, P. L., The properties of metallic materials at low temperatures, Volume 1 of *A Series of Monographs on Metallic and Other Materials*, Chapman and Hall Ltd., London (1952).

Timmerhaus, K. D., and R. Reed, *Cryogenic Engineering: Fifty Years of Progress*, ISBN-10: 038733324X; ISBN-13: 978-0387333243, International Cryogenics Monograph Series, Springer, New York, NY (2007), pp. 1–374.

Tsuneto, T., and M. Nakahara, *Superconductivity and Superfluidity*, Cambridge University Press Edition 2, ISBN: 052102093X, ISBN-13: 9780521020930, Cambridge (2005).

Ventura, G., and L. Risegari, *The Art of Cryogenics: Low-Temperature Experimental Techniques*, ISBN-10: 0080444792; ISBN-13: 978-0080444796, Elsevier Science, New York, NY (2007), pp. 1–378.

Weisend, J. G. II (Ed.), *Handbook of Cryogenic Engineering*. Taylor and Francis, Philadelphia, PA (1998).

White, G. K., and P. Meeson, *Experimental Techniques in Low-Temperature Physics*, Series: Monographs on the Physics and Chemistry of Materials, Book 59, ISBN-10: 0198514271; ISBN-13: 978-0198514275, Oxford University Press, 4th Edition, New York (2002).

Internet Links

Chapter 10: Refrigeration Cycles—http://www.saylor.org/site/wp-content/uploads/2013/08/BolesLectureNotesThermodynamicsChapter10.pdf

Cryocooler—http://en.wikipedia.org/wiki/Cryocooler

Cryogenic Society of America, Cold Facts Newsletter—www.cryogenicsociety.org

Superconductivity—http://www.superconductors.org/Uses.htm; http://www.superconductors.org/index.htm

The Basic Refrigeration Cycle—http://www.achrnews.com/articles/the-basic-refrigeration-cycle

Wikipedia: Carnot Cycle—http://en.wikipedia.org/wiki/Carnot_cycle#The_temperature-entropy_diagram

2

Chemistry, Thermodynamics, and Material Processes at Low Temperatures

Murthy Gudipati

CONTENTS

2.1 Introduction

This overview of low temperature materials in the universe is by no means exhaustive or unbiased. However, it is intended to give the reader a quick glance into the low temperature materials present in space, what we know, and what we still need to understand.

When we speak about cryogenic temperatures, we would like to define the temperature as the meaning varies for different communities. The general definition is unconstrained—very low temperature. For the purposes of this section, we will use the cosmic microwave background (CMB) thermal radiation at 2.72548 K [Fixsen 2009] as the lowest limit and the crystalline water-ice sublimation temperature at ~160 K as the higher temperature limit. This range covers a wide variety of bodies in the solar system and beyond into the interstellar medium. Although sub-Kelvin temperatures are reached in laboratories on Earth, their relevance to astrophysics and planetary sciences (space conditions in general) is not immediately evident.

Our sun is one of the few billion stars in our galaxy and our Milky Way galaxy is one among the several billion galaxies in the universe that are visible both to the naked eye and to instruments in our possession such as space telescopes. Recent developments in exoplanet research increased the potential for the existence of habitable solar systems like ours in our galaxy [Kerr 2013; Schilling 2007; Schwarzschild 2014]. Thus, understanding materials in our solar system that occur under cryogenic conditions would help in understanding many other such solar systems as well as the interstellar medium.

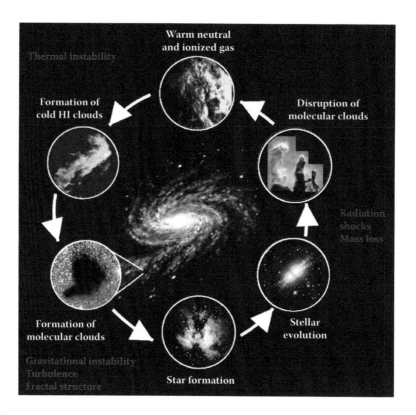

FIGURE 2.1

The birth and death cycle of solar systems and the role of interstellar medium in transporting the cryogenic materials in this cycle. Except for the closer vicinity of stars, star-forming regions, and protoplanetary disks, the remainder of the material is typically under cryogenic temperatures. (From http://soral.as.arizona.edu/HEAT/science/LifeCycle.png.)

The birth and death cycle of solar systems like ours, and the role of the interstellar medium in the evolution of the cryogenic materials in this cycle [van der Tak 2012], are shown in Figure 2.1. Most of the galaxies or the solar systems in them are at cryogenic temperatures. Away from a star or a protostar (a star in birth), radiation flux per unit area falls by the inverse-square of the distance from the star, resulting in cryogenic temperatures at farther distances where most of the volatiles condense, forming a part of the outer solar system and interstellar matter. This makes understanding of material properties in cryogenic temperatures an important aspect of space sciences, particularly astrophysics and planetary sciences. This chapter reviews the physical, chemical, and material processes at cryogenic temperatures pertinent to the two aforementioned areas of research.

2.2 Material in the Solar System and Interstellar Medium at Very Low Temperatures

The cosmic abundances of elements [Arnett 1996] are shown as a bar graph in Figure 2.2. Hydrogen (H) is the most abundant element, followed by helium (He), oxygen (O), carbon (C), nitrogen (N), etc., and molecules formed by these elements, such as H_2, O_2, H_2O, CO,

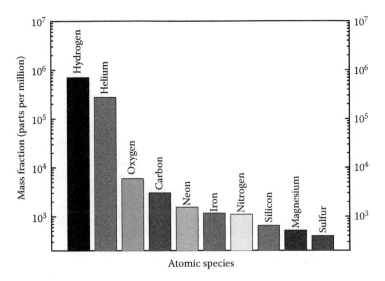

FIGURE 2.2
The abundances of the most predominant elements [Arnett 1996] in our solar system, which are somewhat similar to the galactic abundance. While helium (He) and neon (Ne) are inert elements and cryogenic liquids, the rest of the elements form the basis of a wide variety of molecules in the universe.

N_2, NO, NH_3, etc., are the key determiners of cosmic cryogenic chemical composition. While H_2 exists as a liquid and a solid at very low temperatures, He needs even lower temperatures to reach the liquid phase, and only when doped with impurities under special conditions can solid He be made in the laboratory [Kim and Chan 2012; Pratt et al. 2011]. Solid *para*-H_2 [Fajardo and Tam 1998; Momose and Shida 1998] is used as a quantum solid to study the spectroscopy of trapped atoms and molecules at temperatures typically below 5 K. In space, liquid hydrogen is proposed to exist at extremely high pressures and temperatures at the cores of giant gas planets Jupiter and Saturn [Saumon and Guillot 2004; Trachenko et al. 2014], but not as a liquid or solid at low temperatures.

Thus, low temperature environments in the universe can be approximated as locations distant from any star (or star-forming region), taking our Sun as an example. A temperature versus distance plot of our solar system and a protostar (a star in the formation) is shown in Figure 2.3. It is generally accepted that water-ice formation is the starting point of the low temperature front (also known as the snow-line), assumed to be around 100 K [van Dishoeck 2014] in astronomical time and extremely high vacuum ($\sim 10^{-20}$ mbar or a few molecules per m³), for ice to sublime to water vapor compared to ~160 K in the laboratory.

After molecular hydrogen (H_2), by far the most abundant molecules are CO and water (H_2O) in interstellar space [Aikawa et al. 1999], which can also be easily inferred from the cosmic abundance of elements shown in Figure 2.2. Other molecules with significant abundances include CO_2, C_2H_6, NH_3, SO_2, CH_3OH, etc., all of which are seen both in interstellar ice grains [Gibb et al. 2004; Oberg et al. 2011] and in comet outgassing [Bockelee-Morvan et al. 2000; de Val-Borro et al. 2013]. Under pressure, water becomes a solid at noncryogenic temperature, as exemplified by the formation of ice on Earth, but most of the solar system bodies without atmosphere and the medium between stars (interstellar medium) have water-dominated ice either on the surface or near subsurface. Thus, in the context of space sciences, ice becomes the first target of a solid at cryogenic temperatures. CO, N_2, and CH_4 form ice at temperatures below 35 K, making them abundant only on interstellar ice grains

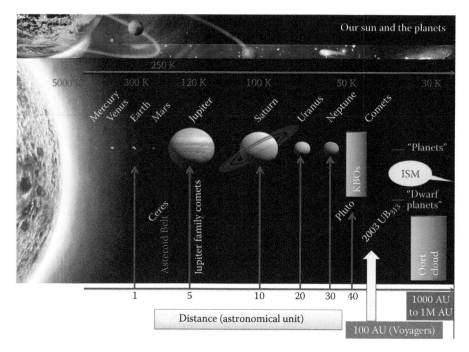

FIGURE 2.3

Temperature versus distance plot of our solar system and potentially other habitable protostars (a star in the formation phase). The Kuiper Belt and Oort cloud contain reservoirs of comets, but comets have eccentric orbits that vary significantly. Material exists under cryogenic temperatures at farther distances from the Sun (several astronomical units—AU—the distance between the Sun and the Earth) and in completely shadowed (from sunlight) surfaces of poor thermal conductivity and no atmosphere, such as on the permanently shadowed regions of the Moon, comets, and asteroids. (Photo courtesy of NASA/JPL [modified].)

(~10 K) and outer solar system icy bodies (beyond Neptune—trans-Neptunian objects or TNOs; Kuiper Belt objects [KBOs], including Pluto), and in the Oort cloud. Comets originate from KBOs and the Oort cloud. A recent book co-edited by the author gives a comprehensive overview of the cryogenic ices in the solar system [Gudipati and Castillo-Rogez 2013], and a review by Clark et al. [2014] provides more details on cryogenic materials in space. KBO comets are scattered into the inner solar system and form so-called short-period comets, whose aphelion is closer to Jupiter's orbit. The interior temperature of a comet is expected to be <50 K, and the comet interior contains these volatile hydrocarbons, as evident from outgassing from comets (comet tails) during a comet's approach closer to the Sun in its eccentric orbit. At the time of this writing, the Rosetta spacecraft is in a close encounter with comet 67P/Churyumov-Gerasimenko (Figure 2.4) studying its surface, interior, coma, and tail composition. Due to the extreme temperature changes a comet undergoes during its eccentric orbits around the Sun and due to the fact that the density of a comet is ~470 ± 45 kg/m^3 [Sierks et al. 2015], making 70% of its volume porous voids, cometary material composition and its physical properties are still a puzzle. Prior to the Philae lander's deployment, it was expected that comets were loosely bound materials with very low tensile strength and high porosity. Because of this, the consensus was that any attempt to land on a comet should be exercised carefully in order not to get sunk into this low-tensile material. However, the Philae lander unexpectedly bounced several times before landing on comet CG/67P, indicating that the surface was much harder than would be expected from a fluffy low-tensile ice/mineral/organic mix [Faber et al. 2015].

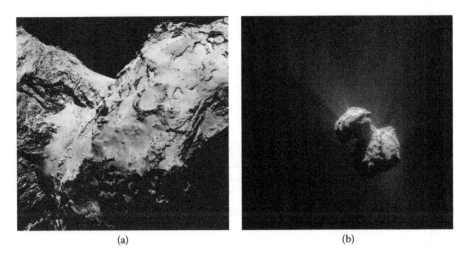

(a) (b)

FIGURE 2.4
The Rosetta spacecraft is presently in a close encounter with comet 67P/Churyumov-Gerasimenko (CG/67P). (a) Irregular shape of the comet with smooth and rugged terrain. (b) Jets of water, carbon dioxide, and other volatiles coming out of the comet from the sunlit surface. (Photo courtesy of ESA/NASA/JPL.)

KBOs contain Pluto and even bigger icy objects such as Eris, Makemake, Haumea, etc., which have been shown to have methane ice on the surface [Schaller and Brown 2007]. Traveling from KBOs toward the Sun, we encounter Neptunian, Uranian, Saturnian, and Jovian systems. Icy bodies in these systems are observed to consist of mostly crystalline water-ice surface [Fraser and Brown 2010; Grundy et al. 2006]. Table 2.1, compiled from data

TABLE 2.1

The Leading Composition of the Surfaces of Objects That Are at Cryogenic Temperatures and Covered with Various Forms of Frozen Molecular Ices

Distance from the Sun (AU)	Object	Surface Ice Composition
1	Earth	H_2O
1.5	Mars	H_2O, CO_2
5	Jovian System	
	Io	SO_2,
	Europa	H_2O, Salts
	Ganymede	H_2O
	Callisto	H_2O
10	Saturnian System	
	Mimas, Enceladus, Tethys, Dione, Rhea, Hyperion, Pohebe, Rings	H_2O
	Iapetus	H_2O, CO_2, Hydrocarbons
20	Uranian System	
	Ariel, Umbriel	H_2O
	Miranda	H_2O, NH_3
	Titania, Oberon	H_2O, Hydrocarbons
30	Neptunian System	
	Triton	N_2, CH_4, CO, CO_2, H_2O
40	Kuiper Belt Objects (KBOs)	
	Pluto, Eris, Makemake, etc.	N_2, CH_4, CO, H_2O

Source: Compiled from Roush, T. L., *Journal of Geophysical Research Planets*, 106, 33315–33323, 2001.

published by Roush [2001], gives a quick overview of major surface composition of objects that are at cryogenic temperatures and covered with various forms of frozen molecular ices. Excellent review articles by Clark et al. [2013] and de Bergh et al. [2013] give detailed data about icy cryogenic surfaces in the outer solar system. Except for Io (dominated by sulfur) and Titan (dominated by hydrocarbons), most of the solar system icy objects are covered with water-ice, and as the surface temperatures become cold enough, N_2 and CH_4 ice are formed on objects such as Pluto and Triton [Cruikshank 2005; Stern 2015]. The Jovian moon Io has a mean surface temperature of ~110 K with sulfur-dominated cryovolcanism. Titan, the largest moon of Saturn, is one of the three bodies with significant atmosphere (Venus, Earth, and Titan) in our solar system. With its surface temperature of ~94 K and about 1.5 bar surface pressure, Titan's surface harbors liquid hydrocarbon lakes as well as a solid hydrocarbon surface (Figure 2.5).

At even closer distances, Mars has polar H_2O and CO_2 ice, which sublime and recondense with seasonal variations [Kieffer et al. 2006] with temperatures reaching as low as ~120 K (Figure 2.6). Finally, the Moon, the closest object to Earth, has on its surface the coldest temperature on its permanently shadowed crater surface (~30 K) [Colaprete et al. 2010; Lanzerotti et al. 1981] and can potentially harbor a significant amount of water that can be utilized for future human explorations to the Moon.

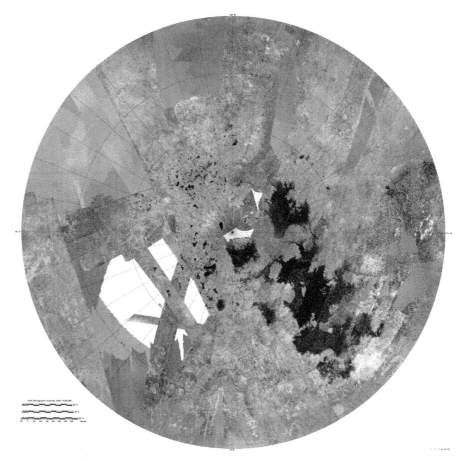

FIGURE 2.5
Cassini mission rendered radar surface mapping of Titan, revealing that Titan's surface harbors liquid hydrocarbon lakes as well as solid hydrocarbon surface. (Photo courtesy of NASA/JPL.)

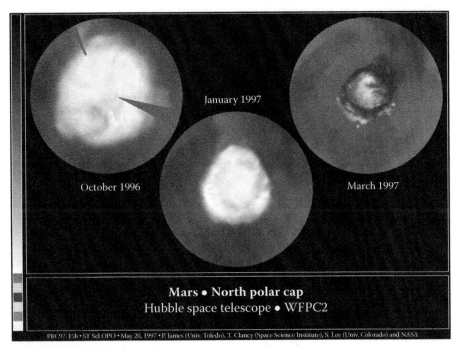

FIGURE 2.6
Northern polar cap images of Mars taken at different seasons from the Hubble Space Telescope (HST). Snow and a warm-up cycle can clearly be seen between winter 1996 and summer 1997. When the surface temperature reaches ~150 K, only water snow covers the surface. At lower temperatures (~120 K), carbon dioxide condenses, forming a CO_2 ice layer on the top. (Photo courtesy of NASA/JPL.)

This section shows that most of the accessible (for robotic and human reach) low temperature objects in our solar system exist at temperatures at and above 30 K, mostly in solid form, except for the liquid hydrocarbon lakes on Titan. Further, minerals and soil on the Moon, Mars, comets, asteroids, and other objects in the solar system can reach similar extremely low temperatures as their frozen volatile ice counterparts. Understanding their physical and chemical properties under radiation, to be discussed in the next section, is an important and critical component of solar system exploration, as robotic or human missions to these objects need to intimately interact with these objects and their surroundings.

2.3 Energy Sources for the Transformation of Low Temperature Materials (Interstellar versus Planetary)

Energy in the form of electromagnetic radiation (photons) and accelerated particles (electrons, ions, subnuclear particles, etc.) form the basis of the radiation environment in the universe as we know it today. The most commonly encountered radiation in our solar system is in the form of photons from the Sun, solar wind (electrons and ions), and cosmic rays (ions). The energy content of this radiation can span anywhere between 10^{20} eV (galactic cosmic rays) to 10^{-10} eV (long wave ~1000 m), as depicted in Figure 2.7. This radiation, constantly impinging on various solar system and interstellar low temperature objects,

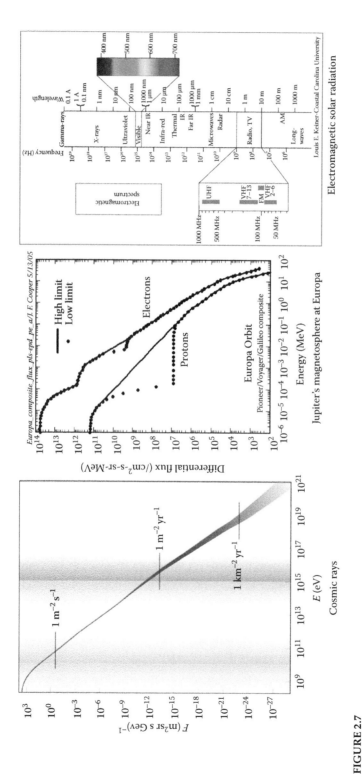

FIGURE 2.7

The energy content of radiation in the universe can span anywhere between ~10^{20} eV (galactic cosmic rays) to ~10^{-10} eV (long wave ~1000 m). The higher the energy content, the deeper it penetrates through planetary surfaces. An analysis of energy content and flux during radiation processing of surfaces of cryogenic solar system bodies is critical for material integrity. (Parts taken from Wikipedia, with permission, and from *Icarus* 149(1), Cooper, J. F., Johnson, R. E., et al., Energetic ion and electron irradiation of the icy galilean satellites, Copyright 2001, with permission from Elsevier.)

causes physical and chemical changes in these materials, some of which are critical for life on Earth (such as photosynthesis). It is generally accepted that energy is one of the four essential necessities for life (water, organics, minerals, and energy). However, the same energy in large doses could also destroy life. Cryogenic temperatures are not expected to be conducive for life, but pockets of warm liquid reservoirs, such as expected under the subsurface of Europa [Khurana et al. 1998; Pappalardo et al. 1999; Schmidt et al. 2011], could be habitable.

2.4 Physical Properties of Water-Ice at Low Temperatures

With water being the dominant molecule on many surfaces, and minerals being present in the form of either dust grains or rocks with little change in their physical properties at lower temperatures, we focus on the physical properties of water-ice in this section. Water-ice on Earth exists in the atmosphere (cirrus clouds), in polar regions as ice caps, and at higher latitudes and elevations (mountains) as snow and glaciers. It exists as very fluffy soft material (fresh snow) and very hard material (ice sheets and ice cores) formed at higher pressures (such as deep beneath the surface). However, most of the ice we find on Earth is not cold enough to be representative of ice on other solar system bodies. For example, ice on Mars reaches temperatures close to 120 K, Jovian icy moons reach similar temperatures, Saturnian moons reach ~100 K, and so on. Cometary ice is expected to be ~30 K in the interior, with extremely high porosity (close to 70%), but it is still a puzzle whether the cometary interior is made of amorphous or crystalline ice. Thermal properties of ice have an important influence on the physical and chemical transformations within these ices. Thermal properties of loosely bound highly porous materials with large voids are known to be drastically different from strongly bound low-porosity materials. An example of such highly porous materials is aerogel [Kwon et al. 2000]. A few laboratory studies have been attempted for amorphous water-ice and its thermal conductivity as well as porosity comes close to aerogels, of the order of a few mW m^{-1} K^{-1}. However, literature values for amorphous ices at low temperatures and low pressures vary by about three to four orders of magnitude ranging between 200 and 0.01 mW m^{-1} K^{-1} [Bar-Nun and Laufer 2003; Klinger 1980; Kouchi et al. 1992], indicating the need for further accurate experimental work.

Tensile strengths of low temperature ices are also important for in situ lander missions to these bodies, as demonstrated by the recent Philae landing. Typically, tensile strength decreases with high porosity, similar to the decrease in the thermal conductivity. Powders of cryogenic ices and organics should have lower tensile strengths compared with the crystalline ice blocks. Here again, there is a need for more laboratory work. One of the few available studies on the tensile strength of amorphous ice gives a value of ~10^3–10^5 dynes cm^{-2} [Bar-Nun and Laufer 2003; Greenberg et al. 1995], several orders below the tensile strength of crystalline ice, found to be ~10^7 dynes cm^{-2} [Petrovic 2003]. The Philae lander of the Rosetta spacecraft found that the surface of the comet CG/67P is much harder than expected [Faber et al. 2015; Hand 2014], which caused it to bounce a few times before finally ending in a rugged, shadowed part of the comet (Figure 2.8). A recent laboratory study [Lignell and Gudipati 2015] indicated that when amorphous ice at higher temperatures converts into harder crystalline ice, organic molecules are expelled—simulating the observations made by the Rosetta VIRTIS instrument [Capaccioni et al. 2015].

Other physical properties include heat capacity, bulk modulus, shear and compressive strength, speed of sound, hardness, viscosity, density, etc. While H_2O ice still needs to be

FIGURE 2.8
A photograph taken by the Philae lander (Rosetta mission) in a shaded region on comet 67P/Churyumov-Gerasimenko. (Photo courtesy of ESA/NASA/JPL.)

adequately studied for all its aforementioned physical properties, other ices in the solar system such as N_2, CH_4, SO_2, CO_2, etc., and mixtures of these ices with mineral grains and organic polymers (such as seen on cometary surfaces and on Iapetus) are also poorly studied. Thus, there is a critical need to study the cryogenic physical properties of a wide variety of molecular ices at macroscopic and microscopic levels.

Extremely low temperatures also pose challenges for materials that can be used for various missions. While many parts of the spacecraft or lander can be protected from cryogenic temperatures through thermal packaging, some parts that interact with the bodies such as drills, anchors, optical equipment for subsurface in situ observations, etc., must be in a position to operate effectively at that temperature, making their fabrication at room temperature extremely difficult. For example, a comet subsurface cryogenic sample analysis (return) mission should have all the exposed parts cryogenically qualified to a minimum of 50 K. Ideally materials for such an application must have the lowest possible thermal expansion or contraction coefficients, no changes in their tensile strengths and brittleness, etc. While the mechanical parts need to retain their functional and material integrity at extremely low temperatures, these and other parts of the spacecraft must also retain their chemical integrity under low temperature and high-radiation conditions such as on Europa, one of the Galilean moons of Jupiter.

2.5 Chemistry at Cryogenic Temperatures

Chemical processes at temperatures below 160 K are mostly caused by radiation (Figure 2.7) that supplies the necessary energy to overcome reaction barriers, barrierless chemical reactions, and light-atom (typically hydrogen) tunneling. While the first two categories dominate a majority of chemical processes in the solar system, tunneling becomes important at extremely low temperatures (<30 K) and over long periods of time [Arnaut et al. 2006; Goumans 2011; Goumans and Andersson 2010]—both appropriate for interstellar conditions. In general, chemical reactions at low temperatures occur at a lower rate than reactions at higher temperatures, mainly due to (thermal) reaction barriers that become larger than the ambient thermal energy available as the temperature lowers [Cavalli et al. 2014]. In general, lower temperatures lower the molecular mobility (including rotations and diffusion), slowing or stopping bimolecular reactions and diffusion-controlled chemical pathways. However, radiation-induced monomolecular chemistry such as molecular dissociation and ionization as well as barrierless reactions will continue to be dominant even at very low temperatures. Our recent work, utilizing newly developed techniques to look at the composition of cryogenic materials using laser ablation and laser ionization time-of-flight mass spectrometry, showed that indeed ionization and other radiation-induced chemistry resulting in complex prebiotic organic molecules could occur even at 10 K—a temperature predominant in the dense molecular clouds (DMSs) of the interstellar medium [Gudipati and Yang 2012; Henderson and Gudipati 2015]. At higher temperatures (such as the solar system temperature >30 K), the radiation-induced chemical processes become even more important. This applies to both the solar system body and the spacecraft parts that are exposed to radiation. Chemical integrity at low temperatures is critical for the functioning of many spacecraft parts. These processes must be taken into account when developing new spacecraft shields or shieldless low-mass CubeSat-like spacecraft.

2.6 Evolution of Complex Molecules in the Coldest Places in the Universe

Atoms and molecules in space (Figure 2.2 and Table 2.1) are constantly subjected to radiation and thermal processing. Whether considering minerals such as silicates, organics such as the polycyclic aromatic hydrocarbons (PAHs) that seem to be abundant in the universe, or other molecular materials that form the basis of surfaces on solar system objects, all these materials are subjected to chemical processing. For example, radiation can cause photoionization (ejection of an electron from the molecule or material), resulting in the formation of an electron–hole pair. This charge separation can initiate a wide variety of chemical evolution processes such as charge-mediated chemistry, Coulomb force–driven long-range effects on the surface, photocurrents, etc. This can equally affect salts such as $MgSO_4$, $NaCl$, or the corresponding acids expected to be the predominant on Europa's surface, which is bombarded with radiation and ions [Carlson et al. 2009; Dalton 2003; Hand and Carlson 2015; Hoerst and Brown 2013; Munoz-Iglesias et al. 2014]. Neutral atoms can be ejected into the atmosphere through photo-attachment (electron attachment to ions or molecules) processes. Molecular dissociations (breaking of a covalent bond) also play a key role in ejecting neutral and ionized volatiles into the tenuous atmosphere of airless bodies. On the Moon, solar radiation and cosmic ray-driven production of H, O, and other neutral and charged species (H^+, O^+, etc.) forms the basis for its atmosphere. Similar processes occur on Europa, Titan, Ganymede, etc. We found that when water-ice containing organic impurities is subjected to photon or electron irradiation, ionization is the primary process [Gudipati 2004; Gudipati and Allamandola 2004], followed by a wide range of complex chemical processes that could form the basis for the composition of prebiotic materials on Earth (Figure 2.9).

2.7 Kinetic versus Thermodynamic Processes at Very Low Temperatures

Most of the radiation-driven processes are kinetically controlled, resulting in a larger amount of local excessive energy. However, processes that occur at thermodynamic equilibrium play a leading role at astronomical timescale. Such processes include chemical

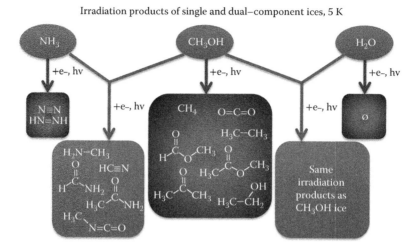

FIGURE 2.9
Complex prebiotic molecules that are detected in the laboratory using in situ laser ablation time-of-flight mass spectrometry from simulated cometary and interstellar ice composition under radiation processing with electrons and UV photons. Boxes show complex organics produced from the initial simple ices and ice mixtures (shown by connecting lines and arrows). Although oxygen species such as O, O_2, H_2O_2, O_3, etc., are produced from pure H_2O ice, their mass spectral signals are strongly overlapped by H_2O signals. Most of the complex organic molecules detected in this laboratory study have also been found on comet tails and ice grains in the interstellar medium [Henderson and Gudipati 2015]. Other ices pertinent to the outer solar system, such as CH_4, CO, N_2, CO_2, SO_2, etc., need to be further studied.

reactions or material phase transitions with a barrier that is comparable with the thermal history of the body [Mastrapa et al. 2013]. Tunneling of a proton (hydrogen) could also lead to chemical reactions at very low temperatures [Hama and Watanabe 2013; Henkel et al. 2014; Herbst 1994]. Kinetically controlled processes include photon-, electron-, or ion (cosmic rays or local magnetospheric radiation)-induced sputtering of the surface [Brown et al. 1986; Johnson et al. 2013], and dust and micrometeorite bombardment [Pham et al. 2009; Turrini et al. 2014]. Penetration depths of such energetic processing are determined by the energy and mass of the particle. Cosmic rays with the highest energies can penetrate several meters below a soil, rock, or ice surface. The rest of the ions, dust, and micrometeorites only affect the first few millimeters of the surface [Cooper et al. 2001].

2.8 Material Properties for Cryogenic Solar System Explorations

Solar system exploration has always pushed the science and engineering capabilities to (and beyond) the limits. In situ explorations to low temperature parts of our solar system pose challenges that not only require material integrity at extremely low temperatures (~30 K), but also must be compatible with very low-sunlight conditions, whether on the permanently shadowed craters of the moon, or Europa's surface, or a comet's shadowed surface. For this reason, many in situ missions at present have very short lifetimes of performance (such as the Philae lander on comet CG/67P), unless local energy sources are available. Future in situ explorations to these extremely cold objects must take these aspects into account: performance under extreme temperature variations, sufficient hardness to penetrate through surfaces as hard as a rock, energy efficient, lightweight, etc. Long-term (more than a few hours to a day) in situ instrumentation would need even more stringent requirements, especially for high-radiation environments such as on Europa, forcing spacecraft and instrumentation to be high-radiation hard and low temperature stable simultaneously. Radiation-induced chemical degradation of the material surface and interior over a long period of time is expected to be a critical issue for lander missions to bodies without atmosphere (such as asteroids and Jovian satellites like Europa).

2.9 Conclusions and Future Outlook

Solar system explorations, as they expand and reach the extreme corners with extreme temperature and radiation conditions, pose challenging demands for new materials that are thermal and radiation hard, lightweight, high tensile strength, etc. At the same time, understanding the thermal, mechanical, and chemical properties of extremely low temperature solar system bodies—from the permanently shadowed craters of the Moon to comets and KBOs—is equally critical. The next several decades will fill in these knowledge and technology gaps, enabling human and robotic explorations of our solar system.

Acknowledgments

This work was enabled through partial funding from JPL's DRDF and R&TD funding for infrastructure of the "Ice Spectroscopy Laboratory" at JPL, NASA Spitzer Science Center, and NASA funding through the Planetary Atmospheres and Cassini Data Analysis programs. This research was carried out at the Jet Propulsion Laboratory, California Institute of Technology, under a contract with the National Aeronautics and Space Administration.

The authors would like to thank Will Grundy, Lowell Observatory, Flagstaff, AZ; Victoria (Maria) Iglesias, JPL/Caltech, Pasadena, CA; and Gabriel Tobie, Université de Nantes, Cedex 3, France, for reviewing this chapter and providing valuable technical comments and suggestions.

References

Aikawa, Y., Umebayashi, T., et al. (1999). Evolution of molecular abundances in proto-planetary disks with accretion flow. *Astrophysical Journal* **519**(2): 705–725.

Arnaut, L. G., Formosinho, S. J., et al. (2006). Tunnelling in low-temperature hydrogen-atom and proton transfers. *Journal of Molecular Structure* **786**(2–3): 207–214.

Arnett, D. (1996). *Supernovae and Nucleosynthesis: An Investigation of the History of Matter, from the Big Bang to the Present*. Princeton, NJ: Princeton University Press.

Bar-Nun, A. and Laufer, D. (2003). First experimental studies of large samples of gas-laden amorphous 'cometary' ices. *Icarus* **161**(1): 157–163.

Bockelee-Morvan, D., Lis, D. C., et al. (2000). New molecules found in comet c/1995 o1 (Hale-Bopp)—investigating the link between cometary and interstellar material. *Astronomy & Astrophysics* **353**(3): 1101–1114.

Brown, W. L., Lanzerotti, L. J., et al. (1986). Sputtering of ices by high-energy particle impact. *Nuclear Instruments & Methods in Physics Research Section B-Beam Interactions with Materials and Atoms* **14**(4–6): 392–402.

Capaccioni, F., Coradini, A., et al. (2015). The organic-rich surface of comet 67p/Churyumov-Gerasimenko as seen by Virtis/Rosetta. *Science* **347**(6220): aaa0571.

Carlson, R. W., Calvin, W. M., et al. (2009). Europa's surface composition. In: R. T. Pappalardo, W. B. McKinnon and K. Khurana (Eds). *Europa*. Tucson, Arizona: The University of Arizona Press: 283–328.

Cavalli, S., Aquilanti, V., et al. (2014). Theoretical reaction kinetics astride the transition between moderate and deep tunneling regimes: The F plus HD case. *Journal of Physical Chemistry A* **118**(33): 6632–6641.

Clark, R., Carlson, R., et al. (2013). Observed ices in the solar system. In: M. S. Gudipati and J. Castillo-Rogez (Eds). *The Science of Solar System Ices*. New York: Springer: 3–46.

Clark, R. N., Swayze, G. A., et al. (2014). Spectroscopy from space. In: G. S. Henderson, D. R. Neuville and R. T. Downs (Eds). *Spectroscopic Methods in Mineralology and Materials Sciences. Reviews in Mineralogy and Geochemistry*, v. 78. Chantilly, Virginia : Mineralogical Society of America: 399–446.

Colaprete, A., Schultz, P., et al. (2010). Detection of water in the lcross ejecta plume. *Science* **330**: 463–468.

Cooper, J. F., Johnson, R. E., et al. (2001). Energetic ion and electron irradiation of the icy Galilean satellites. *Icarus* **149**(1): 133–159.

Cruikshank, D. P. (2005). Triton, Pluto, centaurs, and trans-neptunian bodies. *Space Science Reviews* **116**(1–2): 421–439.

Dalton, J. B. (2003). Spectral behavior of hydrated sulfate salts: Implications for Europa mission spectrometer design. *Astrobiology* **3**(4): 771–784.

de Bergh, C., Schaller, E. L., et al. (2013). The ices on transneptunian objects and centaurs. In: M. S. Gudipati and J. Castillo-Rogez. *The Science of Solar System Ices*. New York: Springer: 107–146.

de Val-Borro, M., Kuppers, M., et al. (2013). A survey of volatile species in Iort cloud comets c/2001 q4 (neat) and c/2002 t7 (linear) at millimeter wavelengths. *Astronomy & Astrophysics* **559**:A48.

Faber, C., Knapmeyer, M., et al. (2015). A method for inverting the touchdown shock of the Philae lander on comet 67p/Churyumov-Gerasimenko. *Planetary and Space Science* **106**: 46–55.

Fajardo, M. E. and Tam, S. (1998). Rapid vapor deposition of millimeters thick optically transparent parahydrogen solids for matrix isolation spectroscopy. *Journal of Chemical Physics* **108**: 4237–4241.

Fixsen, D. J. (2009). The temperature of the cosmic microwave background. *The Astrophysical Journal* **707**(2): 916.

Fraser, W. C. and Brown, M. E. (2010). Quaoar: A rock in the Kuiper belt. *Astrophysical Journal* **714**(2): 1547–1550.

Gibb, E. L., Whittet, D. C. B., et al. (2004). Interstellar ice: The infrared space observatory legacy. *The Astrophysical Journal Supplement Series* **151**(1): 35.

Goumans, T. P. M. (2011). Hydrogen chemisorption on polycyclic aromatic hydrocarbons via tunnelling. *Monthly Notices of the Royal Astronomical Society* **415**(4): 3129–3134.

Goumans, T. P. M. and Andersson, S. (2010). Tunnelling in the O plus CO reaction. *Monthly Notices of the Royal Astronomical Society* **406**(4): 2213–2217.

Greenberg, J. M., Mizutani, H., et al. (1995). A new derivation of the tensile-strength of cometary nuclei—Application to comet Shoemaker-Levy-9. *Astronomy and Astrophysics* **295**(2): L35–L38.

Grundy, W. M., Young, L. A., et al. (2006). Distributions of H2O and CO2 ices on Ariel, Umbriel, Titania, and Oberon from IRTF/SpeX observations. *Icarus* **184**(2): 543–555.

Gudipati, M. S. (2004). Matrix-isolation in cryogenic water-ices: Facile generation, storage, and optical spectroscopy of aromatic radical cations. *Journal of Physical Chemistry A* **108**(20): 4412–4419.

Gudipati, M. S. and Allamandola, L. J. (2004). Polycyclic aromatic hydrocarbon ionization energy lowering in water ices. *Astrophysical Journal Letters* **615**: L177–L180.

Gudipati, M. S. and Castillo-Rogez, J., Eds. (2013). *The Science of Solar System Ices*. Astrophysics and Space Science Library. Springer: New York.

Gudipati, M. S. and Yang, R. (2012). In-situ probing of radiation-induced processing of organics in astrophysical ice analogs—novel laser desorption laser ionization time-of-flight mass spectroscopic studies. *Astrophysical Journal Letters* **756**(1): L24.

Hama, T. and Watanabe, N. (2013). Surface processes on interstellar amorphous solid water: Adsorption, diffusion, tunneling reactions, and nuclear-spin conversion. *Chemical Reviews* **113**(12): 8783–8839.

Hand, E. (2014). Planetary science Philae probe makes bumpy touchdown on a comet. *Science* **346**(6212): 900–901.

Hand, K. P. and Carlson, R. W. (2015). Europa's surface color suggests an ocean rich with sodium chloride. *Geophysical Research Letters* **42**(9):3174–3178.

Henderson, B. L. and Gudipati, M. S. (2015). Direct detection of complex organic products in ultraviolet (ly alpha) and electron-irradiated astrophysical and cometary ice analogs using two-step laser ablation and ionization mass spectrometry. *Astrophysical Journal* **800**(1): 66.

Henkel, S., Ertelt, M., et al. (2014). Deuterium and hydrogen tunneling in the hydrogenation of 4-oxocyclohexa-2,5-dienylidene. *Chemistry: A European Journal* **20**(25): 7585–7588.

Herbst, E. (1994). Tunneling in the C2H-H2 reaction at low-temperature. *Chemical Physics Letters* **222**(3): 297–301.

Hoerst, S. M. and Brown, M. E. (2013). A search for magnesium in Europa's atmosphere. *Astrophysical Journal Letters* **764**(2): L28.

Johnson, R., Carlson, R., et al. (2013). Sputtering of ices. In: M. S. Gudipati and J. Castillo-Rogez. *The Science of Solar System Ices*. Springer New York. **356**: 551–581.

Kerr, R. A. (2013). Exoplanets; Kepler snags super-Earth-size planet squarely in a habitable zone. *Science* **340**(6130): 262.

Khurana, K. K., Kivelson, M. G., et al. (1998). Induced magnetic fields as evidence for subsurface oceans in Europa and Callisto. *Nature* **395**: 777–780.

Kieffer, H. H., Christensen, P. R., et al. (2006). CO2 jets formed by sublimation beneath translucent slab ice in Mars' seasonal south polar ice cap. *Nature* **442**(7104): 793–796.

Kim, D. Y. and Chan, M. H. W. (2012). Absence of supersolidity in solid helium in porous vycor glass. *Physical Review Letters* **109**(15): 155301.

Klinger, J. (1980). Influence of a phase-transition of ice on the heat and mass balance of comets. *Science* **209**(4453): 271–272.

Kouchi, A., Greenberg, J. M., et al. (1992). Extremely low thermal-conductivity of amorphous ice—relevance to comet evolution. *Astrophysical Journal* **388**(2): L73–L76.

Kwon, Y. G., Choi, S. Y., et al. (2000). Ambient-dried silica aerogel doped with TiO_2 powder for thermal insulation. *Journal of Materials Science* **35**(24): 6075–6079.

Lanzerotti, L. J., Brown, W. L., et al. (1981). Ice in the polar-regions of the moon. *Journal of Geophysical Research* **86**(NB5): 3949–3950.

Lignell, A. and Gudipati, M. S. (2015). Mixing of the immiscible: Hydrocarbons in water-ice near the ice crystallization temperature. *The Journal of Physical Chemistry A* **119**(11): 2607–2613.

Mastrapa, R. E., Grundy, W., et al. (2013). Amorphous and crystalline H2O-ice. In: M. S. Gudipati and J. Castillo-Rogez. *The Science of Solar System Ices*. New York: Springer: 371–408.

Momose, T. and Shida, T. (1998). Matrix-isolation spectroscopy using solid parahydrogen as the matrix: Application to high-resolution spectroscopy, photochemistry, and cryochemistry. *Bulletin of the Chemical Society of Japan* **71**(1): 1–15.

Munoz-Iglesias, V., Prieto-Ballesteros, O., et al. (2014). Conspicuous assemblages of hydrated minerals from the H2O-MgSO4-CO2 system on jupiter's Europa satellite. *Geochimica et Cosmochimica Acta* **125**: 466–475.

Oberg, K. I., Boogert, A. C. A., et al. (2011). The Spitzer ice legacy: Ice evolution from cores to protostars. *Astrophysical Journal* **740**(2): 109.

Pappalardo, R. T., Belton, M. J. S., et al. (1999). Does Europa have a subsurface ocean? Evaluation of the geological evidence. *Journal of Geophysical Research* **104**(E10): 24015–24055.

Petrovic, J. J. (2003). Review: Mechanical properties of ice and snow. *Journal of Materials Science* **38**(1): 1–6.

Pham, L. B. S., Karatekin, O., et al. (2009). Effects of meteorite impacts on the atmospheric evolution of Mars. *Astrobiology* **9**(1): 45–54.

Pratt, E. J., Hunt, B., et al. (2011). Interplay of rotational, relaxational, and shear dynamics in solid 4He. *Science* **332**(6031): 821–824.

Roush, T. L. (2001). Physical state of ices in the outer solar system. *Journal of Geophysical Research-Planets* **106**: 33315–33323.

Saumon, D. and Guillot, T. (2004). Shock compression of deuterium and the interiors of Jupiter and Saturn. *Astrophysical Journal* **609**(2): 1170–1180.

Schaller, E. L. and Brown, M. E. (2007). Detection of methane on Kuiper belt object (50000) Quaoar. *Astrophysical Journal* **670**(1): L49–L51.

Schilling, G. (2007). Habitable, but not much like home. *Science* **316**(5824): 528.

Schmidt, B. E., Blankenship, D. D., et al. (2011). Active formation of 'chaos terrain' over shallow subsurface water on Europa. *Nature* **479**(7374): 502–505.

Schwarzschild, B. (2014). Earth-size exoplanets in habitable orbits are common. *Physics Today* **67**(1): 10–12.

Sierks, H., Barbieri, C., et al. (2015). On the nucleus structure and activity of comet 67p/Churyumov-Gerasimenko. *Science* **347**(6220): aaa1044.

Stern, S. A. et al. (2015). The pluto system: Initial results from its exploration by new horizons. *Sci* **350**(6258):aad1815

Trachenko, K., Brazhkin, V. V., et al. (2014). Dynamic transition of supercritical hydrogen: Defining the boundary between interior and atmosphere in gas giants. *Physical Review E* **89**(3): 032126.

Turrini, D., Combe, J. P., et al. (2014). The contamination of the surface of Vesta by impacts and the delivery of the dark material. *Icarus* **240**: 86–102.

van der Tak, F. (2012). The first results from the Herschel-HIFI mission. *Advances in Space Research* **49**(10): 1395–1407.

van Dishoeck, E. F. (2014). Astrochemistry of dust, ice and gas: Introduction and overview. *Faraday Discussions* **168**: 9–47.

3

Solids and Fluids at Low Temperatures

Steve Vance, Thomas Loerting, Josef Stern, Matt Kropf, Baptiste
Journaux, Corey Jamieson, Morgan L. Cable, and Olivier Bollengier

CONTENTS

3.1 Introduction

The fundamental properties of low temperature materials have garnered greater interest with the advent of space exploration. Space technology must survive fluctuations and lows in temperature rarely encountered on Earth. Of deeper scientific interest and continuing mystery is the range of new thermodynamic properties and rheologies (e.g., elasticity, tensile strength, viscosity) of planetary materials encountered on the Moon and Mars, comets, and a multitude of icy worlds. Properties of industrial metals and plastics have been discussed in recent chapters published elsewhere [Van Sciver, 2012], including low temperature heat capacity, thermal contraction, electrical and thermal conductivity, magneto-resistance in metals, and solid–liquid phase changes [Mehling and Cabeza, 2008]. This chapter describes materials and their relevant properties, which we are just beginning to understand in the detail that future exploration requires. It ends with a brief note about applications of piezoelectric materials at low temperatures.

3.2 Solid Materials

At very low temperatures, water and other solid materials that are present on icy satellites have mechanical and chemical properties that resemble those of rocks and metals. This analogy makes them interesting as key to icy satellite geology, but also possibly as industrial materials.

The chemical and physical properties of a solid are greatly influenced by the conditions under which it was formed due to variations in the molecular arrangement. Solids may be *crystalline*, with ordered and repeating fundamental units at the molecular level. This generally gives rise to anisotropic physical properties, higher and sharply defined melting points, and higher thermal conductivities when compared to analog *amorphous* solids that do not have a long-range order [Eucken, 1911; Hendricks and Jefferson, 1933; Lonsdale, 1937]. The degree and type of crystallinity may depend on several factors including pressure, rate and temperature of freezing or deposition, thermal history, specific molecular makeup/existence of impurities, or the presence of bond-disrupting processes [Johannessen et al., 2007; Sestak et al., 2011]. The crystalline state is energetically favored, but kinetically driven formation processes may result in amorphous structures if the atoms do not have time to orderly arrange to the lower-energy crystalline state [Zallen, 1983]. If water is cooled at a rate greater than 10^6 K s^{-1}, the crystalline state does not have time to form and an amorphous solid is formed [Bruggeller and Mayer, 1980].

The geometry of crystalline solids of single repeating units can be classified by one of 14 symmetrically unique lattice structures, called *Bravais lattices*. However, much crystalline material, such as salts or molecular solids, is composed of multiple fundamental units, which break the symmetry of the 14 Bravais lattices and must be defined with additional symmetry elements. These elements are called *point groups*, and the combination of the Bravais lattices and the point groups leads to 230 potential *space groups* that provide the necessary symmetry elements to describe a 3D crystalline solid [Sands, 1975].

Substances often occur in several different crystalline phases—a phenomenon known as *polymorphism*—which depends on the conditions of formation. For example, on distant, frigid bodies of the outer solar system such as Pluto and Triton (a moon of Neptune), nitrogen ice can exist in two phases—a cubic α-phase or a hexagonal β-phase—with a transition temperature between them at 35.6 K. Their differing molecular structures change the electronic and vibrational environment of the material. The difference can be identified in spectroscopic analyses of the surface and has been used to elucidate the surface temperature of these bodies [Grundy et al., 1993; Quirico et al., 1999].

Polymorphism is one of water's anomalous properties: 17 distinct crystalline phases are known at present [Petrenko and Whitworth, 2002; Falenty et al., 2014], labeled by Roman numerals following the chronological order of their discovery. The first new phase (besides hexagonal I_h), described in 1900, was named ice II [Tammann, 1900]; the most recent one, ice XVI, was discovered in 2014 [Falenty et al., 2014]. Other phases may exist [e.g., Wilson et al., 2013; Algara-Siller et al., 2015].

Specific crystal or amorphous structures are identified by analyzing the glass transition temperature or associated thermodynamic quantities; x-ray scattering or electron diffraction patterns; or NMR splitting pattern to determine bond correlation [Stachurski, 2011].

The process of crystallization can be understood as the balance between the enthalpic benefit of bond formation and the entropic drive to disorder. Upon cooling, the kinetic energy of molecules and atoms decreases, diminishing the thermodynamic contribution of entropy. At low temperatures, when the Gibbs free energy is enthalpically dominated, solidification becomes thermodynamically spontaneous. However, not every thermodynamically favored cooling process immediately results in solidification. Even if the phase transformation of the bulk sample is considered spontaneous, below a critical size the transformation of the molecular units may not be. Nucleation sites, if present, often enable crystal formation by allowing additional transitions to take place [De Yoreo and Vekilov, 2003]. If nucleation sites do not exist, liquids may be cooled below the thermodynamically determined solid–liquid phase boundary—a condition known as *supercooling*. For example, water normally freezes at 273 K at 1 atmosphere of pressure, but can be supercooled at the time scale of milliseconds down to 232 K while remaining liquid [Sellberg, 2014].

3.2.1 Crystalline Ices

The phase diagram shown in Figure 3.1 depicts the stability fields of liquid and solid phases of water. That is, the phase diagram merely shows the one phase that is thermodynamically most stable at a specific combination of pressure P and temperature T. Some ice phases, namely ices IV, IX, XII, XIII, XIV, and XVI, cannot be found in Figure 3.1 because they are never the thermodynamically most stable phase. In spite of this metastable nature, such phases readily form and can be isolated and characterized. They do not convert to more stable ice phases at laboratory time scales, and probably also not at astrophysically relevant time scales, provided the temperature and pressure conditions do not change. Crystalline phases can be stable in a P–T field or metastable at all P–T conditions. Some stable ice phases share a phase boundary with liquid water (ices I_h, III, V, VI [e.g., Choukroun and Grasset, 2007 and references therein] and ice VII [Pistorius et al., 1963; Mishima and Endo, 1978; Datchi

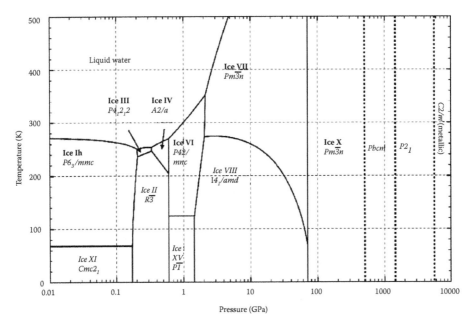

FIGURE 3.1

Phase diagram of H_2O from 0.01 to 10000 GPa and 0 to 500 K, showing stable water phases and their space group when crystalline. Proton disordered phases are written in bold, and proton-ordered phases are written in italics. Phase boundaries are plotted with a thick black line and predicted phase boundaries predicted at ultra-high pressures using computational simulations in dotted lines. (Courtesy of Baptiste Journaux, coauthor of this chapter.)

et al., 2000]), and are disordered [Kuhs, 2007]. Other ice phases are proton ordered and only exist in thermodynamic equilibrium with other ice polymorphs, namely ices II [Kamb, 1964; Kamb et al., 1971; Arnold et al., 1968; Fortes et al., 2003, 2005], VIII [Whalley et al., 1966; Kuhset al., 1984; Jorgensen et al., 1984, 1985; Pruzan et al., 1990, 1992; Besson et al., 1994, 1997; Pruzan et al., 2003; Song et al., 2003; Singer et al., 2005; Knight et al., 2006; Yoshimura et al., 2006; Somayazulu et al., 2008; Fan et al., 2010], XI [Singer et al., 2005; Knight et al., 2006; Fan et al., 2010; Tajima et al., 1984; Matsuo et al., 1986; Fukazawa et al., 1998, 2002; Kuo et al., 2005], XII [Kuhs et al., 1998; Lobban et al., 2000; Salzmann et al., 2006a, 2006b, 2008; Knight and Singer, 2008], XIV [Salzmann et al., 2006a, 2006b; Tribello et al., 2006], and XV [Kuhs et al., 1984; Fan et al., 2010; Knight and Singer,2005; Kuo and Kuhs, 2006; Salzmann et al., 2009]. Such high-pressure ice phases may be found in the mantles of the icy moons, for example, Ganymede, which is covered by an 800-km-thick layer of water ice [Vance et al., 2014], as well as Callisto [Schubert et al., 2004] and Titan [Tobie et al., 2005; Fortes et al., 2007] (Figure 3.2). It is also expected that high-pressure ices play a key role in ocean planets [Leger et al., 2004; Sotin et al., 2007; Grasset et al., 2009], roughly Earth-sized exoplanets with deep oceans that constitute an interesting analog to the icy satellites (Figure 3.2).

The distinction between proton-ordered and proton-disordered ice phases is shown in Figure 3.3. In all ice phases, the oxygen atoms occupy lattice positions, that is, the oxygen atoms are ordered over long ranges. The hydrogen atoms occupy lattice positions only in

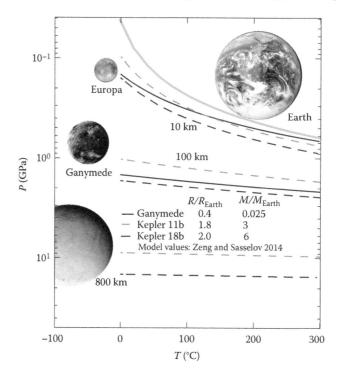

FIGURE 3.2
Pressures in exoplanet oceans span the multi-GPa range of pressures where liquids are possible. A Europa-depth ocean on Earth would behave like Ganymede's ocean in terms of having high-pressure ices. This figure shows profiles of pressure (GPa) and temperature (°C) in the upper mantles of selected objects (modified from Vance, S. et al., *Astrobiology*, 7(6), 2007), beginning at the estimated depth of the seafloor. The overlying ocean is assumed to be at a constant temperature; in general, seafloor temperature will be elevated by more than 40°C in deep oceans. Known exoplanets, albeit very hot ones, are modeled as super Europa objects with seafloor depths like Earth's (10 km), Europa's (100 km), and Ganymede's (800 km). (Courtesy of Steve Vance, principal author of this chapter.)

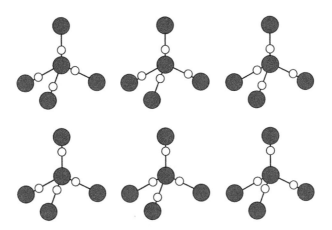

FIGURE 3.3
Six different proton configurations (white circles) allowed by the Bernal–Fowler ice rules. In proton-ordered ices, only one of these is observed, whereas in proton-disordered ices, all of these configurations are observed with equal probability. The tetrahedral coordination around a central oxygen atom (filled circles) is the motif found in all crystalline and amorphous ices (except for the ultrahigh-pressure phases at >100 GPa, ice X and the post-ice X phases). (Adapted from Fuentes-Landete, V., *Proceedings of the International School of Physics "Enrico Fermi," Volume 187: Water: Fundamentals as the Basis for Understanding the Environment and Promoting Technology,* IOS and Bologna: SIF, Amsterdam, 2015.)

the case of the proton-ordered ice phases. For a given network of oxygen atoms, in principle a larger number of ordered proton configurations are possible. Experimentally, however, only a single type of ordering has been observed so far. Ice I_h, the most common form of ice on Earth, is a proton-disordered form of ice. The hydrogen atoms obey the Bernal–Fowler ice rules [Bernal and Fowler, 1933], but they are randomly distributed within the ice crystal. As a result, ice I_h represents a frustrated crystal, which does not have zero configurational entropy at 0 K. The residual entropy at 0 K has become famous as the Pauling entropy ΔS_P [Pauling, 1935] and amounts to approximately 3.41 J K^{-1} mol^{-1}. This entropy can be released if one successfully achieves the transformation to the proton-ordered state. The ordered counterpart of ice I_h is known as ice XI, and its possible ferroelectric nature is currently being questioned [Parkkinen et al., 2014]. Order–disorder pairs can be easily recognized in the phase diagram in Figure 3.1 because the pair is separated by a phase boundary parallel to the pressure axis. This is so because there is barely any volume difference between proton-ordered and proton-disordered form, that is, $\Delta V \approx 0$, but the two ices differ by the Pauling entropy, that is, $\Delta S \approx \Delta S_P$. The slope of the phase boundary dP/dT then goes to infinity, that is, parallel to pressure axis, according to the Clausius–Clapeyron equation $dP/dT = \Delta S/\Delta V$.

Solid–solid transitions between polymorphs are possible by either a rearrangement of the O-lattice (density-driven, e.g., by pressurization) or H-ordering/disordering while preserving the geometry of the O-atoms (entropy-driven, e.g., by cooling a disordered phase to form an ordered one). In the case of density-driven transitions, there is a finite volume change ΔV, but the entropy change may be close to zero $\Delta S \approx 0$ if the transition is from a proton-disordered to a proton-disordered phase. Such phase boundaries can be identified easily in Figure 3.1, since they are almost parallel to the temperature axis. With increasing pressure, the transition sequence $I_h \rightarrow III \rightarrow V \rightarrow VI \rightarrow VII \rightarrow X$ can be identified. All these ice phases, except for ice X, are proton-disordered and of increasing density, starting at 0.92 g cm^{-3} and ending at 2.50 g cm^{-3}. High density is accommodated by improving packing, bending hydrogen bonds, and forming new hydrogen-network topologies. The oxygen networks differ in terms of topology and ring structure. While ices I_h and I_c consist entirely of six rings, ice V has 4-, 5-, 6-, 8-, 9-, 10-, and 12-ring structures [Herrero and Ramírez, 2013]. Some structures show ring threading; for example, ice IV and others show two interpenetrating ice networks (self-clathrates), for example, ices VI, VII, VIII, and X. For the latter ultrahigh-pressure ice X (>100 GPa [Benoit et al., 1996; Pruzan et al., 2003]), the molecular nature disappears due to the symmetrization of hydrogen position between the oxygen atoms.

Three low-density variants of crystalline ice can be produced at ambient pressure: hexagonal ice I_h, cubic ice I_c, and ice XI. While ice I_h is the abundant polymorph of solid H_2O on Earth, ice I_c can occasionally be found in clouds [Murray et al., 2005; Mayer and Hallbrucker, 1987; Whalley, 1983]. Ice I_c forms upon heating of amorphous ice in a vacuum or at ambient pressure. It may thus be present on comets after they have experienced temperatures above the crystallization temperature of ~150 K. It also forms upon heating of high-pressure forms of ice at/below ambient pressure and upon condensation of water vapor on particles at ~140–200 K. Both are very similar in density at P_{atm}, also appearing to be identical when probing the short-range molecular environment (e.g., Raman or mid-infrared). However, they can be distinguished when examining the long-range order (e.g., x-ray diffraction or neutron diffraction) [Kuhs et al., 1987]. Ices I_h and I_c are polytypical relative to each other: identical layers—differing stacking order (hexagonal rings, ABCABC with hexagonal symmetry for I_h, ABAB with fcc symmetry for I_c) [Kuhs and Lehmann, 1986; Kuhs et al., 1987; Guinier et al., 1984; Röttger et al., 1994]. I_c is not obtained as a single crystal, but only in the form of small crystallites with roughly hexagonal stacking faults (quantified by "cubicity index") [Kuhs et al., 1987; Kohl et al., 2000; Hansen et al., 2007]. Ice XI, the proton-ordered form of ice I_h, is only stable at $T < 72$ K. A proton-ordered form of cubic ice I_c might also exist, as has been suggested by *in situ* IR experiments and *ab initio* simulations [Geiger et al., 2014].

As mentioned, a crystalline phase may be transformed into another by ordering/disordering the protons while almost entirely preserving the O-atom topology. These order–disorder pairs will be found in the same P region of the phase diagram (I_h–XI [Singer et al., 2005; Knight et al., 2006; Fan et al., 2010; Tajima et al., 1984; Matsuo et al., 1986; Fukazawa et al., 1998, 2002; Kuo et al., 2005], III–IX [Fan et al., 2010; Kuhs et al., 1998; Lobban et al., 2000; Whalley et al., 1968; LaPlaca et al., 1973; Nishibata and Whalley, 1974; Minceva-Sukarova et al., 1984; Londono et al., 1993; Knight et al., 2006], V–XIII [Kuhs et al., 1998; Lobban et al., 2000; Salzmann et al., 2006a, 2008; Knight et al., 2008; Martin-Conde et al., 2006; Noya et al., 2008], VI–XV [Kuhs et al., 1984; Fan, 2010; Knight and Singer, 2005; Kuo and Kuhs, 2006; Salzmann et al., 2009], VII–VIII [Whalley et al., 1966; Kuhs et al., 1984; Jorgensen et al., 1984, 1985; Pruzan et al., 1990, 1992, 2003; Besson et al., 1994, 1997; Song et al., 2003; Singer et al., 2005; Knight et al., 2006; Yoshimura et al., 2006; Somayazulu et al., 2008; Fan et al., 2010], and XII–XIV [Salzmann et al., 2006a, 2006c; Tribello et al., 2006; Martin-Conde et al., 2006; Noya et al., 2008; Köster et al., 2015]), the ordered phase at lower and the disordered phase at higher temperatures (as the transition connected to the process of disordering an ordered phase is of entropic nature). Six configurations are allowed by the Bernal–Fowler ice rules in the local tetrahedral hydrogen-bond geometry known as Walrafen pentamer. In a fully disordered phase, all six are populated with the same probability (averaged over space/time) [Bernal and Fowler, 1933], whereas in an ordered phase, only one configuration is found. Often, the ordered low temperature phase cannot be accessed due to geometric constraints and kinetic limitations at such low temperatures. Point defects (Bjerrum L/D or ionic defects) in the ice lattice may enhance the reorientational mobility in the ice lattice. By using dopants incorporated into the ice lattice in ice I_h (e.g., HCl, HBr, HF, NH_3, KOH, etc. [Tajima et al., 1984; Matsuo et al., 1986; Hobbs, 1974; Gross and Svec, 1997]) and also at a higher pressure in ice VII (e.g., NaCl, LiCl, RbI [Frank et al., 2006, 2013; Klotz et al., 2009; Journaux et al., 2015]), such point defects may be introduced deliberately. These extrinsic defects may enhance the mobility related to rearrangement of protons and/or rotation of H_2O molecules, thereby overcoming kinetic limitations and allowing access to the proton-ordered low temperature phase [Köster et al., 2015]. However, this may also slow down the dynamics. Empirically it has been found that KOH doping accelerates dynamics in ice I_h whereas HCl doping has been found to accelerate the dynamics in the high-pressure ices V, VI, and XII [Salzmann et al., 2006a, 2006b, 2009]. The reasons why one dopant is effective and the other is not are still unclear, but are certainly related to the dynamics of pairs of point defects within the ice lattices [Burton and Oliver 1935; Mayer and Pletzer 1986]. Some ice phases do not form an order–disorder pair; new ice phases related to them through an order–disorder transition may be found in the future. For example, the ordered counterparts of I_c and IV, as well as the disordered one of II, still await discovery [Lokotosh and Malomuzh, 1993; Zabrodsky and Lokotosh, 1993].

The solubility of dopants in ice is generally assumed to be low. Ice I_h tends to reject any impurities during freezing [Gross and Svec, 1997]. For concentrated solutions, for example, freezing aqueous solutions present during sea ice formation, rejected salts may be incorporated in the bulk ice sheet at the grain boundaries as liquid brine inclusions [Weeks and Akley, 1986]. For the H_2O–NaCl system, the partition coefficient K_d (NaCl) is estimated around 2.7×10^{-3} and $3.2 (\pm 0.2) \times 10^{-3}$ [Gross et al., 1977, 1987]. This incompatible behavior of salts in ice I_h has a notable exception with ammonium fluoride, which forms a solid solution by substituting with H_2O molecules up to 5.2 mol% [Gross and Svec, 1997], most likely due to the isomorphism between the crystal lattices of NH_4F and ice I_h. Such incorporation should be distinguished from that occurring in clathrate hydrates as the ice solid solution keeps the crystallographic structure of the pure H_2O phase. While most substances are completely insoluble in ice, some dopants can be incorporated in the ice crystal by replacing water molecules. In cases of HCl, KOH, NH_3, HBr, or HF, the solubilities are in the ppb and ppm range. In spite of the low solubility, the influence on H-dynamics can be huge.

Recent high-pressure experiments have shown that ice VII can incorporate up to 1.6 mol% of NaCl, 5 mol% of CH_3OH [Frank et al., 2006, 2013], and up to 16.7 mol% of LiCl [Klotz et al., 2009]. Using DFT *ab initio* calculations, Klotz et al. [2009] predict incorporation in ice VII at interstitial (face-centered) and lattice sites (H_2O substitution) for Li^+ and Cl^-, respectively (Figure 3.4). The presence of 16.7 mol% of Li^+ and Cl^- solutes in the ice VII cubic Pn-3m lattice increases the volume of the phase by 8%, increasing the density by 0.2 g cm^{-3}. At low temperatures, the presence of Li and Cl ions inhibits the transition to proton-ordered ice VIII down to 80 K. Strong electrostatic interactions between the incorporated ions and the polar water molecules seem to disadvantage long-range ordering of the H_2O orientations,

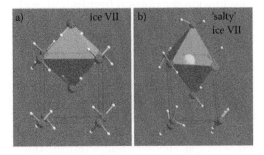

FIGURE 3.4
(a) Proton-disordered ice VII crystallographic structure in one possible proton configuration with the octahedral interstitial face-centered void. (b) Snapshot of a possible configuration of "salty" ice VII (LiCl·6H_2O) derived from *ab initio* calculations illustrating the lattice distortion and H_2O misorientation induced by the incorporation of Li^+ (large circle) and Cl^- (medium circle) ions. (Modified from Klotz, S., *Nat. Mater.*, 8, 2009.)

as it may generate "crystal plasticity"* even at very low temperature. At present, the effect of neutral solutes on the transition to a proton-ordered phase remains unknown.

Incorporating substantial amounts of ionic and molecular solutes may affect the conductivity, volume, density, and thermodynamic stability of water ice. Such modifications of the physical properties of the solid are of interest for planetary scientists, as large water-rich planetary bodies may contain high-pressure ice mantles with thicknesses of hundreds of kilometers. As far as we know, no studies exist of solutes incorporated into other high-pressure ice phases of interest for icy moons and large H_2O-rich planets (e.g., ice III, V, VI, and X).

In the negative pressure regime, phases resembling the geometries of naturally occurring, cage-like clathrate structures are predicted to be thermodynamically stable [Stevenson et al., 1999]. Experiments have not yet been possible on ice at negative pressures. However, one empty clathrate structure could be prepared experimentally and is now known as ice XVI [Kuhs et al., 2014].

At ultrahigh pressures exceeding a few Mbar, currently not accessible experimentally, simulations predict new phases of ice [Wang et al., 2011; Militzer and Wilson, 2010; Hermann et al., 2012; Umemoto and Wentzcovitch, 2011; Sanloup et al., 2013].

3.2.2 Amorphous Ices

Just as polymorphism is regarded to be one of H_2O's anomalous properties, so is polyamorphism: the existence of more than one amorphous solid phase [Mishima et al., 1984, 1985].

One of solid water's amorphous low-density variants, most likely the most abundant form of water in the universe (amorphous solid water, ASW), is produced by the deposition of gaseous water or chemical reaction of atomic H, O, and OH on a very cold solid substrate. Experimentally ASW was produced by deposition of $H_2O_{(g)}$ on a cooled copper rod in early experiments [Ioppolo et al., 2010]. ASW naturally occurs on comets, satellites, interstellar dust, and in cold dark star-forming clouds in space [Mayer and Pletzer, 1986; Ehrenfreund et al., 2003; Ioppolo et al., 2010]). Depending on the conditions of formation (e.g., the temperature and general character of the substrate or the flow rates of the deposit gases [Stevenson et al., 1999; Cartwright et al., 2008; Bossa et al., 2012], the produced ASW may be highly microporous, resulting in specific surface areas of several hundred and even more than 1000 m^2/g [Baragiola, 2003]. These micropores will collapse when temperatures are raised, forming a much more compact variant of ASW with a specific surface area of less than 1 m^2/g [Baragiola, 2003; Mitterdorfer, 2014]. During formation of microporous noncollapsed ASW and in the presence of trace gases (e.g., in the cold regions of dense interstellar clouds), gas molecules can be trapped inside the pores [Ehrenfreund et al., 2003; Mitterdorfer et al., 2011]. Chemical reactions of trapped molecules in the micropores are promoted, and ASW may thus play a pivotal role in the very early stages of planet building [Ehrenfreund et al., 2003]. Furthermore, when pore collapse is induced while guest molecules are trapped, clathrate hydrates may form as a result [Mitterdorfer et al., 2011; Faizullin et al., 2014].

As shown in Figure 3.5, low-density forms of amorphous $H_2O_{(s)}$ are experimentally obtainable in three ways: cryo-deposition of gaseous H_2O (ASW) [Burton and Oliver, 1935], cryo-deposition of micrometer-sized droplets (also on a cooled substrate; hyper-quenched glassy water, HGW) [Mayer, 1985; Kohl et al., 2005] and decompression of higher-density forms of amorphous solid water at elevated temperatures (low-density amorphous ice, LDA) [Mishima et al., 1985]. All of them behave like glassy solids and seem to transform to a deeply supercooled, ultraviscous liquid upon heating, which is known as low-density

---

* Molecular crystal with dynamical disorder in the orientation of its constituent molecules, in which flip rates range from picoseconds to nanoseconds.

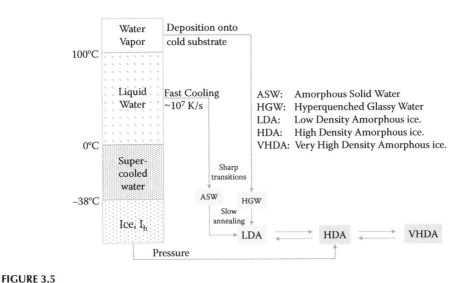

FIGURE 3.5

Preparation routes for amorphous ices, starting from water vapor (leading to ASW), from liquid water (leading to HGW) and from crystalline ice (leading to HDA). Note that the acronyms ASW, HGW, and LDA represent the same structural state (commonly called "LDA") after annealing at >110 K for a few minutes. This is indicated by the dashed arrows. The three polyamorphic forms LDA, HDA, and VHDA can reversibly be interconverted by compression/decompression experiments. (Modified from Fuentes-Landete, V., *Proceedings of the International School of Physics "Enrico Fermi," Volume 187: Water: Fundamentals as the Basis for Understanding the Environment and Promoting Technology*, IOS and Bologna: SIF, Amsterdam, 2015.)

liquid (LDL) water. The temperature of this transformation, known as glass-to-liquid transition, depends on the heating rate, found to be 124 K for slow heating (10 K/h) [Handa et al., 1986], 136 K for "common" heating (10 K/min), and 170 K for very fast heating rates (10^5 K/s) [Sepúlveda et al., 2012]. ASW/HGW/LDA have been shown by DSC at "common" rates to exhibit very similar glass transition temperatures (T_g) of 136–137 K [Hallbrucker et al., 1989; Johari et al., 1987; Elsaesser et al., 2010]. Mechanical indentation experiments have indicated softening and penetration of the material at about 143 K [Johari, 1998].

To date, two higher-density amorphous solid phases are known: high-density amorphous ice (HDA) and very high-density amorphous ice (VHDA), with sub-states of HDA differing in the degree of structural relaxation (unannealed HDA = uHDA [Mishima et al., 1984; Nelmes et al., 2006], expanded HDA = eHDA [Winkel et al., 2008; Salzmann et al., 2006d], and relaxed HDA = rHDA [Salzmann et al., 2006d]). HDA (uHDA) was the first high-density amorphous ice to be produced (in 1984) by compressing I_h to $P > 1$ GPa at 77 K [Mishima et al., 1984] (see Figure 3.5). VHDA can be formed by heating uHDA to temperatures below crystallization at pressures >0.8 GPa [Loerting et al., 2011]. eHDA behaves like a glassy solid and transforms upon heating into a deeply supercooled ultraviscous liquid—HDL water. The glass-to-liquid transition temperature was determined to be 116 K (10 K/min) at ambient pressure [Amann-Winkel et al., 2013].

Isotope substitution neutron diffraction experiments have shown ASW/HGW/LDA to be of highly similar structure [Bowron et al., 2006], as is also the case for uHDA–eHDA [Loerting et al., 2011]. However, uHDA and eHDA differ in their thermal stability (connected to their intrinsic states of relaxation), with uHDA transforming to LDA at approximately 110 K and eHDA at about 132 K at ambient pressure [Winkel et al., 2011].

However, the general question of whether HDA may be considered glassy (or possibly rather an assembly of nanoscaled polycrystallites) has remained [Loerting et al., 2009].

Studies have shown uHDA to first relax and then transform to LDA at P_{atm} [Handa et al., 1986]. At 1 GPa however, there appears to be a reversible glass–liquid transition at about 140 K [Andersson, 2011]. High-pressure experiments have presented dielectric relaxation times of VHDA of 100 s as early as 122 K at 1 GPa [Andersson, 2006]. Dilatometric analysis in the range of 0.1–0.3 GPa has located the onsets of reversible volume changes (attributed to possible glass transitions) to be between 134 and 142 K (at 0.1 and 0.3 GPa, respectively) [Seidl et al., 2011]. DSC measurements on eHDA at P_{atm} have indicated an onset of a glass transition at 116 K, presented as an endothermic feature that could not be found in the case of uHDA [Amann-Winkel et al., 2013]. All these findings suggest transformation of amorphous ices to ultraviscous, deeply supercooled liquids prior to crystallization at <160 K. At 140 K and 0.07 GPa, experimental evidence for spontaneous liquid–liquid phase separation and the formation of an interface between two ultraviscous liquids differing by 25% in density was obtained [Winkel et al., 2011]. At higher temperatures, these deeply supercooled liquids crystallize inevitably and rapidly, and so the thermodynamic connection between amorphous ices/deeply supercooled water and stable/supercooled water above 230 K remains unclear. At astrophysical time scales, water between about 160 and 240 K will always be crystalline.

3.2.3 Rheological Considerations

Durham and Stern [2001] reviewed the rheological properties of water ices, which influence planetary geology, including tectonics, crater formation and relaxation, and global thermal evolution.

Other low temperature materials produce frozen volatiles that are thought to be abundant on icy satellites and hypothesized exoplanets. Ammonia has been identified as a likely constituent in icy bodies beyond Jupiter, which implies a lower ice temperature and a strongly temperature-dependent viscosity [Croft et al., 1988; Durham et al., 1993; Arakawa and Maeno, 1994; Hogenboom et al., 1997; Leliwa-Kopystynski et al., 2002; Fortes et al., 2003]. The thermal conductivity of ammonia-rich (10%–30%) ice is two to three times lower than that of pure water ice, and the loss tangent is about 100 times greater [Lorenz and Shandera, 2001]. Intrinsic absorption in the ice matrix might be responsible for the latter. A recent study of the rheology of ammonia-water slurries as a function of temperature and strain rate show the development of yield stress-like behaviors, shear-rate dependence, and thixotropic behavior, even at relatively low crystal fractions [Carey et al., 2015]. Light and heavy alkanes familiar to the petroleum industry cover the surface of Titan and populate its atmosphere. Formation of methane and ethane clathrates may play a critical role in the evolution of Titan's atmosphere [Choukroun et al., 2010]. Such work has implications for icy satellites where cryovolcanism may exist, such as Triton and Titan.

Rheologies of other frozen volatiles have been studied in connection with other planets and a few locations on Earth: methane clathrate [Stern et al., 1996; Durham et al., 2003], CO2 [Durham et al., 1999], and N2 and CH4 [reviewed by Eluszkiewicz and Stevenson 1990]. Both brittle and ductile behaviors were observed for solid nitrogen and methane; the maximum strengths were determined to be 9 and 10 MPa, respectively, in the brittle failure mode [Yamashita et al., 2010]. These low strengths suggest that H_2O ice or other stronger materials may underlay solid N_2 and CH_4 to generate the topography observed on Triton.

The interaction of materials with different rheologies, melting points, and mixing behaviors has led planetary scientists to consider cryovolcanism as a class of geological process not found on Earth (Figure 3.6) [e.g., Kargel, 1994; Gaidos, 2001; Fagents, 2003; Sotin et al., 2005; Porco et al., 2006; Fortes et al., 2007; Desch et al., 2009; Cooper et al., 2009].

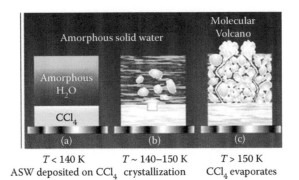

FIGURE 3.6
Amorphous solid water at various low temperatures. (Courtesy of AIP. With permission.)

3.3 Liquid Materials at Low Temperatures

Low temperature liquids are of interest for industrial and laboratory applications, and they exist in a surprising number of places throughout the solar system. Estimates of the thermal evolution of small icy planetary objects in the solar system indicate that natural radiogenic heating of their interiors alone would lead to internal liquid water oceans lasting millions of years [Hussmann et al., 2006]. Temperatures in the young outer solar system beyond Jupiter were below the freezing temperature of volatile species such as alkanes and ammonia, so these would have been included in distant worlds such as Pluto.

3.3.1 Structure of Liquids

Low temperature liquids display a range of viscosity, density, and thermal expansion. These properties determine the efficiency of thermal and material transport of extraterrestrial oceans [e.g., Vance and Goodman 2009; Vance et al., 2014].

Related to the existence of more than one amorphous phase of solid water is the hypothesis of there being more than one liquid phase of water in the supercooled region [Finney et al., 2002]. The liquid phases would be thermodynamically continuously connected to the amorphous solid phases and would correspondingly be of different density (LDL and HDL). At about 0.2 GPa, these two liquids may exist in thermodynamic equilibrium, that is, with a sharp interface between the two liquids of pure H_2O. So, the one-component system water at ambient conditions might in fact be a fluctuating mixture of two liquids, which may separate under low temperature and high pressure conditions. This, in turn, may indicate the possibility of a second critical point in the regime of these two different liquid phases [Stanley et al., 2000]. This second critical point cannot be accessed experimentally because of fast crystallization at its predicted location, and so has remained a virtual point up to now.

3.3.2 Properties of Water and Aqueous Systems

Aqueous solutions are stable in liquid form at an elevated pressure (250 MPa) to temperatures as low as 160 K [Mishima, 2011]. In the pure phase, liquid water exhibits ordering behavior that causes a maximum density at 4°C anomalous to the behavior of other simple fluids (Figure 3.7). The anomalous nature of water becomes more pronounced in

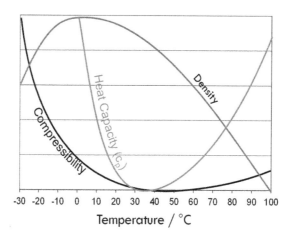

FIGURE 3.7
Compressibility, heat capacity, and density versus temperature. (Adapted from http://www.lsbu.ac.uk/water/.)

the supercooled state below 0°C. Response functions such as heat capacity or isothermal compressibility show a power-law increase, where a singular temperature of about 228 K was identified [Speedy and Angell, 1976]. These anomalies can be rationalized if the single-component system water experiences a spontaneous liquid–liquid separation into a HDL and LDL [Debenedetti, 2003]. The occurrence of polyamorphism (see Section 3.2.2) and the experimental evidence for spontaneous phase separation at 140 K and 0.07 GPa into the two ultraviscous liquids LDL and HDL [Winkel et al., 2011] support this theory. In the presence of sufficient amounts of solutes, the anomalies of water disappear and concentrated aqueous solutions behave as simple liquids.

Low temperature eutectic brines, the last occurring phase in freezing aqueous systems, represent a likely ocean composition in the ice-covered worlds [e.g., Zolotov and Kargel, 2009]. The array of possible compositions comprises a multicomponent space of low temperature liquids that requires a systematic and extensible representation of possible thermodynamic states. Such a systematic framework for computing thermodynamics of aqueous systems has been available for decades, based on the theoretical formulation of Kenneth Pitzer [1991]. The FREZCHEM implementation of the Pitzer thermodynamic framework (e.g., Marion et al., 2012) balances the activity of water (excess chemical potential) of aqueous solutions to compute the properties of aqueous systems in the presence of ice I. The underlying Gibbs energy is computed as the sum of contributions from dissolved cations (c) and anions (a), with pressure- and temperature-dependent coefficients describing the cation–anion interactions (B_{ca} and C_{ca}). Higher-order interactions become relevant for higher concentrations.

$$\frac{G^{ex}}{w_w RT} = f(I) + 2\sum_c \sum_a m_c m_a \left[B_{ca} + \left(\sum_c m_c z_c \right) C_{ca} \right]$$

$$+ \sum_{c<c'} \sum m_c m_{c'} \left[2\Phi_{cc'} + \sum_a m_a \psi_{cc'a} \right] + \sum_{a<a'} \sum m_a m_{a'} \left[2\Phi_{aa'} + \sum_c m_c \psi_{caa'} \right]$$

$$+ 2\sum_a \sum_c m_a m_c \lambda_{ac} + 2\sum_n \sum_a m_n m_a \lambda_{na} + 2\sum_{n<n'} \sum m_n m_{n'} \lambda_{nn'} + \sum_n m_n^2 \lambda_{nn} + \cdots$$

The Pitzer framework is useful for representing the multicomponent thermodynamics of water, but limited in its application to other worlds by Earth-centric data sets as illustrated by available density measurements for $MgSO_4$ shown in Figure 3.8.

Sound speed and specific heat capacity measurements in solutions at low temperature provide a sensitive probe of thermodynamic behaviors. Both are related to Gibbs free energy through the second derivatives in pressure and temperature, respectively:

$$\left(\frac{\partial \rho}{\partial P}\right)_{T,m} = \frac{1}{c^2} + \frac{T\alpha^2}{C_P}$$

$$\left(\frac{\partial C_P}{\partial P}\right)_{T,m} = -T\frac{\partial^2 V}{\partial T^2}$$

$$V = \frac{m}{\rho} = \left(\frac{\partial G}{\partial P}\right)_{T,m}$$

$$C_P = -T\frac{\partial^2 G}{\partial T^2}$$

Sound speed measurements in water [Vance and Brown, 2010; Lin and Trusler, 2012] and aqueous solutions [Chen and Millero 1977; Chen et al., 1978; Millero et al., 1985; Vance and Brown 2013] and nonaqueous systems at low temperatures provide a sensitive measure of water's thermodynamic properties, which can be used to systematically predict the behavior of multicomponent systems in low temperature settings.

Developing large thermodynamic frameworks is made possible by modern computers that can solve large inverse problems associated with sparse multiproperty thermodynamic data sets [Lemmon et al., 2013] and forward problems associated with *ab initio* molecular dynamics calculations [e.g., Hassanali et al., 2011].

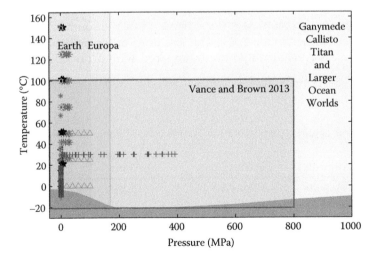

FIGURE 3.8
Pressures and temperatures of density measurements in the $MgSO_4(aq)$ system illustrate that studies typically explore convenient laboratory conditions and conditions relevant to Earth's oceans. (Modified from *Cosmochim. Acta*, 110, Vance, S., J.M. Brown, 176–189, Copyright 2013, with permission from Elsevier.)

3.3.3 Properties of Nonaqueous Systems

Nonaqueous liquids play important roles in the chemistry and transport of materials at cryogenic temperatures. The best example of these interactions can be found on Titan, a moon of Saturn with a thick atmosphere and liquid hydrocarbon lakes. Methane and ethane form the liquid phase and are involved in an active cycle similar to the hydrologic cycle on Earth—these hydrocarbons form clouds, precipitate onto the surface, and pool in lakes at the North and South poles [Brown et al., 2010].

Titan also contains a vast inventory of organic molecules generated by photochemistry in the atmosphere. These species form solid aerosol particles that fall onto the surface, forming dunes in the equatorial regions [Lorenz et al., 2006]. Interactions of liquid hydrocarbons with solid phases include pluvial (via rainfall) and fluvial (via river) transport of particulates into the lakes [Lorenz et al., 2008], and may also generate landforms similar to limestone or gypsum karst structures on Earth [Malaska and Hodyss, 2014; Cornet et al., 2015].

In fact, the conditions present on Titan may be much more ubiquitous throughout the universe than those on Earth. The most common type of stable star in the universe is the M-dwarf (red dwarf), which is smaller and less luminous than the Sun (a G-dwarf). Planets in a safe orbit around such a star (1 AU) would receive a similar amount of radiation as Titan does—just enough to maintain liquid methane and ethane on the planet's surface. Since M-dwarfs are 10–100 times more abundant than G-dwarfs, it may be the case that Titan-like worlds are much more common than Earth-like worlds [Lunine, 2009]. Studies of the physical properties of nonaqueous liquids such as methane and ethane, therefore, may have implications for the exploration of many worlds.

The viscosities and densities of liquid methane, ethane, and propane have been reported at cryogenic temperatures [Haynes, 1973; Diller and Saber, 1981; Goodwin et al., 1973; Diller, 1982; Goodwin, 1977], as these are relevant for the liquefied natural gas industry. Thermal conductivities for these liquid hydrocarbons have also been measured [Younglove and Ely, 1987]. Values relevant to Titan surface conditions are shown in Table 3.1.

A recent study reported the complex dielectric constants of liquid methane and ethane at 90 K [Mitchell et al., 2015]. These loss tangents suggest that the northern lake Ligeia Mare of Titan is almost entirely comprised of methane, while the southern lake Ontario Lacus is more ethane-rich [Brown et al., 2008]. The loss tangent for methane ($2.86 \pm 1.01 \times 10^{-5}$) from Mitchell et al. is considerably lower compared with a previous study (1.14×10^{-3})

TABLE 3.1

Physical Properties of Liquid Hydrocarbons at Titan Surface Temperature (95 K)

Hydrocarbon	Viscosity (Pa·s)	Density (g/cm³)	Thermal Conductivity (W/m·K)	Reference(s)
Methane	1.784×10^{-4}	0.4458	0.2155	Haynes [1973] and Hanley et al. [1977]
Ethane	1.073×10^{-3}	0.6468	0.261	Diller and Saber [1981], Goodwin et al. [1973], and Younglove and Ely [1987]
Propane	5.211×10^{-3}	0.7234	0.226	Diller [1982], Goodwin [1977], and Younglove and Ely [1987]

[Paillou et al., 2008], although the authors note that the Paillou et al. sample contained ~5% heavier hydrocarbons and is probably higher due to this contamination. Work with liquid alkane mixtures [Sen et al., 1992] indicates that the dielectric constant varies linearly with volume fraction as predicted by the Clausius–Mossotti equation [Hill et al., 1969]:

$$(\varepsilon'_m - 1) / (\varepsilon'_m + 2) = \Sigma_i 4\pi\upsilon_i\rho_i N_A \alpha_i / 3M_i$$

where ε'_m is the dielectric constant of the mixture, and for the ith component of the mixture, υ_i is the volume fraction, ρ_i is the mass density, α_i is the electric polarizability, and M_i is the molecular weight.

Interactions of liquid hydrocarbons with solid organics at cryogenic temperatures have also been explored. Simulated Titan aerosols, termed "tholins," have been generated using various energy sources (cold plasma discharge, UV irradiation, etc.) and exposed to liquid hydrocarbons. These complex organics are relatively insoluble in liquid hydrocarbons, although adsorption onto the surface of these particles may enable chemistry at a gas–liquid interface as they fall through the atmosphere of Titan (for a review of Titan tholin formation and reactivity, see Cable et al., 2012).

Although tholins are not very soluble in liquid hydrocarbons, other organic species such as acetylene and HCN are believed to comprise approximately 1% and 2% of the Titan lakes, respectively [Cordier et al., 2009]. Acetylene is capable of polymerization on Titan [Raulin, 1987] and can serve as an energy source for methanogenic bacteria [McKay and Smith, 2005].

Less soluble species also have implications for Titan surface physical properties. Liquid ethane readily forms a co-crystal with solid benzene at 90 K, similar to a hydrated mineral on Earth [Cable et al., 2014; Vu et al., 2014]. These co-crystals may be present in evaporitic deposits around Titan lakes, and can affect evaporite properties such as particle size and dissolution rate. The crystal structure of the benzene–ethane co-crystal has recently been obtained at 90 K using synchrotron powder diffraction [Maynard-Casely et al., 2016].

Understanding the interaction of nonaqueous liquids with spacecraft materials is significant, both for interpreting data from the Huygens lander [Niemann et al., 2005; Lebreton et al., 2005] and for designing future *in situ* Titan missions. Methane behaves like a viscous non-Newtonian fluid just below its melting point (90.6 K), and adheres to materials such as aluminum, stainless steel, and PTFE with similar affinity [Kirichek et al., 2012]. Exposure of hydrocarbons to materials such as aluminum, steel, and titanium at cryogenic temperatures has been extensively explored in the liquefied natural gas industry, and no chemical reactivity has been reported [Kaufman, 1975].

Lorenz [2015] studied the effects of heat rejection for a Titan lake lander or submersible. For a lander with heat leaks on the order of 100 W/m² into the hydrocarbon liquid, experimental simulations indicate the formation of nitrogen bubbles (exsolution) that may interfere with scientific measurements. Such heat leaks should be avoided by using thick insulation and rejecting as much heat as possible into the atmosphere.

3.3.3 Phase Transitions

Solid–liquid phase boundaries define multiple curves that are difficult to treat by extrapolation. Details of the melting point as a function of composition and temperature determine where liquids are stable in natural settings on Earth and other planets (e.g., Figures 3.9 and 3.10). Careful measurements of phase boundaries and crystallization kinetics in aqueous systems are enabled by high-resolution Raman spectroscopy [e.g., Bollengier et al., 2013; Cable et al., 2014].

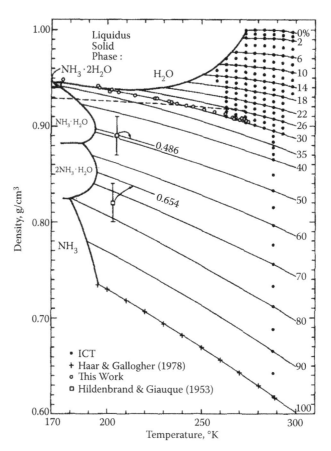

FIGURE 3.9

Phase diagram for ammonia at atmospheric pressure, showing a eutectic temperature of 170 K at 29 wt% and peritectic transitions separating different hydration states of ice at ~55 wt%, 185 K and 80 wt%, 180 K. (Reprinted from *Icarus*, 73(2), Croft, S.K., J.I. Lunine, J. Kargel, Equation of state of ammonia water liquid: Derivation and planetological applications, 279–293, Copyright 1988, with permission from Elsevier.)

3.4 Material Properties of Piezoelectric Materials at Cryogenic Temperatures

Piezoelectric materials are used for investigation of material thermodynamic properties, as described in the preceding text. They are also used as actuators in analytical instruments at cryogenic temperatures to investigate quantum physical phenomena at the nanometer scale, providing precision-controlled displacements for specimen stages used with scanning probes [Eriksson, 1996]. Piezoelectric materials are also used to actively tune superconducting radio frequency (SRF) cavities, enabling maintained resonance at high electromagnetic field intensity. Electro-motive and magneto-motive materials are also used at cryogenic temperatures as the active material in MEMS and NEMS sensor devices to investigate quantum and low temperature physical phenomena at the nanometer scale, for example [Collin et al., 2011]

The electro-mechanical properties of piezoelectric ceramics result from complex interdependency of both mechanical and electrical material properties, and are highly

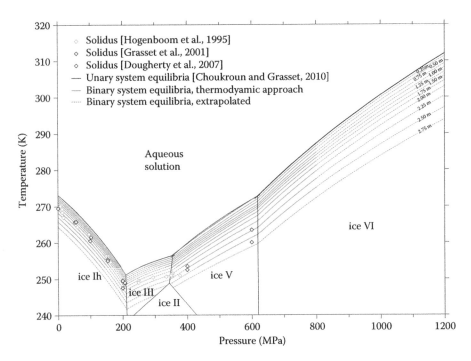

FIGURE 3.10
Suppression of the melting temperature of ices for increasing concentrations of $MgSO_4$ (aq), computed from available solidus data and Gibbs free energies for the fluid phase constructed from b-spline-based thermodynamic equations of state. (Courtesy of Olivier Bollengier, University of Washington, Seattle, Washington.)

temperature dependent. Advancing cryogenic instruments and sensors relying on piezoelectric elements requires characterizing their electromechanical coupling at low temperatures.

Lead zirconate tinate (PZT) is one of the most common piezoelectric materials utilized after quartz. Varying the percentages of constituent materials in PZT affects the crystallographic composition of the piezoelectric crystal. In pure bulk PZT crystals, compositions near the morphotropic boundary (between tetragonal and rhombohedral crystallographic phases) show the largest temperature dependence of pertinent electrical and mechanical material properties (dielectric constant and dissipation factor, respectively) [Zhuang et al., 1989]. Temperature–function relationships for PZT films have since been phenomenologically derived for several compositionally and crystallographically differing PZT films of varying thickness [Wolf and Trolier-McKinstry, 2004]. The variability in electro-mechanical coupling factors in differing composition of PZT films correlates with the differing values of Curie temperature. Furthermore, the relative electrical permittivity decreases with temperature. Decreasing permittivity diminishes the requisite electric field necessary to align poled domains (and thus create strain). Finally, the thermal expansion mismatch between the PZT film and substrate creates both strain and residual stress during and after temperature cycling [Wolf and Trolier-McKinstry, 2004], which further degrades the film's performance through low temperature cycling.

Piezoelectric stack actuator performance at low temperatures has been characterized in the development of active tuning of SRF cavities. In these studies, commercially available actuators are tested both electrically (dielectric loss) and mechanically (displacements) at cryogenic temperatures. In conjunction, thermal properties were also measured, specifically heating due to dielectric loss. In general, Fouaidy's results coincide with the prior measurements and phenomenological derivations [Wolf and Trolier-McKinstry, 2004; Zhuang et al., 1989], showing a marked decrease in electro-mechanical conversion at low temperatures. The capacitance of the actuators decreases with temperature and mechanical performance. However, hysteresis in the mechanical response to applied voltages decreases to a negligible value at temperatures below 4 K. Despite approximately a 95% reduction in mechanical performance between the range of 300 K and 1.8 K, displacements on the order of several micrometers have been observed at the lowest temperatures. As temperatures decrease and mechanical performance diminishes, dielectric heating of the actuator diminishes. Despite significant degradation near 1.8 K, the commercial actuators perform in the range viable for deployment as active tuning devices for SRF cavities, with the benefits of reduced thermal contributions from dielectric heating and increased accuracy due to minimal response hysteresis [Fouaidy et al., 2005].

Neutron flux (~2.1 × 10^{15} N/cm^2) has been used to verify the efficacy of piezoelectric stack actuators for use in SRF cavities. These studies indicated negligible degradation due to radiation, and are thus viable in the radiation environment [Martinet, 2006]. Over the anticipated lifetime of an SRF cavity (>>3 Giga cycles), the effects of dynamic preloading on the electrical properties of the actuator stacks is predominantly linear throughout the range of temperatures (~2–300 K) [Fouaidy et al., 2007]. Increased preloading force corresponds to increased capacitance, and thereby mechanical performance. However, the proportionality of this linear correlation, or sensitivity, decreases with lower temperatures, with a 1000 N increase in preloading force corresponding to an increase in capacitance of 426 nF for actuators at 300 K, but only an increase of 16 nF for actuators at 2 K. Unlike the hysteresis in mechanical response to applied voltage at low temperatures, the change in electrical properties (capacitance) shows significant hysteresis during loading and unloading the piezoelectric actuators at $T = 2$ K [Fouaidy et al., 2007].

Piezoelectric material properties affecting electro-mechanical performance have been studied at cryogenic temperatures under various external conditions. In general, there is good agreement that the maximum mechanical displacement from a given voltage decreases with temperature. The degree of performance degradation is largely affected by the composition and associated crystallographic morphology of the piezoelectric device. Both bulk and thin-film exhibit reduced mechanical performance at a lower temperature, with more interfering phenomena in thin-film devices (thermal expansion mismatch, etc.). Despite the decrease in performance, piezoelectric material properties provide consistent mechanical response at cryogenic temperatures, with the added benefits of reduced dielectric heating and hysteresis. For applications in sensitive instrumentation and small-scale sensors, the loss in maximum displacement capabilities has little effect on the viability of the piezoelectric material.

3.5 Summary/Conclusions

The frontiers of physics and coupled human–robotic exploration beyond Earth drive progress in understanding the properties of low temperature materials. Planetary low-temperature materials comprise a large space of volatile compositions, involving complex rheologies, and multiple crystalline, amorphous, and liquid phases. Low temperature technologies enable innovative measurement techniques for probing fundamental properties such as sound speeds and phase boundaries, while fast computers enable detailed thermodynamic frameworks that will be necessary for future planetary exploration.

3.6 Acknowledgments

Some of the research reported in this chapter was conducted at the Jet Propulsion Laboratory (JPL), California Institute of Technology (Caltech), under a contract with the National Aeronautics and Space Administration (NASA). T.L. would like to acknowledge continuing support of his work by the Austrian Science Fund FWF (recently through START award Y391 and bilateral project I1392) and the European Research Council (ERC Starting Grant SULIWA). The authors would like to thank Jeffrey D. Hein, Robert P. Hodyss, and Victoria (Maria) Iglesias, JPL/Caltech, Pasadena, CA, for reviewing this chapter and providing valuable technical comments and suggestions.

References

Algara-Siller, G., O. Lehtinen, F.C. Wang, R.R. Nair, U. Kaiser, H.A. Wu, A.K. Geim, I.V. Grigorieva. Square ice in graphene nanocapillaries, *Nature*, 519 (2015), pp. 443–445.

Angell, C.A. Supercooled water, *Annu. Rev. Phys. Chem.*, 34 (1983), pp. 593–630.

Amann-Winkel, K., C. Gainaru, P.H. Handle, M. Seidl, H. Nelson, R. Bohmer, T. Loerting. Water's second glass transition, *Proc. Natl. Acad. Sci. U. S. A.*, 110 (2013), p. 17720.

Andersson O., A. Inaba. Dielectric properties of high-density amorphous ice under pressure, *Phys. Rev. B.*, 74 (2006), p. 184201.

Andersson O. Glass–liquid transition of water at high pressure, *Proc. Natl. Acad. Sci. U. S. A.*, 108 (2011), p. 11013.

Arakawa, M., N. Maeno. Effective viscosity of partially melted ice in the ammonia-water system, *Geophysi. Res. Lett.*, 21(14) (1994), pp. 1515–1518.

Arnold G.P., E.D. Finch, S.W. Rabideau, R.G. Wenzel. Neutron-diffraction study of ice polymorphs. III. Ice Ic, *J. Chem. Phys.* 49 (1968), p. 4365.

Baragiola R.A., in: V. Buch, J.P. Devlin (Eds.), *Water in Confining Geometries*, Heidelberg: Springer-Verlag (2003).

Benoit M., M. Bernasconi, P. Focher, M. Parrinello. New high-pressure phase of ice, *Phys. Rev. Lett.*, 76(16) (1996), pp. 2934–2936, doi:10.1103/PhysRevLett.76.2934.

Bernal J.D., R.H. Fowler. A theory of water and ionic solution, with particular reference to hydrogen and hydroxyl ions, *J. Chem. Phys.*, 1 (1933), p. 515.

Besson J.M., M. Kobayashi, T. Nakai, S. Endo, P. Pruzan. Pressure dependence of Raman linewidths in ices VII and VIII, *Phys. Rev. B.*, 55 (1997), p. 11191.

Besson J.M., P. Pruzan, S. Klotz, G. Hamel, B. Silvi, R.J. Nelmes, J.S. Loveday, R.M. Wilson, S. Hull. Variation of interatomic distances in Ice VIII to 10 GPa, *Phys. Rev. B.*, 49 (1994), p. 12540.

Bollengier, O., M. Choukroun, O. Grasset, E. Le Menn, G. Bellino, Y. Morizet, L. Bezacier, A. Oancea, C. Taffin, G. Tobie. Phase equilibria in the $H_2O–CO_2$ system between 250–330K and 0–1.7 GPa: Stability of the CO_2 hydrates and H_2O-ice VI at CO_2 saturation, *Geochim. Cosmochim. Acta.*, 119 (2013), pp. 322–339.

Bossa J.-B., K. Isokoski, M.S.d. Valois, H. Linnartz. Thermal collapse of porous interstellar ice, *Astron. Astrophys.*, 545 (2012), p. A82.

Bowron D.T., J.L. Finney, A. Hallbrucker, I. Kohl, T. Loerting, E. Mayer, A.K. Soper. The local and intermediate range structures of the five amorphous ices at 80 K and ambient pressure: A Faber-Ziman and Bhatia-Thornton analysis, *J. Chem. Phys.*, 125 (2006), p. 194502.

Brown, R. H., L.A. Soderblom, J.M. Soderblom, R.N. Clark, R. Jaumann, J.W. Barnes, C. Sotin, B. Buratti, K.H. Baines, P.D. Nicholson. The identification of liquid ethane in Titan's Ontario Lacus, *Nature*, 454 (2008), pp. 607–610.

Brown, R., J.P. Lebreton, H.J. Waite, (Eds.) *Titan from Cassini-Huygens*. Netherlands: Springer (2010).

Bruggeller, P., E. Mayer. Complete vitrification in pure liquid water and dilute aqueous solutions, *Nature*, 288 (1980), p. 569.

Burton E.F., W.F. Oliver. X-ray diffraction patterns of ice, *Nature*, 135 (1935), p. 505.

Cable, M.L., T.H Vu, R. Hodyss, M. Choukroun, M. J. Malaska, P. Beauchamp. Experimental determination of the kinetics of formation of the benzene-ethane co-crystal and implications for Titan, *Geophys. Res. Lett.*, 41(15) (2014), pp. 5396–5401.

Cable, M.L., S.M. Horst, R. Hodyss, P.M. Beauchamp, M.A. Smith, P.A. Willis. Low-temperature microchip nonaqueous capillary electrophoresis of aliphatic primary amines: Applications to Titan chemistry, *Chem. Rev.*, 112 (2012), pp. 1882–1909.

Carey, E., K. Mitchell, M. Choukroun, F. Zhong. Laboratory studies on the rheology of cryogenic slurries with implications for icy satellites, *Bull. Amer. Phys. Soc.*, 60(4) (2015), p. R2.00006.

Cartwright J.H.E., B. Escribano, C.I. Sainz-Diaz. The mesoscale morphologies of ice films: Porous and biomorphic forms of ice under astrophysical conditions, *Astrophys. J.*, 687 (2008), p. 1406.

Chen, C.T., F.J. Millero. Speed of sound in seawater at high-pressures, *J. Acoust. Soc. Am.*, 62(5) (1977), pp. 1129–1135.

Chen, C.T., L.S. Chen, F.J. Millero. Speed of sound in NaCl, $MgCl_2$, Na_2SO_4, and $MgSO_4$ aqueous-solutions as functions of concentration, temperature, and pressure, *J. Acous. Soc. Am.*, 63(6) (1978), pp. 1795–1800.

Choukroun, M., O. Grasset. Thermodynamic model for water and high-pressure ices up to 2.2 GPa and down to the metastable domain, *J. Chem. Phys.*, 127(12) (2007), p. 124506.

Choukroun, M., O. Grasset, G. Tobie, C. Sotin. Stability of methane clathrate hydrates under pressure: Influence on outgassing processes of methane on Titan, *Icarus*, 205(2) (2010), pp. 581–593.

Collin E., T. Moutonet, J.S. Heron, O. Bourgeois, Y.M. Bunkov, H. Godfrin. A tunable hybrid electro-magnetomotive NEMS device for low temperature physics, *J. Low Temp Phys.*, 162(2011), p. 653.

Cooper, J., P. Cooper, E. Sittler, S. Sturner, A. Rymer. Old faithful model for radiolytic gas-driven cryovolcanism at Enceladus, *Planet. Space Sci.*, 57(13) (2009), pp. 1607–1620.

Conde, M.M., C. Vega, G.A. Tribello, B. Slater. The phase diagram of water at negative pressures: Virtual ices, *J. Chem. Phys.*, 131 (2009), pp. 034510-1–034510-8.

Cordier, D., O. Mousis, J.I. Lunine, P. Lavvas, V. Vuitton. An estimate of the chemical composition of Titan's lakes, *Astrophys. J. Lett.*, 707(2) (2009), pp. L128–L131.

Cornet, T., D. Cordier, T. Le Bahers, O. Bourgeois, C. Fleurant, S. Le Mouelic, N. Altobelli. Dissolution on Titan and on Earth: Towards the age of Titan's karstic landscapes, *J. Geophys. Res.*, (2015), doi10.1002/2014JE004738.

Diller, D.E., J.M. Saber. Measurements of the viscosity of compressed gaseous and liquid ethane, *Phys. A: Stat. Mech. Appl.*, 108(1) (1981), pp. 143–152.

Diller, D.E. Measurements of the viscosity of saturated and compressed liquid propane, *J. Chem. Eng. Data*, 27(3) (1982), pp. 240–243.

Croft, S.K., J.I. Lunine, J. Kargel. Equation of state of ammonia water liquid: Derivation and planetological applications, *Icarus*, 73(2) (1988), pp. 279–293.

Datchi, F., P. Loubeyre, R. Letoullec. Extended and accurate determination of the melting curves of argon, helium, ice (H_2O), and hydrogen (H_2), *Phys. Rev. B.*, 61(10) (2000), pp. 6535–6546.

Debenedetti P. G. Supercooled and glassy water, *J. Phys.: Condens. Matter*, 15 (2003), p. R1669.

Desch, S., J. Cook, T. Doggett, S. Porter. Thermal evolution of Kuiper belt objects, with implications for cryovolcanism, *Icarus*, (2009) doi:10.1016/j.icarus.2009.03.009.

De Yoreo, J.J., P.G Vekilov. Principles of crystal nucleation and growth, *Rev. Mineral. Geochem.*, 54 (2003), pp. 57–93.

Durham, W., S. Kirby, L. Stern. Flow of ices in the ammonia-water system, *J. Geophys. Res.: Solid Earth*, 98(B10), (1993), pp. 17667–17682.

Durham, W. B., S.H. Kirby, L.A. Stern. Steady-state flow of solid CO_2: Preliminary results, *Geophys. Res. Lett.*, 26(23) (1999), pp. 493–3496.

Durham, W., S. Kirby, L. Stern. Rheology of water ice—Applications to satellites of the outer planets, *Ann. Rev. Earth Planet. Sci.*, 29 (2001), pp. 295–330.

Ehrenfreund, P., H.J. Fraser, J. Blum, J.H.E. Cartwright, J.M. Garcia-Ruiz, E. Hadamcik, A.C. Levasseur-Regourd, S. Price, F. Prodi, A. Sarkissian. Physics and chemistry of icy particles in the universe: Answers from microgravity, *Planet. Space Sci.*, 51 (2003), p. 473.

Elsaesser, M.S., K. Winkel, E. Mayer, T. Loerting. Reversibility and isotope effect of the calorimetric glass → Liquid transition of low-density amorphous ice, *Phys. Chem. Chem. Phys.*, 12(3) (2010), pp. 708–712.

Eluszkiewicz, J., D. Stevenson. Rheology of solid methane and nitrogen—Applications of Triton, *Geophys. Res. Lett.*, (ISSN 0094-8276) (1990), p. 17.

Eucken, A. The change in heat conductivity of solid metalloids with temperature, *Ann. Phys.*, 34 (1911), pp. 185–222.

Fagents, S.A. Considerations for effusive cryovolcanism on Europa: The post-Galileo perspective, *J. Geophys, Res. Planets*, 108(E12) (2003), p. 5139.

Faizullin, M.Z., A.V. Vinogradov, V.N. Skokov, V.P. Koverda. Fomation of gas hydrate during crystallization of ethane-saturated amorphous ice, *Russ. J. Phys. Chem.*, 88 (2014), p. 1706.

Falenty, A., T.C. Hansen, W.F. Kuhs, Formation and properties of Ice XVI obtained by emptying a type sII clathrate hydrate, *Nature*, 516 (2014), p. 231.

Fan X., D. Bing, J. Zhang, Z. Shen, J.-L. Kuo. Predicting the hydrogen bond ordered structures of Ice I_h, II, III, VI and Ice VII: DFT methods with localized based set, *Comp. Mater. Sci.*, 49 (2010), p. S170.

Finney, J., A. Hallbrucker, I. Kohl, A. Soper, D. Bowron. The local and intermediate range structures of the five amorphous ices at 80 K and ambient pressure: A Faber-Ziman and Bhatia-Thornton analysis. *Phys. Rev. Lett.*, 88(22) (2002), p. 225503.

Fortes, A., P. Grindrod, S. Trickett, L. Vocadlo. Ammonium sulfate on Titan: Possible origin and role in cryovolcanism, *Icarus*, 188(2007), pp. 139–153.

Fortes, A.D., I.G. Wood, J.P. Brodholt, L. Vocadlo. Ab initio simulation of the Ice II structure, *J. Chem. Phys.*, 119 (2003), p. 4567.

Fortes A.D., I.G. Wood, M. Alfredsson, L. Vocadlo, K.S. Knight. The incompressibility and thermal expansivity of D_2O Ice II determined by powder neutron diffraction, *J. Appl. Cryst.*, 38 (2005), p. 612.

Fouaidy, M., G. Martinet, N. Hammoudi, F. Chatelet, S. Blivet, A. Olivier, H. Saugnac. Proceedings from Particle Accelerator Conference (2005), p. 728.

Fouaidy, M., M. Saki, N. Hammoudi, L. Simonet. Electromechanical characterization of piezoelectric actuators subjected to a variable preloading force at cryogenic temperature. pp. 1–15, Report No. CARE-Note-2006-007-SRF, HAL Id: in2p3-00148455, https://hal.archives-ouvertes.fr /in2p3-00148455/document.

Frank, M., R.E. Aarestad, H.P Scott, V.B. Prakapenka. A comparison of ice VII formed in the H_2O, NaCl-H_2O, and CH_3OH-H_2O systems: Implications for H_2O-rich planets, *Phys. Earth Planet. Interiors*, 215(C) (2013), pp. 12–20. doi:10.1016/j.pepi.2012.10.010.

Frank M., C. Runge, H. Scott, S. Maglio, J. Olson, V. Prakapenka, G. Shen. Experimental study of the NaCl–H2O system up to 28 GPa: Implications for ice-rich planetary bodies, *Phys. Earth Planet. Interiors*, 155(12) (2006), pp. 152–162. doi:10.1016/j.pepi.2005.12.001.

Fuentes-Landete, V., C. Mitterdorfer, P.H. Handle, G.N. Ruiz, J. Bernard, A. Bogdan, M. Seidl, K. Amann-Winkel, J. Stern, S. Fuhrmann, T. Loerting. Crystalline and amorphous ices. In: P. G. Debenedetti, M. A. Ricci, and F. Bruni (Eds.), *Proceedings of the International School of Physics "Enrico Fermi,", Volume 187: Water: Fundamentals as the Basis for Understanding the Environment and Promoting Technology*. Amsterdam: IOS and Bologna: SIF (2015), pp. 173–208.

Fukazawa, H., S. Ikeda, M. Oguro, T. Fukumura, S. Mae. Deuteron ordering in KOD-doped ice observed by neutron diffraction, *J. Phys. Chem. B.*, 106 (2002), pp. 6021.

Fukazawa, H., S. Ikeda, S. Mae. Incoherent inelastic neutron scattering measurements on ice XI; the proton-ordered phase of ice I_h doped with KOH, *Chem. Phys. Lett.*, 282 (1998), p. 215.

Gaidos, E.J. Cryovolcanism and the recent flow of liquid water on Mars, *Icarus*, 153(1) (2001), pp. 218–223.

Geiger P., C. Dellago, M. Macher, C. Franchini, G. Kresse, J. Bernard, J.S. Stern, T. Loerting. Proton ordering of cubic ice I_c: Spectroscopy and computer simulations, *J. Phys. Chem. C.*, 118(20) (2014), pp. 10989–10997.

Goodwin, R.D. Provisional thermodynamic functions of propane, from 85 to 700 K at pressures to 700 bar, *Nat. Bur. Stand.* (U.S.) Interagency Report NBSIR 77–860 (1977).

Goodwin, R.D., H.M. Roder, G.C Straty. Thermophysical properties of ethane, from 90 to 600 K at pressures to 700 bar, *Nat. Bur. Stand.* (U.S.), Technical Note 684 (1973).

Grasset, O., J. Schneider, C. Sotin. A study of the accuracy of mass-radius relationships for silicate-rich and ice-rich planets up to 100 Earth masses, *Astrophys. J.*, 693(1) (2009), p. 722.

Gross G.W., P.M. Wong, K. Humes. Concentration dependent solute redistribution at the ice–water phase boundary. III. Spontaneous convection. Chloride solutions, *J. Chem. Phys.*, doi:10.1063/1.434704, 67, (1977), pp. 5264–5274.

Gross G., A. Gutjahr, K. Caylor. Recent experimental work on solute redistribution at the ice/water interface. Implications for electrical properties and interface processes, *J. Phys.*, 48 (C1), (1987), pp. C1-527–C1-533. https://hal.archives-ouvertes.fr/jpa-00226318/document

Gross G.W., R. K. Svec. Effect of ammonium on anion uptake and dielectric relaxation in laboratory-grown ice columns, *J. Phys. Chem. B.*, 101(32) (1997): 6282–6284. doi:10.1021/jp963213c.

Grundy, W.M., B. Schmitt, E. Quirico. The temperature-dependent spectra of α and β nitrogen ice with application to Triton, *Icarus*, 105(1) (1993), pp. 254–258.

Guinier A., G.B. Bokii, K. Boll-Dornberger, J.M. Cowley, S. Durovic, H. Jagodzinski, P. Krishna, P.M. De Wolff, B.B. Zvyagin, et al. Nomenclature of polytype structures. Report of the International Union of Crystallography Ad hoc Committee on the Nomenclature of Disordered, Modulated and Polytype Structures, *Acta Crystallogr.*, A40 (1984), p. 399.

Hallbrucker, A., E. Mayer, G. Johari. Glass-liquid transition and the enthalpy of devitrification of annealed vapor-deposited amorphous solid water: A comparison with hyperquenched glassy water, *J. Phys. Chem.*, 93(12)(1989), pp. 4986–4990.

Handa Y.P., O. Mishima, E. Whalley. High-density amorphous ice. III. Thermal properties, *J. Chem. Phys.*, 84 (1986), p. 2766.

Hanley, H.J.M., W.M. Haynes, R.D. McCarty. The viscosity and thermal conductivity coefficients for dense gaseous and liquid methane. *J. Phys. Chem. Ref. Data*, 6(597) (1977), pp. 597–609.

Hansen, T.C., A. Falenty, W.F. Kuhs. Modelling ice I_c of different origin and stacking-faulted hexagonal ice using neutron powder diffraction data, *Phys. Chem. Ice, Proc. Int. Conf.*, 11th (2007) 201.

Hassanali, A.A., J. Cuny, V. Verdolino, M. Parrinello. Aqueous solutions: State of the art in ab initio molecular dynamics, *Phil. Trans. Royal Soc. London A: Math., Phys. Eng. Sci.*, 372(2011) (2014), p. 20120482.

Haynes, W.M. Viscosity of saturated liquid methane, *Physica*, 70 (1973), pp. 410–412.

Hendricks, S.B., M.E. Jefferson. On the optical anisotropy of molecular crystals, *J. Opt. Soc. Am.*, 23 (1933), pp. 299–307.

Hermann, A., N.W. Ashcroft, R. Hoffmann. High pressure ices, *PNAS.*, 109(3) (2012), pp. 745–750. doi:10.1073/pnas.1118694109/-/DCSupplemental/Appendix.pdf.

Herrero, C.P., R. Ramírez,Topological characterization of crystalline ice structures from coordination sequences, *Phys. Chem. Chem. Phys.*, 15(2013), p. 16676.

Hill, N.E., W.E. Vaughan, A.H. Price, M. Davis. *Dielectric Properties and Molecular Behavior.* New York, NY: Van Nostrand (1969).

Hobbs P.V. *Ice Physics.* Oxford: Clarendon Press (1974).

Hogenboom, D.L., J.S. Kargel, J.P. Ganasan, L. Lee. Magnesium sulfate-water to 400 MPa using a novel piezometer: Densities, phase equilibria, and planetological implications, *Icarus*, 115(2) (1995), pp. 258–277.

Hogenboom, D.L., J.S. Kargel, G.J. Consolmagno, T.C. Holden, L. Lee, M. Buyyounouski. The ammonia-water system and the chemical differentiation of icy satellites, *Icarus*, 128(1) (1997), pp. 171–180.

Hussmann, H., F. Sohl, T. Spohn. Subsurface oceans and deep interiors of medium-sized outer planet satellites and large trans-neptunian objects, *Icarus*, 185 (2006), pp. 258–273.

Ioppolo, S., H.M. Cuppen, C. Romanzin, E.F.v. Dishoeck, H. Linnartz. Water formation at low temperatures by surface O_2 hydrogenation I: Characterization of ice penetration, *Phys. Chem. Chem. Phys.*, 12 (2010), p. 12065.

Johannessen, B., P. Kluth, D.J. Llewellyn, G.J. Foran, D.J. Cookson, M.C. Ridgway. Amorphisation of embedded Cu nanocrystals by ion irradiation, *Appl. Phys. Lett.*, 90(7) (2007), p. 073119.

Johari, G.P. Liquid state of low-density pressure-amorphized ice above its T_g, *J. Phys. Chem. B.*, 102 (1998), p. 4711.

Johari G. P., A. Hallbrucker, and E. Mayer. The glass–liquid transition of hyperquenched water, *Nature*, 330, pp. 552–553, doi:10.1038/330552a0.

Jorgensen, J.D., R.A. Beyerlein, N. Watanabe, T.G. Worlton. Structure of D_2O Ice VIII from in situ powder neutron diffraction, *J. Chem. Phys.*, 81 (1984), p. 3211.

Jorgensen, J.D., T.G. Worlton. Disordered structure of D2O ice VII from in situ neutron powder diffraction, *J. Chem. Phys.*, 83 (1985), p. 329.

Journaux B., I. Daniel, H. Cardon, S. Petitgirard, J. Perrillat, R. Caracas, and M. Mezouar. Experimental determination of salt partition coefficients between aqueous fluids, ice VI and ice VII: Implication for the composition of the deep ocean and the geodynamics of large icy moons and water rich planets, EGU2015-9503-3, EGU General Assembly 2015, Vienna, April 2015. http://meetingorganizer.copernicus.org/EGU2015/EGU2015-9503-3.pdf.

Kamb, B. Ice II: A proton-ordered form of ice, *Acta Crystallogr.*, 17 (1964), p. 1437.

Kamb, B., W.C. Hamilton, S.J. LaPlaca, A. Prakash. Ordered proton configuration in ice II, from single-crystal neutron diffraction, *J. Chem. Phys.*, 55 (1971), p. 1934.

Kargel, J. Cryovolcanism on the icy satellites, *Earth Moon Planets*, 67(1) (1994), pp. 101–113.

Kaufman, J. G. *Properties of Materials for Liquefied Natural Gas Tankage*. Baltimore, MD: American Society for Testing and Materials (1975).

Kirichek, O., A.J. Church, M.G. Thomas, D. Cowdery, S.D. Higgins, M.P. Dudman, Z.A. Bowden. *Cryogenics*, 52(7–9) (2012), pp. 325–330.

Klotz S., L. Bove, T. Strässle, T. Hansen, A. Saitta. The preparation and structure of salty ice VII under pressure, *Nat. Mater.*, 8 (2009), pp. 405–409.

Knight, C., S.J. Singer. Prediction of a phase transition to a hydrogen bond ordered form of ice VI, *J. Phys. Chem. B.*, 109 (2005), p. 21040.

Knight, C., S.J. Singer. A reexamination of the Ice III/IX hydrogen bond ordering phase transition, *J. Chem. Phys.*, 125 (2006), p. 064506/1.

Knight, C., S.J. Singer, J.-L. Kuo, T.K. Hirsch, L. Ojamae, M.L. Klein. Hydrogen bond topology and the Ice VII/VIII and I_h/XI proton ordering phase transitions, *Phys. Rev. E.*, 73 (2006), p. 056113/1.

Knight, C., S.J. Singer. Hydrogen bond ordering in ice V and the transition to ice XIII, *J. Chem. Phys.*, 129 (2008), 164513/1.

Kohl, I., E. Mayer, A. Hallbrucker. The glassy water–cubic ice system: A comparative study by x-ray diffraction and differential scanning calorimetry, *Phys. Chem. Chem. Phys.*, 2 (2000), p. 1579.

Kohl, I., L. Bachmann, A. Hallbrucker, E. Mayer, T. Loerting. Liquid-like relaxation in hyperquenched water at ≤140 K, *Phys. Chem. Chem. Phys.*, 7 (2005), p. 3210.

Köster, K.W., V. Fuentes Landete, A. Raidt, M. Seidl, C. Gainaru, T. Loerting, R. Böhmer. Dynamics enhanced by HCl doping triggers full Pauling entropy release at the ice XII-XIV transition. *Nat. Commun.*, 6 (2015), pp. 1–7.

Kuhs, W.F.E., J.L. Finney, C. Vettier, D.V. Bliss. Structure and hydrogen ordering in ices VI, VII, and VIII by neutron powder diffraction, *J. Chem. Phys.*, 81 (1984), p. 3612.

Kuhs, W.F., M.S. Lehmann. *Water Science Reviews 2*. In: F. Franks (Ed.), Cambridge: Cambridge University Press, (1986), p. 1.

Kuhs, W.F., D.V. Bliss, J.L. Finney. High-resolution neutron powder diffraction study of ice Ic, *J. Phys., Colloq.* (1987), p. C1.

Kuhs, W.F., C. Lobban, J.L. Finney. Partial H-ordering in high pressure ices III and V, *Rev. High Press. Sci. Technol.*, 7 (1998), p. 1141.

Kuhs, W.F. *Physics and Chemistry of Ice*. Royal Society of Chemistry (2007). http://books.google.com/books?id=sFp56qUc2HMC&pg=PA249&dq=intitle:Physics+and+Chemistry+of+Ice+Proceeding&hl=&cd=1&source=gbs_api

Kuo, J.-L., M.L. Klein, W.F. Kuhs. The effect of proton disorder on the structure of ice-I_h: A theoretical study, *J. Chem. Phys.*, 123 (2005), pp. 134505/1.

Kuo, J.-L., W.F. Kuhs. A first principles study on the structure of ice-VI: Static distortion, molecular geometry, and proton ordering, *J. Phys. Chem. B.*, 110 (2006), p. 3697.

LaPlaca, S.J., W.C. Hamilton, B. Kamb, A. Prakash. On a nearly proton-ordered structure for ice IX, *J. Chem. Phys.*, 58 (1973), p. 567.

Lebreton, J.-P., O. Witasse, C. Sollazzo, T. Blanquaert, P. Couzin, A.M. Schipper, J.B. Jones, D.L. Matson, L.I. Gurvits, D.H. Atkinson. An overview of the descent and landing of the Huygens probe on Titan, *Nature*, 438 (2005), pp. 758–764.

Leger, A., F. Selsis, C. Sotin, T. Guillot, D. Despois, D. Mawet, M. Ollivier, A. Labeque, C. Valette, F. Brachet. A new family of planets? "Ocean Planets". *Icarus*, 169(2) (2004), pp. 499–504.

Leliwa-Kopystynski, J., M. Maruyama, T. Nakajima. The water-ammonia phase diagram up to 300 MPa: Application to icy satellites, *Icarus*, 159 (2002), pp. 518–528.

Lemmon E.W., M.L. Huber, M.O. McLinden, NIST standard reference database 23: Reference fluid thermodynamic and transport properties-REFPROP, Version 9.1, National Institute of Standards and Technology, Standard Reference Data Program, Gaithersburg (2013).

Lin, C., Trusler. The speed of sound and derived thermodynamic properties of pure water at temperatures between (253 and 473) K and at pressures up to 400 MPa, *J. Chem. Phys.*, 136 (2012), p. 094511.

Lobban, C., J.L. Finney, W.F. Kuhs. The structure and ordering of ices III and V, *J. Chem. Phys.*, 112 (2000), p. 7169.

Loerting, T., K. Winkel, M. Seidl, M. Bauer, C. Mitterdorfer, P.H. Handle, C.G. Salzmann, E. Mayer, J.L. Finney, D.T. Bowron. How many amorphous ices are there?, *Phys. Chem. Chem. Phys.*, 13 (2011), p. 8783.

Loerting, T., V.V. Brazhkin, T. Morishita. Multiple amorphous-amorphous transitions, *Adv. Chem. Phys.*, 143 (2009), p. 29.

Lokotosh, T.V., N.P. Malomuzh. Proton ordering in cubic ice, *Khim. Fiz.*, 12 (1993), pp. 897–907.

Londono, J.D., W.F. Kuhs, J.L. Finney. Neutron diffraction studies of ices III and IX on under-pressure and recovered samples, *J. Chem. Phys.*, 98 (1993), p. 4878.

Lonsdale, K. Diamagnetic and paramagnetic anisotropy of crystals, *Rep. Prog. Phys.*, 4 (1937), p. 368.

Lorenz, R.D., S.E. Shandera. Physical properties of ammonia-rich ice: Application to Titan, *Geophys. Res. Lett.*, 28, 2 (2001), pp. 215–218.

Lorenz, R.D., S. Wall, J. Radebaugh, G. Boubin, E. Reffet, M. Janssen, E. Stofan, R. Lopes, R. Kirk, C. Elachi, C., et al. Titan's damp ground: Constraints on Titan surface thermal properties from the temperature evolution of the Huygens GCMS inlet, *Science*, 312, 5774 (2006), pp. 724–727.

Lorenz, R.D., R.M. Lopes, F. Paganelli, J.I. Lunine, R.L. Kirk, K.L. Mitchell, L.A. Soderblom, E.R. Stofan, G. Ori, M. Myers, M., et al. Fluvial channels on Titan: Initial Cassini RADAR observations, *Planet. Space Sci.*, 56, 8 (2008), pp. 1132–1144.

Lorenz, R.D. Heat rejection in the Titan surface environment: Potential impact on science investigations, *J. Thermophys. Heat Trans.*, 1–9 (2015), pp. 1.T4608.

Lunine, J.I. Saturn's Titan: Jonathan I. Lunine. A strict test for life's cosmic ubiquity, *Proc. Am. Phil. Soc.* 153(4) (2009), pp. 403–418.

Malaska, M.J., R. Hodyss. Dissolution of benzene, naphthalene, and biphenyl in a simulated Titan lake, *Icarus*, 242 (2014), pp. 74–81.

Martin-Conde, M., L.G. MacDowell, C. Vega. Computer simulation of two new solid phases of water: Ice XIII and ice XIV, *J. Chem. Phys.*, 125 (2006), p.116101.

Marion, G., J. Kargel, D. Catling, D., J. Lunine. Modeling ammonia–ammonium aqueous chemistries in the solar system's icy bodies, *Icarus*, 220 (2012), pp. 932–946.

Matsuo T., Y. Tajima, H. Suga. Calorimetric study of a phase transition in D_2O Ice I_h doped with KOD: Ice XI, *J. Phys. Chem. Solids.*, 47 (1986), pp. 165.

Mayer, E., A. Hallbrucker. Cubic ice from liquid water, *Nature*, 325 (1987) 601.

Mayer, E. New method for vitrifying water and other liquids by rapid cooling of their aerosols, *J. Appl. Phys.*, 58 (1985), p. 663.

Mayer, E., R. Pletzer. Astrophysical implications of amorphous Ice—A microporous solid, *Nature*, 319 (1986), p. 298.

Maynard-Casely H., R. Hodyss, M. Cable and T. Vu. A co-crystal between benzene and ethane, an evaporite material for Saturn's moon Titan, *IUCr J.*, (2016), accepted for publication.

McKay, C.P., H.D. Smith. Possibilities for methanogenic life in liquid methane on the surface of Titan, *Icarus*, 178, 1 (2005), pp. 274–276.

Mehling, H., F. Cabeza. *Heat and Cold Storage with PCM*, Springer Science & Business Media, (2008), p. 308.

Militzer, B., H.F. Wilson. New phases of water ice predicted at megabar pressures, *Phys. Rev. Lett.*, 105(19) (2010), p. 195701.

Millero, F.J., C.A. Chen. The speed of sound in mixtures of the major sea salts–a test of Young's rule for adiabatic PVT properties, *J. Solut. Chem.*, 14(4)(1985), pp. 301–310.

Minceva-Sukarova, B., W.F. Sherman, G.R. Wilkinson. A high pressure spectroscopic study on ice III-ice IX, disordered-ordered transition, *J. Mol. Struct.*, 115 (1984), p. 137.

Mishima, O., S. Endo. Melting curve of ice VII, *J. Chem. Phys.*, 68(10) (1978), pp. 4417–4418. doi:10.1063/1.435522.

Mishima, O., L.D. Calvert, E. Whalley. Melting ice I at 77 K and 10 kbar: A new method of making amorphous solids, *Nature*, 310 (1984), p. 393.

Mishima, O., L.D. Calvert, E. Whalley. An apparently first-order transition between two amorphous phases of ice induced by pressure, *Nature*, 314 (1985), p. 76.

Mishima, O. Melting of the precipitated ice IV in LiCl aqueous solution and polyamorphism of water, *J. Phys. Chem. B.*, 115(48) (2011), pp. 14064–14067.

Mitchell, K. L., M.B. Barmatz, C.S. Jamieson, R.D. Lorenz, J.I. Lunine. Laboratory measurements of cryogenic liquid alkane microwave absorptivity and implications for the composition of Ligeia Mare, Titan, *Geophys. Res. Lett.*, 42, 5 (2015), pp. 1340–1345.

Mitterdorfer, C., M. Bauer, T. Loerting.Clathrate hydrate formation after CO_2–H_2O vapour deposition, *Phys. Chem. Chem. Phys.*, 13 (2011), p. 19765.

Mitterdorfer, C., M. Bauer, T.G.A. Youngs, D.T. Bowron, C.R. Hill, H.J. Fraser, J.L. Finney, T. Loerting. Small-angle neutron scattering study of micropore collapse in amorphous solid water, *Phys. Chem. Chem. Phys.*, 16 (2014), p. 16013.

Murray, B.J., D.A. Knopf, A.K. Bertram. The formation of cubic ice under conditions relevant to earth's atmosphere, *Nature*, 434 (2005), p. 202.

Nelmes, R.J., J.S. Loveday, T. Straessle, C.L. Bull, M. Guthrie, G. Hamel, S. Klotz. Annealed high-density amorphous ice under pressure, *Nat. Phys.*, 2 (2006), p. 414.

Niemann, H.B., S.K. Atreya, S.J. Bauer, G.R. Carignan, J.E. Demick, R.L. Frost, D. Gautier, J.A. Haberman, D.M. Harpold, D.M. Hunten, et al. The abundance of constituents of Titan's atmosphere from the GCMS instrument on the Huygens probe, *Nature*, 438 (2005), pp. 779–784.

Nishibata, K., E. Whalley. Thermal effects of the transformation ice III–IX, *J. Chem. Phys.*, 60 (1974), p. 3189.

Noya, E.G., M.M. Conde, C. Vega. Computing the free energy of molecular solids by the Einstein molecule approach: Ices XIII and XIV and a simple model of proteins, *J. Chem. Phys.*, 129 (2008), p. 16.

Paillou, P., K. Mitchell, S. Wall, G. Ruffie, C. Wood, R. Lorenze, E. Stofan, J. Lunine, R. Lopes, P. Encrenaz. Microwave dielectric constant of liquid hydrocarbons: Application to the depth estimation of Titan's lakes, *Geophys Res. Lett.* 35 (2008) L05202.

Parkkinen, P., S. Riikonen, L. Halonen. Ice XI: Not that ferroelectric, *J. Phys. Chem. C.*, 118(2014), p. 26264.

Pauling, L. The structure and entropy of ice and of other crystals with some randomness of atomic arrangement, *J.Am. Chem. Soc.*, 57(12) (1935), pp. 2680–2684.

Petrenko, V.F., R.W. Whitworth. *Physics of Ice*. Oxford University Press, 2002.

Pistorius, C.W.F.T., M.C. Pistorius, J.P. Blakey, L.J. Admiraal. Melting curve of ice VII to 200 kbar. *J. Chem. Phys.*, 38(3) (1963), p. 600. doi:10.1063/1.1733711.

Pitzer, K. *Activity Coefficients in Electrolyte Solutions*. CRC Press, Boca Raton, FL, (1991).

Poole, P.H., F. Sciortino, U. Essmann, H.E. Stanley. Phase behaviour of metastable water, *Nature*, 360 (1992), pp. 324–328.

Pruzan, P., J.C. Chervin, B. Canny. Determination of the D_2O ice VII–VIII transition line by Raman scattering up to 51 GPa, *J. Chem. Phys.*, 97 (1992), pp. 718–721.

Pruzan, P., J.C. Chervin, E. Wolanin, B. Canny, M. Gauthier, M. Hanfland. Phase diagram of ice in the VII–VIII–X domain. Vibrational and structural data for strongly compressed ice VIII, *J. Raman Spectrosc.*, 34 (2003), p. 591.

Pruzan, P., J.C. Chervin, M. Gauthier. Raman spectroscopy investigation of ice VII and deuterated ice VII to 40 GPa. Disorder in ice VII, *Europhys. Lett.*, 13 (1990), p. 81.

Quirico, E., S. Douté, B. Schmitt, C. de Bergh, D.P. Cruikshank, T.C Owen, T.R Geballe, T.L Roush. Composition, physical state, and distribution of ices at the surface of Triton, *Icarus*, 139 (1999), pp. 159–178.

Raulin, F. Organic chemistry in the oceans of Titan, *Adv. Space Res.*, 7(5) (1987), pp. 71–81.

Röttger, K., A. Endriss, J. Ihringer, S. Doyle, W.F. Kuhs. Lattice constants and thermal expansion of H_2O and D_2O Ice I_h between 10 and 265 K, *Acta Crystallogr.*, B50 (1994), p. 644.

Salzmann, C.G., P.G. Radaelli, A. Hallbrucker, E. Mayer, J.L. Finney. The preparation and structures of hydrogen ordered phases of ice, *Science*, 311 (2006a), p. 1758.

Salzmann, C.G., A. Hallbrucker, J.L. Finney, E. Mayer. Raman spectroscopic study of hydrogen ordered ice XIII and of its reversible phase transition to disordered ice V, *Phys. Chem. Chem. Phys.*, 8 (2006b), p. 3088.

Salzmann, C.G., A. Hallbrucker, J.L. Finney, E. Mayer. Raman spectroscopic features of hydrogen-ordering in ice XI, *Chem. Phys. Lett.*, 429 (2006c), p. 469.

Salzmann, C.G., T. Loerting, S. Klotz, P.W. Mirwald, A. Hallbrucker, E. Mayer. Isobaric annealing of high-density amorphous ice between 0.3 and 1.9 GPa: In situ density values and structural changes, *Phys. Chem. Chem. Phys.*, 8 (2006d), p. 386.

Salzmann, C.G., P.G. Radaelli, E. Mayer, J.L. Finney. Ice XV: A new thermodynamically stable phase of ice, *Phys. Rev. Lett.*, 103 (2009), p. 105701/1.

Salzmann, C.G., P.G. Radaelli, J.L. Finney, E. Mayer. A calorimetric study on the low temperature dynamics of doped ice V and its reversible phase transition to hydrogen ordered ice XIII, *Phys. Chem. Chem. Phys.*, 10 (2008), p. 6313.

Sands, D.E. *Introduction to Crystallography*. Dover Publications (1975).

Sanloup, C., S.A. Bonev, M. Hochlaf, H.E. Maynard-Casely. Reactivity of xenon with ice at planetary conditions. *Phys. Rev. Lett.*, 110(26) (2013), p. 265501.

Schubert, G., J. Anderson, T. Spohn, W. McKinnon. Interior composition, structure and dynamics of the Galilean satellites. In: F. Bagenal, et al. (Eds.), *Jupiter: The Planet, Satellites and Magnetosphere* (2004), pp. 281–306.

Seidl M., M.S. Elsaesser, K. Winkel, G. Zifferer, E. Mayer, T. Loerting. Volumetric study consistent with a glass-to-liquid transition in amorphous ices under pressure, *Phys. Rev. B*, 83 (2011), p. 100201.

Sellberg, J.A. et al. Ultrafast x-ray probing of water structure below the homogeneous ice nucleation temperature, *Nature*, 510 (2014), p. 381.

Sen, A.D., V.G. Anicich, T. Arakelian. Dielectric constant of liquid alkanes and hydrocarbon mixtures, *J. Phys. D: Appl. Phys.*, 25 (1992), pp. 516–521.

Sepúlveda, A., E. Leon-Gutierrez, M. Gonzalez-Silveira, C. Rodríguez-Tinoco, M.T. Clavaguera-Mora, J. Rodríguez-Viejo. Glass transition in ultrathin films of amorphous solid water, *J. Chem. Phys.*, 137 (2012), p. 244506.

Sestak, J., J.J. Mares, P. Hubik. *Glassy, Amorphous, and Nano-Crystalline Materials, Thermal Physics, Analysis, Structure and Properties*. Springer, New York, (2011).

Singer, S.J., J.-L. Kuo, T.K. Hirsch, C. Knight, L. Ojamaee, M.L. Klein. Hydrogen-bond topology and the ice VII/VIII and ice I_h/XI proton-ordering phase transitions, *Phys. Rev. Lett.*, 94 (2005), p. 135701/1.

Somayazulu, M., J. Shu, C.-s. Zha, A.F. Goncharov, O. Tschauner, H.-k. Mao, R.J. Hemley. In situ high-pressure x-ray diffraction study of H_2O Ice VII, *J. Chem. Phys.*, 128 (2008), p. 064510/1.

Song, M., H. Yamawaki, H. Fujihisa, M. Sakashita, K. Aoki. Infrared investigation on ice VIII and the phase diagram of dense ices, *Phys. Rev. B.*, 68 (2003), p. 014106/1.

Sotin, C., O. Grasset, A. Mocquet. Mass-radius curve for extrasolar earth-like planets and ocean planets, *Icarus*, 191(1) (2007), pp. 337–351.

Speedy, R.J., C.A. Angell. Isothermal compressibility of supercooled water and evidence for a thermodynamic singularity at–45°C, *J. Chem. Phys.*, 65 (1976), pp. 851–858.

Stachurski, Z.H. On structure and properties of amorphous materials, *Materials*, 4 (2011), pp. 1564–1598.

Stanley, H.E., S.V. Buldyrev, M. Canpolat, O. Mishima, M.R. Sadr-Lahijany, A. Scala, F.W. Starr. The Puzzling Behavior of Water at Very Low Temperature, Proceedings—International Meeting on Metastable Fluids, *Phys. Chem. Chem. Phys.*, 2 (2000), p. 1551.

Stevenson, K.P., G.A. Kimmel, Z. Dohnalek, R.S. Smith, B.D. Kay. Controlling the morphology of amorphous solid water, *Science*, 283 (1999), p. 1505.

Tajima, Y., T. Matsuo, H. Suga. Calorimetric study of phase transition in hexagonal ice doped with alkali hydroxides, *J. Phys. Chem. Solids*, 45 (1984), p. 1135.

Tammann, G. Ueber die Grenzen des festen Zustandes IV, *Ann. Phys.*, (1900), p. 1.

Tobie, G., O. Grasset, J.I. Lunine, A. Mocquet, C. Sotin. Titan's internal structure inferred from a coupled thermal-orbital model. *Icarus*, 175(2) (2005), pp. 496–502.

Tribello, G.A., B. Slater, C.G. Salzmann. A blind structure prediction of ice XIV, *J. Am. Chem. Soc.*, 128 (2006), p. 12594.

Umemoto, K., R.M. Wentzcovitch. Two-stage dissociation in $MgSiO_3$ post-perovskite, *Earth Planet. Sci. Lett.*, 311(3) (2011), pp. 225–229.

Van Sciver, S.W. *Helium Cryogenics*, International Cryogenics Monograph Series, 17 Springer Science (2012), doi10.1007/978-1-4419-9979-5_2.

Vance, S., J.M. Brown. Layering and double-diffusion style convection in Europa's ocean, *Geochim. Cosmochim. Acta*, 110 (2013), pp. 176–189.

Vance, S., J. Goodman. *Oceanography of an Ice Covered Moon*. Europa: Arizona University Press (2009), pp. 459–482.

Vance, S., J. Harnmeijer, J. Kimura, H. Hussmann, B. deMartin, J.M. Brown. Hydrothermal systems in small ocean planets, *Astrobiology*, 7(6) (2007), pp. 987–1005.

Vance, S., M. Bouffard, M. Choukroun, C. Sotin. Ganymede's internal structure including thermodynamics of magnesium sulfate oceans in contact with ice, *Planet. Space Sci.*, 96 (2014), pp. 62–70.

Vu, T.H., M.L. Cable, M. Choukroun, R. Hodyss, P. Beauchamp. Formation of a new benzene-ethane co-crystalline structure under cryogenic conditions, *J. Phys. Chem. A*, 118(23) (2014), pp. 4087–4094.

Wang, Y., H. Liu, J. Lv, L. Zhu, H. Wang, Y. Ma. High pressure partially ionic phase of water ice, *Nat. Commun.*, 2 (2011), p. 563.

Weeks, W.F., S.F. Ackley. *The Geophysics of Sea Ice*. In: Norbert Untersteiner (Ed.), 9 164. NATO ASI Series. Springer US(1986).

Whalley, E., Cubic ice in nature, *J. Phys. Chem.*, 87 (1983), p. 4174.

Whalley, E., J.B.R. Heath, D.W. Davidson. Ice IX: An antiferroelectric phase related to ice III, *J. Chem. Phys.*, 48 (1968), p. 2362.

Whalley, E., D.W. Davidson, J.B.R. Heath. Dielectric properties of ice VII. Ice VIII: A new phase of ice, *J. Chem. Phys.*, 45 (1966), p. 3976.

Wilson, H.F., et al. Superionic to superionic phase change in water: Consequences for the interiors of Uranus and Neptune, *Phys. Rev. Lett.*, 110 (2013), p. 151102.

Winkel, K., M.S. Elsaesser, E. Mayer, T. Loerting. Water polyamorphism: Reversibility and (dis)continuity, *J. Chem. Phys.*, 128 (2008), p. 044510.

Winkel, K., E. Mayer, T. Loerting. Equilibrated high-density amorphous ice and its first-order transition to the low-density form, *J. Phys. Chem. B*, 115 (2011), p. 14141.

Wolf, R.A., S. Trolier-McKinstry. Temperature dependence of the piezoelectric response in lead zirconate titanate films, *J. Appl. Phys.*, 95(3) (2004).

Yamashita, Y., M. Kato, M. Arakawa. Experimental study on the rheological properties of polycrystalline solid nitrogen and methane: Implications for tectonic processes on Triton. *Icarus J.*, 207(2) (2010), pp. 972–977, doi:10.1016/j.icarus.2009.11.032.

Yoshimura, Y., S.T. Stewart, M. Somayazulu, H.-k. Mao, R.J. Hemley. High-pressure x-ray diffraction and Raman spectroscopy of ice VIII, *J. Chem. Phys.*, 124 (2006), p. 024502/1.

Younglove, B.A., J.F. Ely. Thermophysical properties of fluids. II. Methane, ethane, propane, isobutane, and normal butane, *J. Phys. Chem. Ref. Data*, 16, 4 (1987), pp. 577–798.

Zabrodsky, V.G., T.V. Lokotosh. A basic state and collective excitations in proton ordered phase of cubic ice, *Ukr. Fiz. Zh.*, 38 (1993), pp. 1714–1723.

Zallen, R. *The Physics of Amorphous Solids*. New York, NY: John Wiley and Sons (1983).

Zhuang, Z.Q., M.J. Haun, S.-J. Jang, L.E. Cross. *Ultrasonics, Ferroelectrics, and Frequency Control, IEEE Transactions on*, 36(4) (1989), p. 413.

Zolotov, M.Y., J. Kargel. *On the Chemical Composition of Europa's Icy Shell, Ocean, and Underlying Rocks*. Europa: University of Arizona Press (2009), pp. 431–458.

4

Characterization Methods for Low Temperature Materials

Yoseph Bar-Cohen, W. Kinzy Jones, and Robert Cormia

CONTENTS

4.1 Introduction

Developing and using materials that are applicable at low temperatures as well as assuring their performance and durability requires physical and chemical characterization techniques that can be used to analyze them before, after, and possibly during their operation in service as well as in simulated conditions [Reed and Clark, 1983]. Various techniques are used to examine the surface and bulk properties of the material. Some of the techniques are nondestructive like those described in Chapter 5. The destructive test methods, and the ones that are used to remotely examine the surface nondestructively, are described in this chapter. Generally, image analysis is widely used to examine the physical structure of the surface of the material or structure that needs to be characterized. For materials and structures that are easily accessible to the surface, there are many applicable methods. Analysis of structures below the surface is done by techniques that provide insight and gauging of the internal characteristics, including stress and strain, and chemical composition.

Many analytical techniques are used to examine low temperature materials. While these techniques are covered widely in the literature, this chapter reviews some of the key ones in order to provide completeness to this book. The methods and their underlying principles are described in the following sections.

4.2 Properties of Materials

Conducting proper characterization requires knowledge and understanding of the relevant properties that are affected by the temperature [White, 1979]. Some of the properties that need to be taken into account are related to refrigeration, storage at low temperatures, and thermal properties such as heat transfer, thermal capacity, and thermal contraction coefficient (TCC) [Van Sciver, 2012]. Generally, the lowest temperatures at which the materials are investigated are in the range of 10^{-3} K.

4.2.1 Heat Capacity and Specific Heat Capacity

Heat capacity is the measure of the energy needed to raise the temperature of a material by one degree; it is related to other thermodynamic state variables. In other words, it is the ratio of the amount of heat energy that is transferred to an object and the resulting increase in its temperature. Its density, that is, the specific heat capacity, is the heat capacity per unit mass. The heat capacity is the temperature derivative of either the entropy or the internal energy. Another heat capacity measure that is commonly used is the volumetric heat capacity and it is the value per volume. Generally, there are a lot of experimental data on the heat capacity of solids at low temperatures. For most solids, the heat capacity drops by three or four orders of magnitude when the temperature is reduced from about room temperature to the level of single-digit Kelvin.

4.2.2 Thermal Contraction

When cooling a material, its physical dimensions are reduced [Corruccini and Gniewek, 1961]. When the temperature is lowered from room temperature to the level of liquid helium, the contraction can reach as much as a few tenths of a percent in volume. The effect can be critical to structures that are made of several materials having different TCCs, and it can be as high as an order of magnitude. Since most structures and systems are constructed at room temperature, the differences in TCC can cause thermal stresses and affect the seal of vacuum systems as well as the structure/system thermal insulation. Also, if an assembly has moving parts, the TCC differences can prevent part movement. It is interesting to note that, when cooled, some materials expand rather than contract. Taking advantage of this property of materials, it is possible to design thermally stable composite materials that do not expand or shrink as a function of temperature.

4.2.3 Thermal Conductivity

Thermal conductivity is the property of a material to conduct heat, and it is a temperature-dependent parameter. The inverse of thermal conductivity is called thermal resistivity, and it indicates the ability of a material to resist the flow of heat. A pure crystalline material can exhibit different thermal conductivity along its various crystallographic axes, and it is the result of having differences in phonon coupling along each of these axes. Thermal conductance is the quantity of heat that passes in a unit time through an area and thickness having one Kelvin difference between the two opposite faces through which the heat passes. Materials of high thermal conductivity allow heat transfer at a higher rate than materials of low thermal conductivity. Therefore, materials of high thermal conductivity are widely used in heat sink applications and, in contrast, materials with low thermal conductivity are used as thermal insulators. Generally, at about room temperature, the thermal conductivity of pure metals linearly increases with the drop in temperature. However, at low temperatures around the temperature of liquid helium, the value of the thermal conductivity tends to be constant.

For metallic materials, thermal conductivity is based on both phonons and electrons. At room temperatures, phonons dominate, but for pure metals at very low temperatures, electron transfer dominates. The less the obstacles to electron movement, the higher the thermal conductivity (the obstacles may include impurity atoms, grain boundaries, dislocations, and even lattice vibrations). At low temperatures, the lattice vibrations have decreased amplitude, so the mean free path for the electrons increases and, therefore, very pure annealed metals with few grain boundaries will have higher thermal conductivity.

4.2.4 Electrical Resistivity and Conductivity

Electrical resistivity is a measure of the resistance of a material to the flow of an electric current, and it is a related to how readily a material allows the travel of electric charge. The inverse of the electrical resistivity is the electrical conductivity or specific conductance and it is the measure of a material's ability to conduct an electric current. According to the Wiedemann–Franz law [Kittel, 2005], the thermal conductivity in metals approximately tracks the material electrical conductivity, since freely moving valence electrons transfer both electric current and heat energy. For nonmetallic materials, this correlation of the electrical and thermal conductance does not hold due to the increased role of phonon carriers for heat transfer.

4.2.4.1 Semiconductors

A semiconductor is a pure crystalline material that has electrical conductivity with a value between those of a conductor and an insulator. The low conductivity in semiconductors is attributed to the limited number of charge carriers more than to impurity or phonon scattering. The electrical conductivity of semiconductor materials decreases with the reduction in temperature and is opposite to the response of metallic conductors. The properties of semiconductors are explained by the movement of electrons and holes inside a lattice and can be made to have variable resistivity. The electrical properties of a semiconductor material can be modified by controlled addition of impurities. The conduction of current is done by charge carriers consisting of moving free electrons and "holes" [Berman, 1976]. Adding impurity atoms to a semiconducting material, that is, "doping," greatly increases the number of charge carriers. A doped semiconductor that contains mostly free holes is called "p-type," and a semiconductor with mostly free electrons is called "n-type." The n- and p-types are the key elements of electronic devices including diodes and transistors.

4.2.4.2 Superconductors

A superconductor is an element, intermetallic alloy, or compound that conducts electricity without resistance when it is cooled below a characteristic critical temperature [Rose-Innes and Rhoderick, 1978]. Generally, the electrical resistivity of a metallic conductor decreases as the temperature is lowered, but in superconductors, at the critical temperature, the resistance drops abruptly to zero. In ordinary conductors such as copper or silver, this decrease is limited by impurities and other defects. An electric current flowing through a loop of superconducting wire can continue indefinitely with no need for a power source. The quantum mechanical phenomenon of superconductivity was discovered in Leiden, Holland, on April 8, 1911 by Heike Kamerlingh Onnes [Laesecke, 2002]. The phenomenon is characterized by the Meissner effect, which is the complete ejection of magnetic field lines from the interior of the superconductor as it transitions into the superconducting state.

One of the applications that can benefit from superconductivity is the formation of magnetic levitation and the operation of transport vehicles such as trains. These trains are made to float on strong superconducting magnets and, therefore, virtually eliminate friction with the tracks. In contrast to superconducting magnets, conventional electromagnets waste much of the electrical energy as heat and they are physically much larger. The feasibility of operating such trains has been demonstrated in Japan where in December 2003 a test train reached a speed of 581 km/h (361 mph).

Electric generators that are made with superconducting wire are far more efficient than conventional generators that are wound using copper wire. The efficiency of superconductor-based generators is above 99%, and this is achieved with generators that are half the size of conventional ones.

Generally, many metals, alloys, and certain ceramic materials exhibit superconductivity. The discovery that some cuprate–perovskite ceramic materials have a critical temperature above 90 K (–183°C) led to the term "high-temperature superconductors." These superconducting materials have a critical temperature higher than the liquid nitrogen boiling temperature of 77 K.

4.2.5 Mechanical Properties

When designing and operating materials at cryogenic temperatures, it is critical to take into account their mechanical properties. This includes the mechanical support of cryogenic systems: they need to have components that are operational at both low and ambient temperatures. The ultimate stress that a material can be subjected to indicates the load level that causes failure. Generally, most structural materials are characterized by their uniaxial stress limits. For ductile materials, the yield stress has to be accounted for in their design and it indicates the load that is required to cause 0.2% permanent strain deformation. For applications in cyclic loading, the level of load that the material can be subjected to needs to be lower. While the yield and ultimate tensile stresses increase for most materials, there may be a decrease in ductility and toughness. Some metals undergo a ductile-to-brittle transition. They may become more notch sensitive, so one needs to eliminate stress concentrations in the design. Depending on the metal, the increase in tensile strength may be only a few percent to over 100%, where Al 6061-T6 shows an increase of about 40% while the ultimate strength of the 300 series CRES increases about 150%.

4.3 Materials and Processes Analyses

Materials developed for use in low temperature applications require testing at various stages of their life from research and development to their retirement from use. The tests are typically performed during and after fabrication, during service, as well as after failure to determine the cause and prevent future occurrences. The methods of testing that are used to characterize materials include chemical and mechanical, and they may be destructive and/or nondestructive. Tests may include elemental analysis, determining the heat treatment condition, metallography, microstructure and phase analysis, as well as the determination of how the material was processed and formed, and if it met its necessary specification(s). Failure analysis includes determining the cause of fracture, including corrosion, creep, fatigue, wear, deformation, and contamination. The test may consist of analyzing the cross-section where the sample is prepared by mounting it into a mold, polishing and etching the surface, as well as possibly using replication techniques. The tests may include microstructure evaluation, grain size determination, and analysis of the heat treatment, as well as identification of porosities or flaws, and hardness testing. Microscopic tests and imaging may be performed to determine the microstructure, carburization and decarburization, plating thickness, carbide precipitation, intergranular corrosion, alpha case, sensitization, surface contamination, nodularity and nodule count, as well as eutectic melting.

The test methods that are used help determine the material properties, confirm reaching required property goals/specifications, assure the quality of the tested materials and their capability to sustain the service conditions, as well as monitor the material properties during service and determine the causes of failure. Various methods of testing materials and structures are described in the following sections of this chapter as well as some that are also used to test high-temperature materials and are described in detail in the study by Bar-Cohen and Cormia [2014]. The simplest test method is visual inspection, and it may involve unaided examination, simple imaging with photography or video viewing, or the use of an optical microscope.

To determine the mechanical properties of test materials, various configuration coupons are made and tested. The tests are done under various conditions including low temperatures, vacuum, etc., and the sample stress versus strain curve, plastic deformation, form of failure, and many other characteristics are examined. Depending on the application, one may also want to perform fatigue and fracture toughness testing. For materials with a ductile-to-brittle transition, one may want also to perform Charpy testing (also known as Charpy V-notch test, which is a standardized high-strain-rate test for determining the amount of energy absorbed by a material during fracture) [Meyers and Chawla, 2008]. Also, tribology measurements are performed to examine friction, lubrication, and wear characteristics.

4.4 Imaging and Visualization Analyzers

Visual inspection of test samples and structures is the simplest and fastest method of obtaining information about the surface, shape, anomalies, colors, and many other physical features. To enable viewing objects beyond the normal capability of the human eye, various tools are used, including magnifying glasses and microscopes. Generally, there are three categories of microscopes: optical, electron, and scanning probe microscopes. Optical and electron microscopes use the interactions of electromagnetic radiation/electron beams with the material and collect information regarding how the specimen reflects, refracts, diffracts, and/or scatters radiation. The beam can be focused on the sample using wide-field irradiation or raster scanning via a fine beam. Scanning probe microscopes use the interaction of a probe tip and the surface of the tested material, including tunneling current, magnetic attraction, and lateral force and stiction. A general illustration and a summary of the techniques are shown in Figure 4.1 and Table 4.1.

4.4.1 Optical Microscopes

Visual inspection is the most basic method of examining as well as characterizing materials and structures. Such methods are used to inspect grain boundaries and identify phases, as well as to monitor the propagation of cracks, distinguish between brittle and ductile failure, as well as assess the presence of fatigue failure.

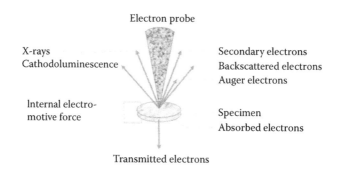

FIGURE 4.1
Illustration of the general beam characteristics that serve as the basis of many of the analytical and imaging techniques described in this chapter. (Courtesy of Charlie H. Nielsen, Vice President, Director SM/SA/IB Divisions, JEOL USA, Inc.)

TABLE 4.1

Image Analysis Techniques

Acronym and Name		Analysis/ Information	Strengths	Challenges
Optical microscopes		The most basic method of examining as well as characterizing materials and structures	Relatively inexpensive, simple to operate, high sensitivity, and 3-D imaging	The resolution is limited by the diffraction limit of the visible light. Surface preparation is needed for certain tasks
AFM	Atomic force microscopy	Contact and noncontact surface morphology	Straightforward analysis with digital data files	Solid samples only. Irregular surfaces can be difficult to test
SEM	Scanning electron microscopy	Secondary electron detectors (SED). Backscattered electron detectors (BSD)	Fast imaging of conductive materials	Insulators are a bit more difficult to analyze Cannot test liquids
TEM	Transmission electron microscopy	High-energy electron transmission	Very high spatial resolution. Some 3D information	Sample preparation can be tedious/ expensive
EDX	Energy-dispersive x-ray analysis	Elemental composition analysis	Very fast, semiquantitative	Not sensitive to elements with low Z-number
FIB	Focused ion beam	Surface topography	Fast analysis of semiconductor defects/features	Technology is more expensive than SEM
EMP	Electron microprobe	Elemental composition analysis	Fast, sensitive, semiquantitative	

To enhance the ability to examine materials, one can use optical methods with increasing capability starting from a simple magnifying glass. Further enhancement of the visualization is done by using optical microscopes with greater magnification, resolution, and stereoscopic viewing [Murphy, 2001]. Generally, optical microscopes provide detailed viewing of microstructural characteristics of tested materials (Figure 4.2). These include the grain boundaries, phase structure, and others, but the viewing is restricted by the diffraction limit and dictated by the wavelength of visible light, which is about 0.3 µm. To increase the image contrast, there are many methods that are used, including contrasting colors, polarization, and phase contrasting. Also, to improve the visualization resolution and contrast, one can use confocal microscopes, which use an aperture with a pinhole for examining sample areas that are larger than the focal plane. The pinhole light source is located in the optically conjugate plane in front of the detector, and it eliminates out-of-focus light [Minsky, 1961; Price and Jerome, 2011] and significantly increases the image resolution. To ease on the scanning and viewing of samples, there are microscopes that are designed as inverted wide-field type where the sample is placed onto an XYZ stage and scanned while viewing various sections of its area.

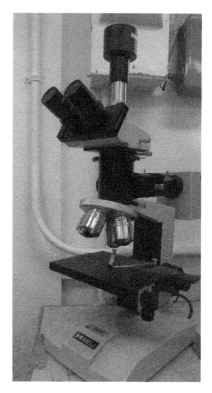

FIGURE 4.2
An example of an optical microscope (Model ML 7500, Meiji Techno Co., Japan). Photographed at JPL, Pasadena, CA.

4.4.2 Scanning Electron Microscopy

The scanning electron microscope (SEM) is one type of electron microscope (Figure 4.3) [Reimer, 1998]. SEM creates images of the topmost surface of a sample by rastering a focused, collimated electron beam over the sample; for an example of an SEM image, see Figure 4.4. The image that is created using the interaction of the electrons with atoms that are located on the sample surface contains information about the material elements, electrical conductivity, and surface topography. When an accelerated electron beam hits the sample, multiple interactions occur, including backscattered electrons, secondary electrons, and characteristic x-rays and Auger electrons. Backscattered electrons are electrons that are backscattered from the incident beam due to interaction with the atoms in the sample. These backscattered electrons are high-energy electrons, with an energy range from the accelerating voltage of the beam to a few hundred volts, and their special resolution is a few microns. The amount of electrons backscattered is dependent on the atomic number of the materials and accelerating voltage. Secondary electrons are low-energy electrons from the interaction of the incident beam and typically have energy less than 50 eV, so they can only escape the sample from a few nanometers in depth and their special resolution is based primarily on the spot size of the incident beam. When a high-energy electron beam impacts an atom, if the energy of the incident electron is more than the binding energy of the bound electron, the incident electron knocks out the bound electron and an outer bound electron is transferred into the excited energy shell, leading to the emission of a characteristic x-ray for conservation of energy. Since the electron

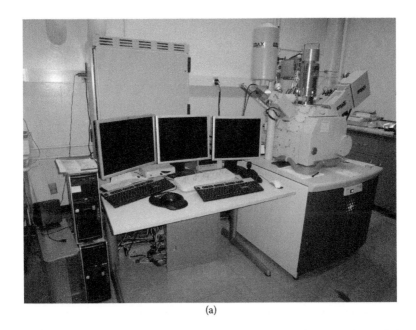

(a)

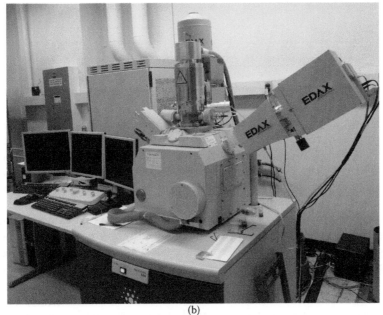

(b)

FIGURE 4.3
(a) and (b) Two views of an SEM model FEI Nova Nanosem 600 (made by EDAX). Photographed at JPL, Pasadena, CA.

binding energies are quantized and unique for each element, characteristic x-rays have very specific energies and can be utilized for chemical analysis (see Section 4.3).

SEM imaging systems use multiple detectors that collect the various emitted signals from the surface. Secondary electrons, due to their low energy, can be attracted to an electron detector with a small attraction bias, where the high-energy backscattered electrons

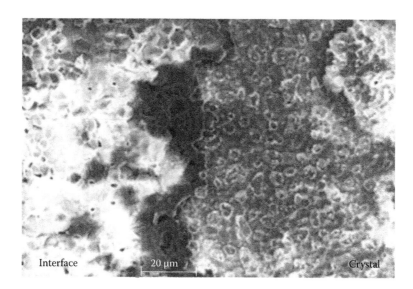

FIGURE 4.4
An SEM image of the solder to the electrode of a piezoelectric actuator. (Courtesy of David Braun, Jet Propulsion Laboratory, Pasadena, CA.)

have line of sight travel to the detector and are unaffected by the bias of the detector. By reversing the bias, secondary electrons can be repelled from the detector, allowing only backscattered electrons to be detected. Recent developments include a solid-state backscatter detector for improved resolution or the use of four backscatter detectors around the sample to quantify sample topography.

In older SEM instruments, the electron beam is generated thermionically from an electron gun using a tungsten filament cathode. Tungsten is widely used in such filaments, since it has a very high melting point and very low vapor pressure and it is relatively inexpensive. However, this type of source limits the magnification to 50,000×. Other types of thermionic electron emitters that have high brightness, such as lanthanum hexaboride (LaB_6), allow a smaller initial spot size with the same beam current, improving both resolution and source lifetime. Field emission guns (FEG), using a cold cathode made of single-crystal tungsten emitter etched to an atomic scale point, produce the smallest initial spot size and reach magnifications of 400K×. The newest source, the hot filament FEG, can produce magnifications of 1 million× and subnanometer resolution. An SEM that uses FEG as the emission source is known as an FE-SEM. A summary of the capabilities and applications of SEM is given in Table 4.2.

The magnification of an SEM can be controlled over a range of six orders of magnitude reaching as much as one million times in FE-SEM systems. The latter is three orders of magnitude higher than the best optical microscopes. The size of the electron beam interaction volume depends on the electron's energy as well as the atomic number and density of the tested material. Generally, the tested samples are mounted rigidly onto a holder. Standard SEM imaging requires a reasonable electric conductivity of the sample surface and adequate grounding to avoid accumulation of electrostatic charges. To produce a conductive surface for imaging materials that have low conductivity (polymers and ceramics) a deposit of an ultrathin coating (10–20 nm) of an electrically conducting material is sputter coated on the sample. Gold is the primary metal, and graphite

TABLE 4.2

Summary of the Capabilities and Applications of SEM

Method Provides	Analysis Capabilities	Applications
• High-magnification digital images • Uses secondary and backscattered electrons	• Solid materials, both conductive and insulating, particles, fibers, plastics, and polymers • Environmental samples using special vacuum equipment	• Materials characterization • Failure analysis • Process development • Quality assurance (QA) and quality control (QC) • Metallurgy

can be utilized to minimize any effect on x-ray analysis. Additionally, an FE-SEM can operate at low voltages with the same resolution and an accelerating voltage can be determined where the electrons absorbed by the sample and scattered (secondary and backscattered) are equal, leading to no charging of nonconductive samples. Helium and nitrogen cooling stages are readily available for SEMs from multiple sources for cryogenic analysis of materials.

4.4.3 Energy-Dispersive X-Ray Spectroscopy

Energy-dispersive x-ray spectroscopy (EDS or EDX) is used to perform elemental analysis and consists of examining the interaction with excited x-rays [Jenkins et al., 1995]. EDX systems are widely used as an addition to SEM instruments. x-ray spectroscopy is based on the fact that each element in a tested material has a unique atomic structure that generates a unique set of x-ray peaks characteristic to the element. Characteristic x-rays are generated from the interaction of a high-energy beam of electrons with the atom. If the incident electron has an energy higher than the binding energy of the bound electron, the collision can knock the bound electron out of its shell, creating an electron vacancy. The electrons from outer shells fill the inner bound electron vacancies and characteristic x-rays are generated for conservation of energy. The electron shells are historically labeled as the K, L, M, and N shells. If a K electron is ejected and replaced by an L electron, a Kα x-ray is emitted, but if it is replaced by a M shell electron, a Kβ x-ray is emitted. The probability of which electron would be replaced is well established, so if a Kα is observed, there must be a Kβ with about 20% of the intensity. Likewise, L series and M series transitions also occur because if the L electron fills the K shell, it produces a vacancy in the L shell, which is filled from the M or N shells, producing L series x-rays, and so forth. EDS consists of measuring the energies of the emitted x-rays. Since the energy of the x-ray is a characteristic of the difference in energy between the related two atomic shells, and the atomic structure of a specific element, this method allows measuring of the elemental composition of the material. It captures the entire x-ray spectrum at the same time and provides chemical analysis for a range of a few microns in an SEM to a few nanometers when used with a TEM. EDS uses a silicon PIN (consisting of a p-type semiconductor, undoped intrinsic semiconductor region, and an n-type semiconductor region) diode detector that is cryogenically chilled to minimize noise. As the x-ray travels through the detector, it produces electron–hole pairs that are detected so that the signal from a single x-ray event is converted into a current pulse based on the incident x-ray energy. The individual pulses are characterized and registered into a multichannel analyzer, resulting in a spectrum of intensity versus wavelength. The energy-dispersive detector has a resolution of 130 eV, so Kα x-ray, though only having an energy spread of less than 2 eV, produces a spectrum that is 130 eV, measured at half the maximum width of the pulse. Typically the 1–10 KeV range is used for EDS analysis. Although peak identification software is readily available with well-established correction software for computational evaluation of multicomponent samples, observing the shape of the spectrum can be helpful in insuring correct identification of the elements under analysis. K radiation is symmetrical in shape with energy width determined by the detector, and also the presence of the Kβ peak. L lines are asymmetrical with extra peaks on the high-energy side of the Lα peak, and M lines also look symmetrical but have a wider energy spread (300 eV) because they are composed of two M peaks of near equal intensity. A summary of the capabilities and applications of EDX is given in Table 4.3.

4.4.4 Focused Ion Beam

The focused ion beam (FIB) system uses an ion beam of Ga to raster scan the surface of a test material similar to the electron beam raster of SEM [Giannuzzi and Stevie, 2004]. The FIB ion source is a liquid metal ion system. Gallium is heated to above its melting point (29.7°C) and it flows onto a tungsten tip that has a radius of 2–5 μm. Field emission optics forms a 2- to 5-nm liquid gallium emission tip called a Taylor cone. The Ga^+ ions are extracted from the source and electrostatically (not electromagnetically as in an electron lens) focused onto a beam-defining aperture to change the beam current and size, then an objective lens is utilized to focus the resulting beam and scan coils to raster the ion beam. The resulting secondary electrons (or ions) are used to create an image of the material surface. By milling small holes in the material surface using the ion beam, cross-section images are produced. Multibeam FIB systems can also be incorporated with an SEM column, gas injectors to deposit nanoscale metal patterns, and nanomanipulators to obtain more accurate control, and used to create images with much higher magnifications and resolution. The dual-beam FIB is an excellent and rapid tool to produce thin sections for TEM analysis, where sample preparation is measured in a few hours, not days [Brandon and Kaplan, 2008]. Electron backscatter diffraction (EBSD) [Stojakovic, 2012] is a technique that allows the determination of the crystallographic orientation of a polycrystalline sample for a 2D analysis in an SEM and for 3D analysis in a dual-beam FIB by ion milling the surface of the sample and measuring the crystal morphology and orientation through the sample. Electrons diffract according to Bragg's law (see Section 5.1), producing a pattern of intersecting bands called Kossel cones. The backscattered electrons are observed from a CCD 2D camera. The sample is tilted to a high angle (70°) relative to the incident beam to maximize the backscattered electrons. The arrangement of bands is a function of crystal orientation, and the symmetry of the lattice is reflected in the pattern, with the angles between the bands directly related to angles between crystallographic planes. The EBSD patterns are collected from diffraction regions within the top 50 nm on the surface. Therefore, the sample must be free of deformation on the surface, so standard metallographic sample preparation techniques are not suitable without electropolishing, etching, vibratory polishing, or ion milling.

TABLE 4.3

Summary of the Capabilities and Applications of EDX

Method Provides	Analysis Capabilities	Applications
• Qualitative and quantitative analysis of elements that are heavier than boron • Spatial mapping	Solid materials, particles, fibers, and biological specimens	Materials characterization, failure analysis, process development, metallurgy, and QA/QC

Table 4.4 is a summary of FIB capabilities and applications, including such applications as failure analysis.

4.4.5 Transmission Electron Microscopy

Transmission electron microscopy (TEM) is an imaging technique that is used to examine the interaction characteristics of a transmitted electron beam after passing through an ultrathin material [Williams and Carter, 2004]. The TEM sample fits within the magnetic lens system and is therefore limited to a disc 3 mm in diameter. The formed image is magnified and focused onto an imaging charged coupled device (CCD) that operates as a camera. Similar to SEM, TEM is used to produce images with a resolution significantly higher than that of optical microscopes (over 1 million×) and sub-0.2-nm point image resolution; with recent advances in electron optics, it can examine fine details such as single atoms. TEM can be used to observe phase changes, crystal orientation, electronic structure, defects such as dislocations, stacking faults, nanoscale precipitates and sample-induced electron phase shift, as well as for absorption-based imaging [Pennycook and Nellist, 2011]. TEM can perform electron diffraction and EDS. Additionally, the electron source can be an SEM (making a scanning transmission electron microscope [STEM]), which allows the beam to be focused in a nanometer diameter spot and providing x-ray spectroscopy from a nanoscale volume of materials (unlike the scattering in an SEM that limits the analysis volume to a few microns). Because the sample is so thin, there is no beam spreading and the x-ray generation volume is almost identical to the incident beam size. Cryogenic sample holders are available. A summary of the capabilities and applications of TEM is given in Table 4.5.

4.4.6 Electron Probe Micro-Analyzers

Electron probe micro-analyzers (EPMA) use a micro-beam of electrons to perform nondestructive chemical analysis of minute solid samples [Goldstein et al., 2003; Reed, 2005]. The significance of EPMA is its ability to perform precise quantitative elemental analyses of very small areas, as small as 1–2 μm, using wavelength-dispersive x-ray analysis. Wavelength-dispersive x-ray detector uses Bragg diffraction (Bragg's law: $\lambda = 2d \sin \theta$) from a single crystal (fixed interplanar spacing, d) to separate the various x-ray wavelengths emitted from the sample. The beam current is higher than that in an SEM (in the μA range vs. nA in the SEM) due to the need for greater x-ray generation because of the inefficiency of diffraction

TABLE 4.4

Summary of FIB Capabilities and Applications

Method Provides	Analysis Capabiltiies	Applications
High-resolution imaging while ion milling through a material	Solid materials, metals and alloys, semiconductors, and ceramics	Semiconductor failure analysis, materials science, and preparing samples for TEM

TABLE 4.5

Summary of the Capabilities and Applications of TEM

Method Provides	Analysis Capabilities	Applications
• Very high resolution image analysis • 3D tomography	Solid materials that are properly prepared	Materials characterization, failure and semiconductor analysis, and process development

compared with nondispersive spectroscopy. While SEM can generate 3D images of objects, analysis by EPMA requires the use of flat polished areas that are examined in 2D. The beam current is kept constant, the sample is maintained at the fixed point on the diffracting circle, the diffracting crystal is bent as the goniometer changes theta, the x-ray take-off angle remains constant, and the detector must stay on the diffracting circle, making the mechanical mechanism for the detector extremely complex. Multiple crystals, such as LiF, α-quartz, potassium hydrogen phthalate (KAP), and lead stearate, and multiple goniometers must be used to cover the full range of the periodic table. Similar to SEM, the image resolution is far higher than possible by optical imaging. The electron beam has a higher accelerating potential (typically 30 KeV) than an SEM to maximize x-ray production. A sample holder with a moving stage containing three vertical axes is a standard fixture in EPMA systems. High vacuum is needed as with all electron optic instruments. EPMA can contain additional detectors that perform EDS and cathodoluminescence analysis, when the electron beam interaction generates photons in the infrared to ultraviolet frequency range. Another feature that is included in EPMA is a high-powered visible light microscope that allows direct optical observation of the tested samples. Due to the small depth of field in the optical microscope, it insures that the sample is located exactly on the x-ray diffracting focusing circle.

4.4.7 Scanning Probe Microscopy

Scanning probe microscopy (SPM) forms images of surfaces and surface interactions using a physical probe that scans a very small area (10 nm to 20 μm) of the tested material [Binnig et al., 1986]. The image of the surface is obtained by mechanically moving the probe in a line or square raster over the sample surface and recording the probe–surface interaction as a function of position. Use of multiple scanning probe tips allows simultaneous imaging of several surface interactions. The resolution of the SPM technique depends on the specific method that is used and it can be as high as atomic levels. This capability results from the use of piezoelectric actuators that move the probe and provides atomic-level precision and accuracy. A summary of the capabilities and applications of SPM is given in Table 4.6.

4.4.8 Scanning Tunneling Microscopy

Scanning tunneling microscope (STM) is a method of imaging surfaces at the atomic level [Binnig and Rohrer, 1986]. STM is often operated in high vacuum using a conducting tip that is brought very close to the examined surface with bias voltage applied between the tip and the sample. The method is based on the phenomenon of quantum tunneling that occurs when a particle tunnels through a barrier that classically cannot be passed. The STM system can cause electrons to tunnel between the tip and the test sample. The resulting tunneling current depends on the position of the tip, the applied voltage, and the local electron density of the sample. An image is formed by displaying the current as the tip is rastered across the material surface.

TABLE 4.6

Summary of the Capabilities and Applications of SPM

Method Provides	Analysis Capabilities	Applications
Family of SPMs performing surface morphology, lateral and magnetic force, as well as electrostatic surface characteristics	Solid surfaces including metals, alloys, ceramics and glasses, polymers, and biopolymers	Metal finish, process development, and QA/QC

TABLE 4.7

Summary of the Capabilities and Applications of ASM and SPM

Method Provides	Analysis Capabilities	Applications
• Quantitative surface topology • Force interaction (magnetic, electrostatic, and lateral friction)	Solid materials, metals, ceramics and glasses, as well as softer materials such as plastics, polymers, and biopolymers	Surface finish, magnetic media, metallurgy, and polymer engineering

An STM consists of a scanning tip, piezoelectric actuated scanner that controls the XYZ of the tip, vibration isolation system, and a computer for control, display, and data acquisition. Even though it is an imaging technique that can provide lateral resolutions of 0.1-nm resolution and depth resolution of 0.01 nm when used in UHV, it is very sensitive to the test conditions and requires extremely clean and stable surfaces, a sharp tip, excellent vibration control, and sophisticated electronics. The high resolution of STM allows both imaging and manipulation of individual atoms on compliant materials. The extreme sensitivity of the tunneling current to the height of the tip is critically dependent on having effective vibration isolation. Originally, magnetic levitation was used to suppress vibrations, but today, mechanical springs or gas spring systems are used. It is interesting to note that tests with STM can be performed in the air at 1 atm, as well as in various gases and liquids, with temperature ranging from close to zero Kelvin to several hundred degrees Celsius.

4.4.9 Atomic Force Microscopy

Atomic force microscopy (AFM), also known as scanning force microscopy (SFM), is a high-resolution SPM [Eaton and West, 2010]. The image is produced by scanning the surface with a mechanical probe and its movement is controlled by piezoelectric actuators allowing for accurate and precise movements. It normally uses a probe that is made of silicon or silicon nitride cantilever consisting of a sharp tip (a tip radius of curvature on the order of 20–50 µm) to scan the sample surface. The tip is brought toward the sample surface, and the atomic force between the tip and the sample causes both attraction and deflection of the cantilever. The forces that are measured include mechanical contact forces, van der Waals forces, capillary forces, lateral forces, and electrostatic forces. The signal is produced by a laser beam that is reflected from the cantilever surface and recorded by a photodiode array having two or more position-sensitive detectors. Other methods that can be used to measure the deflection include optical interferometry, capacitive sensing, and piezoresistive AFM cantilevers. Under ideal conditions, the AFM resolution can reach fractions of a nanometer. It is an effective imaging, measuring, and manipulation tool for materials at the nanometer scale. A summary of the capabilities of ASM and SPM is given in Table 4.7.

4.5 Materials and Metallurgical Analyzers

Besides methods of viewing the material using such tools as optical and electron microscopes, there are many analytical tools that are used to determine the elemental composition and other characteristics of the test materials including the use of x-ray diffraction (XRD) analysis and x-ray fluorescence (XRF) spectrometry. The various methods are covered in this section.

4.5.1 XRD Analysis

XRD analysis is used to examine the crystallography of materials as well as obtain information about the chemical composition and physical properties of materials [Cullity and Stock, 2001]. Bragg's law ($\lambda = 2d \sin \theta$), where λ is the wavelength of the x-rays, d is the interplanar spacing between an (hkl) plane, and θ is the angle of the diffracted beam, states the requirement of diffraction of an x-ray beam and can be utilized in a number of configurations. The most common is a diffractometer where a single-wavelength x-ray source is diffracted at multiple angles of 2θ corresponding to the d values found from the crystallographic structure. The diffracted beam intensity is dependent on the atom location, atomic number, and a number of other factors. Every material has a unique set of d values and I/I_0 (measured for a (hkl) plane to the strongest peak), which has been catalogued by the International Center for Diffraction Data (formerly known as JCPDS). The data historically was called the Hanawalt Search Method, and currently there are more than 800,000 unique material data sets, available in printed books, search manuals, or searchable databases. Given the ability to provide an absolute characterization of materials, a miniature XRD system was included in the suite of instruments used on the Mars Science Lab mission's Curiosity rover that landed on Mars in August 2012. A summary of the capabilities and applications of XRD is given in Table 4.8.

4.5.2 XRF Spectrometry

Bombarding materials with high-energy x-rays or gamma rays results in fluorescent emission (i.e., secondary emission) of x-rays and this emission is used to characterize materials [Jenkins, 1999]. The bombardment causes emission of electrons with binding energies smaller than the energy of the x-rays. Following this excitation, electrons from outer, lower-energy orbitals drop to fill the generated holes and this leads to energy release in the form of photons that are equal to the difference in energy between the related two electron orbitals. The emitted radiation, which has the characteristics of the atoms that are present, is measured by proportional counters or various types of solid-state detectors. XRF instruments are widely used for elemental analysis of ceramics, glass, and metals. An XRF instrument was also used in NASA's Mars Science Lab mission that landed on Mars in 2012 [Blake et al., 2012]. A summary of capabilities and applications of XRF is given in Table 4.9.

TABLE 4.8

Summary of the Capabilities and Applications of XRD

Method Provides	Analysis Capabilities	Applications
Crystallography information (identification of crystal and material phases) and crystal lattice structure	Metals, alloys, glasses and ceramics, particles, powders, polymers, and biopolymers	Crystal structure and material phase determination, process development, and QA/QC

TABLE 4.9

Summary of the Capabilities and Applications of XRF

Method Provides	Analysis Capabilities	Applications
• Elemental composition of inorganic elements in a variety of materials, thin films, and solids • Qualitative and semiquantitative data with the use of standards	Solid materials and thin films, ceramics and glasses, powders, particles, plastics, and polymers	Determination of trace elements, alloy composition, thin-film analysis, material validation, and QA/QC

4.5.3 X-Ray Absorption Spectroscopy

X-ray absorption spectroscopy (XAS) is used to determine the local geometric and/or electronic structure of tested materials [von Bordwehr, 1989]. Tests are done using a synchrotron radiation source of intense and tunable x-rays, and the samples are tested in gas, solution, or solid form. A narrow, parallel, monochromatic x-ray beam radiates through the test material causing intensity decrease due to absorption; the decrease depends on the types of atoms and the material density [http://www.chem.ucalgary.ca/research/groups/faridehj/xas.pdf]. At certain energies, the absorption increases drastically and involves absorption edges on the spectrum. Each such edge occurs when the energy of the incident photons is just sufficient to cause excitation of a core electron of an absorbing atom, producing a photoelectron. The energies of the absorbed radiation at these edges correspond to the binding electron energies in the K, L, M, etc., shells of the absorbing elements. The emitted photoelectron wave from an absorbing atom is backscattered by neighboring atoms, and the two interfere constructively and destructively and are measured as maxima and minima after the edge. A summary of the capabilities and applications of XAS is given in Table 4.10.

4.5.4 X-Ray Photoelectron Spectroscopy

X-ray photoelectron spectroscopy (XPS) is a surface analysis tool [Watts and Wolstenholme, 2003]. It provides semiquantitative elemental composition and chemical bonding state information (Table 4.11) and it is different from the Auger electron spectroscopy (AES) method. The latter provides composition data that are more quantitative and easier to interpret chemical state information (Table 4.12). Every element (except H and He) has a set of characteristic binding energies and photo-ionization cross-sections. The semiquantitative nature of XPS is based on a combination of theoretically calculated photo-ionization cross-sections such as the Scofield table [Scofield, 1976] and well-characterized standards. Applications of XPS include characterization of surface passivation and conversion coatings, surface segregation and oxidation thickness, and approximate alloy composition. Contamination analysis is one of the important applications of XPS.

TABLE 4.10

Summary of the Capabilities and Applications of XAS

Method Provides	Analysis Capabilities	Applications
Determining the local geometric and/or electronic structure of materials. Has unique sensitivity to the local structure compared with x-ray diffraction	Amorphous solids and liquids, metals and alloys, solid solutions, organometallics, powders, and some plastics	Determining molecular bonding environments and fine structure, especially in amorphous and some crystalline materials

TABLE 4.11

XPS Capabilities for Materials Characterization

XPS Analysis Provides	Limitations
• Semiquantitative (with calibration standards) • Chemical bonding state information • Point analysis, line scan, and x-ray mapping • Depth composition profiles (conductive and nonconductive materials)	• Cannot make measurements in liquids • Cannot detect hydrogen • Sensitive to about 0.5 atom % • Minimum beam area is about 50 μm

TABLE 4.12

AES Capabilities for Materials Characterization

Method Provides	Analysis Capabilities	Applications
• Surface composition • Some chemical bonding state information • High spatial resolution • Line scan and X-Y elemental mapping • Depth composition profiles • Secondary electron detector (SED) images • EDX (on some instruments)	• Metals and alloys • Carbon composites • Some ceramics	• Grain boundary analysis • Corrosion and defect analysis • Failure analysis

4.5.5 Low-Energy Electron Diffraction

Low-energy electron diffraction (LEED) is used to determine the surface structure of crystalline materials by bombarding the surface with a collimated beam of low-energy electrons ranging from 20 to 200 eV and analyzing diffracted electrons as spots on a fluorescent screen [VanHove et al., 2012]. Using the LEED technique, diffraction patterns are examined to determine information about the symmetry of the surface structure. Also, the intensity of the diffracted beams as a function of the energy of the incident electron beam is used to obtain accurate information about the atomic positions on the surface of the tested material.

4.5.6 Neutron Diffraction

Neutron diffraction, which is also known as elastic neutron scattering, is used to determine the atomic and/or magnetic structure of materials [Lovesey, 1984; Squires, 1996]. This technique is highly sensitive to certain light atoms and can be used to distinguish between isotopes. A thermal or cold neutron beam irradiates the tested material, and the formed diffraction pattern provides information about the structure of the material. Similar to the XRD method, the diffraction of the scattered neutron beam follows Bragg's law and provides complementary information. Neutrons interact with the nucleus of the atom rather than the electrons, and the diffraction intensity depends on the interacted isotope. For some light atoms, the diffraction intensity is very strong even in the presence of elements with a large atomic number.

The main disadvantage of neutron diffraction is the limited availability of the radiation source. Neutrons are produced by a reactor, and to select a specific neutron wavelength, components such as crystal monochromators and filters are used. Alternatively, a spallation source can be used where the energies of the incident neutrons are sorted by a series of synchronized aperture elements that filter the neutron pulses with the desired wavelength. Samples in the form of polycrystalline powder are tested using crystals that are much larger than the ones used in XRD. Since neutrons carry a spin, they interact with the magnetic moment of electrons, allowing analysis of the microscopic magnetic structure of tested materials.

4.5.7 Auger Electron Spectroscopy

Auger electron spectroscopy (AES) is a method of analyzing the surface composition of materials, and its characterization capabilities are summarized in Table 4.12 [Briggs and Seah, 1996]. This method is especially useful for grain boundary analysis. AES can provide point analysis of areas as small as 50 Angstroms, making it ideal for characterizing inclusions, defects, cracks, etc. The Auger effect is a multielectron interaction starting with an incident electron that creates a hole that is filled by an electron from a higher-level orbital and imparts transition energy onto a third electron that is emitted.

An AES instrument generally consists of an ionizing energy source in the form of a high-voltage electron beam that is focused on a test material. The emitted electrons are deflected around the electron gun passing through an aperture to the back of the analyzer and directed into an electron multiplier. Secondary electrons that are ejected from the surface are measured using a cylindrical mirror analyzer or hemispherical analyzer. The relatively low energy of the secondary electrons limits the information volume to the outer 50–100 Angstrom surface layer, and the spatial resolution of AES can be as low as 10–25 Angstroms depending on the source. By varying the supplied voltage, Auger data are measured, and using an ion gun, the depth profile is measured. The Auger effect produces a characteristic signature for each element, which is somewhat dependent on chemical bonding state. AES is capable of point, line, and mapping analysis, and the speed of AES allows for simultaneous ion etching and analysis to provide depth composition profiles.

4.5.8 Raman Spectroscopy

Raman spectroscopy is a characterization method that is used to study vibrational, rotational, and other low-frequency modes [Smith and Dent, 2005]. This method is widely used to characterize materials, especially carbon, measure temperature, and find the crystallographic orientation of a sample [Gardiner, 1989]. It is interesting to note that a Raman analyzer was included in the suite of instruments on the Curiosity rover that landed on Mars in August 2012.

Raman spectroscopy is applied to perform analysis of inelastic scattering (also known as Raman scattering) of monochromatic light, and the radiation source is a laser in the visible, near infrared, or near ultraviolet range. The laser beam interacts with molecular vibrations, phonons, or other excitation forms and results in energy shift of the laser photons providing information about the tested material. The test is done by illuminating a spot of a material, and the scattering is optically acquired through a monochromator. The elastic Rayleigh scattering of wavelengths that are close to the laser light is filtered out and the rest of the acquired light is dispersed onto a detector. The weak intensity of the dispersed light that results from inelastic scattering is difficult to separate from the high-intensity Rayleigh scattering. In recent years, the detectors that used to be based on photomultipliers, which required long acquisition times, have been significantly improved by using Fourier transform (FT) spectroscopy and CCD detectors.

Generally, the Raman effect occurs when light impinges on a molecule and interacts with the electron cloud to create excitation from the ground state to a virtual energy state. When the molecule relaxes, it emits a photon and returns to a different rotational or vibrational state. The energy difference between the original state and the new one leads to a frequency shift from the excitation wavelength in the emitted photon. If the final vibrational state of the molecule is more energetic than the initial state, then a Stokes shift to a lower frequency takes place balancing the total energy. If the final vibrational state is less energetic than the initial state, then an anti-Stokes shift occurs and the emitted photon is shifted to a higher frequency. To exhibit a Raman effect, a molecule needs to have a change in the polarization potential with respect to the vibrational coordinate. The level of the polarizability change determines the Raman scattering intensity, and the pattern of shifted frequencies is determined by the sample rotational and vibrational states. A summary of the capabilities and applications of Raman spectroscopy is given in Table 4.13.

TABLE 4.13

Summary of the Capabilities and Applications of Raman Spectroscopy

Method Provides	Analysis Capabilities	Applications
Vibrational, rotational, and low-frequency modes in a system, including molecular and crystalline bonding environments	Organic and amorphous materials, some ceramics and glasses, plastics, and powders	Structural determination of nanocarbon materials, some ceramics and glasses, process development

4.5.9 Electron Microprobe

Electron microprobes (EMP) are used to determine the chemical composition of solid materials in small volumes (typically $\leq$ 10–30 μm^3). These probes are also known as an electron microprobe analyzer (EMPA) or electron probe micro-analyzer (EPMA) [Reed, 1997; Goodhew et al., 2000]. Similar to SEMs, tested materials are bombarded by an electron beam and the determined wavelengths of the excited x-rays are used to perform elemental analysis. Improvements in the capability of EMP instruments have led to significant accuracies in measuring minute concentration of trace elements.

4.5.10 Electron Energy Loss Spectroscopy

Electron energy loss spectroscopy (EELS) is a method of analyzing materials by exposing them to an electron beam with a narrow range of kinetic energies [Brydson, 2001]. The electrons undergo inelastic scattering loss of energy with their beam path deflected slightly and randomly. Inelastic interactions include phonon excitations where the inner-shell ionizations are particularly useful for detecting the elemental components of a tested material. The energy loss is measured by an electron spectrometer and analyzed to determine the types of atoms and the atomic number of each type that interacts with the electron beam. The scattering angle can also be measured and provide information about the dispersion relation of the excitation source of inelastic scattering. Generally, the EELS method is more effective for testing materials having low atomic numbers, since the excitation edges tend to be better defined.

4.5.11 Inductively Coupled Plasma Mass Spectrometry

Inductively coupled plasma mass spectrometry (ICP-MS) is a spectrometry method with a significantly high sensitivity and it can detect concentrations of metals and several nonmetals at levels of parts per trillion [Montaser, 1998]. The operation of the ICP-MS method is based on inducing coupled plasma onto the test material causing ionization, and the ions generated are separated and quantified by a mass spectrometer. Generally, ICP-MS has much greater analytical precision, sensitivity, and speed than the atomic absorption techniques. However, some ions interfere with the detection of others and may affect the accuracy of the measurements.

4.5.12 Particle-Induced X-Ray Emission

Particle-induced x-ray emission (PIXE) is another technique of elemental analysis also known as proton-induced x-ray emission [Johansson et al., 1995]. The method involves radiating a test material using an ion beam, causing atomic interactions that release electromagnetic radiation with wavelengths in the spectrum range of x-rays. The wavelengths that are emitted are related to the specific elements that are present in the tested material. By focusing the radiating beam on a 1 μm diameter spot, the method has been improved significantly, allowing for the performance of microscopic analysis. This latter analytical method is known as microPIXE, and it is effective in determining the distribution of trace elements in a wide range of materials.

4.5.13 Atomic Spectroscopy

Optical atomic spectroscopy is an elemental analysis method of identifying and quantifying free atoms in the gas phase [Cullen, 2003]. The tested material elements are converted by atomization into gaseous atoms or elementary ions, resulting in ultraviolet and visible absorption, emission, or fluorescence of atomic species that are measured. A summary of the capabilities and applications of atomic spectroscopy is given in Table 4.14. Generally, there are three types of atomic spectroscopy:

Atomic absorption (AA): In a heated gas, atoms absorb radiation in wavelengths that are characteristic of electron transitions from ground to higher excited states. The related atomic absorption spectrum consists of resonance lines that are the result of this transition.

Atomic emission (AE): Heating by an electric arc, flame, plasma, or spark excites atoms to higher orbitals, and the excitation is accompanied by emission or radiation of a photon.

Atomic fluorescence (AF): Irradiation with an intense source having wavelengths that are absorbed by atoms or ions in a flame can cause fluorescence. The observed radiation results from mostly resonance fluorescence due to returning to the ground state from excited states.

4.5.14 Static Secondary Ion Analysis

Static secondary ion analysis (SSIMS) is a very sensitive surface analysis technique, and information about this method is given in Table 4.15. Also, a summary of the SSIMS method's capabilities and applications is given in Table 4.16. The SSIMS uses an ion beam, typically argon, which is rastered over a surface to cause fragmenting of the surface species that are ejected into vacuum. The fragmented molecules are analyzed using time of flight (TOF) or quadrupole mass analyzers to obtain information about their mass and relative intensity. SSIMS is part of an ion microprobe family that is used to detect trace elements in steel and other alloys and is much more sensitive than electron microprobe techniques. Depth profiles using SSIMS can be created fairly fast, but commercially, its use can be very expensive even with the use of molecular spectral libraries.

TABLE 4.14

Summary of the Capabilities and Applications of Atomic Spectroscopy

Method Provides	Analysis Capabilities	Applications
Determination of elemental composition by electromagnetic or atomic or mass spectrum	Trace elements in solution, identification of some molecular species, metal and ceramic particles, and some powders	Qualitative and quantitative trace element analysis, testing for impurities and contaminants in aqueous solutions/mixtures

TABLE 4.15

SSIMS Capabilities for Materials Characterization

SSIMS Features	SSIMS Information	Advantages/Disadvantages
• Surface sensitive • Detects positive ions very well • Analyses fragments of organic molecules • Identifies contamination • Destructive technique, since the surface is destroyed by the ion beam	• Molecular ions • Inorganic ions • Elemental concentration (% of volume) • Depth concentration profiles • Chemical identification (using spectral libraries)	• Very sensitive technique (parts per billion/trillion) • Fragment ions provide species identification • Surface sensitivity provides contamination analysis • Requires calibration for concentration profiles

TABLE 4.16

Summary of the Capabilities and Applications of SSIMS

Method Provides	Analysis Capabilities	Applications
• Molecular fingerprint information—ion masses • Qualitative/semiquantitative analysis of some elements	• Organic films • Thin films • Silicon wafers/semiconductors	• Semiconductor wafer testing • Characterizing thin organic films • Identifying contamination

4.6 Summary/Conclusions

Understanding and determining the properties of materials at low temperatures requires multidisciplinary expertise including materials science as well as chemical, electrical, and mechanical engineering. The methods of characterization need to account for the processing techniques that can involve multiple steps and require precision control of the temperature, pressure, and processing time. Key properties that need to be characterized include elemental content, particle/grain size, surface structure, chemical purity, and crystallography.

Combining image, surface, structural, and chemical testing with physical properties measurements allows the development of better materials by understanding the fundamental material structure and property relationships. Knowledge of the changes in chemistry, kinetics, phase, and physical properties helps materials engineers in the selection and engineering of materials. Further, grain boundary engineering and characterization help develop high-performance metal alloys, ceramics, and composites that maintain the strength of the test material. Composite materials, in particular, require extensive engineering to ensure that they maintain compatibility in interfacial bonding, thermal contraction, and with all the material phases, work at the desired operating temperature.

Structural analysis using XRD, TEM, and FE-SEM provides materials engineers with insights into the atomic and chemical bonding of a material, the structure and chemistry of the different phases, and how the macrostructure and physical properties of a material are dependent on micro/nanostructured domains. Physical property measurements are as important as materials characterization, and support the development and optimization of the performance of low temperature materials. Strength, modulus of elasticity, stiffness, thermal contraction, electrical and thermal conductivity, and optical (phonon) properties are measurements made in materials laboratories and play an important role in materials selection.

Acknowledgments

Some of the research reported in this chapter was conducted at the Jet Propulsion Laboratory (JPL), California Institute of Technology (Caltech), under a contract with the National Aeronautics and Space Administration (NASA). The author would like to thank Donald Lewis, JPL/Caltech, Pasadena, CA, and Arvind Agarwal, Florida International University, Miami, FL, for reviewing this chapter and providing valuable technical comments and suggestions.

References

Berman, R., *Thermal Conduction in Solids*, Clarendon Press, Oxford (1976).

Binnig, G., and H. Rohrer, Scanning tunneling microscopy, *IBM Journal of Research and Development*, Vol. 30, No. 4 (1986).

Binnig, G., C. F. Quate, and C. Gerber, Atomic force microscope. *Physical Review Letters*, Vol. 56, No. 9, doi:10.1103/PhysRevLett.56.930 (1986), pp. 930–933.

Blake, D. F., D. Vaniman, C. Achilles, R. Anderson, D. Bish, T. Bristow, C. Chen, S. Chipera, J. Crisp, D. Des Marais, R. T. Downs, J. Farmer, S. Feldman, M. Fonda, M. Gailhanou, H. Ma, D. Ming, R. Morris, P. Sarrazin, E. Stolper, A. Treiman, and A. Yen, Characterization and calibration of the CheMin mineralogical instrument on Mars science laboratory, *Space Science Review*, doi:10.1007/s11214-012-9905-1 (2012).

Brandon, D., and W. Kaplan, *Microstructural Characterization of Materials*, ISBN978-0-470-02785-1, Wiley, New York, NY (2008).

Briggs, D., and M. P. Seah, *Practical Surface Analysis, Auger and X-Ray Photoelectron Spectroscopy*, ISBN-10: 0471953407, ISBN-13: 978-0471953401, John Wiley & Sons, Berlin, Germany (1996), pp. 1–674.

Brydson, R., *Electron Energy Loss Spectroscopy*, ISBN-10: 1859961347, ISBN-13: 978-1859961346, Garland Science, New York, NY (2001), pp. 1–160.

Corruccini, R. J., and J. J. Gniewek, *Thermal Expansion of Technical Solids at Low Temperatures*, NBS Monograph 29, U.S. Government Printing Office, Washington, DC (1961).

Cullen, M., *Atomic Spectroscopy in Elemental Analysis*, Blackwell, Oxford, UK, ISBN-10: 0849328179, ISBN-13: 978-0849328176 (2003), pp. 1–310.

Cullity, B. D., and S. R. Stock, *Elements of X-Ray Diffraction*, 3rd Edition, ISBN-10: 0201610914, ISBN-13: 978-0201610918, Prentice Hall, Upper Saddle River, NJ (2001), p. 664.

Eaton, P., and P. West, *Atomic Force Microscopy*, ISBN-10: 0199570450, ISBN-13: 978-0199570454, Oxford University Press, Cary, NC (2010), pp. 1–288.

Gardiner, D. J. *Practical Raman Spectroscopy*, Springer-Verlag, Berlin, Germany, ISBN 978-0-387-50254-0 (1989).

Giannuzzi L. A., and F. A. Stevie (Eds.), *Introduction to Focused Ion Beams: Instrumentation, Theory, Techniques and Practice*, ISBN-10: 0387231161, ISBN-13: 978-0387231167, Springer, New York, NY (2004), pp. 1–376.

Goldstein, J., D. E. Newbury, D. C. Joy, C. E. Lyman, P. Echlin, E. Lifshin, L. Sawyer, and J. R. Michael, *Scanning Electron Microscopy and X-Ray Microanalysis*, 3rd Edition , ISBN-10: 0306472929, ISBN-13: 978-0306472923, Springer, New York, NY (2003), p. 689.

Goodhew, P. J., J. Humphreys, and R. Beanland, *Electron Microscopy and Analysis*, 3rd Edition, ISBN: 0748409688, ISBN-13: 978-0748409686, Taylor & Francis Group; Boca Raton, FL (2000), pp. 1–254.

Jenkins, R., *X-Ray Fluorescence Spectrometry*, 2nd Edition, ISBN-10: 0471299421, ISBN-13: 978-0471299424, Wiley-Interscience, New York, NY (1999), pp. 1–232.

Jenkins, R., R. W. Gould, and D. Gedcke, *Quantitative X-Ray Spectrometry*, 2nd Edition, ISBN-10: 0824795547, ISBN-13: 978-0824795542, Marcel Dekker, New York, NY (1995), pp. 1–484.

Johansson, S. A. E., J. L. Campbell, and K. G. Malmqvist, *Particle-Induced X-Ray Emission Spectrometry (PIXE)*, ISBN-10: 0471589446, ISBN-13: 978-0471589440, Wiley-Interscience, New York, NY (1995), pp. 1–451.

Kittel, C., *Introduction to Solid State Physics*, 5th Edition, Wiley, New York (1976).

Laesecke, A. Through measurement to knowledge: The inaugural lecture of Heike Kamerlingh Onnes (1882), *Journal of Research of the National Institute of Standards and Technology*, Vol. 107, No. 3, doi:10.6028/jres.107.021 (2002), pp. 261–277.

Lovesey, S. W., *Theory of Neutron Scattering from Condensed Matter; Volume 1: Neutron Scattering*, ISBN 0-19-852015-8, Clarendon Press, Oxford, United Kingdom (1984).

Meyers, M. A., and K. K. Chawla, *Mechanical Behavior of Materials* ISBN-10: 0521866758; ISBN-13: 978-0521866750, Cambridge University Press, London, UK (2008), p. 882.

Minsky, M., Microscopy apparatus, US Patent 3,013,467 (December 19, 1961).

Montaser, A. *Inductively Coupled Plasma Mass Spectrometry*, ISBN-10: 0471186201, ISBN-13: 978-0471186205, Wiley-VCH, Berlin, Germany (1998), pp. 1–1004.

Murphy, D. B., *Fundamentals of Light Microscopy and Electronic Imaging*, ISBN-10: 047125391X, ISBN-13: 978-0471253914, Wiley-Liss, Wilmington, DE (2001), pp. 1–360.

Pennycook, S. J., and P. D. Nellist (Eds.) *Scanning Transmission Electron Microscopy: Imaging and Analysis*, ISBN-10: 1441971998, ISBN-13: 978-1441971999, Springer, New York, NY (2011), pp. 1–774.

Price, R. L., and W. G. Jerome (Eds.) *Basic Confocal Microscopy*, ISBN-10: 0387781749, ISBN-13: 978-0387781747, Springer, New York, NY (2011), pp. 1–313.

Reed, R. P., and A. F. Clark, *Materials at Low Temperatures*, American Society of Metals (ASM), Metals Park, OH (1983).

Reed, S. J. B., *Electron Microprobe Analysis*, 2nd Edition, ISBN-10: 052159944X, ISBN-13: 978-0521599443, Cambridge University Press, New York, NY (1997), pp. 1–350.

Reed, S. J. B., *Electron Microprobe Analysis and Scanning Electron Microscopy in Geology*, 2nd Edition, Cambridge University Press, New York, NY (2005).

Reimer, L., *Scanning Electron Microscopy: Physics of Image Formation and Microanalysis*, Springer Series in Optical Sciences, New York, NY, ISBN-10: 3540639764, ISBN-13: 978-3540639763 (1998), pp. 1–541.

Rose-Innes, C., and E. H. Rhoderick, *Introduction to Superconductivity*, 2nd Edition, International Series in Solid State Physics, Vol. 6, Pergamon Press, New York, NY (1978).

Scofield, J. H., Hartree-Slater subshell photoionization cross-sections at 1254 and 1487 eV, *Journal of Electron Spectroscopy and Related Phenomena*, Vol. 8 (1976), pp. 129–137.

Smith, E., and G. Dent, *Modern Raman Spectroscopy: A Practical Approach*, ISBN-10: 0471497940, ISBN-13: 978-0471497943, Wiley, New York, NY (2005), pp. 1–222.

Squires, G. L. *Introduction to the Theory of Thermal Neutron Scattering*, 2nd Edition, ISBN 0-486-69447-X, Dover Publications Inc., New York, NY (1996).

Stojakovic, D., Electron backscatter diffraction in materials characterization, *Processing and Application in Ceramics*, Vol. 6, No. 1 (2012), pp. 1–13.

VanHove, M. A., W. H. Weinberg, and C.-M. Chan, *Low-Energy Electron Diffraction: Experiment, Theory and Surface Structure Determination*, ISBN-10: 3642827233, ISBN-13: 978-3642827235, Springer Series in Surface Sciences, New York, NY (2012).

Van Sciver, S. W., Low-Temperature Materials Properties, Chapter 2 in *Helium Cryogenics*, International Cryogenics Monograph Series, doi 10.1007/978-1-4419-9979-5_2, ISBN 978-1-4419-9978-8, Springer LLC (2012), pp. 17–58.

von Bordwehr, R. S., A history of the x-ray absorption fine structure, *Annales de Physique*, Vol. 14, No. 4, doi: 10.1051/anphys:01989001404037700 (1989), pp. 377–465.

Watts, J. F., and J. Wolstenholme, *Introduction to Surface Analysis by XPS and AES*, 2nd Edition, ISBN-10: 0470847131, ISBN-13: 978-0470847138, Wiley, New York, NY (2003), pp. 1–224.

White, G., *Experimental Techniques in Low Temperature Physics*, 3rd Edition, Clarendon Press, Oxford, UK (1979).

Williams, D. B., and C. B. Carter, *Transmission Electron Microscopy: A Textbook for Materials Science* (4-Vol Set), ISBN-10: 030645324X, ISBN-13: 978-0306453243, Springer, New York, NY (2004), pp. 1–703.

5

NDE and Structural Health Monitoring for Low Temperature Applications

Yoseph Bar-Cohen, Victor Sloan, Terry D. Rolin, and Tom Guettinger

CONTENTS

5.1 Introduction

Nondestructive testing (NDT) and nondestructive evaluation (NDE) methods are used throughout the life of materials and structures from the initial steps of fabrication and throughout the operating life of the structures in service [Bar-Cohen, 2000]. Such methods are applied to inspect materials, parts, safety-critical components, structures, and systems. Similar methods are even used to perform diagnosis of our body and are also applied throughout the world to detect concealed weapons and explosives on passengers in airports. Generally, NDE methods require removal of the test articles from service; the structure is scanned to detect the presence of flaws as well as determine the material properties. To test structures while they are in service, structural health monitoring (SHM) methods are used where sensors are employed to detect defects that are forming. Flaws that may emerge in service include corrosion and fatigue cracks. The use of SHM methods allows extending the life of structures and safely operating them, since they are monitored while they are aging and they are removed from service only if a cause for concern is identified. The NDE methods that are generally used include ultrasonics, visual, infrared, thermal, electrical, electromagnetic and eddy current, radiographic, microwave, and spectroscopy. The SHM methods that are used rely on the availability of sensor technologies that are applicable to the operating conditions of the specific materials and structures as well as the required information that needs to be determined. The application of NDE and SHM methods at low temperatures is mostly done at the lab and under outdoor conditions in the field, including aircraft structures that are operated under cold conditions, as well as drilling facilities and pipelines in such cold places as Alaska in the winter.

5.1.1 NDE and NDT Methods

NDT and NDE methods are used to prevent premature failure of structures due to the formation of such defects as cracks, corrosion, and other field-related flaws that may cause degradation of the structure integrity. The methods that are widely used in the field include liquid penetrant, visual testing, eddy current, magnetic particles, radiography, and various ultrasonic methods. Each of these methods have advantages and disadvantages; they can be relatively expensive, generate inconclusive findings, and may not be feasible to automate or integrate into a system, nor generate prompt information about the initiation and formation of damage. The methods that involve the use of liquids (such as liquid penetrant, magnetic particles, as well as ultrasonic testing using couplants) are not applicable at low temperatures and particularly below the freezing point of specific liquids. For performing ultrasonic tests at low temperatures that can reach cryogenic levels, the method of electromagnetic acoustic transduction (EMAT) testing is widely used for flaw detection and material property characterization [Hirao and Ogi, 2003]. This method does not require direct contact with the surface of the tested object, since it is activated by induction and therefore does not require a couplant. It allows for determination of material properties from the measurement of the elastic wave velocity or the attenuation. To excite ultrasonic waves by EMAT, a coil and magnet structure are used and the method can generate various complex wave patterns and polarizations that are difficult to realize compared to testing by fluid coupled piezoelectric probes.

5.1.2 Structural Health Monitoring

Structural health monitoring (SHM) methods are used for monitoring, inspection, damage assessment, and residual life prediction, and can be applied for real-time sensing. SHM methods are also used to assess the condition of structures and monitor for early detection of damage, thus increasing their reliability, safety, and efficiency as well as allow for repairs or, if it is determined necessary, provide information for making decisions regarding part replacement. The test system uses sensors to extract from the measurements information about damage and related features and to determine the integrity of the tested structure. A structure that is equipped with monitoring sensors is generally called a smart structure. Various techniques and disciplines are involved with the use of SHM systems, including structural dynamics modelling and simulations, materials science, fatigue and fracture analysis, materials and flaws characterization, sensors and actuators, and microelectronics.

SHM methods are applied either to conduct continuous monitoring or to periodically perform specific tests. The application of continuous monitoring requires the use of reliable and robust systems that can be operated for a long period of time in the field in harsh environments. In many systems, the condition of the tested structure is monitored to determine the state of integrity and to detect deviation from reference properties. Generally, such systems are connected to the Internet and allow remote monitoring where the system acquires data and periodically transmits it to a host computer either in raw form or preanalyzed.

Searching the Internet for NDE studies and research that were conducted at low temperatures shows a limited number of documented reports. One study that can be referenced is a PhD dissertation covering composite sandwich structures that were tested using acoustic emission at temperatures as low as −60°C (−140°F) [Deng, 2007]. This experimental and numerical study describes the application of acoustic emission NDE to determine the fatigue behavior of composite sandwich beams related to hull material of a ship that plies the polar regions. For this purpose, 4-point static and load-controlled fatigue bending tests were performed with sandwich beams made of unidirectional carbon/epoxy and woven glass/epoxy skin materials. The effect of low temperatures on the flexural static behavior of the beam was determined from the increase in stiffness, strength, and elastic limit, and the decrease in the displacement as a function of temperature. At −60°C, it was observed that core shear was the main failure mode under fatigue loading conditions and brittle type core shear failure increases at lower temperatures. The acquired AE data provided insight into the characteristics of damage progression and the phenomena responsible for it. Furthermore, using finite element analysis revealed the effects of skin stiffness on the crack initiation location under fatigue loading conditions. Given the limited documented information that was identified in the literature, this chapter is focused mostly on the results of NDE studies at low temperatures that were conducted by the authors.

5.1.3 Electromagnetic Acoustic Transduction

Electromagnetic acoustic transduction (EMAT) is an NDE method that involves the use of electromagnetic excitation to generate ultrasonic waves and the use of the reverse transformation to produce electric signals that are applied to determine the response. In contrast to the conventional ultrasonic tests that involve the use of piezoelectric transducers, the EMAT method does not require coupling material. EMAT can be used to inspect most ferrous and nonferrous metals and can be applied at various temperatures including cryogenic ones. EMAT can be used to excite ultrasonic guided waves and perform high-speed material analysis where the complete circumference and volume of tubular materials can be inspected. Electromagnetic fields are used to generate and receive ultrasonic waves that travel in a circumferential direction within the pipe material. This transmitted wave is sensitive to corrosion losses on both the internal diameter and the outer diameter of the pipe, and it travels between the transmitting and the receiving EMAT transducer. The EMAT method has many of the advantages of electromagnetic inspection methods including the ability to inspect irregular surfaces. This method provides qualitative information about the severity of corrosion damage.

The EMAT method can be used to perform echo polarization tests that employ birefringence measurements (Figure 5.1). It allows for studying the fast and slow axes of material velocity before and after the exposure of the material to cryogenic temperatures where the material residual stress component changes are recorded. The velocity axial characteristics are attributed to the crystallographic structure resulting from the process of rolling the sheet metal.

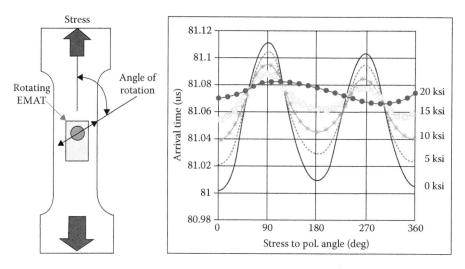

FIGURE 5.1

Electromagnetic acoustic transduction (EMAT) ultrasonic shear waves are polarized perpendicular to their propagation direction. Rotating the polarization relative to the stress can be used to find and measure the principal stresses in a tested part.

5.2 Cryogenic NDT and Transition Detection Methods

For planetary applications, liquid nitrogen or helium is used in computer-controlled vacuum insulated cryogenic chambers that simulate the environment to which space vehicle components are exposed. The rate of temperature change and the dwell time constant at any temperature can also be programmed to match specific mission profiles. Various forms of excitation can be introduced that simulate the conditions to which a space vehicle component may be exposed. These forms of excitation include vibration, ultraviolet light, magnetic fields, mechanical shock, and vacuum. Real-time NDT techniques are quite useful in furthering the understanding of the effects of the exposure to low temperatures as well as determining how the material properties are impacted.

The ability to identify and quantify changes in the microstructure of metal alloys is valuable in determining fatigue life and materials behavior in service. For example, certain metals, after being subjected to cryogenic temperatures, have shown large increases in their fatigue life. Moreover, improved component life has also been documented when such low temperature exposure has been used in aerospace, vehicles, wind energy, and turbine power plants and transmissions and can reduce the total operation cost [Jordine, 1996]. However, the mechanisms of microstructure changes are not yet fully understood (Chapter 3). Such changes are currently evaluated in a semi-quantitative manner either by visual inspection of microscopic images of the microstructure or by destructive examination of the test material, as well as by characterizing precipitation hardening and phase transformation of materials (Chapter 4).

Novel NDT methods that can determine material property changes, which occur when test specimens are exposed to cryogenic temperatures, have been conceived and developed by the coauthor Victor Sloan [Sloan, 2014a and 2014b]. The developed methods are based on the use of EMAT and ultrasound transmission (UT) for real-time recognition. These methods can determine the time and temperature at which phase transformations of materials take place and identify the residual stress changes as well as provide a guide in removing these stresses. The methods are used to measure changes in microstructure and validate the initial assertion of improved fatigue life or material property characteristic changes under exposure to cryogenic temperatures. A photo made by time delay photography showing the EMAT probe and test part viewed through the window of a cryogenic LN2 chamber is shown in Figure 5.2. The sensor is magnetically attached to a tested crankshaft measuring changes in ultrasonic transit time.

Generally, temperature gradients that are created during manufacture of parts can produce nonuniform dimensional changes. Also, when metal castings are cooled and solidified, compressive stresses are developed in the lower-volume sections of parts, which cool first and therefore develop tensile stresses in the sections with larger volume, which are last to cool. As a result, shear stresses can develop between the sections having different volume. Also, the surface of structures cools first and the core last, resulting in stresses that are developed particularly when phase change takes place during cooling. The fatigue life of the structure can be predicted using finite element analysis (FEA) and corroborated by dynamic stress and strain measurements.

The EMAT method that was used is based on the effect of the material microcrystalline structure on the transmission of ultrasound. This effect allows for the evaluation of the kinetics and completion of the time-temperature transformation between structural phases. The measurements are based on the difference between the Young's moduli of the phases at the transformation temperature [Ray et al., 2003]. Since the velocity of longitudinal sound waves is a function of the square root of the elastic constants, a change in velocity is expected during the transformation. Hence, the velocity of the wave should be between the values of the two phases. In addition to velocity changes, sound attenuation is also a highly sensitive parameter that may also give insight into the mechanism of the transformation. Generally, the velocity of the longitudinal sound wave (V_l) is related to the elastic constant. In Hooke's law, the elastic constant relates to the linear stress and strain but these are fundamentally functions of the thermodynamic state of the material and the interatomic binding forces in the microcrystalline

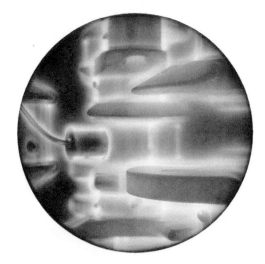

FIGURE 5.2
A photograph of an EMAT test where the probe and the test parts are viewed through the window of a cryogenic LN2 chamber. (Courtesy of Victor Sloan.)

lattice structure. The longitudinal velocity is related to the Young's modulus (*E*), density (P), and Poisson's ratio (σ) by

$$V_1 = \left[\frac{E}{\rho} \frac{1-\sigma}{(1+\sigma)(1-2\sigma)} \right]^{1/2}$$

To utilize this to nondestructively assess the transformation within an alloy, the sound wave velocity is determined from the transit time through a structure, where the time (τ) is related to longitudinal velocity (*V*) and the path length (*L*) by

$$\tau = \frac{L}{V}$$

Since changes in length are small during the phase transformation, changes in the time of flight of ultrasonic waves reflect changes in the properties of the product and provide information about the progress of the phase transformation. Hence, measuring the ultrasonic transit time provides a nondestructive control parameter for optimizing the treatment duration and temperature while reducing excessive processing or the need for multiple treatment cycles, both of which reduce the desirable mechanical properties [Sahoo, 2011]. Performing EMAT, while subjecting the tested material to cryogenic conditions, allows for obtaining the needed data from the changes in the wave velocity and amplitude. The time and temperature are recorded, as well as changes in velocity or amplitude, and thus phase transformation is determined (see Figure 5.3).

5.2.1 EMAT Test of Transonic Wind Tunnel Turbine Blades

To demonstrate the capabilities of the EMAT method, tests were performed on transonic wind tunnel turbine blades made of 2014 aluminum that had been removed from the NASA Ames Research Center (located at Moffett Field, California). These blades (Figure 5.4) are part of the 3.35 × 3.35 m² (11 × 11 ft) Unitary Plan Wind Tunnel (UPWT). This wind tunnel facility (Figure 5.5) has been used for testing many structures including several generations of commercial and military aircraft, as well as NASA space vehicles including models of the Space Shuttle, Mercury, Gemini, and Apollo capsules. The NASA UPWT design is a closed-return, variable-density tunnel with a fixed geometry and ventilated test section having a flexible

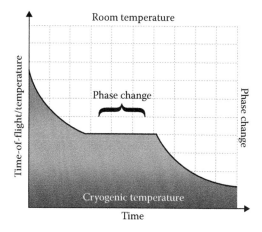

FIGURE 5.3
The time of flight and temperature as a function of the duration of exposure to cryogenic temperature.

FIGURE 5.4
CAD drawing of turbine blade used in the NASA Ames Research Center 3.35 × 3.35 m² (11 × 11 ft) transonic wind tunnel.

FIGURE 5.5
The NASA Ames Research Center Unitary Plan Wind Tunnel (UPWT) complex. (Courtesy of NASA Ames Research Center, Moffett Field, CA.)

wall nozzle. It is one of three separate test sections powered by a common drive system. Its aluminum turbine blades are used in a three-stage, axial-flow compressor powered by four wound rotors with variable-speed induction motors, producing airflow at a speed of up to 1.5 Mach. The interchangeability of models among the test sections allows testing across a wide range of conditions. The blades of this wind tunnel were monitored by EMAT while they were subjected to a temperature cycle from +23.3°C to –154.4°C and back to +23.3°C (+74°F to –310°F and back to +74°F) (Figure 5.6). For comparison reference, 2014 aluminum alloy coupons that were etched and 180 grit sanded were included in the test processes.

An example of one of the hysteresis graphs is shown in Figure 5.7 presenting the transit time versus temperature for the cooling and reheating cycle to which the turbine blades were subjected. The cooling was done using liquid nitrogen in the chamber, and the cooling and heating cycle took 16 h. This hysteresis chart plots the ultrasonic time of flight of transmission through the thickness of the aluminum turbine blade. During the cooling stage bringing the temperature to approximately –86°C (–123°F) over 3.5 h in the ramp-down of the test process, one will note a spike on the hysteresis curve. This spike indicates that a stress component change has taken place in the turbine blade. The cooling process continues, reducing the temperature in the cycle from –86°C (–123°F) to –190°C (–310°F). During the reheating back to +23.3°C (+74°F) one can notice that the spike did not appear at –86°C (–123°F). This indicates that the stress component previously identified at this temperature was stabilized and the slope of the hysteresis curve was uninterrupted.

FIGURE 5.6
The EMAT setup and the tested turbine blades in the cryogenic chamber. (Courtesy of Victor Sloan.)

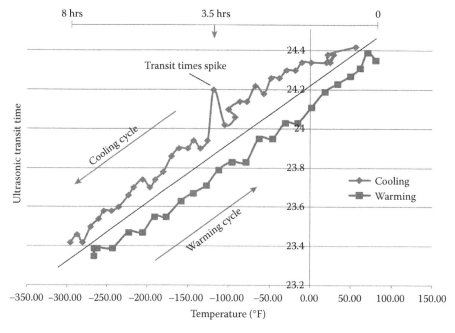

FIGURE 5.7
Hysteresis graph—EMAT ultrasonic transit time versus temperature in degrees F (NASA turbine blade).

Prior to the cryogenic tests, x-ray diffraction (Pulstec X360n) was performed to measure the change in the atomic lattice structure of the crystal of the aluminum coupons, which were etched and 180 grit sanded (Figure 5.8). The Sigma (x) MPa residual stress and Tau (xy) MPa shear stress were determined by making x-ray diffraction measurements at 0 and 90° to the Z axis. Following the cryogenic tests, these x-ray stress tests were repeated. The residual stress comparisons indicated that the atomic lattice structure had changed following the cryogenic exposures (Figure 5.9).

Debye ring (3D)—The circular patterns shown in Figure 5.9 are called Debye rings. They are produced by the diffraction of x-rays from the crystal lattice of the samples, as determined by Bragg's law. The angles of diffraction and the size of the resulting Debye rings are characteristic of the metal being tested.

FIGURE 5.8
The Pulstec X360n x-ray diffraction test of a NASA turbine blade along the 0 degree position. The operator shown in the photo is the co-author, Victor Sloan. (Courtesy of NASA.)

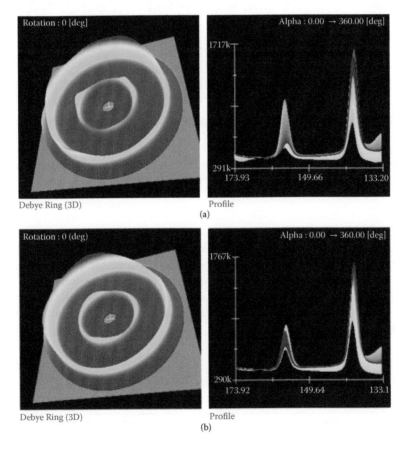

FIGURE 5.9
X-ray diffraction measurements showing the diffraction Debye ring/profile images of aluminum 2014. (a) Virgin 2014 aluminum coupon—0° (Sigma (x) –125 MPa; Tau (xy) 7 MPa; FWHM 1.78°). (b) Virgin 2014 aluminum coupon—90° (Sigma (x) –103 MPa; Tau (xy) 12 MPa; FWHM 1.83°).   (*Continued*)

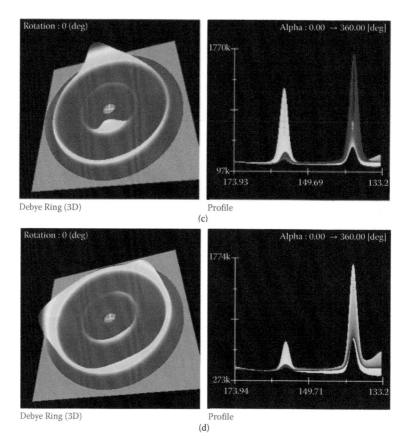

FIGURE 5.9 (*Continued*)
X-ray diffraction measurements showing the diffraction Debye ring/profile images of aluminum 2014. (c) 2014 aluminum coupon after cryogenic test—0° (Sigma (x) –64 MPa; Tau (xy) 14 MPa; FWHM 1.71°). (d) 2014 aluminum coupon after cryogenic test—90° (Sigma (x) –46 MPa; Tau (xy) 49 MPa; FWHM 1.70°).

Full width at half maximum of diffraction (FWHM)—Crystallite size is indicated by the broadening of a particular peak in a diffraction pattern associated with a particular planer reflection from within the crystal unit cell. The size is inversely related to the FWHM of an individual peak, where the narrower the peak, the larger the crystallite size.

Profile—The profile of a diffraction pattern is defined by the peak position, intensity, and Bragg's angle.

X-Ray Diffraction Stress Measurements at Cryogenic Temperatures

Virgin Material Coupon A	Coupon C after Cryogenic Exposure
0° reference x-ray sensor position	0° reference x-ray sensor position
Sigma (x) –125 MPa; Tau (xy) 7 MPa; FWHM 1.78	Sigma (x) –64 MPa; Tau (xy)14 MPa; FWHM 1.71
90° reference x-ray sensor position	90° reference x-ray sensor position
Sigma (x) –103 MPa; Tau (xy) 12 MPa; FWHM 1.83	Sigma (x) –46 MPa; Tau (xy) 49 MPa; FWHM 1.70

A room temperature (+23.3°C [+74°F]) birefringence stress test was also performed prior to the cryogenic exposure. EMAT tests using a dual polarized shear wave sensor attached to the turbine blades were conducted. The ultrasonic transit times through the aluminum 2014 coupons were measured to compare the fast and slow transit time axes of the turbine blades. The birefringence test results documented the change in stress conditions at various points when comparing before and after the cryogenic process (Figure 5.10).

In the tests that were conducted in the aluminum coupons, it was noted that the residual stress was relieved. This was confirmed by x-ray diffraction measurements taken before and after the cryogenic exposure as well as by comparison with through-thickness EMAT averaged birefringence transit time tests.

5.2.2 EMAT Test of North American Eagle F-104 Fighter Jet

The use of EMAT in ground vehicles at cryogenic temperatures was found to be also quite effective. A study similar to that reported in Section 5.2.1 was conducted on the North American Eagle F-104 fighter jet [www.landspeed.com] that is owned and operated by Ed Shadle and Keith Zanghi (Figure 5.11). This jet aircraft was used in an effort to break the world landspeed record of Mach 1.0. The jet's aluminum 7075-T7351 steering wheel (Figure 5.12) and aluminum 6061 braking systems (Figure 5.13) must maintain their material properties

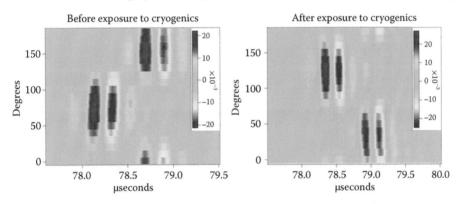

FIGURE 5.10
Time-of-flight measurements before and after exposure to cryogenic temperatures.

FIGURE 5.11
The North American Eagle F-104. (Courtesy of Keith Zanghi, North American Eagle, Inc.)

Brake Rotor Left

	Before	After	Difference
FlatnessBase	0.0000318	0.0000161	0.0000
Diameter_ID	3.7506789	3.7510161	−0.0003
Roundness_ID	0.0006173	0.0006042	0.0000
Diameter_ID1	4.998383	4.9988825	−0.0005
Roundness_ID1	0.0030368	0.0036521	−0.0006
Z Value_Point1	3.2492196	3.2494454	−0.0002
Z Value_Point4	3.2494936	3.249331	0.0002
Z Value_Point7	3.2491024	3.2490035	0.0001
Z Value_Point10	3.2489308	3.2492066	−0.0003
Z Value_Point2	3.5245799	3.5243502	0.0002
Z Value_Point5	3.5210045	3.5199586	0.0010
Z Value_Point8	3.5181677	3.5173298	0.0008
Z Value_Point11	3.5199165	3.5199725	−0.0001
Z Value_Point3	3.5214485	3.5208506	0.0006
Z Value_Point6	3.5136998	3.5116718	0.0020
Z Value_Point9	3.5086057	3.5069966	0.0016
Z Value_Point12	3.5101573	3.510207	0.0000
X Value_OD	0.00002644	0.0003587	−0.0001
Y Value_OD	0.0003004	0.0003755	−0.0001
Diameter_OD	23.9515193	23.9526332	−0.0011
Roundness_OD	0.0009237	0.0008118	0.0001
Flatness1	0.0151252	0.0155093	−0.0004

Brake Rotor Right

	Before	After	Difference
FlatnessBase	0.0000393	0.0000193	0.0000
Diameter_ID	3.7516514	3.751878	−0.0002
Roundness_ID	0.0007424	0.000832	−0.0001
Diameter_ID1	5.00026	5.0007771	−0.0005
Roundness_ID1	0.00014751	0.0013203	0.0002
Z Value_Point1	3.2490052	3.2492106	−0.0002
Z Value_Point4	3.2494122	3.2494081	0.0000
Z Value_Point7	3.2497353	3.2496072	0.0001
Z Value_Point10	3.2492326	3.2493356	−0.0001
Z Value_Point2	3.6021837	3.6017012	0.0005
Z Value_Point5	3.6341153	3.6329279	0.0012
Z Value_Point8	3.6465903	3.6455247	0.0011
Z Value_Point11	3.6155509	3.6146996	0.0009
Z Value_Point3	3.5621998	3.5611868	0.0010
Z Value_Point6	3.6227449	3.6206038	0.0021
Z Value_Point9	3.648326	3.6460155	0.0023
Z Value_Point12	3.5911781	3.5896399	0.0015
X Value_OD	−0.0018968	−0.0019216	0.0000
Y Value_OD	0.0007701	0.0008203	−0.0001
Diameter_OD	23.9427926	23.9440392	−0.0012
Roundness_OD	0.0015054	0.0014181	0.0001
Flatness1	0.0261686	0.0268404	−0.0007
Flatness2	0.0155771	0.0161554	−0.0006

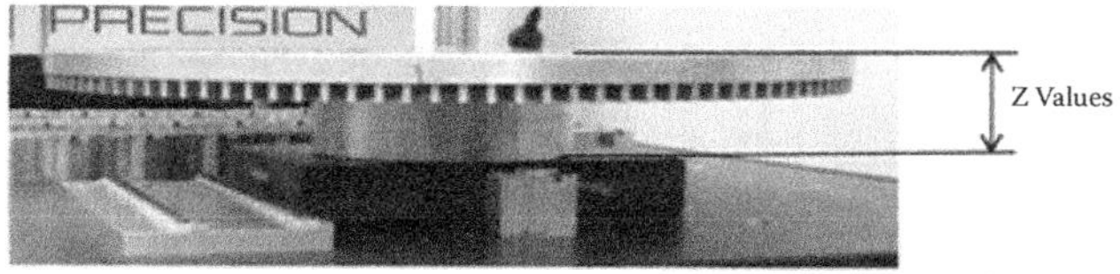

Aluminum 6061 brake rotor

Z Values

(a)

FIGURE 5.12
Using Zeiss CMM computer measurements of the shape for the brake rotor and the steering wheel before and after the cryogenic exposure: (a) the brake rotor. *(Continued)* (Courtesy of Victor Sloan.)

Eugen F 05.31.2011

24" Wheel	Before	After	Difference
Flatness1	0.000572	0.000584	0.0000
Diameter_ID	3.000668	3.000748	-0.0001
Roundness1	0.000251	0.000245	0.0000
X Value_Pin	-1.87668	-1.87674	0.0001
Diameter_Pin	0.250095	0.250103	0.0000
Roundness2	0.000268	0.000277	0.0000
X Value_OD	6.94E-05	8.64E-05	0.0000
Y Value_OD	9.04E-05	9.61E-05	0.0000
Diameter_OD	6.998888	6.999025	-0.0001
Z Flange_Point1	-1.01239	-1.01251	0.0001
Z Flange_Point2	-1.01267	-1.01276	0.0001
Z Flange_Point3	-1.01132	-1.01141	0.0001
Z Flange_Point4	-1.01095	-1.01103	0.0001
Z Value_Point5	-2.48978	-2.48992	0.0001
Z Value_Point6	-2.48927	-2.48936	0.0001
Z Value_Point7	-2.49329	-2.49341	0.0001
Z Value_Point8	-2.49373	-2.49383	0.0001
RawDataCurve	0.009661	0.008896	0.0008
FilterDataCurve	0.002527	0.002602	-0.0001
Diameter_Circle4	22.52557	22.5261	-0.0005
Roundness3	0.00075	0.000346	0.0004
X Value_Circle4	-0.00079	-0.00076	0.0000
Y Value_Circle4	-0.00063	-0.00061	0.0000

Zeiss CMM computer measuring machine
before and after cryogenic NDT shape data

Aluminum 7075 steering wheel

(b)

FIGURE 5.12 (*Continued*)
Using Zeiss CMM computer measurements of the shape for the brake rotor and the steering wheel before and after the cryogenic exposure: (b) the steering wheel. (Courtesy of Victor Sloan.)

FIGURE 5.13

The F-104 aluminum brake rotors in preparation for cryogenic exposure and EMAT testing. (Courtesy of Victor Sloan.)

and shape geometry while exposed to severe stresses that are imposed at a high speed. Any change in residual stress in these components during operation could dramatically impact the success of the attempt to operate and steer at high speeds, since the material properties and/or shape geometry may change. If the aluminum materials have a high level of residual stress, they may distort or even fail as a result of change in residual stress components as a function of the exposure to time, pressure, vibrations, and cycles of use.

Using cryogenic exposure and EMAT testing, high residual stress components were identified in the 7075-T7351 aluminum steering wheel (Figure 5.12). The cryogenic exposure process and the EMAT recording are listed in Table 5.1. Reviewing the ultrasonic transit time shifts has shown indications of a change in material stress component at various temperatures and time periods (Figure 5.14). After the residual stresses were cryogenically relieved, the wheel bearing bore dimensions significantly changed as the original machined tight fit of 0.038 mm (0.0015″)T changed to a loose fit of 0.0127 mm (0.0005″)L. This result means that the aluminum wheel would sustain distortion when exposed to the high stresses that are incurred on the wheel at Mach 1.0 speed. This is important information, since the residual stress component change would cause material distortion and failure of the bearing and wheel assembly. A new 7075-T7351 wheel was then made and

TABLE 5.1

EMAT Ultrasonic Scanning Recording Periods

Time (Hours)	Scan No.	Differential (Hours)	Time (Seconds)	Temperature
0	21	4	14,500	−0 to −150°F
4	22	4	29,000	−150 to −300°F
8	23	4	43,500	−300°F Soak
12	24	4	58,000	−300°F Soak
16	25	4	72,500	−300°F Soak
20	26	4	87,000	−300 to 150°F
24	27	4	101,500	−150 to 60°F

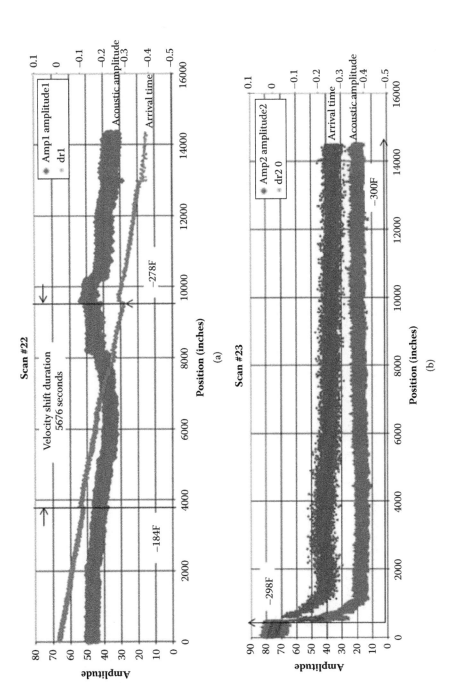

FIGURE 5.14
EMAT velocity and conductivity changes of aluminum 7075 wheel. (a) Test at –101.1°C (–150°F) ramped down to –184.4°C (–300°F). (b) Test at –184.4°C (–300°F) dwell temperature. *(Continued)*

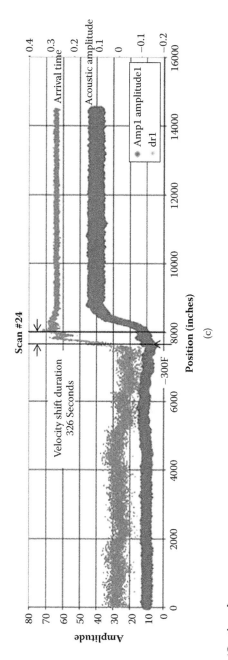

FIGURE 5.14 (*Continued*)
EMAT velocity and conductivity changes of aluminum 7075 wheel. (c) Test at –184.4°C (–300°F) soaking.

cryogenically tested prior to the machining of the final wheel bearing insert installation and provided for a stress-free wheel assembly.

The dimensions of the F-104's steering wheel and brake systems were taken before and after the cryogenic tests using a Zeiss Contura computer measuring machine (Figure 5.12). Deviations from the original measurements (Figure 5.15) were documented: a change in shape geometry occurred in the materials following the cryogenic exposure indicating a change in the stress components (Figure 5.16).

As a summary of the test, it has been shown that 7075-T7351 aluminum is a good choice of alloy material to sustain the intended operation of the F-104 rocket car at a high speed. Traditional rubberized wheels could not be used, as they would disintegrate at Mach 1.0 speed. Due to the original heat treating and machining processes of the wheel assembly,

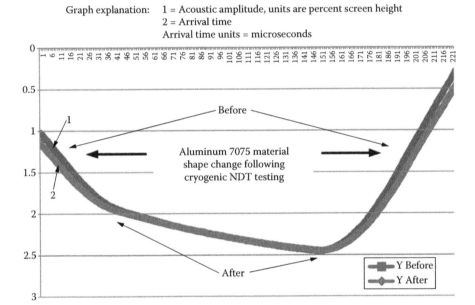

FIGURE 5.15
The cross-section shape change after cryogenic exposure of the 61 cm (24 inch) North American Eagle 7075-T7351 aluminum wheel.

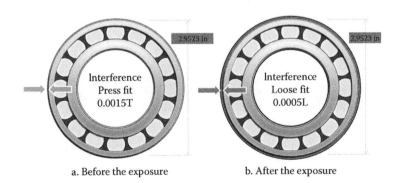

a. Before the exposure b. After the exposure

FIGURE 5.16
Interference fit between the wheel bearing and the wheel assembly changed after the cryogenic exposure.

residual stresses were developed within the wheel structure and they were nondestructively identified by the EMAT tests during the cryogenic exposure. If the original wheel assembly was exposed to the intended Mach 1.0 speed, it is quite likely that the bearing and steering wheel would have failed due to the residual stresses.

5.2.3 Signal-to-Noise Reduction of Low Temperature Eddy-Current NDE

Eddy current is one of the NDE methods that can be operated at cryogenic temperatures. The method uses electromagnetic induction where a probe that contains a wire coil is used to test conductive materials for near-surface flaws and material properties. An alternating current flows through the coil and generates an oscillating magnetic field. If the probe and its magnetic field are brought close to a conductive material such as a metal test piece, circular flow of electrons (shaped as swirling water in a stream) is induced in the metal and the flow is known as eddy current. In response, the flowing eddy current through the metal generates its own magnetic field, which interacts with the coil and its field through mutual inductance. Changes in metal thickness or defects such as near-surface cracking interrupt or alter the amplitude and pattern of the eddy current and the resulting magnetic field. This in turn affects the movement of electrons in the coil by varying its electrical impedance. The eddy-current instrument determines and plots the changes in the impedance amplitude and phase angle, which are then used by a trained operator to identify changes in the test piece. Figure 5.17 shows the illustration of the interaction of the coils induced and the resulting magnetic fields when it is brought close to the surface of a conductive material.

5.2.3.1 The Material Parameters That Affect Eddy-Current Measurements

Changes in the tested material properties (i.e., conductivity or permeability) affect the eddy-current flow in the metal specimen, resulting in a change in the secondary magnetic field that is created by the current flow. The variation in the secondary field strength causes a shift in flux distribution and strength of the primary magnetic field around the coil. This, in turn, changes the coil's electrical parameters, which is a cascade event.

One of the historical arguments against using eddy-current testing is that it is too sensitive. The signal that is measured during inspection might be created by a combination of the conductivity and magnetic permeability changes.

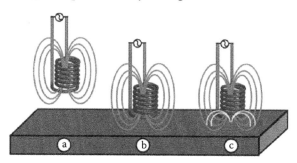

FIGURE 5.17
Illustration of the interaction of the coils induced and the resulting magnetic field when the coil is brought close to the surface of a conductive material. (a) An alternating current flowing through the coil at a chosen frequency generates a magnetic field around the coil. (b) When the coil is placed close to an electrically conductive material, eddy current is induced in the material. (c) If a flaw in the conductive material disturbs the eddy current circulation, the magnetic coupling with the probe is changed and a defect signal can be read by measuring the coil impedance variation.

Conductivity is an electrical property which determines how well electrons move through a material. Generally metals are classified as conductors. Theoretically, eddy-current tests can be conducted on any piece of metal. Resistivity is another term that describes the electrical properties of a material, and it is the inverse of the conductivity.

Permeability is a magnetic property, and not all metals have significant levels of permeability. The permeability value of a metal suggests how it alters a magnetic field moving through it. Metals with high permeability values (carbon steels) restrict the inspectable area to the upper skin layer with a given probe. Fortunately, a large percentage of aircraft structures and surfaces are made from nonferromagnetic materials. Their relative permeability is assumed to be one. This means that as the primary magnetic field moves through these materials, their inherent permeability does not alter the magnetic field energy level.

5.2.3.2 Maximizing the Signal-to-Noise Ratio at Cryogenic Temperatures

In defect detection, it is essential to choose inspection parameters that maximize the signal-to-noise (S/N) ratio, that is, the response from the defect divided by the average response from the competing grain noise. On the other hand, for material characterization it is often useful to choose inspection parameters that maximize the backscattered grain noise.

The signals generated during an eddy-current testing (ECT) inspection allow for the localization and dimensioning of defects in the material being examined. A noise-free ECT signal from a defect generally has an "eight"-shaped Lissajous figure in the impedance plane having the key characteristics of the phase angle and amplitude. However, in practice, ECT signals are corrupted by several forms of noise, which can interfere with the analysis of the acquired signal, reducing the correct/misreading ratio and increasing the duration of the inspection.

Generally, the material noise can produce significant distortions in the Lissajous figure and may cause misreading of the phase angle and amplitude. In some cases, the distortions are so extensive that making an evaluation is not possible. To operate eddy-current tests in cryogenic temperatures at low noise levels, a novel method was conceived named Victor noise reduction (VNR). In this patent-registered technique [Sloan 2014a; Sloan 2014b], the material is cycled through cryogenic temperatures at specific temperature–time profiles to reduce/eliminate lattice defects. Then, a final heat-treatment process restores the material to its original hardness. This process reduces the material noise caused by nonuniformities in the magnetic properties of the material that is being inspected.

Using the VNR method, the reduction of the material noise enables one to detect small flaws. The effectiveness of this process is shown in the next section.

5.2.3.3 NDT of the Cryogenic Processing Effects on Structure-Critical Fasteners

Aircraft structures are commonly assembled using a large number of fasteners, bolts, and rivets. Most of these are not fracture-critical, that is, if they crack or shear, the structure of the aircraft and its performance are not in danger. However, there are a few structure-critical bolts and they require inspection prior to the assembly of the aircraft or during regular maintenance intervals. An example of a typical shoulder bolt is shown in Figure 5.18. The critical area is the "shoulder," that is, the segment between the head of the bolt and the threads. This area is often inspected for flaws using eddy-current testing. In a few instances, the steel alloys and heat treatment (e.g., 4XXX series steel, PH17-4 and PH18-8 hardened steel) of these bolts result in a very large eddy-current noise-background, which impedes the signal interpretation and flaw-discrimination.

FIGURE 5.18
Typical structure-critical shoulder bolts.

For the initial feasibility evaluation, a semi-automated system, ETC-2000 (Figure 5.19a), was used to scan the shoulder of a variety of bolts using a custom-made probe (Figure 5.19b). The probe is an anisotropic drive-receive type that is operated at 3 MHz. Later, a programmable table-top system was developed to scan the bolts' shoulder (Figure 5.19c).

Figure 5.20 shows the eddy-current responses from two example heat-treated bolts (labeled A and B) prior to cryogenic processing. The bolts were scanned at 30 rpm

(b)

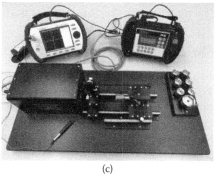

(a)   (c)

FIGURE 5.19
The eddy-current inspection system. (a) The ETC-2000 setup, (b) the specialized probe, and (c) the bolt inspection system.

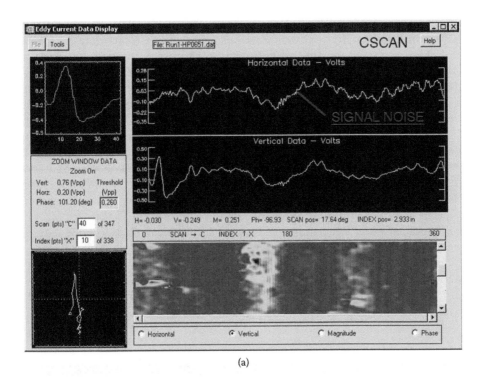

(a)

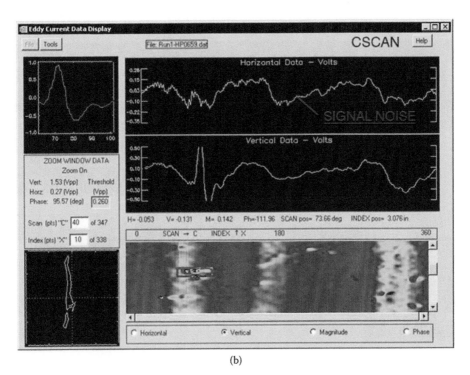

(b)

FIGURE 5.20
Eddy-current signal responses from two bolts prior to cryogenic treatment (a) Bolt A prior to VNR, (b) Bolt B prior to VNR.

with an index rate of 0.005. The bottom-right section of the data image plots the vertical signal amplitude of the eddy-current response in a C-scan display format. The horizontal line graphs above the colored section represent the voltage levels along a horizontal axis of the C-scan. The smaller square windows on the left side display signal details of the response contained in a small red box drawn on the C-scan image. The C-scans of both bolts display a large mix of colors, which are indicative of the eddy-current noise recorded during the inspection and also shown in the line graphs. It is quite difficult to extract signals from flaws or to evaluate the responses with respect to flaw severity.

Both bolts subsequently underwent the VNR cryogenic process. Figure 5.21 shows the signal responses from the bolts after the VNR treatment. The cryogenic process significantly reduced the material noise responses such that the remaining indications could be evaluated. The indication highlighted by the small red box in the C-scan of Bolt A exhibits the phase response (lower left window of Figure 5.21a) typical of a rejectable flaw, possibly an inclusion. The phase response of the flaw marked in the C-scan of Bolt B is indicative of a small nonrejectable surface pit.

The VNR process successfully reduced the eddy-current signal response noise generated by minute discontinuities in the material of the bolts, resulting in an improved signal-to-noise ratio and hence in an improved response evaluation.

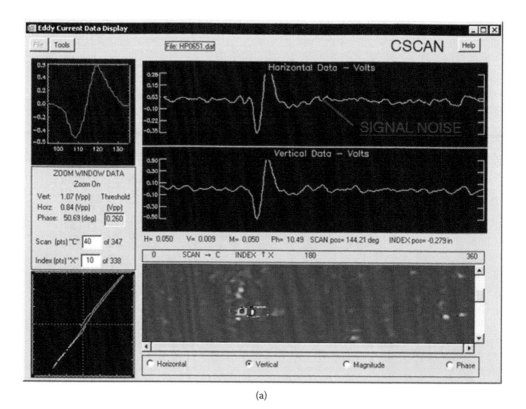

(a)

FIGURE 5.21

Signal responses from both bolts after the VNR process: (a) Bolt A after VNR.   *(Continued)*

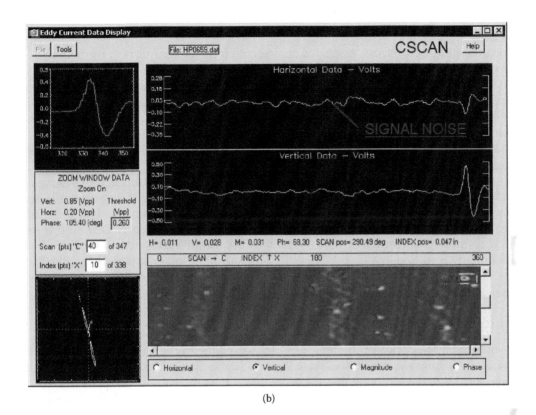

(b)

FIGURE 5.21 (Continued)
Signal responses from both bolts after the VNR process: (b) Bolt B after VNR.

5.3 Low Temperature NDE Case Study

Backscatter radiography and terahertz scanning are additional NDE methods that can be used to examine materials at low temperatures, and two examples of testing cryogenic tanks are described in this section [Ussery, 2008]. These methods provide invaluable data concerning tank insulation voiding and fracture behavior before, during, and after cryogenic loading. Given the recent progress and improvements in imaging technology, investigating the *in situ* material behavior is being considered. This direction is driven in part by the costs required to purchase novel technologies, which are expensive and must include the necessary training to become competent with data interpretation. These costs are particularly important to the aerospace industry, which is driven to do more with less. For example, aerospace avionics assemblies that are flight qualified involve many hours of costly testing including vibration, shock, and thermal cycling. These costs are not easily recovered, particularly when destructive testing or analysis reveals that a particular flight article is not the root cause of a failure. This section describes a case study of analyzing material behavior during cryogenic liquid exposure. The equipment that was used is a real-time x-ray radiography system that is standard in many NDE laboratories.

5.3.1 Problem Description

During the shuttle era, NASA employed engine cut-off (ECO) sensors near the bottom of the external liquid oxygen and liquid hydrogen tanks (Figure 5.22). The purpose of these sensors is to detect when the tanks are nearly depleted of oxidizer or fuel. The ECO sensors are monitored both prior to launch and while the vehicle is ascending prior to tank separation. If a "dry condition" is detected, the information is relayed back to a point sensor box, which automatically commands an engine shutdown or no start in prelaunch scenarios.

To avoid engine failure from mismatched fuel/oxidizer ratios, the tanks were never allowed to be completely "dry." In 2006, during an acceptance test procedure (ATP), an ECO sensor was monitored for thermal response during liquid nitrogen exposure and, during the last thermal response test, the sensor began showing intermittent electrical behavior. This behavior was manifested as a resistance change that was found to mimic a dry condition, despite the sensor being visibly submersed in cryogen. During this same period, a shuttle ECO sensor exhibited an intermittent condition while on the launch pad, subsequently resulting in a launch scrub. This occurred despite the tank being filled with liquid cryogen. The launch scrub coupled with the acceptance test failure kicked off a large-scale investigation into the cause of the erratic sensor behavior.

5.3.1.1 NDE Limitations

The intermittent behavior of the tank ECO sensor while on the launch pad was clearly a serious issue. However, the sensors were internal to the tank and could only be monitored via electrical signals. Although other sensors inside the tank showed the presence of cryogen, launch commit criteria prevented the launch. There was no NDE technique available to examine the erratic sensor due to its location deep within the tank. However, the similarity in erratic performance of the ATP failure to the tank failure suggested that the behavior could be repeated in a lab environment. With this knowledge, the suspect flight-qualified sensor was subsequently removed from the tank and sent for inspection. In many cases, location is the limiting factor of an NDE technique. The investigator needs to look at subassemblies buried within top-level assemblies. In these cases, breaking configuration of the top-level

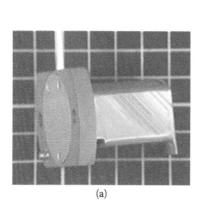 

(a)   (b)

FIGURE 5.22
(a) The ECO sensor that failed the acceptance test by exhibiting erratic resistance upon exposure to liquid nitrogen. (b) The internal construction of an ECO sensor including the tiny wire that changes resistance when exposed to cryogens.

assembly requires additional acceptance testing, which incurs significant costs. This clearly drives the need for even greater technological advances in NDE equipment and methods.

5.3.1.2 The Developed Test Method

Because this particular ECO was flight qualified, no one wanted the sample destroyed unless necessary. A destructive study of a nonflight-qualified ECO was conducted and this investigation helped point investigators toward weak points in the ECO design. One weak point found was the construction of a loaded terminal post. In order to complete the ECO electrical circuit, two terminal posts are depressed against a conducting surface through use of a loaded Belleville spring washer. The washer is loaded by swaging the terminal post against the washer. Nondestructive examination of several nonflight washers showed Belleville washer depression down to 0.46–0.48 mm (18–19 mils) from a nominal nonloaded height of 0.51–0.58 mm (20–23 mils). This amount of compression resulted in hundreds of pounds of contact load, which ensured low contact resistance [Kichak, 2005]. Washer heights were measured using the computer numerically controlled table of an x-ray radiography instrument down to a precision of ±0.127 mm (±0.5 mils). These data were then compared to the same data acquired from both the ATP failure and the suspect tank sensor. Measurement of the right washer on the ATP failure showed the right washer height to be 0.5 mm (19.7 mils) from a baseline washer height of 0.51 mm (20.0 mils) [Rolin, 2006]. The baseline washer height was obtained by destructively removing the washer. The data revealed that there was very little compression on the right side. This translated to minimal contact and, therefore, to intermittent electrical contact. The same measurements were performed on the flight sensor. The left washer shown in Figure 5.23 exhibited a washer height of 0.58 mm (23.0 mils) from a baseline washer height of 0.58 mm (23.0 mils), that is, essentially no compression.

These findings clearly indicated that loss of Belleville washer deflection was at the heart of the issue. What was not known was how this compression was affected by exposure to cryogen. In other words, was the washer improperly compressed during manufacture or did the washer lose compression upon repeated exposure to cryogen? Modeling analysis showed that there was little chance for loss of washer compression and it should not be expected. However, due to the criticality of these sensors, actual testing was requested. To test behavior of the washer deflection, various ECO sensors were chosen, radiographed, and then measured using the computer numerically controlled (CNC) table of the test system.

These samples were then exposed to a variety of cryogens, radiographed, and measured again. This process was repeated several times. The results from this test are shown in Table 5.2;

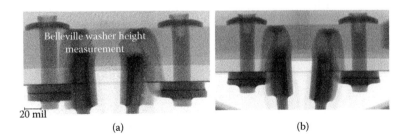

(a)　　　　　　　　　　　　　(b)

FIGURE 5.23
(a) This radiograph image shows the ECO sensor that failed the acceptance test by exhibiting erratic resistance upon exposure to liquid nitrogen. An NDE washer height measurement compared to the baseline revealed very little compression. The scale bar in the lower left corner is not calibrated. (b) This radiograph shows the ECO sensor removed from the external tank associated with the launch scrub. The left washer demonstrated no compression. The right one showed about one mil of compression beyond baseline.

TABLE 5.2

Belleville Washer Deflection with Various Cryogen Exposures Measured via NDE

Test article	Thermal cycling post ATP	LSWH before cycle, inches (+/- 0.0005)	LSWH after cycle, inches (+/- 0.0005)	Delta, inches (+/ 0.0007)	Additional thermal cycles	LSWH before cycle, inches (+/- 0.0005)	LSWH after cycle, inches (+/- 0.0005)	Delta, inches (+/- 0.0007)	Total LSWH delta, inches (+/- 0.0007)	RSWH before cycle, inches (+/- 0.0005)	RSWH after cycle, inches (+/- 0.0005)	Delta, inches (+/- 0.0007)	Additional thermal cycles	RSWH before cycle, inches (+/- 0.0005)	RSWH after cycle, inches (+/- 0.0005)	Delta, inches (+/- 0.0007)	Total RSWH Delta, inches (+/- 0.0007)
A	1 LH2	0.0212	0.0214	0.0002	–	–	–	–	–	0.0205	0.0216	0.0011	–	–	–	–	–
B	1 LH2	0.0205	0.0206	0.0001	–	–	–	–	–	0.0202	0.0201	-0.0001	–	–	–	–	–
C	1 LH2	0.0164	0.0164	0.0000	–	–	–	–	–	0.0191	0.0188	-0.0003	–	–	–	–	–
D	1 LH2	0.0222	0.0220	-0.0002	–	–	–	–	–	0.0217	0.0225	0.0008	–	–	–	–	–
E	1 LH2	0.0223	0.0220	-0.0003	–	–	–	–	–	0.0227	0.0224	-0.0003	–	–	–	–	–
F	3 LH2	0.0182	0.0187	0.0005	–	–	–	–	–	0.0186	0.0187	0.0001	–	–	–	–	–
G	3 LH2	0.0187	0.0184	-0.0003	–	–	–	–	–	0.0187	0.0186	-0.0001	–	–	–	–	–
H	3 LH2	0.0187	0.0188	0.0001	–	–	–	–	–	0.0187	0.0187	0.0000	–	–	–	–	–
I	3 LH2	0.0184	0.0183	-0.0001	–	–	–	–	–	0.0185	0.0185	0.0000	–	–	–	–	–
J	3 LH2	0.0183	0.0191	0.0008	–	–	–	–	–	0.0188	0.0191	0.0003	–	–	–	–	–
K	1 LN2	0.0214	0.0228	0.0014	1 LH2	0.0228	0.0226	-0.0002	0.0012	0.0228	0.0224	-0.0004	1 LH2	0.0224	0.0221	-0.0003	-0.0007
L	1 LN2	0.0218	0.0225	0.0007	1 LH2	0.0225	0.0224	-0.0001	0.0006	0.0213	0.0219	0.0006	1 LH2	0.0219	0.0220	0.0001	0.0007
M	1 LN2	0.0223	0.0222	-0.0001	–	–	–	–	–	0.0233	0.0231	-0.0002	–	–	–	–	–
N	8 LN2	0.0193	0.0201	0.0008	–	–	–	–	–	0.0193	0.0200	0.0007	–	–	–	–	–
O	8 LN2	0.0193	0.0200	0.0007	–	–	–	–	–	0.0200	0.0201	0.0001	–	–	–	–	–
P	8 LN2	0.0194	0.0199	0.0005	–	–	–	–	–	0.0194	0.0199	0.0005	–	–	–	–	–
Q	8 LN2	0.0195	0.0198	0.0003	–	–	–	–	–	0.0199	0.0199	0.0000	–	–	–	–	–

Note: LSWH, Left swage washer height; RSWH, right swage washer height; E-LSWH, average of two measurements (0.0220 and 0.0226). F-LSWH, note on TPS was that one measurement revealed 0.0188; K-LSWH, two measurements averaged (0.0230 and 0.0226)/severely mishandled; K-RSWH, average of two measurements (0.0220 and 0.0222); L-LSWH, average of three measurements (0.0226, 0.0226, and 0.0222).

they indicate that the model was not correct because several samples did indicate the loss of deflection. Washer movement was verified using scanning electron microscopy (SEM), and photos are shown in Figure 5.24. Several of these sensors had their Belleville washer removed and the terminal post examined. During this examination, dimpling in the witness mark area was found. This dimpling was indicative of relaxation, movement up the post, tightening of the washer onto the post, relaxation, and more movement up the post upon repeated thermal cycles. The total width measured using the SEM data was well correlated with the final measurement of washer displacement using the radiography data. In addition, each dimple corresponded to the number of cryogenic cycles seen by the terminal post.

These measurements pushed investigators to monitor Belleville washers visually while in cryogenic conditions. To accomplish this task, an open-top Styrofoam container was placed into the real-time x-ray machine. A sensor was positioned inside this container and liquid nitrogen was added. The x-ray machine was turned on, and the sensors were monitored (Figure 5.25). Measurements of the Belleville washer height were made and tracked while the sensors were immersed and during warm-up. Although movement of Belleville washer height was detectable, measurements were below the measurement uncertainty. Despite measurement uncertainty, the experiment demonstrated that NDE equipment could be used to monitor in real time a test article in a cryogenic environment. Following this study, operation of relays, switches, and cable mating interfaces have been

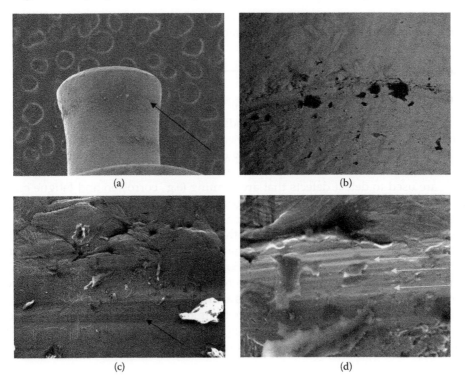

(a)   (b)

(c)   (d)

FIGURE 5.24
(a) This SEM image shows a terminal post at magnification of 55× after the Belleville washer was removed. Note the witness marks indicative of post to washer contact (black arrow). (b) This image shows the same post at 200×. (c) This image shows the witness mark at 1100× with vertical abrasion marks (black arrow) indicating possible washer fit at first swaging of post or collateral damage from removing the washer. (d) This image shows the witness mark at 5000×. There is dimpling in the witness mark area (arrows) indicative of thermal cycling.

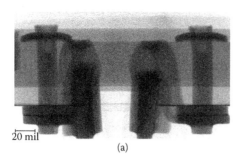
(a)

(b)

FIGURE 5.25

These radiographs show terminal posts both during and after submersion in liquid nitrogen. The internals were monitored in real time during immersion and subsequent boil-off to look for thermal expansion and contraction anomalies. (a) This radiograph shows the ATP failed sensor while in liquid nitrogen. (b) This radiograph shows the ATP failed sensor after reaching ambient temperature. Although there was no detectable change, the experiment demonstrated that liquid nitrogen submersed samples could be safely radiographed with good resolution. (From Rolin, T.D., and R. Abrams, MSFC LH2 ECO Sensor DPA Report, PowerPoint presentation, EI42 EEE Parts Engineering and Analysis Team, Marshall Space Flight Center Internal Report, Huntsville, AL (2006), pp. 1–23)

Note: The scale bar in the lower left corner of each image is not calibrated.

investigated at cryogenic temperatures and the results were critical to understanding the components' behavior in cryogenic environments.

5.4 Conclusions

Nondestructive testing and evaluation methods are used in every step of the life of materials and structures, from fabrication through the entire operating life. These methods are used to test the integrity of materials, parts, safety-critical components, structures, and systems as well as to determine the material properties. To perform real-time tests in service, SHM methods are used to detect defects that are forming (e.g., corrosion and fatigue cracks) as well as to monitor material properties changes. The NDE methods that are generally used include ultrasonics, visual, infrared, thermal, electrical, electromagnetic, radiographic, microwave, and spectroscopy. The SHM methods that are used rely on the availability of sensor technologies that are applicable to the operating conditions, the specific materials and structures, as well as the required detection information. While conventional NDE and SHM methods that are employed at close to room temperatures can be used at cryogenic temperatures, the methods that involve fluids are not applicable. At cryogenic temperatures, ultrasonics using EMAT is one of the effective techniques, and related studies were covered in this chapter. Also covered in this chapter is the use of eddy current in a noise reduction method as well as a case study of the use of x-ray radiography.

Acknowledgments

Some of the research reported in this chapter was conducted at the Jet Propulsion Laboratory (JPL), California Institute of Technology, under a contract with the National Aeronautics and Space Administration (NASA). The part that covers the case study reports research conducted at the George C. Marshall Space Flight Center (NASA/MSFC). Some of this work was done in collaboration with the NASA Engineering and Safety Center (NESC). The author would like to thank Dave Arms, Level III NDT Inc., Santa Clara, CA; Tod Sloan, University of Colorado School of Medicine, and University of Texas at Dallas, School of Behavior and Brain Sciences; and Russell (Buzz) Wincheski, NASA Langley Research Center, Hampton, VA, for reviewing this chapter and for providing valuable technical comments and suggestions.

References

Bar-Cohen, Y. (Ed.), *Automation, Miniature Robotics and Sensors for Nondestructive Evaluation and Testing,* Volume 4 of the *Topics on NDE (TONE) Series,* American Society for Nondestructive Testing, Columbus, OH, ISBN 1-57117-043 (2000), p. 481.

Deng, F., Acoustic emission applications to composite sandwich structures at room and low temperatures, Wayne State University PhD dissertation, Publication Number 3292471 (2007), p. 102.

Hirao, M., and H. Ogi, *EMATS for Science and Industry,* ISBN 1-4020-7494-8, Kluwer Academic Publishers, Boston, MA (2003), p. 372.

Jordine, A., Increased life of carburized race car gears by cryogenic treatment, *International Journal of Fatigue,* Vol. 18, No. 6, (1996), p. 418.

Kichak B., *Sensor Reliability Testing Status Report,* NASA Engineering and Safety Center (NESC) Technical Consultation Report RP-05-127 (2005).

Ray, B. C., S. T. Hasan, and D. W. Clegg, Effects of thermal shock on modulus of thermally and cryogenically conditioned kevlar/polyester composites, *Journal of Materials Science Letters,*Vol. 22 (2003), pp. 203–204.

Rolin, T. D., *Failure Analysis Memo EI42 (FA2006-016)—Nondestructive and Destructive Failure Analysis of External Tank (ET) Liquid Level Engine Cut-off (ECO) Sensor 3478 (JN06-016), EI42 EEE Parts Engineering and Analysis Team,* Marshall Space Flight Center, Huntsville, AL, Failure Analysis memo number FA2006-16 (2006), pp. 1–17.

Rolin, T. D., and R. Abrams, MSFC LH2 ECO Sensor DPA Report, PowerPoint presentation, EI42 EEE Parts Engineering and Analysis Team, Marshall Space Flight Center Internal Report, Huntsville, AL (2006), pp. 1–23.

Sahoo, B.N., *Effect of Cryogenic Treatment of Cemented Carbide Inserts on Properties & Performance Evaluation in Machining of Stainless Steel,* Masters Thesis, Department of Mechanical Engineering, National Institute of Technology, Rourkela, India (2011), p. 73.

Sloan, V., Cryogenic transition detection, U.S. Patent number 8,894,279 (November 25, 2014a).

Sloan, V., Cryogenic nondestructive testing (NDT) and material treatment, U.S. Patent number 8,920,023 (December 31, 2014b).

Ussery, W., *Application of Terahertz Imaging and Backscatter Radiography to Space Shuttle Foam Inspection,* Lockheed Martin Space Systems Company, ASNT Digital Imaging XI symposium (2008).

Internet Reference

EMAT for operation at cryogenic temperatures, http://cryogenicndt.com/EMAT_Velocity_Shift_Testing.php

6

Refrigeration Systems for Achieving Cryogenic Temperatures

Ronald G. Ross, Jr.

CONTENTS

6.1 Introduction: Achieving Cryogenic Temperatures

A sort of workingman's definition of cryogenic temperatures is temperatures below around 123 K, which equals –150°C or –238°F. In this temperature range and below, a number of physical phenomena begin to change rapidly from room-temperature behaviors, and new phenomena achieve greatly increased importance. Thus, study at cryogenic temperatures typically involves a whole set of new temperature-specific discipline skills, operational constraints, and testing methodologies. One of these special attributes of cryogenics is the science and engineering of achieving cryogenic temperatures, both in the laboratory as well as in a sustained "production" environment. The latter can extend from a hospital magnetic resonance imaging (MRI) machine, to a long-wave instrument on a space telescope, to a night-vision scope on a military battlefield. A number of technologies can provide the cooling required for these and other applications; the choice generally depends on the desired temperature level, the amount of heat to be removed, the required operating life, and a number of operational interface issues such as ease of resupply, sensitivity to noise and vibration, available power, and so on.

This chapter provides an overview of the common means of achieving cryogenic temperatures for useful exploitation, including both passive systems involving the use of liquid and frozen cryogens, as well as active cryorefrigeration systems—commonly referred to as cryocoolers. Separate subsections articulate the basic operating principles and engineering aspects of the leading cryocooler types: Stirling, pulse-tube, Gifford-McMahon (GM), Joule–Thomson (JT), and Brayton. Because this field is very extensive, the goal of the chapter is to provide an introductory description of the available technologies and summarize the key decision factors and engineering considerations in the acquisition and use of cryogenic cooling systems.

After summarizing the details of the refrigeration systems themselves, the remaining 40% of the chapter is devoted to reviewing the critical aspects of cryogenic cooling system design and sizing—including load estimation and margin management—and cryocooler application and integration considerations. Key integration topics include thermal interfaces and heat sinking, structural support and mounting, vibration and electromagnetic interference (EMI) suppression, and interface issues with electrical power supplies. The final subsection touches on techniques for measuring the performance of cryogenic refrigeration systems.

6.2 Passive Cooling Systems: Liquid and Frozen Cryogens, and Radiators in Space

For many years, the use of stored cryogen systems has provided a reliable and relatively simple method of cooling over a wide range of temperatures—from below 4 K for liquid helium, to 77 K for liquid nitrogen, up to 150 K for solid ammonia. These systems rely on the boiling or sublimation of the low temperature fluid or solid cryogens to provide cooling of the desired load. For solid cryogens, the temperature achieved may be modulated to a modest extent by varying the backpressure on the vented gas from atmospheric pressure down to a hard vacuum.

In most cases, stored-cryogen cooling technology is fairly well developed with proven design principles and many years of experience in the trade. The advantages of these systems are temperature stability, freedom from vibration and electromagnet interference, and negligible power requirements. The disadvantages are the systems' limited life or requirement for constant replenishment, the inability to smoothly control the cryogenic load over a broad range of temperatures, and the high weight and volume penalty normally associated with long-life, stand-alone systems.

In systems where the temperature stability and heat transfer associated with cooling with a liquid cryogen are advantageous, one can often extend the useful life of the cryogen or greatly minimize the need for replenishment by adding in a mechanical refrigerator with the cryogenic dewar to either recondense the boiled-off vapor and return it to the dewar, or to simply intercept a significant fraction of the parasitic thermal load entering the dewar.

The use of stored cryogens such as liquid nitrogen or liquid helium has often been the preferred method for cryogenic cooling of a wide variety of devices—from a laboratory apparatus to an MRI machine in a hospital setting. Cryogenic liquids can be used for cooling in a number of different states, including normal two-phase liquid-vapor (subcritical), low-pressure liquid-vapor (densified), and high-pressure, low temperature single-phase (supercritical) states. Subcritical fluids such as low-pressure helium have long been the cooling means of choice for very low temperature (1.8 K) sensors for space astronomy missions.

Solid cryogens are mostly used below their triple point where sublimation occurs directly to the vapor state. They provide several advantages over liquid cryogens including elimination of phase-separation issues, providing higher density and heat capacity, and yielding more stable temperature control, which is desirable for many applications.

6.2.1 Available Temperatures from Various Cryogens

The detailed thermodynamic properties of common cryogens are available in the literature for those designing cryogenic systems. However, it is useful to provide a brief overview of the practical operating temperature ranges and properties of common cryogens, along with an introduction to the thermodynamic operating regimes for their solid, liquid, vapor, and gas states.

Figure 6.1 shows the operating temperatures attainable with 10 common cryogens that can be used to directly cool cryogenic loads or other components. Each cryogen is represented by a bar that extends from its minimum operating temperature as a solid, based on sublimation at a vapor pressure of 0.10 torr, to its maximum operating temperature—its critical point, which is the maximum temperature at which a cryogen can exist as a two-phase liquid vapor. Within each bar, the region of solid phase is denoted by the shaded area defined at its maximum temperature by the cryogen's triple point, which is the maximum temperature at which a cryogen can exist as a solid. Above this, the cryogen's boiling point at a pressure of one atmosphere is noted by the dashed line. The use of 0.10 torr to define the lowest achievable temperature is for convenience, as the temperature can be lowered if the ability to pull a stronger vacuum is available.

6.2.2 Thermodynamic Principles of Cryogen Coolers

A modest familiarity with thermodynamics fundamentals is useful for understanding the limitations and constraints of stored-cryogen system operating states. Figure 6.2 provides an expanded description of the key fluid parameters shown in Figure 6.1 via an idealized temperature–entropy (T-S) diagram for a pure cryogenic fluid. Since entropy is defined as

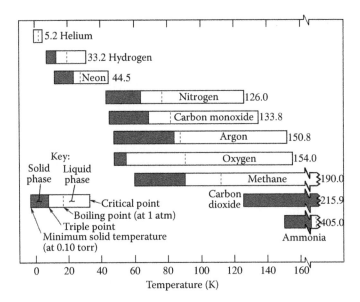

FIGURE 6.1
Operating temperature ranges for common expendable coolants.

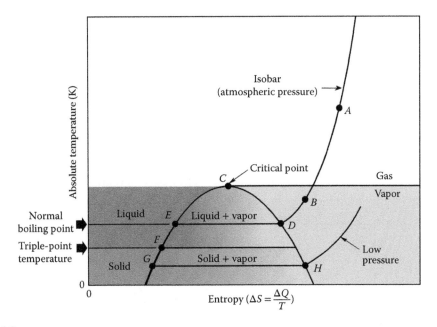

FIGURE 6.2
Idealized temperature–entropy diagram for a cryogenic fluid.

the heat transferred divided by the temperature at which the change occurs, the T-S chart is useful to not only visualize the boundaries between fluid states, but also quantify the amount of heat transferred when a fluid undergoes a change of state.

Starting with the point *C*, the apex of the dome is called the critical point, and the conditions at that point are called the critical pressure, critical temperature, and so on. When the fluid is at or above the critical temperature, it can never exist in the liquid state, but will remain as a single-phase, homogeneous gas. Fluids stored under these conditions are sometimes called cryo-gases.

The line described by curve *ABD* in Figure 6.2 shows the path of a gas being cooled at constant atmospheric pressure. The horizontal line drawn at *C* represents the dividing line between a vapor and a gas. While they are technically in the same state, the points along the line *DE* represent a liquid and vapor mixture at constant temperature and pressure—point *D* being 100% saturated vapor and point *E* being 100% saturated liquid. As an example, for water, this *DE* line would be at 100°C, the boiling point of water at a pressure of one atmosphere. The change in energy from point *E* to *D* is the heat of vaporization. When a liquid is heated along this line, point *E* is also called the bubble point because it is where the first vapor bubbles appear.

Further cooling of the liquid from *E* to *F* reduces the vapor pressure, and eventually the liquid freezes into a solid. Point *F* is defined as the triple point (or melting point), where the fluid exists as solid, vapor, and liquid. For water, this would be the temperature of an ice/water mixture, that is, 0°C.

For a cryogen that is below the triple point temperature, such as point *G*, any addition of heat will cause the solid to sublimate, as opposed to melting to a liquid. For the conditions of point *G*, the heat of sublimation is given by the change in energy from *G* to *H*.

Figure 6.3 expands on Figure 6.2 by describing the dependence of boiling-point or sublimation temperature on external pressure for common cryogens. Also noted is the triple point where the cryogen transitions to a solid. This plot also indicates the temperature and pressure where external contaminant gases, such as water vapor, will begin to condense on cryogenic surfaces such as low-emittance shields and MLI. Preventing such condensation is a critical issue for managing radiant parasitic loads on low-emittance shields and cryogenic surfaces. This topic of emittance degradation from contaminant films is covered later in this chapter in Section 6.4.3.5.

6.2.3 Cooling with Liquid Cryogens

Over the years, many liquid cryogenic systems have been developed, fabricated, and operated in both ground environments and in space. They cover a wide range of cryogen fluids

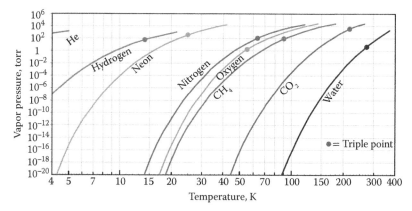

FIGURE 6.3
Boiling-point temperature of common gases as a function of external pressure.

and construction features in terms of stored volume, pressure and temperature limitations, and relative efficiency in terms of the parasitic heat leaks. Many of these systems utilize liquid helium for achieving temperatures between 1.4 and 4 K or liquid nitrogen for achieving temperatures around 77 K. To achieve temperatures below 4.2 K requires that liquid helium be stored under partial vacuum conditions. At pressures from 10 to 40 torr, temperatures 1.4–1.8 K are achievable with liquid helium.

6.2.3.1 Engineering Aspects of Liquid Cryogen Systems

6.2.3.1.1 Typical Dewar Construction Features

As shown in Figure 6.4, liquid cryogen systems typically involve a nested storage tank concept whereby the inner tank, which holds the liquid cryogen, is suspended inside an outer vacuum shell with low-conductivity structural supports. These structural supports are typically made of low-conductivity tubes, struts, or tension bands to achieve high structural efficiency and minimum conductivity between the two tanks. The gap between the two tanks is then evacuated and filled with multilayer insulation (MLI). In addition, a high-efficiency dewar may also contain one or more strategically placed vapor-cooled shields (VCS) that are cooled by the evaporating cryogen as it vents from the inner tank.

The goal of the gap construction is to prevent gaseous conduction and radiation between the outer and the inner tanks and to achieve maximum thermal benefit from the evaporating cryogen. Although the heat of vaporization of the cryogen is the primary cooling force in the system, there is also considerable benefit associated with extracting the available heat from the vapor as it rises up in temperature from the cryogen temperature to the external vent temperature. This is accomplished by piping the venting gas through the vapor cooled shields, which serve to intercept much of the radiant energy coming through the MLI layers from the outer tank. The VCS can also be attached to the support struts or plumbing to further reduce conductive heat leaks.

An extensive summary of the construction features for representative stored cryogen tanks fabricated for space applications—and in some cases airborne use—has been assembled by Donabedian and is available in the *Spacecraft Thermal Control Handbook, Vol II: Cryogenics* [Donabedian, 2003b] and in Chapter 15 of the earlier *Infrared Handbook* [Donabedian, 1993]. These reviews include tanks built for a variety of uses, fluids, and pressures, and contain designs for both subcritical and supercritical cryogen storage. As

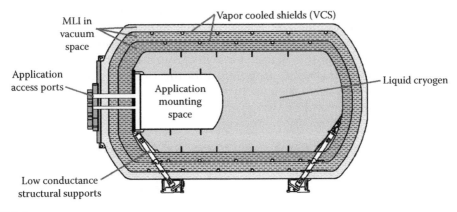

FIGURE 6.4
Example liquid cryogen dewar construction features.

part of that work, a general performance, figure-of-merit database was also developed to serve as a convenient means of comparing and evaluating the performance of the various stored cryogen systems. The figure of merit is defined as the net heat leakage of the fluid (Q) divided by the surface area of the pressure vessel (A) in units of W/m². This Q/A parameter varies from as high as 0.1 to as low as 0.003 W/m² depending on the temperature and properties of the fluid stored, the tank design, and the external temperature.

6.2.3.1.2 Multilayer Insulation

High-quality MLI has been found to play a particularly critical role in achieving high-efficiency stored cryogen systems, and as a result, has been a focus for much research over the years within the cryogenic community. Lockheed-Martin Palo Alto, in particular, spent years testing and optimizing MLI for space cryogen dewars and has assembled a substantial database on performance attributes, lessons learned, and preferred practices (see, e.g., Johnson, 1974; Nast, 1993).

Because MLI is also a key element of solid cryogen systems and cryocooler-cooled systems, a separate subsection of this chapter (Section 6.4.3) has been devoted to a detailed summary of cryogenic MLI performance attributes and measured data. Its particular focus is on the use of MLI at cryogenic temperatures, which places greatly increased emphasis on the conduction properties of MLI.

6.2.3.1.3 Porous Plugs

For liquid helium systems, when the temperature is dropped below about 2 K, a point referred to as the lambda point, liquid helium undergoes a phase transition and becomes a "superfluid" with very special properties. These properties include infinite thermal conductivity, zero viscosity, and zero entropy. The substance, which looks like a normal liquid, will flow without friction along any surface and circulate over obstructions and through pores in containers that attempt to hold it. The porous plug was invented to contain superfluid helium in a cryogen dewar while allowing for evaporative cooling. Its micron-level pore sizes are required to separate the liquid and vapor phases to ensure that liquid does not escape before its heat of vaporization can be utilized. Selzer et al. provided a summary of the physics behind the porous plug's operation as part of their paper on its original development at Stanford University in 1970 [Selzer et al., 1971].

6.2.3.1.4 Zero Boil-Off Systems

In the last 10 years or so, the concept of zero boil-off (ZBO) systems utilizing mechanical refrigerators combined with high-efficiency cryogenic tanks has been pursued to provide long-term storage of cryogenic fluids while minimizing storage volume and refrigerator power. These ZBO systems are being used for many ground-based systems and examined to provide liquid oxygen, methane, or hydrogen for future space planetary-mission propulsion and life-support systems. Helium, hydrogen, and deuterium ZBO systems are also being considered for future space-based lasers.

6.2.3.2 Liquid Cryogen Cooler Development History and Availability

As noted earlier, liquid cryogenic systems have been developed, fabricated, and operated for many years, in both ground environments and space [Ross, 2007]. They cover a wide range of cryogen fluids and construction features in terms of temperatures, stored volume, and thermal isolation systems. Many present-day systems are used for cooling superconductor electronics and magnetics in applications such as MRI systems and for cooling spacecraft instruments utilizing liquid helium to achieve temperatures below 2 K. Although not really a "cooling" application, much of the same technology is also used in liquid cryogen "storage" systems for liquid helium, hydrogen, oxygen, and nitrogen.

A number of manufacturers provide generic liquid cryogen dewars for laboratory and commercial applications. In space, most dewars have been custom built for particular missions by contractors such as Lockheed Martin in Palo Alto, CA, and Ball Aerospace and Technology Corp. (BATC) in Boulder, CO.

The tanks developed for NASA's Gemini and Apollo programs for storage of supercritical oxygen and hydrogen for their 7- to 14-day missions were the earliest space-qualified tanks for operation in near-zero-gravity environments. More advanced and larger tanks were developed by the Beech Aircraft Corp. Boulder Division (now BATC) for longer life. They were designed for general-purpose storage of LO_2, LH_2, LN_2, and LHe for extended periods.

Nearly all actual liquid cryogen cooling applications in space have involved the use of superfluid helium to achieve temperatures below 2 K for cooling long-wavelength infrared (IR) sensors in space telescopes. The first of these was the IRAS dewar in 1983, followed by COBE in 1989, Spitzer in 2003, GPB in 2004, and XRS in 2005. For most other temperatures, space cryogen-cooled missions used solid cryogens, which are discussed next. A summary of the history of cryogen systems used in space is provided by Ross [2007].

6.2.4 Cooling with Solid Cryogens

A second efficient way to use stored cryogens is in the frozen state. As shown in Figure 6.5, the normal operating regime of a solid cryogen cooler is below its triple-point temperature. In this region, the addition of heat causes conversion of the solid directly into vapor through the process of sublimation, bypassing the liquid state. Operating below the triple point also eliminates the problems of fluid management and phase separation associated with fluid systems. From an efficiency point of view, working with the solid phase provides greater density (and thus lower storage volume) and higher heat content per unit mass of cryogen. Other advantages of a solid cooler are relative simplicity, absence of moving parts, absence of noise and vibration, excellent temperature stability, and no power requirements.

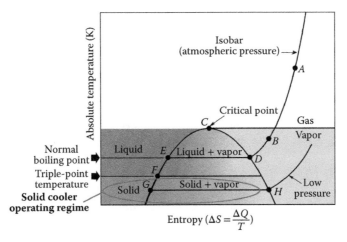

FIGURE 6.5
Solid cryogen operating regime; heat of sublimation = $T'(S_H - S_G)$.

The primary limitations or disadvantages include a limited number of suitable cryogens, very large mass and volume required for large heat loads or long design lives, the need for very significant ground servicing facilities and manpower support, and safety implications associated with venting toxic or flammable vapors or having a clogged vent—causing an explosion.

For space applications, the overboard vent of a solid cooler must also be designed to maintain the necessary pressure level on the cryogen and to not create unacceptable torques on the spacecraft. If the vent rate is significant, the gas may have to be vented along an axis passing directly through the center of gravity of the satellite and/or thrust compensation devices must be used to prevent attitude disturbances. Sustaining the desired operating temperature also requires that the back pressure on the sublimating cryogen be maintained at the fixed value associated with the design temperature.

This important interrelationship between the temperature and pressure of a frozen cryogen is presented in the vapor pressure–temperature plot shown in Figure 6.3. As noted, the sublimation temperature is strongly dependent on the vapor pressure maintained above the solid. Thus, given a fixed back pressure, a solid cryogen can be designed to maintain a very stable temperature.

6.2.4.1 Thermodynamics of Solid Cryogens

Table 6.1 tabulates the key thermophysical properties of a number of cryogens traditionally used in solid coolers, including their operating temperature range. The cryogens are listed in descending order of minimum practical operating temperature (the table's fourth column) based on a reasonably achievable back pressure of 0.1 torr.

As shown earlier in Figure 6.3, the minimum operating temperature depends on the minimum back pressure that can be sustained during cooler operation. If the heat load is small enough and the applied vacuum is high enough, then lower pressures, and thus lower temperatures, can be attained. The maximum temperature (the fifth column) is the highest temperature at which the solid phase can exist, which is its triple-point temperature. By proper design of the back pressure imposed by the vent gas system, any temperature between these two extremes can be achieved. The operating temperature range of a solid cooler is thus much wider than the range for a stored liquid cryogen cooler.

TABLE 6.1

Thermal Properties of Selected Solid Cryogens

Cryogen	Heat of Sublimation (J/g)	Density of Solid at Melting Point (kg/m³)	Temperature Range (K)	
			0.1 Torr	Triple Point
Ammonia (NH_3)	1719	822	150	195
Carbon dioxide (CO_2)	574	1562	125	216
Methane (CH_4)	569	498	60	90
Oxygen (O_2)	227	1302	48	55
Argon (A)	186	1714	48	84
Carbon monoxide (CO)	293	929	46	68
Nitrogen (N_2)	225	1022	43	63
Neon (Ne)	106	1439	14	25
Hydrogen (H_2)	508	80.4	8	14

6.2.4.2 Engineering Aspects of Solid Cryogen Coolers

6.2.4.2.1 Solid Cryogen Dewar Construction Features

As shown in Figure 6.6, solid cryogen dewar systems are fundamentally similar to liquid cryogen dewars in their structural support and thermal insulation systems. Both include a nested storage tank concept whereby the inner tank, which holds the solid cryogen, is suspended inside an outer vacuum shell with low-conductivity structural supports. As with liquid cryogen dewars, these structural supports are typically made using low-conductivity tubes, struts, or tension bands to achieve high structural efficiency and minimum conductivity between the two tanks. The gap between the two tanks is evacuated and filled with MLI, and for high-efficiency dewars, one or more VCS may be strategically placed inside the gap.

As with liquid cryogen dewars, the goal of the gap construction is to prevent gaseous conduction and radiation between the outer and inner tanks and to achieve maximum thermal benefit from the evaporating cryogen. Although the heat of sublimation of the solid cryogen is the primary cooling force in the system, there is also considerable benefit associated with extracting the available heat from the vapor as it rises up in temperature from the cryogen temperature to the external vent temperature. This is accomplished by piping the venting gas through the vapor cooled shields, which can serve to intercept much of the radiant and conducted heat coming through the MLI layers from the outer tank. The VCS can also be attached to the support struts or plumbing to further reduce conductive heat leaks.

Two key differences between liquid- and solid-cryogen dewars are the need for a thermally conductive matrix and freezing coils in a solid cryogen dewar. Because the solid cryogen evaporates selectively near the interface with the cryogenic load, it tends to withdraw away from this critical thermal interface as it is depleted and requires a means to keep it thermally connected to the load. This is accomplished by filling the cryogen tank with a high-conductivity open-cell metallic matrix or foam to provide a known thermal bridge to the load interface during all stages of evaporation of the solid cryogen.

The second addition to a solid cryogen dewar is a means to freeze the cryogen in place within the foam-filled dewar. This is accomplished by providing a cooling coil within the dewar that can be connected to an external coolant source sufficiently cold to freeze the cryogen. A second use of the cooling coil can be to keep the cryogen frozen during system integration and test periods.

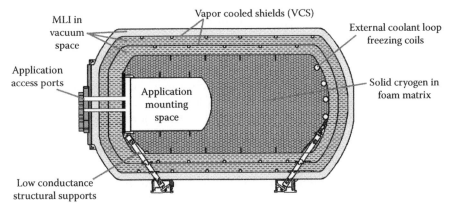

FIGURE 6.6
Example solid cryogen dewar construction features.

This process of freezing the cryogen, or precooling it to temperatures well below its freezing point, requires great care and experience to avoid damaging the dewar. Like water ice, the frozen state of the cryogen often has a different density, and thus occupies a different volume than the liquid state. In the solid state, it is also likely to have a different coefficient of thermal expansion (CTE) than the metallic dewar. If not fully accommodated in the dewar design and filling procedure, these expansions or contractions of the cryogen can lead to rupturing portions of the dewar.

A compounding issue with solid cryogens is their ability to relocate themselves within a dewar by evaporative transport from warmer regions to cooler regions; this involves the physics of cyropumping and can lead to a solid cryogen filling in the space critically needed for expansion during a subsequent warm-up operation. A very traumatic failure of the NICMOS solid nitrogen dewar occurred due to this cause during preparation for use on the Hubble Space Telescope [Miller, 1998a, 1998b].

6.2.4.3 *Solid Cryogen Cooler Development History and Availability*

The first operational long-life solid cooler used in space was a single-stage carbon dioxide system developed by Lockheed Martin and launched aboard an Air Force satellite (STP-72-1) on October 20, 1972 [Nast and Murray, 1976]. Since that time nearly a dozen cooler designs, both single- and two-stage, have been used to cool sensors to temperatures over a range of 10 K to 65 K, with operational lifetimes from 10 months to 2.5 years.

Two-stage designs have been used several times to optimize the overall cooler performance and minimize cooler mass by using a high-temperature cryogen such as carbon dioxide or ammonia (both of which have higher operating temperatures and high heat content) to provide a shield for cryogens that have lower operating temperatures, such as hydrogen, neon, and methane. In some cases, methane has been used as the shield cryogen for low-heat-content cryogens such as neon.

With the rapid development of long-life mechanical refrigerators and the relatively high cost of designing and servicing solid coolers, mechanical cryocoolers have increasingly become the cooler of choice for space missions that historically would have used solid cryogen coolers. An overview of the history of cryogenic coolers in space is provided by Ross [2007].

6.2.5 Radiation to Deep Space

For space-borne applications, cryogenic temperatures as low as 40–60 K can also be achieved using very carefully designed radiant cooler systems radiating into deep space. Although the effective radiation temperature in space is approximately 3 K, achieving these 40–60 K temperatures is generally limited to sophisticated cryoradiators on spacecraft well separated from the much warmer environment of Earth orbit. In Earth orbit, practical cryoradiator temperatures are closer to 80 K and above.

When striving for cryogenic temperatures above these levels, radiant cooling to deep space can provide an effective and cost-effective means of cooling, although even then, elaborate shields from the sun and Earth, and from the warm environment of the supporting spacecraft are required.

The advantage of cryoradiators is relatively stable long-term performance without the need for power, or concerns about mechanical wearout, electronics failures, or depletion of a stored cryogen supply. Countering this attractiveness is the relatively challenging design associated with achieving sufficiently low parasitic thermal loads and maintaining sufficient structural robustness to survive the launch loading environment. To achieve useful performance, constraints are also typically required on the spacecraft's geometric configuration and orbital attitudes.

As with nearly all cryogenic applications there is a strong competition between structural robustness and thermal isolation (minimum thermal conductivity). This invariably leads to highly optimized structural/thermal designs often involving mechanical mechanisms for latching or unlatching supplementary structural supports used only to survive launch. Added to this is the difficulty of isolating from direct solar and Earth reflected solar (albedo) radiation, which typically requires Earth and sun shades; these too can often end up with mechanical deployment mechanisms. Lastly, isolating the radiator and cold plumbing from the warm spacecraft requires careful application of low-emittance surfaces and cryo MLI. At cryogenic temperatures, such surfaces and MLI can perform much more poorly than they do in room-temperature applications, so these contribute to additional engineering challenges in the design process.

The bottom line is that a significant number of cryoradiators have been successfully used in space since the early 1970s [Nast and Murray, 1976], but a modest fraction of these have had significant schedule and cost growths associated with meeting the design challenges, and each design tends to be a new custom design for each new spacecraft and mission. For higher temperatures, like the 150–170 K temperatures needed for space optics, design criticality is much less severe, and cold radiators in this temperature range—just above cryogenic temperatures—have provided very effective long-term cooling of space instruments. See, for example, the 12-year space radiator performance on the Atmospheric Infrared Sounder (AIRS) instrument [Ross, Johnson, and Elliott, 2014].

Crawford [2003] and Donabedian [2003c] provide excellent reviews of the more detailed design principles developed for space cryoradiators over the past 40 years. Donabedian also presents comparative performance data for nearly two dozen flight designs. Their chapters in the *Spacecraft Thermal Control Handbook, Vol II: Cryogenics* are an excellent starting point for those wishing to more carefully examine the design options for space cryoradiators.

6.3 Active Refrigeration Systems: Stirling, Pulse Tube, GM, JT, and Brayton

For cryogenic applications where stored cryogens such as liquid nitrogen and liquid helium are not readily available or are inconvenient to use, mechanical refrigerators, or cryocoolers, are often the preferred design solution. The primary considerations that differentiate mechanical refrigerators from stored cryogen cooling systems are the issues of cryogen storage, resupply and safety for cryogen systems, and the requirement for electrical power and a means of heat rejection for cryocoolers. Because cryocoolers, or cryorefrigerators, are typically driven by electrical powered compressors, means must be available to provide both the electrical power and the means to reject the resulting heat dissipation. The power dissipation issue is particularly important because the resulting heat reject temperature strongly affects the thermodynamic efficiency of the cryocooler. A second aspect of the electrically driven compressor is the strong likelihood of measurable levels of equipment vibration, EMI, and audible noise that may interact negatively with the intended cryogenic application. Achieving low levels of vibration and noise has been an important focus in the cryocooler development cycle industry, and is an important distinguishing attribute of certain cryocooler types and constructions. Another key advantage of a cryocooler is the ability of a single unit to provide cooling over a broad range of temperatures, many with closed-loop temperature control.

Some of the most important applications for cryocoolers include achieving high vacuum levels with cryopumps in semiconductor processing facilities, cooling IR detectors and superconducting devices in a broad range of military, space, and laboratory instruments, and reliquefying cryogens to provide a ZBO recapture of the cryogen in systems using liquid helium or nitrogen. Key decision factors include the cooling system operational cost, complexity, and reliability/maintainability.

To meet these broad needs, a wide range of cryocoolers has been developed, and these coolers use a number of different thermodynamic cycles. In general, the size (cooling capacity) and available cooling temperature range of mechanical cryocoolers span many orders of magnitude—from room temperature down to 1 K and below, and from micro-watts to kilowatts of cooling power. The most common mechanical refrigeration cycles include Stirling, pulse tube (PT), GM, JT, and reverse-Brayton cycles. The attributes of each of these types of coolers are discussed in the subsections that follow, including a brief description of their thermodynamic cycle, operational features, representative perfor-mance, and general commercial availability.

In addition to these five cooler types, there are a number of lesser known cycles such as adiabatic demagnetization and the dilution cycle that are used primarily for achieving ultra-low temperatures below 1 K. The reader is referred to the literature on sub-Kelvin coolers for further information on these specialty cooler types.

To achieve the lowest temperatures, typically 30 K and below, cryocoolers generally employ two or more linked stages, where an upper stage (higher temperature) cooler is used to provide a low temperature heat rejection path for a lower-temperature stage. Although many multiple-stage coolers employ the same thermodynamic cycle for each of the linked stages, there is sometimes an advantage to linking different types of coolers using different thermodynamic cycles. These are generally referred to as hybrid coolers and include combinations such as a Stirling, PT, or GM upper stage with a JT bottom stage. Such a cooler can take advantage of the high efficiency of the Stirling cycle for higher-temperature precooling and also capture the remote-coldhead low-vibration attributes of a JT system for the final interface with the application load.

6.3.1 Cryocooler Cycle Types and Efficiency Measures

6.3.1.1 Cryocooler Classifications and Practical Systems

All mechanical refrigerators generate cooling by basically expanding a gas from a high pres-sure to a low pressure. The primary distinguishing feature between cycles is how the com-pression is accomplished, what pressure ratio is used, what method of expansion is used to achieve the cold temperature, how well and where heat is rejected, and how well thermody-namic efficiency is maintained using heat exchangers, regenerators, and recuperators.

Probably the most fundamental distinction between cryocooler types is the nature of the refrigerant flow within the cryocooler: either alternating flow (ac systems) or continuous flow (dc systems). This distinction is also denoted as *regenerative systems* versus *recuperative systems* based on the type of heat recovery heat exchanger that is applicable: regenerators for an alternating flow (ac system), or recuperators for a continuous flow (dc system).

In an ac-type cooler system, a regenerative heat exchanger stores and releases energy to the alternating refrigerant stream using a regenerator made of, for example, fine mesh screens or densely packed particles with good specific heat properties. In a dc-type system, a recuperative heat exchanger exchanges energy between two opposing streams of flowing gas or liquid using a counterflow heat exchanger referred to as a recuperator. Of the common cooler types, Stirling, PT, and GM use regenerative (ac flow) cycles, while JT and turbo-Brayton systems use recuperative (dc) flows. A key distinguishing feature of such systems is that the compressor generally must be quite close to the cold-end expander in a regenerative ac-flow cooler, and can be very remote (many meters away) for a recuperative dc-flow cooler. This has important implications on managing the compressor's heat dissipation and possible vibration and noise in close proximity to the cryogenic load. One exception is the GM cooler; it uses a regenerative refrigeration cycle, but uses a constant flow dc compressor that can be remotely located. To do this a GM cooler chops the dc flow into an ac flow within the coldhead itself, remote from the GM compressor.

6.3.1.2 The Carnot Cycle and Efficiency References for Cryocoolers

As background before delving into the details of the various cryo refrigerator types, it is useful to touch briefly on the standard measure of cryocooler efficiency: the percent of Carnot coefficient of performance. The coefficient of refrigeration performance (COP) for any refrigerator is defined as the ratio of the extracted heat to the applied work, that is

$$\text{COP}_{\text{Cooler}} = \frac{\text{cooling power}}{\text{input power}} \tag{6.1}$$

Next, let us examine the Carnot-cycle refrigerator, which has the highest efficiency of any refrigeration cycle, and thus serves as a reference for all other refrigeration cycles. As shown in Figure 6.7 (a temperature–entropy [T–S] diagram of the cycle), the cycle consists of gas compression on the right, constant-temperature heat rejection at the top, an expansion phase on the left, and constant-temperature heat absorption on the bottom. Heat absorbed

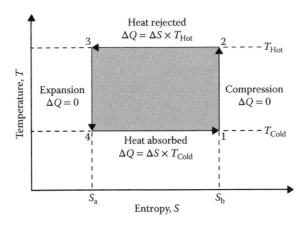

FIGURE 6.7
The Carnot refrigeration cycle.

during the process corresponds to the line with endpoints 1 and 4 (the refrigeration), while the heat rejected to the environment during the process (work done plus heat absorbed) is line 2 to line 3. Note that the compression and the expansion are both done isothermally (at constant temperature), while the expansion and the compression processes are done isentropically (i.e., with no heat transfer).

Applying Equation 6.1 to the Carnot cycle in Figure 6.7, we find that the ratio of cooling power (heat absorbed) to input power (heat rejected minus cooling power) is purely a function of the cold and hot temperatures. Thus, the COP of the Carnot-cycle refrigerator is uniquely defined in terms of the cold-tip temperature (T_{cold}) and heat-sink temperature (T_{hot}) as

$$COP_{Carnot} = \frac{T_{cold}}{T_{hot} - T_{cold}} \tag{6.2}$$

6.3.1.2.1 Cryocooler Efficiency as Percent of Carnot

An important figure of merit for cryocoolers is the thermodynamic COP of the refrigerator expressed as a percentage of the ideal Carnot COP. This efficiency measure is applicable to all the various cryocooler cycles discussed in the remaining subsections of this chapter and is thus defined as

$$\%Carnot\ COP = 100 \times \frac{COP_{Cooler}}{COP_{Carnot}} = 100 \times \frac{(\text{Cooling power @ } T_{cold})\,(T_{hot} - T_{cold})}{\text{Input electrical power} \times (T_{cold})} \tag{6.3}$$

Notice that the percent Carnot COP is strongly dependent on both the hot and cold operating temperatures. Thus, when comparing the efficiency of various cooler candidates it is important to use common reference temperatures (both hot and cold) for the comparison. Figure 6.8 shows such a comparison assembled by Radebaugh, based on the reported performance of a broad number of cryocoolers with data available in the 2004 time frame. This particular comparison is made for cryocooler operation at 80 K with a 300 K reject temperature [Radebaugh, 2004].

From the plot, the relative efficiency trends of the various cryocooler types are evident, as well as their typical range of available cooling powers at 80 K and their required input powers. These trends will be one of the key factors that influence the applications most appropriate for each cryocooler type as they are discussed later in this chapter.

6.3.1.2.2 Dissecting Cryocooler Efficiency

To provide visibility into the principal parameters controlling cryocooler efficiency, it is sometimes useful to separate the overall efficiency or COP of a cryocooler, Equation 6.1, into its two main components: the thermodynamic efficiency of the compressor/expander combination, and the efficiency of the compressor drive motor. Cryocooler compressor motor efficiency is often found to be around 80%, or even less, and can represent a sizable fraction of the inefficiency of a cryocooler. Thus, it can be useful to break it out separately in understanding overall cryocooler efficiency.

Compressor/expander thermodynamic COP, which is a measure of the ability of the cooler to convert work done on the gas into net cooling power to the load, is thus defined as

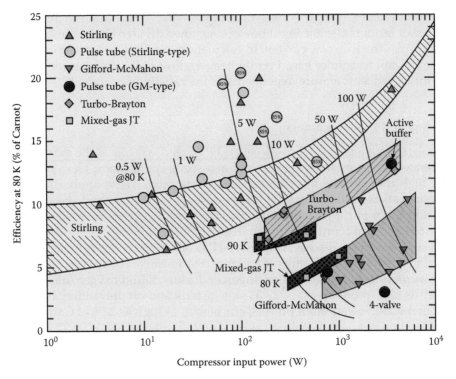

FIGURE 6.8
Efficiency for 80 K cooling for various measured cryocoolers of different types and input powers. (Radebaugh, R. Refrigeration for superconductors, *Proc. IEEE, Special Issue on Applications of Superconductivity*, vol. 92, © 2004 IEEE.)

$$\text{COP}_{\text{Thermodynamic}} = \frac{\text{cooling power}}{\text{input } PV \text{ work}} \approx \frac{\text{cooling power}}{\text{input electrical power} - i^2 R} \tag{6.4}$$

In this expression, the work done on the gas is referred to as the pressure–volume work or *PV*-work and is often approximated as the compressor input electrical power minus the drive-motor $i^2 R$ losses. This is a relatively good approximation because the other compressor loss mechanisms such as windage, mechanical friction, and eddy current forces are minor loss terms in a good compressor compared to its $i^2 R$ losses.

Similarly, because $i^2 R$ losses are generally the dominant loss term in a good motor, cooler motor efficiency can be usefully estimated as

$$\text{Motor efficiency} = \frac{\text{input power} - i^2 R}{\text{input power}} \tag{6.5}$$

There are four principal contributors to high $i^2 R$ losses: (1) low magnetic flux density in the motor's magnet circuit, which requires greater current to generate a given drive force; (2) higher coil resistance for a given number of coil turns; (3) higher operating temperature of the coil and magnet; and (4) excessive capacitive or inductive circulating currents that contribute to $i^2 R$ losses, but do no useful work. Eliminating circulating currents is the same as requiring that the motor have a near-unity power factor, where power factor is defined as the cosine of the phase angle between the input drive voltage and the input drive current. The power factor is also the input power consumed divided by the product of the true rms voltage times the true rms current. In calculating the i^2R losses, a common practice is to estimate the coil resistance based on the temperature of the compressor motor casing using the measured temperature dependence of the resistivity of copper.

6.3.2 Stirling and PT Cryocoolers

Stirling coolers (both mechanical displacer and PT based) are one of the most widely used cryorefrigerator types for small remote and aerospace applications. Here small size and mass and high thermodynamic efficiency are paramount. These applications are often remote from available utility-supplied power, and are often mass and space constrained. Classic examples of Stirling applications include remote cell phone towers, military IR vision sensors, and spacecraft-instrument IR and gamma-ray sensors. However, the development of large commercial-scale Stirling-type PT coolers has recently increased, aimed at efficiency and reliability improvements for large cost-sensitive continuous cooling applications such as cooling high-temperature superconductors, liquid oxygen/nitrogen production, as well as LNG production and LNG storage tank boil-off prevention. For these large coolers with multi-kilowatt input powers, efficiencies as high as 22% of Carnot have been achieved based on net useful cooling capacity on the order of 650 W at 77 K and 8.5 kW total electrical power to the cryocooler.

6.3.2.1 Stirling Thermodynamic Cycle and Operational Features

Stirling-cycle coolers tend to come in two flavors: those using a mechanical displacer to effect the thermodynamic cycle, and those based on a pneumatic PT circuit to achieve the thermodynamic cycle. Both use an oscillating-flow compressor to generate the ac flow needed by the coldhead. However, the PT version replaces the mechanical displacer of the classic Stirling cycle with a pneumatic (no moving part) expander to achieve the desired mass flow/gas pressure phase relationship needed for high thermodynamic efficiency. The benefit of the PT version is lower expander vibration and elimination of complexity and possible mechanical wear associated with the moving displacer. These days, most of the industry is moving to the use of PT expanders.

Because of the direct coupling between the compressor drive frequency and the expander drive frequency, most Stirling-based coolers operate at between 30 and 70 Hz using helium in the 10–35 bar pressure range as the refrigerant gas. This relatively high ac frequency is an advantage for cooling in the temperature range above 80 K, but serves as a disadvantage for obtaining high efficiency at very low operating temperatures (below 20 K), where the reduced specific heat of regenerator materials drastically limits heat storage between cycle phases.

Figure 6.9 schematically shows the thermodynamic cycles of the mechanical Stirling-cycle cryocooler. Generally, for the mechanical-displacer Stirling cycle the regenerator and the displacer are combined in one single unit as noted.

The cycle can be roughly divided into four steps, as follows:

- The cycle starts with the compressor compressing the gas in the expander cold finger. Because the gas is heated by compression, the displacer is used to position the expander's gas pocket at the warm end of the cold finger, which is coupled to a heat sink to dissipate the generated heat.

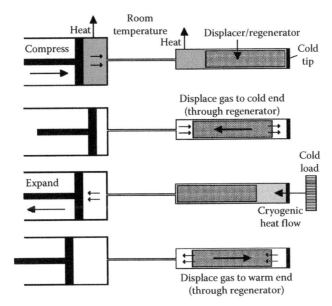

FIGURE 6.9
Schematic of Stirling cooler refrigeration cycle.

- At the completion of the compression phase, the displacer moves to the left to reposition the expander's gas pocket to the cold end of the cold finger to ready it for the upcoming expansion phase. During this part of the cycle, the gas passes through the regenerator entering the regenerator at ambient temperature T_{High} and leaving it with temperature T_{Low}. Thus, the heat storage feature of the regenerator smooths out the cyclic temperature of the gas in the two ends of the expander cold finger.

- Next, the compressor enters its expansion phase, thus expanding and cooling the gas in the expander's cold finger—adjacent to the cryocooler's cold load. This is where the useful cooling power is produced.

- In the final portion of the cycle, the displacer moves to the right to reposition the expander's gas pocket to the hot end of the cold finger to ready it for the upcoming compression phase. During this part of the cycle, the gas again passes through the regenerator entering the regenerator at the cold temperature T_{Cold} and leaving it with temperature T_{Hot}. Thus, the heat storage feature of the regenerator again smooths out the cyclic temperature of the gas as it flows between the two ends of the expander cold finger.

6.3.2.2 PT Stirling Cycle

In contrast to the mechanical driven displacer of the classic Stirling cycle, the PT version of the cooler uses a tuned pneumatic circuit with no moving parts to accomplish the gas position management functions accomplished by the conventional Stirling mechanical displacer.

The key elements of the PT tuned circuit are analogous to the principal elements of an electrical resistance-inductance-capacitance (RLC) phase shifting network. In the PT, the mechanical analogs are the reservoir volume, which provides the capacitance function, and an inertance tube, whose flow resistance provides the resistance function. The inductance or inertia function comes from the inertia of the gas flowing in the inertance tube, thus its name. The design objective of the circuit is to achieve an optimum phase shift (~70°) between the mass flow through the regenerator and the instantaneous pressure from the compressor. This is accomplished by carefully tuning the PT coldhead's RLC parameters: the length and diameter of the inertance tube and volume of the reservoir.

The gas displacing function of the expander is carried out by the PT itself. In addition to being the name of this type of expander, it is the name given to a short hollow tube between the inertance tube circuit and the regenerator. The objective of the hollow PT is to isolate the cold end of the regenerator from the hot gases returning from the inertance tube circuit. It does this by achieving a careful stratification of temperatures along its length and having sufficient volume such that the gas at the hot (inertance) end of the PT never reaches the cold-load interface end during each pressure/expansion cycle. To maintain this strict stratification of temperatures, the PT design must carefully prevent any kind of gas mixing in the PT due to turbulent flow or gravity-induced convection.

The four cyclic phases of the PT cooler are shown in Figure 6.10. In this figure, the displacer function of the PT is noted by a virtual displacer which represents the cold and hot boundaries of the stratified gas plug that oscillates back and forth in the PT during the cooler's operation.

- As with the conventional Stirling cycle, the cycle starts with the compressor, compressing the gas in the expander cold finger. Because the gas is heated by compression, the PT's pneumatic circuit is used to position the expander's gas at the warm end of the regenerator, which is coupled to a heat sink to dissipate the generated heat.

- At the compression phase ends, the PT's pneumatic circuit repositions the expander's gas to the cold end of the regenerator to ready it for the upcoming expansion phase. During this part of the cycle, the gas passes through the regenerator to dampen out the cyclic temperature variations and preserve the temperatures at the two ends of the regenerator.

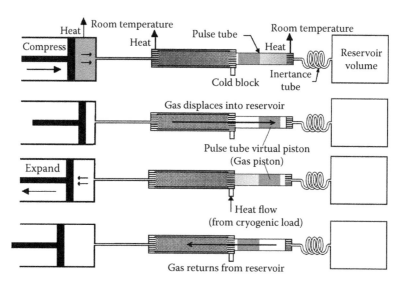

FIGURE 6.10
Schematic of PT cooler refrigeration cycle.

- Next, the compressor enters its expansion phase, thus expanding and cooling the gas in the expander's cold finger—adjacent to the cryocooler's cold-load interface. This is where the useful cooling power is produced.
- In the final portion of the cycle, the PT's pneumatic circuit repositions the expander's gas to the hot end of the regenerator to ready it for the upcoming compression phase. During this part of the cycle, the gas again passes through the regenerator to dampen out the cyclic temperature variations and preserve the temperatures at the two ends of the regenerator.

6.3.2.3 Engineering Aspects of Stirling and PT Cryocoolers

Supporting the requirement for an alternating fluid flow with a frequency of 30–70 Hz, Stirling cooler compressors are invariably piston-type compressors driven either by a rotating crank shaft like a car engine, or by a linear voice-coil motor, like a HiFi loud speaker.

6.3.2.3.1 Rotary Crank Compressor

The advantage of the rotary-crankshaft design is that the displacer can also be driven off the same crank shaft as the piston, thus achieving both gas compression and displacer control with the needed phase relationship between them from the same drive motor. The key disadvantage of the rotary crank design is the life issues associated with piston, displacer, and bearing wear, and contamination of the helium working fluid by outgassing products of the required bearing lubricants. Note that a rotary compressor is essentially a constant-stroke variable-frequency compressor, the frequency being determined by the motor drive speed (rpm). A miniature Ricor K508 rotary-drive Stirling cooler in shown in Figure 6.11.

6.3.2.3.2 Linear Compressors

Over the past 25 years, the vast majority of Stirling coolers have migrated over to the linear compressor configuration to achieve higher-reliability, longer-life designs. An example is the DRS (formerly Texas Instruments) 1.75 W at 80 K dual-piston linear drive Stirling cooler shown in Figure 6.12. This design uses a variable stroke and constant drive frequency, where the linear piston's mechanical resonant frequency is closely aligned with the drive frequency to achieve high drive motor efficiency. Maintaining a close match minimizes the required drive current and results in the drive current being closely in phase with the drive voltage. This minimizes circulating reactive currents that add to the i^2R losses, but does not contribute to work done by the motor.

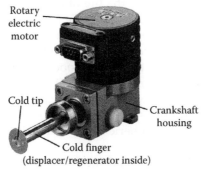

FIGURE 6.11
Miniature Ricor K508 rotary Stirling cooler (500 mW at 80 K).

FIGURE 6.12
1.75 W at 80 K DRS linear-motor dual-piston Stirling cooler.

The primary determiners of the compressor resonant frequency are the moving mass of the compressor piston assembly and the elastic spring constant of the combination of the gas under compression by the piston and the piston suspension springs. The resonant frequency is tuned to the desired value by adjusting these parameters.

To minimize exported vibration caused by the internal moving piston mass, most Stirling compressors are manufactured as a balanced head-to-head pair with two pistons moving in opposition into a common compression chamber. In this way, the momentum of the two pistons is cancelled out to a high degree, leaving a very quiet and relatively vibration-free compressor.

Although early linear compressor designs avoided the bearing wear and lubricant-caused issues of rotary Stirling coolers, they still contained rubbing pistons and displacers, which limited their useful lives to around 10,000 h.

6.3.2.3.3 The Oxford Compressor Design

In the mid-1980s, Steve Werrett, Gordon Davey, and their associates at Oxford University in England attempted to greatly extend the life of a linear Stirling cooler by supporting both the compressor piston and the displacer-regenerator/piston on linear flexure bearings [Werrett et al., 1986; Bradshaw et al., 1986]. These were designed to prevent piston and displacer contact with the cylinder wall while maintaining a tight (~0.0003″) clearance between the piston and the cylinder to achieve good compression efficiency. Figure 6.13 shows the mechanical features of the original 1980s Oxford cooler, including its spiral flexure spring design. This design was highly successful and launched into space to support 80 K cooling of the Improved Stratospheric and Mesospheric Sounder (ISAMS) instrument on board NASA's Upper Atmospheric Research Satellite (UARS) in September 1991 [Ross, 2007].

Based on the demonstrated long life and mechanical simplicity of its flexure bearing design, the Oxford cooler concept was quickly adopted worldwide by nearly all the leading manufacturers of long-life Stirling coolers. Since then, Oxford-style flexure supports

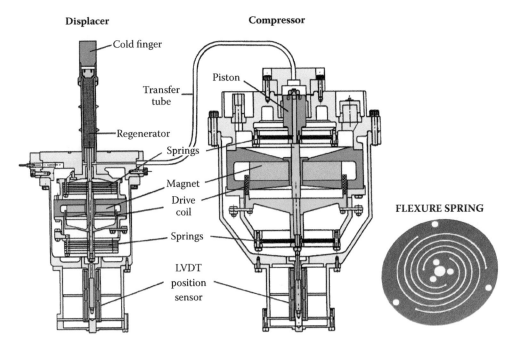

FIGURE 6.13
Construction features of the 1980s Oxford Stirling cooler, which incorporates a flexure-bearing-supported linear-drive compressor and a flexure-bearing-supported linear-driven active displacer.

have been adopted into all sizes of Stirling coolers from the lowest cost "tactical coolers" used in short-life military applications, to large-scale multikilowatt machines targeted at liquefaction of natural gas.

6.3.2.3.4 Mechanical Displacer

The classic mechanical Stirling-cycle expander combines a regenerator with a mechanical piston displacer, often integrated into a single regenerator/displacer unit as shown in Figure 6.13. For rotary-crank driven coolers, such as that shown in Figure 6.11, the regenerator/displacer is driven off the crankshaft, offset from the piston position by around 70°. For linear compressors with mechanical displacers, such as that shown in Figure 6.12, the displacer is generally a passive resonant system like the compressor, but tuned to have its phase shifted from that of the compressor by that needed for good Stirling-cycle efficiency.

With the introduction of the long-life Oxford cooler design in the late 1980s, a greater degree of control over piston/displacer phasing was introduced by embedding a second linear motor in the displacer as shown in Figure 6.13. This displacer motor was then used to provide precise stroke and phase control to the displacer via closed-loop drive electronics. However, the downside of this high-efficiency, long-life design was that the displacer and electronics complexity, mass, and cost increased substantially.

6.3.2.3.5 PT Expander

The first PT research dates from the 1960s with the work of Gifford and Longsworth [Gifford and Longsworth, 1965] and progressed rather slowly over the next 20 years [Radebaugh et al., 1986]. However, in the early 1990s, research with pulse-tube expanders for Stirling cryocoolers made a giant leap forward in terms of efficiency. This was brought about by the introduction of the inertance tube, first introduced by TRW (now Northrop Grumman Aerospace Systems—NGAS) into cryocoolers being developed for space applications. This technology allowed substantially improved Stirling-cycle tuning over that achievable with the use of the existing orifice PT. As noted earlier, the inertance tube introduced the ability to provide three-parameter resistance/inductance/capacitance (RLC-type) tuning, and thus achieved the more extensive phase-angle control required for high cryocooler efficiency. Since the late 1990s, PT expanders have been adopted worldwide as a leading expander type for Stirling-cycle coolers. Figures 6.14 and 6.15 show the features and appearance of a typical single-stage PT cooler utilizing an integral head-to-head Oxford-style linear compressor. PT coldheads are also being adapted for use on GM coolers, as described in Section 6.3.3.

6.3.2.3.6 *Drive Electronics*

A second area of advanced development first introduced by the Oxford cooler and its space-cooler derivatives is advanced solid-state drive electronics for precise control of cooler operation. For Stirling coolers with mechanical displacers, this typically involves precise control of compressor and displacer stroke amplitude and the phase between them; with PT coolers, only compressor stroke needs to be controlled. Taking advantage of the precise control of compressor stroke, many electronics expand this capability to also provide closed-loop control of the cold-tip temperature and active nulling of vibration harmonics in the drive axis [Harvey et al., 2004]. A representative set of modern PT drive electronics is shown in Figure 6.16.

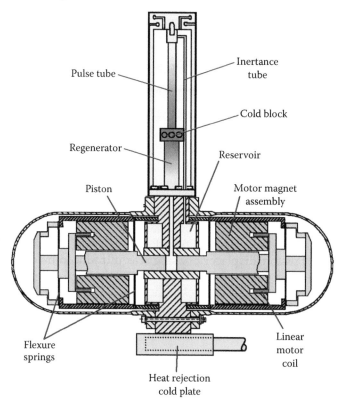

FIGURE 6.14
Schematic of dual-piston, linear-motor PT cooler.

FIGURE 6.15
NGAS dual-piston, linear-motor HEC PT cooler. (From Raab, J. and Tward, M., *Cryogenics*, Vol. 50, Issue 9, September, 2010.)

FIGURE 6.16
NGAS HEC dual-piston, linear-motor PT cooler drive electronics. (From Harvey, D. et al., *Cryogenics*, 44, 6–8, 2004.)

In addition to controlling and managing the power interface, many advanced electronics also provide a digital interface for remote programming of the cooler and feedback of cryo-cooler-related digital data such as cold-tip temperature, stroke level, and vibration level.

A common electrical interface issue with linear-drive coolers is the feedback of large ripple currents at twice the cooler drive frequency into the power supply bus. This and other electrical and mechanical interface considerations are discussed later in this chapter in Sections 6.5.2 and 6.5.3.

6.3.2.4 Stirling and PT Cooler Development History and Availability

Small Stirling-cycle cryocoolers (such as those shown in Figures 6.11 and 6.12) were first used in military/space applications in the early 1970s and first launched into space in 1975

[Ross, 2007]. Since that time, they have become the workhorse of the military and space industry. Starting in the mid 1990s high-efficiency PT coolers (such as that shown in Figure 6.15) emerged and have all but replaced the mechanical displacers of earlier generations of Stirling-cycle refrigerators [Raab and Tward, 2010].

As shown in Table 6.2, Stirling and PT cryocoolers have developed an enviable record in space applications over the past 20 years, with some units having demonstrated lives of greater than 139,000 h (over 15 years) of continuous 24/7 operation [Ross, 2007].

TABLE 6.2

Space Stirling and PT Cryocooler Flight Operating Experience as of October 2013

Cooler/Mission	Hours/Unit	Comments
Ball Aerospace (BATC) Stirling		
HIRDLS (60 K one-stage Stirling)	80,000	Turn on 8/04, ongoing, no degradation
TIRS cooler (35 K two-stage Stirling)	7,000	Turn on 3/6/13, ongoing, no degradation
Fujitsu Stirling (ASTER 80 K TIR system)	119,400	Turn on 3/00, ongoing, no degradation
Mitsubishi Stirling (ASTER 77 K SWIR system)	115,200	Turn on 3/00, ongoing, load off at 71,000 h
NGAS (TRW) Coolers		
CX (150 K Mini PT (2 units))	139,000	Turn on 2/98, ongoing, no degradation
HTSSE-2 (80 K mini Stirling)	24,000	3/99 thru 3/02, mission end, no degradation
MTI (60 K 6020 10cc PT)	119,000	Turn on 3/00, ongoing, no degradation
Hyperion (110 K Mini PT)	111,000	Turn on 12/00, ongoing, no degradation
SABER (75 K Mini PT)	107,000	Turn on 1/02, ongoing, no degradation
AIRS (55 K 10cc PT (2 units))	99,000	Turn on 6/02, ongoing, no degradation
TES (60 K 10cc PT (2 units))	80,000	Turn on 8/04, ongoing, no degradation
JAMI (65 K HEC PT (2 units))	72,000	Turn on 4/05, ongoing, no degradation
GOSAT/IBUKI (60 K HEC PT)	40,700	Turn on 2/09, ongoing, no degradation
STSS (Mini PT (4 units))	30,200	Turn on 4/10, ongoing, no degradation
Oxford/BAe/MMS/Astrium Stirling		
ISAMS (80 K Oxford/RAL)	15,800	10/91 to 7/92, instrument failed
HTSSE-2 (80 K BAe)	24,000	3/99 to 3/02, mission end, no degrad.
MOPITT (50–80 K BAe (2 units))	114,000	Turn on 3/00, lost one disp. at 10,300 h
ODIN (50–80 K Astrium (1 unit))	110,000	Turn on 3/01, ongoing, no degradation
AATSR on ERS-1 (50–80 K Astrium (2 units))	88,200	3/02 to 4/12, no Degrad, satellite failed
MIPAS on ERS-1 (50–80 K Astrium (2 units))	88,200	3/02 to 4/12, no Degrad, satellite failed
INTEGRAL (50–80 K Astrium (4 units))	96,100	Turn on 10/02, ongoing, no degradation
Helios 2A (50–80 K Astrium (2 units))	74,000	Turn on 4/05, ongoing, no degradation
Helios 2B (50–80 K Astrium (2 units))	30,200	Turn on 4/10, ongoing, no degradation
Raytheon ISSC Stirling (STSS (2 units))	30,200	Turn on 4/10, ongoing, no degradation
Rutherford Appleton Lab (RAL)		
ATSR 1 on ERS-1 (80 K Integral Stirling)	75,300	7/91 to 3/00, satellite failed
ATSR 2 on ERS-2 (80 K Integral Stirling)	112,000	4/95 to 2/08, instrument failed
Planck (4 K JT)	38,500	5/09 to 10/13, mission end, no degradation
Sumitomo Stirling Coolers		
Suzaku (100 K 1-stg)	59,300	7/05 to 4/12, mission end, no degradation
Akari (20 K 2-stg (2 units))	39,000	2/06 to 11/11 EOM, 1 Degr., 2nd failed at 13 kh
Kaguya GRS (70 K one-stage)	14,600	10/07 to 6/09, mission end, no degradation
JEM/SMILES on ISS (4.5 K JT)	4,500	Turn on 10/09, could not restart at 4,500 h
Sunpower Stirling (75 K RHESSI)	102,000	Turn on 2/02, ongoing, modest degradation

Presently there are a number of active manufacturers of Stirling and PT cryocoolers located all over the world: in the United States, Europe, Israel, and Asia. Starting originally with modest-size units with a cooling capacity of around 1 W at 80 K, the recent stable of available Stirling-cycle coolers ranges from palm-size units weighing just a few ounces and providing a cooling capacity of 500 mW at 80 K, to units that weigh 350 lb. and provide 650 W of cooling at 80 K.

6.3.3 GM and GM/Pulse Cryocoolers

GM cryocoolers (with both mechanical displacer and PT coldheads) are one of the most widely used coolers for commercial and laboratory use where low cost and operational convenience are important, and lots of electrical power is widely available. The GM cycle is very similar to the Stirling cycle in that its expander is based on an ac oscillating flow, typically using helium in the 10–30 bar range as the refrigerant gas with a working frequency of 1–2.4 Hz.

The one significant difference between Stirling-type coolers and GM coolers is that the GM cooler uses a low-cost high-availability dc flow compressor (typically acquired from a commercial air conditioning application) to provide the primary gas-compression function. The alternating flow needed by the GM expander is then provided by a rotary valve mounted on the GM cooler's coldhead assembly. This valve chops the dc flow into an ac flow by alternately connecting the expander to the high- and low-pressure sides of the compressor at the required oscillatory frequency of 1–2 Hz. This low frequency is particularly useful for obtaining improved efficiency at very low operating temperatures where the reduced specific heat of regenerator materials limits heat storage between cycle phases. The required phase relationship between refrigerant pressure and mass flow is achieved by synchronizing the rotary valve with the motor- or pneumatic-driven motion of the displacer. Because the compressor is a dc-flow device, it can be located remote from the actual cryogenic application, connected only by high-pressure hoses. However, the GM compressor must also use a highly efficient oil separator and a high-quality gas purification trap to prevent compressor oil vapor from reaching the expander.

6.3.3.1 GM Thermodynamic Cycle and Operational Features

GM cryocoolers tend to come in two flavors: those based on the historic GM motor-driven mechanical-displacer expander, and those based on the more recently developed PT expander. Both use the same dc-flow compressor and rotary valve to generate the ac flow needed by the coldhead. However, the PT version replaces the motor-driven mechanical displacer of the GM cycle with a pneumatic (no moving part) expander to achieve the desired mass flow/pressure phase relationship needed for high thermodynamic efficiency. The benefit of the PT version is lower vibration and elimination of mechanical wear in the moving displacer.

Figure 6.17 schematically shows the thermodynamic cycle of the mechanically driven GM refrigerator. Generally, for the mechanical GM cycle the regenerator and the displacer are combined into one displacer/regenerator unit as noted. The cooling cycle for a GM-PT-type cooler is essentially identical, except that the phasing of the gas flow in the cold finger is controlled by the PT's tuned pneumatic circuit instead of by the motion of the mechanical displacer.

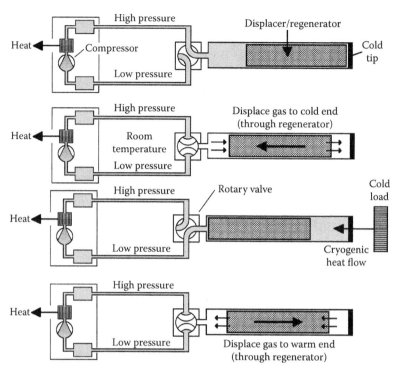

FIGURE 6.17
Schematic of Gifford-McMahon (GM) refrigeration cycle.

The GM cooling cycle can be divided into four steps as follows:

- The cycle starts with the rotary valve connecting the expander to the high-pressure room-temperature gas from the compressor. This fills the expander's gas pocket, which has been previously positioned at the warm end of the cold finger, with high-pressure gas.

- At the completion of the high-pressure filling phase, the displacer moves to the left to reposition the expander's gas pocket to the cold end of the cold finger to ready it for the upcoming expansion phase. During this part of the cycle, the gas passes through the regenerator entering the regenerator at ambient temperature $T_{Ambient}$ and leaving it with temperature T_{Low}. Thus, the heat storage feature of the regenerator retains the temperature gradient between the warm and cold ends of the cold finger and smooths out the cyclic temperature variation of the gas.

- Next, the rotary valve connects the low-pressure suction from the compressor return to the expander, thus expanding and cooling the gas in the expander's cold finger tip—adjacent to the cryocooler's cold load. This is where the useful cooling power is produced.

- In the final portion of the cycle, the displacer moves to the right to reposition the expander's gas pocket to the room temperature end of the cold finger to ready it for the upcoming high-pressure gas-filling phase. Again, during this part of the cycle, the gas passes through the regenerator, and the heat storage feature of the regenerator smooths out the cyclic temperature of the gas as it flows between the two ends of the regenerator.

6.3.3.2 Engineering Aspects of GM Cryocoolers

6.3.3.2.1 Compressor

A key engineering attribute of the GM process is the use of readily available, mature, dc-flow air conditioning compressors for the compression part of the cycle. This allows GM coolers to directly benefit from the years of reliability development and cost reduction of these OEM compressors. In addition, the dc-flow nature of these compressors allows them to be remotely located (up to tens of feet) from the application cold load—a big advantage in many applications. The one downside is probably that these compressors are almost all sized for large loads and draw input powers from 1 to 14 kW. Thus, they have serious electrical power and heat dissipation requirements—typically requiring 220–440 V electrical supplies and facility-based chilled water for cooling. In summary, GM machines are generally associated with modest to large loads in institutional settings, and are not particularly well suited to small, compact, or portable applications.

6.3.3.2.2 Coldhead

Having the GM coldhead operational frequency decoupled from the compressor's drive frequency provides another big advantage for the GM cooler. This allows the coldhead operational frequency to be optimized for maximum coldhead performance without regard to the compressor's operation. The ability to select a low expander frequency of around 1 Hz greatly simplifies the design of the regenerator beds for low temperature applications where the low specific heat of materials is a severe constraint. The low drive frequency allows the use of much larger regenerator particles (on the order of 0.25 mm diameter), which are much easier to package and contain in comparison to ~0.05 mm particles that would be required for, say, a 20 Hz Stirling-cycle coldhead.

The one disadvantage of the GM coldhead, particularly those with the classic mechanically driven displacer, is a modest (some would say high) level of mechanical vibration and audible noise generated directly by the coldhead. However, the recent GM PT coldheads are a big improvement in this regard, leaving the rotary valve as the only noise source in these units; and the rotary valve can be separated away from the coldhead in some units to further reduce vibration [Wang, 2005; Xu et al., 2003].

6.3.3.3 GM Development Status and Typical Performance

GM refrigerators have been the workhorse of the domestic cryogenic cooler industry for many years. Primary applications include cryopumping vacuum chambers used for semiconductor processing, cooling superconductor magnets such as in MRI machines in hospitals, and providing general-purpose cooling in cryogenic laboratories. Another common use these days is in ZBO systems where the GM cooler is used to reliquefy the evaporated gases from liquid nitrogen and liquid helium systems. Where the use of liquid nitrogen and liquid helium were once the preferred cooling means in the past, GM cryocoolers have replaced the stored cryogen systems in many places because of their ability to cool to a wide range of temperatures from 4 to 150 K and at loads as large as 1.5 W at 4.2 K and 600 W at 80 K. Both single- and two-stage machines are widely available (see Figure 6.18), with two-stage machines offering simultaneous cooling of loads at two different temperatures. Leading suppliers of GM machines include Sumitomo Heavy Industries (SHI) in Japan and Cryomech and CTI-Cryodyne in the United States. Figure 6.19 shows representative cooling curves for a variety of GM coolers manufactured by Cryomech. SHI and CTI have their own offerings. In addition, both Cryomech and SHI have units for substantial cooling down to 4.2 K [Wang, 2005; Wang and Gifford, 2003; Xu et al., 2003].

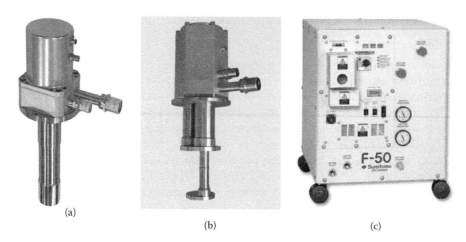

(a)

(b)   (c)

FIGURE 6.18
Example GM cooler components: (a) 200 W at 80 K Cryomech GM expander, (b) 0.5 W at 4.2 K Cryomech two-stage GM PT expander, and (c) example Sumitomo GM compressor.

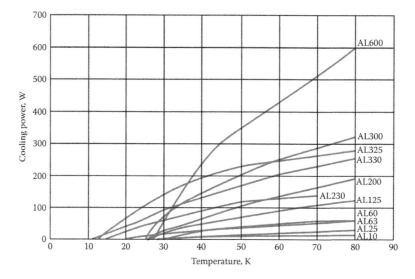

FIGURE 6.19
Representative cooling curves for the family of Cryomech GM refrigerators. Sumitomo has a similar family.

In general, GM cryocoolers have a good mean time to maintenance of around 10,000 h or more, comparable to a commercial air conditioning system. This has been improved to 30,000–45,000 h with some of the latest of PT coldheads.

As shown earlier in Figure 6.8, GM machines tend to be less efficient than Stirling coolers and have large power draws (typically from 1 to 8 kW) and often utilize three-phase electricity at 200–440 V. To manage the rejected heat from their large compressors, most provide facilities for water cooling via user-provided coolant water supplies. Smaller compressor units can also be acquired with interfaces for air cooling and utilizing 120 V single-phase power.

6.3.4 JT Refrigeration Systems

JT-based refrigeration systems are probably the most familiar type of refrigeration system to the general public. A variant of this cycle, referred to as the vapor compression or throttle cycle, is used in nearly all domestic refrigerators and freezers, and residential, commercial, and automotive air conditioning systems. A second major use of the vapor compression cycle is the liquefaction of oxygen and nitrogen for industrial uses. However, today, the use of the JT or throttle cycle is not particularly common for general cryogenic cooling applications. Two specialized uses include the cooling of the small tip of cryogenic surgical probes and as a bottoming cycle for cooling focal planes to 4–6 K in vibration-sensitive space-viewing instruments and telescopes. Previously, JT open-cycle blow-down systems were commonly used to cool IR detectors in many tactical military applications [Longsworth and Steyert, 1988; Bonney and Longsworth, 1990]. However, such applications have greatly diminished in recent years, replaced mostly by small fast-cooldown tactical Stirling cryocoolers.

6.3.4.1 *JT Thermodynamic Cycle and Operational Features*

Fundamentally, the JT cycle is a recuperative cycle that is built on a constant dc flow of high-pressure fluid that is expanded isenthalpically (no heat transfer) to a low pressure through a JT expansion valve [Maytal and Pfotenhauer, 2012]. Except for the open-cycle, fast-cooldown military applications mentioned earlier, most JT cooling systems are closed-cycle systems, meaning that the fluid is circulated in a closed-cycle system, as shown in Figure 6.20.

Such systems incorporate a high-pressure compressor to first pressurize the refrigerant stream to a relatively high pressure—much higher than that of a Stirling-cycle cooler, for example. Here, the heat of compression is extracted via heat exchange to an ambient-temperature heat sink. For the vapor compression or throttle-cycle version of the JT cycle, the refrigerant is chosen so that it is actually liquefied at this temperature and pressure; for the conventional JT cycle, it is generally still a gas, but must be cooled below its inversion temperature before reaching the expansion valve.

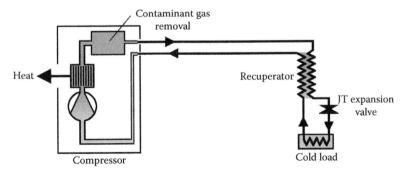

FIGURE 6.20
Basic mechanical setup of the closed JT cycle.

To achieve maximum efficiency, the circulating refrigerant may be passed through a counterflow heat exchanger (recuperator) to utilize the remaining cooling capacity of the spent refrigerant to precool the refrigerant stream entering the JT valve. Typically, the vapor compression or throttle-cycle version of the JT cycle does not incorporate the recuperative heat exchanger.

The refrigerant is next expanded through the JT valve, or throttle valve in the case of a liquid, to where it is used to cool the refrigeration load. Depending on the refrigerant gas and pressures used, the resulting cold refrigerant can be a pure gas or a mixture of gas and liquid. After cooling the load, the refrigerant is circulated back to the compressor for repressurization. For the case where the refrigerant is a liquid following the expansion process, the liquid may be contained in a reservoir or "evaporator" where it is boiled off as it is used to cool the load.

6.3.4.2 Basic Thermodynamics of the JT Cycle

A key feature of the JT effect is that it critically depends on gas properties that deviate from those of an ideal gas; in fact, an ideal gas exhibits no JT cooling effect when expanded. As a result, the JT cycle is highly dependent on the choice of refrigerants and the temperatures and pressures used. Table 6.3 tabulates some of the key properties of common gases used in cryogenic JT cooling systems. When using these gases, conditions close to gas liquefaction temperature typically lead to properties that deviate the most from ideal gas properties, and thus tend to be ideal for the maximum JT cooling effect. For refrigerators using a vapor compression or throttle cycle, the fluid is kept very close to liquefaction and thus very nonideal. Thus, those applications tend to have the highest efficiency.

From a physics point of view, as a nonideal gas expands, the average distance between molecules grows. Because of the attractive part of the intermolecular force, expansion causes an increase in the potential energy of the gas. If no external work is extracted in the process and no heat is transferred (the isenthalpic JT process), the total energy of the gas remains the same because of the conservation of energy. The increase in potential energy thus implies a decrease in kinetic energy and therefore a decrease in temperature of the gas.

Unfortunately, a second mechanism has the opposite effect. During gas molecule collisions, kinetic energy is temporarily converted into potential energy (corresponding to the repulsive part of the intermolecular force). As the average intermolecular distance increases, there is a drop in the number of collisions per time unit, which causes a decrease in average potential energy. Again, total energy is conserved, so this leads to an increase in kinetic energy and an increase in gas temperature upon expansion.

TABLE 6.3

Fluid Properties Important to JT Coolers

Fluid	Normal Boiling Point (K)	Freezing Point (K)	Critical Point (K)	Max. Inversion Temp. (K)
Helium	4.2	1.8	5.2	39
Hydrogen	20.4	13.8	33.2	195
Neon	27.1	24.6	44.5	220
Nitrogen	77.4	63.3	126.2	608
Argon	87.3	83.8	150.7	763
Oxygen	90.2	54.4	154.6	758
Methane	111.7	90.7	190.6	980
Krypton	119.8	115.8	209.4	1054

Below what is referred to as a gas inversion temperature (column 5 in Table 6.3) the former effect dominates, and JT expansion results in gas cooling; above the inversion temperature, the second process dominates, and JT expansion causes the gas to increase in temperature.

The rate of change of temperature for a change in pressure in a JT expansion process is referred to as the *JT coefficient* of a gas. It is commonly expressed in °C/bar or K/Pa and depends critically on the type of gas and the temperature of the gas before expansion. Its pressure dependence is usually only a few percent for pressures up to 100 bar. Figure 6.21 shows the JT coefficient for a number of common gases at atmospheric pressure. For positive JT cooling to take place, the JT coefficient must be positive; thus, the zero crossings in the plot define the *inversion temperatures* of the displayed gases.

6.3.4.2.1 Use of Mixed-Gas Refrigerants

When trying to achieve cryogenic temperatures between 80 and 120 K with conventional gases such as nitrogen, one finds that the efficiency is quite low and the required pressure ratio for good efficiency is impractical to achieve with an inexpensive single-stage compressor. To combat these limitations, experimentation initiated in the 1970s examining the possibility of combining nitrogen with various hydrocarbon gases to yield a mixed gas with substantially improved JT properties that would allow efficient operation with a lower-pressure-ratio single-stage compressor. The primary challenges involved achieving throttle-cycle performance at the higher temperatures while achieving gas–liquid solubilities that prevented expansion valve plugging as the higher temperature constituents drop in temperature. Other constraints on the constituents include oil solubility and flammability. Progress on the development of mixed gases over the years has been quite successful and led to a variety of proven mixed-gas refrigerants for various cryogenic temperature ranges [Boiarski, 1998; Arkhipov et al., 1999; Bradley et al., 2009].

One of the first commercial cryogenic JT coolers, called the Cryotiger, was developed by Ralph Longsworth in 1994 [Longsworth et al., 1995; Longsworth, 1997]. It married the use of a mixed-gas refrigerant with an inexpensive oil-lubricated compressor and GM oil stripping technology to yield a low-cost, relatively long-life JT cryocooler for use in the range of 80 K.

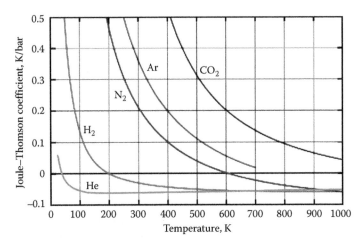

FIGURE 6.21
JT coefficients for various gases at atmospheric pressure.

6.3.4.3 Engineering Aspects of JT Cycle Cryocoolers

6.3.4.3.1 High-Pressure Piston Compressors

Key engineering attributes of the JT process are the need for high pressure ratios, relatively low refrigerant mass flow rates, and the need to prevent blockage of the JT valve by refrigerant contaminants such as water ice or compressor lubricants. For ground-based systems, this typically leads to a high-pressure piston or scroll-type compressor with oil lubrication and the complementary need to have the refrigerant compatible with the oil lubricant. For applications near room temperature (such as domestic refrigeration and air conditioning) the compressor lubricant is allowed to flow freely around the JT circuit as it remains dissolved in the refrigerant. However, for cryogenic applications, the oil from a lubricated compressor must be stripped from the refrigerant stream before entering the expansion valve where it could freeze out and plug the valve. A similar need exists in the implementation of GM compressors, and thus this technology has been well developed for a mean time between maintenance of 10,000 h and more.

6.3.4.3.2 Oxford-Based Linear-Motor Compressors

For long-life applications, the desire for a lubricant-free JT compressor has led to the use of oil-free, linear-motor compressors based on the Oxford-cooler technology used in Stirling and PT coolers. Such compressors are fitted with reed valves to effect a dc flow, but the low piston forces available from such compressors severely limit the allowable JT pressure ratios available to around 3:1; at a cost of increased complexity, this can be overcome by using multiple compressor stages as required.

Initial research on Oxford-based JT compressors was conducted by Bradshaw and Orlowska at England's Rutherford Appleton Laboratory in the late 1980s [Bradshaw and Orlowska, 1991]. This technology, shown in Figure 6.22, utilized a two-stage JT compressor with helium gas to achieve cooling at 4.2 K. When matured, this technology successfully flew on the European Plank mission launched in May 2009 [Bradshaw et al., 1999]. More recently, the Oxford-JT-compressor technology is being used to provide 6 K cooling for the MIRI instrument on the James Webb Space Telescope. This application, scheduled to launch in 2018, uses a single-stage Oxford-style JT compressor with helium that is precooled to ~18 K via a three-stage Oxford-style PT refrigerator [Raab and Tward, 2010; Petach and Michaelian, 2014].

6.3.4.3.3 Sorption Compressors

Another compressor technology that is sometimes used in a JT cooler is a sorption compressor. Such a compressor uses a chemical or physical sorbent material to absorb a refrigerant at low pressure and temperature and then to desorb it at high pressure and high temperature, thus achieving the required compression. Microporous activated carbons, zeolites, and silica gels are some typical physical sorbers, whereas metal hydrides and oxides are well-known chemical sorbers for hydrogen and oxygen, respectively. Here, the driving force is thermal input, either from electrical heaters or from other heat sources. A sorption compressor cell basically consists of a container that is filled with the sorbent material that is fitted with a means of being heated and cooled (thermally cycled) while the gas flow to and from the system is controlled with check valves. As a result, a sorption compressor has no moving parts and generates no vibration that would disturb a sensitive application.

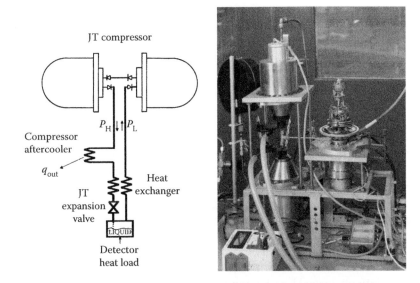

FIGURE 6.22
Oxford-compressor-driven 4 K JT cryocooler at Rutherford Appleton Labs in 1988.

Sorption-based cryogenic JT systems tend to be a speciality technology focused mainly at space observatory applications requiring extremely low levels of vibration. The Planck sorption cooler developed by NASA's Jet Propulsion Laboratory for the European Planck mission is a prominent recent sorption cryocooler development [Wade et al., 2000]. This system uses a metal-hydride compressor to compress hydrogen gas for cooling to 18 K. Another prominent research center for sorption cryocooler technology is the University of Twente in the Netherlands [ter Brake et al., 2011].

6.3.4.3.4 JT Blockage

Experience has shown that water vapor is the most common cause of blockage of a JT valve, even with oil-lubricated compressors. Generally, moisture concentration must be less than around 2 ppm to achieve long blockage-free operational periods [Bonney and Longsworth, 1990]. The probability of blockage tends to increase for smaller systems with low flow rates, as the orifices tend to be smaller and the gas spends a greater time cold prior to reaching the JT valve, thus producing larger ice crystals. The shape of the JT restrictor is also important. A large-ID capillary tube is least prone to plugging, followed by an equivalent round orifice. The annular gap of a needle valve is worst. Systems with high flow rates and relatively large JT restrictions tolerate much higher levels of contamination. For long-lifetime systems, often an in-line getter is incorporated in the system to maintain the necessary low level of contaminant gases to prevent JT blockage over long operating periods [Bradshaw et al., 1999].

6.3.4.3.5 Integration Features

From an integration point of view, the dc-flow nature of a JT system makes it easy to provide cooling at relatively large distances (many meters) from the compressor, thus minimizing exposure of the cryogenic load to compressor-generated vibration and EMI. The long flow path also allows flexibility in packaging and integration. Cooling can be distributed over large areas or multiple coldheads. For applications where saturated liquid is generated, the pooled liquid can provide temperature stability and load leveling for varying heat loads.

6.3.4.3.6 Hybrid Coolers

One means of achieving higher efficiency or lower temperatures is to precool the JT gas stream using a second cooler, either a second JT cooler (as with the Planck cryocooler system), or another active cooler such as a Stirling, PT, or GM, as with the JWST/MIRI cooler [Petach and Michaelian, 2014]. This can significantly reduce the temperature range required for the JT working fluid and significantly (by a factor of 10) reduce the operating pressures required of the JT compressor. The disadvantage of using multiple coolers is a somewhat lower level of reliability that may result from increased system complexity.

6.3.4.4 JT Cryocooler Development History and Availability

An important downside of JT cryo refrigerators is their relative scarcity. The 4–6 K hybrid JT bottoming-stage units developed for the Planck and JWST MIRI space applications are one-of-a-kind custom units costing millions of dollars each and are not easily transferrable to other applications. However, in the area of commercial hybrid 4 K JT coolers, Sumitomo acquired the cryocooler and cryopump business of the Daikin Company in 2005. One of the cryocooler products acquired was the line of GM refrigerators with a JT third stage.

Designed primarily for radio telescope astronomers, these three-stage cryocoolers, an example of which is shown in Figure 6.23, have high cooling powers of up to 5 W at 4.3 K as well as very stable temperatures at the third stage.

In terms of commercial units providing 80 K temperatures, the Cryotiger developed by Ralph Longsworth of APD Cryogenics in the 1995 time frame is one of the few commercial units available [Longsworth et al., 1995; Longsworth, 1997]. The latest reincarnation of this cooler is now being sold by Brooks as their Polycold PCC cooler. This unit (shown in Figure 6.24) uses a separate GM-type compressor with a remote coldhead to achieve cooling powers of around 5 W at 70 K using a mixed-gas refrigerant.

FIGURE 6.23
Sumitomo hybrid GM/JT coldhead for 4 K cooling.

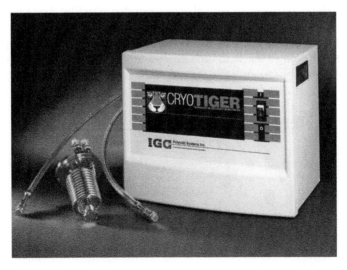

FIGURE 6.24
Cryotiger 70 K JT cryocooler.

Another application of custom JT coolers has been for cooling cryogenic medical catheters and cryosurgical probes. Here, the small size of a JT cold end can be successfully integrated into the medical probe's tip while the compressor unit is external, some distance away [Dobak et al., 1998; Marquardt et al., 1998; Longsworth, 2001].

6.3.5 Brayton Refrigeration Systems

A reverse-Brayton cycle cryocooler is a second type of recuperative or dc-flow cryocooler. However, it differs considerably from the JT cycle by using a high-flow-rate, low-pressure-ratio refrigerant stream to produce cooling. As a result, it generally uses an entirely different type of compressor, a high-speed gas turbine that operates at speeds of 100,000–600,000 rpm. Positive displacement compressors and expanders could also meet the functional needs of the cycle; however, their mass and vibration characteristics tend to make them less desirable.

As shown in Figure 6.25, the reverse-Brayton cycle schematic looks very similar to that of the JT cycle, but with the JT expansion valve replaced with a gas turbine expander. This turbine expander both expands the gas and withdraws some work from the gas; this provides near isentropic expansion as opposed to the isenthalpic expansion (constant energy expansion) used in the JT cycle. Because of the use of the turbine compressor and expander, the cycle is also commonly referred to as the turbo-Brayton cycle.

Unfortunately, like the JT cycle, the turbo-Brayton cycle is not particularly common in general cryogenic cooling applications. Its key advantages, like the JT cycle, are very low generated vibration and the ability of the compressor to be remotely located from the cold load. The low-mass turbine rotors are the only moving parts in the system, and because they are precision balanced and operate at very high rotational speeds (1000–10,000 revolutions per second), the systems generate extremely low levels of vibration. With the use of nonwearing gas bearings, they also tend to have high reliability for long-life applications. This is achieved by avoiding the oil-lubricated compressors and oil stripping issues associated with common JT and GM systems.

The primary disadvantage of the turbo-Brayton cycle is its manufacturing challenges that are driven by the need for a very high construction quality to achieve competitive efficiencies. Both the tiny turbines and the recuperator must be designed and fabricated to very exacting standards if high efficiency is to be realized. Because of the issues with achieving the build precision required with very small turbine parts and high-effectiveness recuperators, the technology tends to be better suited to larger applications requiring several watts of cooling and requiring the unique features of an ultra-low-vibration, long-life refrigerator.

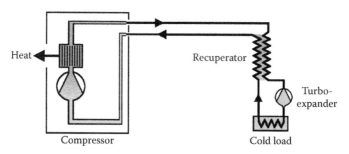

FIGURE 6.25
Turbo-Brayton cryocooler schematic.

Recent specialized uses of the turbo-Brayton technology include building a long-life, ultra-low-vibration cryocooler (the NICMOS cooler) for use on the Hubble Space Telescope, building a low-vibration freezer for the International Space Station, and building helium liquefiers for large industrial-scale liquid helium applications.

6.3.5.1 Engineering Aspects of Brayton Cycle Cryocoolers

Three factors influence the efficiency of a turbo-Brayton cryocooler: the speed of the compressor, the thermal effectiveness of the recuperator, and the precision of the small-scale turbine blades. A fourth key element of the cooler is the specialized electronics used to drive the compressor and control the cooler's operation.

6.3.5.1.1 Compressor

A typical turbo-Brayton compressor comprises a small-diameter, low-mass rotor with a centrifugal impeller at one end. This rotor is driven by a three-phase variable-frequency motor built into the assembly. Radial and longitudinal support for the rotor is commonly provided by self-acting gas bearings. The compressor assembly may also incorporate an integral heat exchanger to reject the heat of compression and motor losses to a suitable thermal interface.

From a practical point of view, the pressure ratio that can be achieved depends primarily on the speed of the compressor. Mechanical features (centrifugal stresses, shaft dynamics, etc.) and gas properties (molecular weight, Mach number, etc.) constrain the maximum pressure ratio that can be achieved in a single compression stage to about 1.8 in neon and 1.26 in helium. Compression efficiency, on the other hand, is driven by the precision of the turbines and their internal losses, both of which result in an increasing proportion of losses as size is decreased.

6.3.5.1.2 Expansion Turbine

The expansion turbine provides the refrigeration to the cycle by expanding the gas from high pressure to low pressure; this reduces the temperature of the gas and also produces shaft work. This work may be used to drive a brake or an alternator; if the work is removed by a brake, the component is referred to as a turboexpander. If the work is converted to electric power, the component is a turboalternator.

To be reasonably efficient, the expansion turbines in small-capacity machines must be extremely small, as shown in Figure 6.26. Rotational speeds are comparable to the turbo compressors, with speeds of 100,000–500,000 rpm. Self-acting gas bearings similarly provide radial and longitudinal support for the rotors, and the rotors must be balanced to a high degree of precision, resulting in vibration levels that are nearly undetectable.

6.3.5.1.3 Recuperator

The thermal effectiveness of the recuperator is also very critical to the refrigeration efficiency of the reverse-Brayton cycle. The recuperator's primary purpose is to efficiently precool the high-pressure gas flow between the compressor and the expansion turbine. Achieving the high thermal effectiveness values (>0.99) that are typically needed to achieve competitive Brayton-cycle efficiencies requires very thoughtfully developed, highly optimized designs, particularly given the low pressure ratio and high flow rate of the reverse-Brayton cycle. The major contributor to ineffectiveness of the recuperator is generally the longitudinal heat conduction between its warm end and its cold end. Thus, a major consideration in the design of

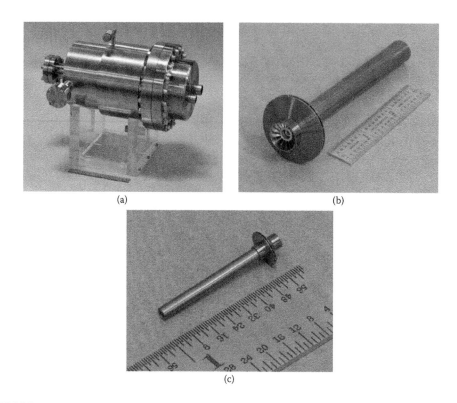

(a)

(b)

(c)

FIGURE 6.26
Fully assembled turboalternater (a) and rotor assemblies for 2 kW class turboalternater (b) and 2 W class turboalternater (c). (From Zagarola, M. et al., *Cryocoolers 17*, ICC Press, Boulder, CO, 2012.)

the devices is the reduction of axial conduction to the lowest practical value. Meeting these
needs frequently results in the recuperator being the highest-mass device in the cooler system.

6.3.5.1.4 Electronics

The electronics of a turbo-Brayton cryocooler perform two functions. First, they convert power
from the available power source to regulated, three-phase ac power at the multi-kilohertz
drive frequency of the centrifugal compressor. To achieve high overall cryocooler electrical
efficiency, a key driver on the power supply design is achieving high electronic conversion
efficiency. A second key function of the cooler's electronics is to provide the control functions
needed to operate the cooler system. A key one of these is to adjust the frequency of the compressor drive (i.e., the turbine speed) to increase or decrease the pressure ratio. This increases
or decreases the available refrigeration to allow control of the cold-load temperature to correct
for possible changes in the heat-rejection temperature or variations in the cold load.

Both of these processes (power conversion and control) tend to be highly specialized
functions that lead to carefully optimized, highly customized electronics [Konkel and
Bradley, 1999].

6.3.5.2 Turbo-Brayton Cryocooler Development History and Availability

Turbo-Brayton cryogenic expanders have been under development for many years,
starting in the late 1950s and early 1960s [Sixsmith, 1984]. In the early 1970s, under DoD

sponsorship, large DoD contracts to General Electric and Garrett AiResearch were made to develop high-capacity 12 K cryocoolers for military space reconnaissance applications [Ross, 2007]. The original system goal for the GE unit was for cooling loads of 1.5 W at 12 K plus 30 W at 60 K, with a 30,000-h lifetime and a maximum power consumption of 4 kW [Sherman, 1982]. The Defense Advanced Research Project Agency (DARPA) initiated a follow-on turbo-Brayton program in 1978 with Garrett AiResearch [Harris et al., 1981]. This system was designed, fabricated, and performance tested in the early 1980s. Although it exhibited satisfactory operation, contamination and other issues prevented performance goals from ever being fully reached.

6.3.5.2.1 Creare Coolers

In the early 1980s, Creare, Inc. in Hanover, NH, began work on critical elements of smaller turbo-Brayton systems for both DoD and NASA applications. These activities eventually led to the development of a successful engineering model cooler [Swift and Sixsmith, 1993; Dolan et al., 1997] and eventual selection of Creare to build a very-low-vibration turbo-Brayton cooler to cool the NICMOS instrument on the Hubble Space Telescope. This cooler was designed to replace the instrument's 65 K solid-nitrogen dewar, which was severely degraded during launch in 1997 [Miller, 1998b]. This NICMOS turbo-Brayton cooler, which was installed by astronauts on orbit in March 2002, worked exceptionally well, providing around 7 W of cooling at 77 K to the NICMOS instrument over the next 7 years [Swift et al., 2008]. The cryocooler refrigerant gas in the NICMOS cooler is neon at a nominal pressure ratio of 1.6:1. The inlet pressure to the compressor is 1.5 atm, and the compressor operates at variable speeds up to 440,000 rpm. Under these conditions, the input power to the compressor is 315 W at a rejection temperature of 280 K [Swift et al., 2008].

Since the NICMOS instrument development and launch, Creare has been developing a number of additional advanced turbo-Brayton technologies focused on future NASA and DoD cooler opportunities, both smaller than and larger than the NICMOS cooler [Zagarola et al., 2012].

6.3.5.2.2 Air Liquide

In Europe, Air Liquide developed, qualified, and delivered a reversed turbo-Brayton cooler for the −80°C MELFI freezer on the International Space Station (ISS) in 2006 [Ravex et al., 2005]. The MELFI cooler, like the Creare NICMOS cooler, is based on high-speed turbo machinery supported on gas bearings, and carrying the compressor and expander turbine wheels. For 840 W of electrical power, the unit is designed to provide 60 W of cooling at 178 K, and for 1000 W input power, to produce 90 W at 178 K. The mass is 8.5 kg.

Air Liquide, like Creare, is also actively expanding its base of turbo-Brayton technologies to support additional future missions on board scientific satellites. One key focus is on turbo machines for applications of around 110 mW in the 2–5 K temperature range.

6.4 Cryogenic Cooling System Design and Sizing

The intent in this section is to cover some important aspects of cryogenic cooling system design that are generic in nature, and thus applicable to any cooling means, passive or active. Section 6.5 will then expand on a number of cryocooler-specific considerations associated with integrating with active cryocoolers. In general, the cryogenic system design process includes a number of iterative steps:

1. Derive a strawman cryogenic system design including rough estimates of all key parameters such as geometric sizes, power dissipations and active cooling loads, and candidate cooling approaches.
2. Estimate the total cooling load over the system's total operating range and life including both active loads and passive parasitic loads.
3. Acquire performance data for the candidate cooling approaches for the full range of projected cooling loads, cooling temperatures, and external environmental temperatures.
4. Iterate the cryogenic load projections with the cooler performance projections to achieve a successful cooling system design.
5. Validate the design with detailed calculations and engineering tests.

Unfortunately, for many cryogenic applications, parasitic loads, which typically carry large uncertainties, often represent 80% of the total load. Thus, it is best to apply large conservative margins for these poorly predictable loads during the design process and to implement a process to "burn down" the uncertainties as the design progresses toward the final implemented system. This process of resolving the uncertainties over time to lower the risk of poor system performance or the cost of excessive conservatism is a critical part of the cryogenic system design process.

6.4.1 Cryogenic Load Estimation and Management

Generally, one of the most important and difficult tasks in cryogenic system design is establishing what the expected cryogenic loads are going to be and what the primary drivers are in determining the loads. Knowing the load drivers allows emphasis to be focused on these specific areas of the design and to perhaps integrate in additional cooling stages to carry some of the loads at higher temperatures, or to refine specific load estimates or reduce them through experiments or technology developments. It is also important to recognize that the total range of loads that can be efficiently provided at a given temperature by a given cryogenic cooling system is generally not much greater than a factor of two. Therefore, in practical terms, the cooling system design must be matched to the load early in the design process with relatively good accuracy, that is, better than a factor of two. Achieving a proper match between system design and load, and maintaining the match over a multiyear project development cycle is an important integration challenge faced by cryogenic system engineers.

To estimate the cryogenic load it is essential to address a number of key issues. These include:

1. Accurately estimate the active part of the cryogenic load and the rough details of the overall cooling system configuration. By active loads we mean the loads other than parasitic conduction and radiation loads. An example would be direct electric power dissipation by focal planes, motors, electronics, etc., at the cryogenic temperature. Other active loads would be the cryogenic load associated with liquefying gases or cooling a fluid as part of the cryogenic application.
2. Estimate any conduction loads associated with connecting the application to the outside world. Connection parasitics include conduction down electronic wires

from outside the cryogenic application or conduction down tubing and pipes from the cryogenic stage to outside the application. Wiring and harness loads can often be large contributors to the cryogenic load if very special low-conductance wiring is not employed.

3. With the proposed system configuration in mind, attempt to accurately estimate the total parasitic conduction loads associated with structurally supporting the application. These are often difficult to estimate without at least a conceptual structural design and estimates of any applied structural loads such as vibration and handling during transportation to the application site. Helpful tips on estimating structural conduction loads are presented in Section 6.4.2.

4. As with the conduction parasitics, attempt to estimate the total parasitic radiation loads absorbed by the application from the external thermal environment. These are often the most challenging to accurately quantify, as radiation loads are a strong function of the surface emittance of application materials, and these emittances can have very large uncertainties and change over the life of an application. Background on the prediction of radiation loads and MLI performance is presented in Section 6.4.3.

6.4.2 Estimating Structural Support Thermal Conduction Loads

For structural support conduction there are four key issues: (1) achieving a high-strength and low-thermal-conductivity structural support design using low-conductivity, high-strength materials such as stainless steel, titanium, fiberglass, and Kevlar; (2) supporting the assembly from an intermediate-temperature support to reduce the ΔT across the support structure; (3) using vacuum insulation systems to avoid gaseous conduction loads; and (4) minimizing the mass and size of the assembly to reduce the structural loads that the supports must carry.

In a relatively mature design, the conduction loads can be computed with good accuracy based on the design details and the relatively well known conductivity properties of common cryogenic structural materials; representative properties of common low-conductance materials are shown in Figure 6.27. The most structurally and thermally efficient structural designs are generally the ones that use structural members in pure axial tension and compression and minimize loads carried in bending.

6.4.2.1 Load Estimating "Rule of Thumb"

Unfortunately, early in an application's conceptual design, details of the design do not yet exist, and what is needed are some generic "rules of thumb" for estimating overall support conductance in terms of the ΔT's involved, the candidate structural materials, and the supported masses. Such a load estimating algorithm has been developed for space applications, based on examining a wide variety of flight-proven space-instrument designs with cryogenic structural supports [Ross, 2004]. The developed rules account for the known relationships between material conductivity and temperature, between launch acceleration level and assembly mass, between launch acceleration loads and required support-member cross-sections, and between support-member cross-section and conductive thermal load.

Although these estimates were generated for space applications with typical launch loading environments, they should also be useful for estimating conductive loads for

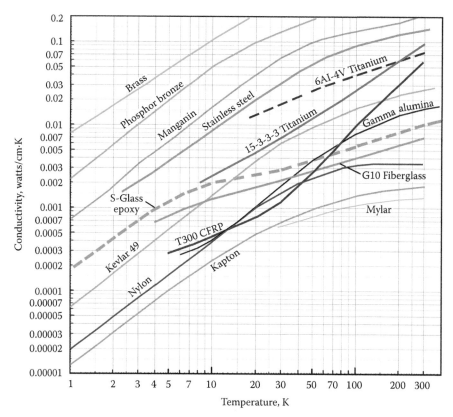

FIGURE 6.27
Thermal conductivity of common cryogenic structural materials as a function of temperature.

systems subject to common transportation loads, and may serve as useful bounds for stationary laboratory facilities, as often considerations of structural robustness may lead to similar design-load levels.

The general relationship derived for predicting cryogenic structural conduction loads is given by

$$Q \approx \textbf{A}\kappa m^{0.66}\Delta T \tag{6.6}$$

where Q is the conductive load being estimated (watts), $\textbf{A}$ is the empirical scaling factor [0.02 (very efficient structures) to 0.27 (less efficient)], κ is the average conductivity of support material in temperature range ΔT (W/cm·K), m is the mass of supported assembly (kg), and ΔT is the differential temperature across support structure (K).

The empirically derived constant $\textbf{A}$ in Equation 6.6 reflects how efficiently the structural materials are used in the design. Thus, a lower value ($\textbf{A} = 0.02$) correlates with axially loaded structures using high-strength materials such as a tension band system, while a higher value ($\textbf{A} = 0.27$) is used for structures using members in bending, like a cantilever-type structure, or one using lower-strength materials. Note that $\textbf{A} = 0.02$ to 0.27 reflects an order of magnitude difference between the structural conduction loads of a highly efficient structural design and those of a satisfactory, but inefficient design.

6.4.2.2 MLI and Gold Plating Lateral Conductivity

Because of the relatively high thermal conductance of MLI blankets when compared with low-conductivity structures, one needs to be particularly sensitive about allowing MLI to thermally bridge between two components or surfaces with substantially different temperatures. This is true for conduction through the thickness of an MLI blanket, and is even more true for heat transfer parallel to the blanket surface. The aluminized layers of MLI have quite high lateral conductivity and can seriously deteriorate the thermal resistance of low-conductance structural supports upon which they might be wrapped. In a similar vein, gold plating to achieve a low-emittance surface on low-conductance structural members can substantially deteriorate their thermal resistance.

6.4.2.3 Vacuum Requirements to Minimize Gaseous Conduction

To achieve low thermal conduction loads it is invariably necessary to eliminate gaseous conduction and convection heat transfer from the ambient environment into the cold load. Thus, nearly all cryogenic applications incorporate vacuum insulation systems and must be mounted in vacuums that achieve levels of at least 10^{-4} torr. For applications striving for very low conduction rates and elimination of condensation on cold low-emittance surfaces, much better vacuum levels are required.

Gaseous conduction (Q) between surfaces surrounded by gas in the free molecular regime is independent of the gap spacing, and linearly dependent on the temperature difference (ΔT) between the surfaces, on the pressure of the gas (P), and on the surface area of the cold object (A). Thus,

$$Q \propto A P \, \Delta T \tag{6.7}$$

Additional parameters that affect the proportionality relate to the specific heat transfer properties of different gases, which can vary somewhat with temperature. Useful estimates for gaseous conduction loads for various vacuum levels are shown in Figure 6.28. These are for free molecular conduction through a He/H$_2$ mixture likely to exist at cryogenic temperatures and for representative temperature differences between surfaces. For more exact calculations for specific system designs, one should appeal to the equations governing thermal conduction through free molecular gases as described in cryogenic heat transfer texts such as Scott [1988].

6.4.3 Estimating Thermal Radiation Loads

Estimating thermal radiation loads often poses the largest challenge to the cryogenic systems designer trying to predict cryogenic loads over multiyear operational periods. This is because radiation loads increase as the fourth power of the radiating temperature—making loads very sensitive to temperature predictions—and are also directly proportional to the surface emittance properties of cryogenic surfaces, which can vary by two orders of magnitude. Low-emittance gold-plated and polished-aluminum surfaces and MLI have very low effective emittance values and are commonly used to minimize radiant heat transfer.

MLI is constructed of layers of aluminized Mylar or Kapton separated by low-conductance spacers or stippled surfaces that are designed to minimize heat conduction between adjacent shield layers. The spacer layers also have the important function of providing for the evacuation of residual gas from between the shield layers when vacuum is

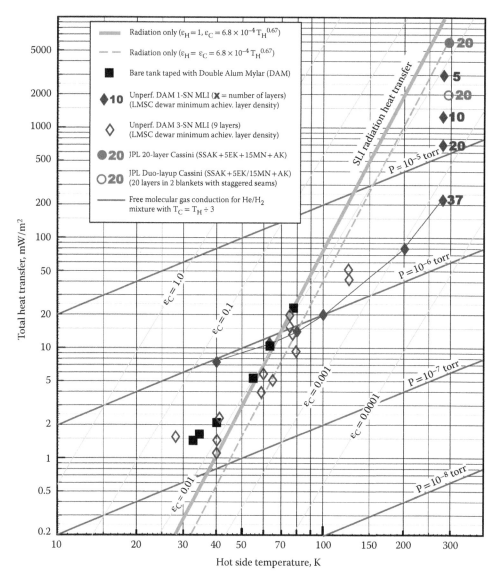

FIGURE 6.28
Radiation heat transfer to a cold polished aluminum body and through various MLI formulations as a function of hot-side temperature. Also shown are contours of constant emittance and background gas pressure.

applied. Although the spacer features minimize the contact between the low-emittance shields, the result is still a relatively high degree of heat conduction through a typical MLI blanket when compared with low-power cryogenic loads.

A key difficulty in estimating thermal radiation loads is that the heat transfer properties of MLI are not easily addressed using analytical models because the effective emittance is strongly affected by a number of physical parameters such as contact pressure between the layers and the level of thermal conductance that results. To eliminate gaseous conduction and surface contamination, MLI and low-emittance shields must also be mounted in a low-pressure vacuum space.

6.4.3.1 Radiation Heat Transfer in Cryogenic Applications

To help understand the behavior of radiation heat transfer in cryogenic applications, it is instructive to examine the heat transfer properties of various surfaces and MLI that have been measured in extensive studies conducted in support of the development of early spacecraft and ground cryogenic applications [Johnson, 1974; Kutzner, 1973]. Heat transfer measurements from these and other studies have been summarized by Nast [1993] and reduced by this author into a single plot (Figure 6.28) so that comparisons and conclusions can be more readily drawn.

At the highest level, Figure 6.28 shows total heat transfer between a hot enclosure of temperature (T_{Hot}) and a cold surface that is enveloped by the hot enclosure. A central feature of the plot is the bold line labeled "SLI Radiation Heat Transfer." This line describes the radiation absorbed by a single-layer insulation (SLI) cold polished-aluminum surface from a facing hot-side surface with temperature (T_{Hot}) that has an emittance of unity (the typical case for a small cold object in a large hot enclosure); it is also assumed that the cold surface is cold enough that reradiation from this surface is negligible compared to the heat absorbed from the hot surface (the usual case).

The SLI bold line follows the relationship ($P \propto T^{4.67}$), where the $T^{4.67}$ term includes the classic ($P \propto T^4$) relationship for radiation heat transfer plus the relationship between the emittance of aluminum and temperature ($\epsilon \propto T^{0.67}$) as described in [Nast, 1993]. In reality, this "emittance" is actually the IR absorptance of the cold surface to the wavelengths emitted by the hot surface. Since emittance equals absorptance at any given temperature, the IR absorptance of the cold surface is computed as the emittance of the cold surface at the hot-side temperature (T_{Hot}).

The resulting $P \propto T^{4.67}$ relationship for radiation heat transfer as a function of temperature is a fundamental part of the classic Lockheed MLI equation widely promoted in the literature. The position of the bold line in the plot corresponds to an emittance (infrared absorptance for room-temperature radiation) of 0.031, the consensus value used for 300 K aluminum MLI surfaces. Lines of constant effective emittance are also drawn on the plot, roughly parallel to the bold line for SLI heat absorptance. The bold SLI line drops faster, reflecting the drop in emittance ($\epsilon \propto T^{0.67}$) with temperature. The dashed line (parallel to the bold line) denotes the heat absorbed by the cold surface when the hot and cold surfaces are closely spaced, parallel, and both have the emittance of polished aluminum.

6.4.3.2 Room-Temperature MLI Performance

Also shown in Figure 6.28 for $T_{Hot} = 300$ K are measured property data for a variety of room-temperature MLI constructions. By room temperature we mean MLI designed for a hot-side temperature near 300 K where heat fluxes are quite large. This is the typical MLI discussed in heat transfer texts and must be carefully distinguished from that used in cryogenic applications with much lower hot-side temperatures.

In general, MLI comes in two distinct designs, that designed to be self-supporting and attached to the outside of an application, and that captured in the internal vacuum space of a dewar. For dewar MLI, the layers are individually stacked with minimal pressure pushing them together; this minimizes the thermal conduction through the MLI stack. In contrast, self-supporting, external MLI, which is usually sewn together, tends to have much higher conduction, and thus has substantially poorer performance than dewar MLI.

Plotted in Figure 6.28 are measured heat transfer data for Lockheed's best (maximum loft) dewar MLI (solid diamonds) [Nast, 2003] and JPL's traditional sewn-through space-craft MLI (solid circular bullet) [Lin et al., 1995]. Note that the 20-layer JPL sewn-through MLI has 10× higher heat transfer than the 20-layer dewar MLI.

As one means of reducing the effective emittance of spacecraft MLI, note that two sewn-through half-blankets with staggered seams (open circular bullet) provides a 3× improvement over a single sewn-through blanket with the same total number of layers [Lin et al., 1995].

6.4.3.3 MLI Performance at Cryogenic Temperatures

Next, consider the MLI data shown in Figure 6.28 for hot-side temperatures of 100 K and below. As the hot-side temperature decreases, the radiation heat transfer drops by $T^{4.67}$ and the conduction term only drops linearly (proportional to T). The net result is that conduction through MLI becomes a critical issue at cryogenic temperatures. This relatively high conductivity at cryogenic temperatures in compounded by any compressive pressure or blanket bending that squeezes the layers together.

For hot-side temperatures below around 100 K, most of the advantage of the multiple layers of even fluffy dewar MLI is lost (see plot data for the 37-layer Lockheed MLI). As a result, Lockheed formulated a special cryo-temperature dewar MLI for hot-side tempera-tures below 100 K (open diamonds). This MLI has only nine aluminized layers and uses three silk nets between each layer to further minimize the conduction between layers. However, even this cryo-MLI leads to only a marginal improvement over just a single low-emittance aluminized surface (black squares) below around 77 K. At the lowest hot-side temperatures, say below 50 K, MLI is often replaced by just bare low-emittance surfaces, as they exhibit comparable or better performance than even the best low-conductivity cryo MLI. If one has to use sewn-through construction for cryo-MLI for an external-surface application, the MLI is likely to act essentially as a single-layer shield (SLI) at cryo tempera-tures. Or, if the MLI does not maintain a smooth surface, its emittance can be much (10×) worse than that of a single well-polished low-emittance surface.

6.4.3.4 Effect of Vacuum Pressure on MLI Conductance

In viewing the measured data for the cryo-MLI shown in Figure 6.28, it is seen that the data begin to diverge and rise above the bold SLI line for hot-side temperatures below about 50 K. When gas conduction effects are examined (see the heat transfer lines shown in Figure 6.28 for various vacuum pressure levels), it is seen that these gas-conduction effects correlate quite well with the observed flattening of the measured MLI heat transfer. This suggests that gas conduction is beginning to dominate the heat transfer at these tem-peratures. In general, operational vacuum environments rarely achieve pressures much below 10^{-6} to 10^{-7} torr, and this pressure correlates quite well with gaseous-conduction heat transfer becoming comparable to and even exceeding radiation heat transfer at hot-side temperatures below 50 K.

6.4.3.5 Effects of Contaminants on Emissivity

The aluminum emittance properties shown in Figure 6.28 are for hyper-clean, smooth polished surfaces. When a surface becomes contaminated with an external high-emittance substance such as a thin film of water ice, the emittance of the surface increases rapidly.

Figure 6.29 shows the sensitivity of surface emissivity to the thickness of common contaminants [Viehmann and Eubanks, 1972]. Note that just a 1-µm deposit of water ice will more than double the emittance (IR absorptance) of a low-emittance surface. Figure 6.3, located earlier in Section 6.2.3, shows data on the allowable maximum pressure levels to prevent condensation of various common gases. Means of computing the rate of buildup of contaminant films are provided by Ross [2003a]. Such computations are useful to gauge the required vacuum level for the interior of vacuum insulation systems and the expected rate at which radiation loads can be expected to build up.

One means of maintaining and improving the vacuum level achievable is to incorporate getters into the vacuum space. A wide variety of getter materials are available that can sorb all active gases such as H_2, O_2, H_2O, CO, CO_2, and N_2 by a chemical reaction under vacuum [Saes, 2007]. Depending on the final application and the production processes, different getter metallic alloys and shapes have been developed. These are used to pump out residual gases that remain in, or develop in the vacuum space over time.

Once contaminated, the well-established approach to deal with radiation shield or MLI degradation is to periodically boil off the contaminants by heating the cryogenic surfaces to near room temperature. However, this deep thermal cycling can be very stressful to other cycled components. To minimize and limit the amount of decontamination cycling, it is prudent that the cryogenic system designer develop a robust system design that is able to accommodate predicted cryogenic load increases due to radiation property degradation over the life expectancy of the application as well as having margin for other degradation mechanisms.

6.4.4 Coldlink Design and Integration Considerations

As part of the cryogenic system design process, there is another set of integration issues that are both thermal and structural. These relate to how the cryogenic cooling source is attached to the cold application. The mechanical attachment to the load is often referred to as the coldlink. This assembly is typically a highly optimized, multidiscipline,

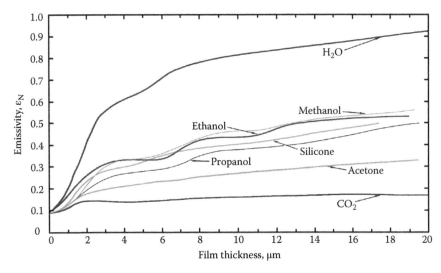

FIGURE 6.29
Emissivity of polished stainless steel to 300 K blackbody radiation versus film thickness.

thermal–mechanical device designed to transmit heat efficiently at cryogenic temperatures, but to also carry out many other functions.

Important coldlink design considerations include:

1. Minimizing the ΔT between the cryogenic load and the cryogen or cryocooler
2. Providing flexibility to accommodate differential motions between the cold load and the cooler cold-load interface
3. Minimizing parasitic conduction and radiative parasitics into the coldlink assembly, including the effects of surface contamination of low-e surfaces
4. Providing for easy attachment and removal of the load from the cooler, or vice versa
5. Providing for active temperature control of the cold load
6. Providing for measurement of the cryogenic load for system diagnostics

Although one's initial focus may be on the coldlink's thermal design issues, the structural issues associated with the coldlink are inseparable and must be fully integrated into the design so that a complete multidiscipline solution of the integrated problem is achieved.

6.4.4.1 Coldlink Thermal Conductance

The primary requirement of any coldlink is to conduct heat between the cold load and the cryocooler interface with minimal thermal drop. One means of doing this is to utilize very-high-conductivity materials in the primary heat conduction path, especially materials that have improved thermal conductivities at cryogenic temperatures. Example materials include high-purity copper and aluminum, silver, and single-crystal materials such as sapphire and silicon. As shown in Figure 6.30, the versions of aluminum and copper that exhibit very high conductivity are those that are exceptionally pure, as reflected in the material's high residual resistivity ratio (RRR). Common materials, such as OFHC copper and 1100 aluminum, have RRR values around 20; these have good conductivity, but not the exceptional conductivity of the hyperpure materials. A point to be drawn from these curves is that hyperpure and single-crystal materials can have very high cryogenic conductivities, but their conductivity is also likely to be extremely sensitive to small levels of impurity elements and crystalline imperfections, such as those acquired from cold working. Measuring the actual conductivity of any high-conductivity materials that are to be used is highly recommended.

As an example use of single-crystal sapphire to achieve exceptionally high conductivity, Figure 6.31 shows the coldlink assembly of the AIRS instrument that conducts heat from the instrument's 58 K focal plane array to its PT cryocoolers [Ross and Green, 1997]. An important point is that when using high-conductivity materials, one must pay particular attention to interface resistances that can often dominate the total thermal resistance of the coldlink assembly. These are the attachment resistances between cooler and flex element, between the flex element and primary conductor, between the primary conductor and load, etc. As shown in Table 6.4, various attachment resistances and flex braid assembly greatly exceed the resistance of the sapphire rod in the AIRS coldlink.

6.4.4.2 Providing Flexibility to Accommodate Differential Motions

A prominent part of nearly all coldlinks is a flex braid or S-link assembly such as that shown at the left end of the coldlink in Figure 6.31. Such a flexible coupling is invariably

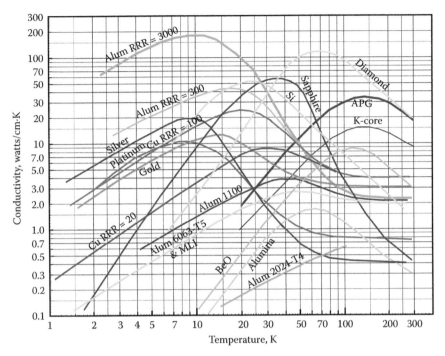

FIGURE 6.30
Thermal conductivity of common high-conductivity materials as a function of temperature.

FIGURE 6.31
AIRS focal plane/cryocooler coldlink assembly with sapphire rod conductor.

TABLE 6.4

Breakdown of AIRS Coldlink Assembly Thermal Resistances

ITEM	Resistance (K/W)
Focal plane to sapphire rod	1.57
Conduction down sapphire rod	0.16
Sapphire rod to moly coupling	0.34
Resistance across shrink-fit joint	0.40
Resistance across flex braid	1.35
Coldblock contact resistance	0.30
Total focal plane/PT thermal resistance	4.12

required to limit structural loads transmitted into the cryocooler or the cryogenic payload due to differential motions between the two. These differential motions can be caused by dimensional differences during assembly, by thermal contractions during cooldown, and by relative motions during transportation or launch. The level of flexibility required is specific to the payload and cooler design; however, typical devices achieve structural stiffnesses around 1–3 N/cm (0.5–1.5 lb./in), and thermal resistances around 1–2 K/W [Sugimura et al., 1995; Williams et al., 1997; Arentz et al., 1995; Kawecki, 1995]. A common design requirement on the flexlink assembly is to maintain the sideloads on the cryo-cooler's cold finger below the level that would cause excessive side deflection, or rubbing and wear of any internal moving elements. This is particularly true for Stirling and GM cryocoolers, which have a tight-tolerance moving displacer internal to the cold finger.

6.4.4.3 Minimizing Coldlink Support Conduction and Radiative Loads

Minimizing parasitic conduction to the coldlink and radiative loads onto its surfaces, including the effects of surface contamination, are an important element of the design of the coldlink assembly. Candidate materials and a load estimation algorithm for achieving low-conductance structural supports were highlighted in Section 6.4.2. However, a key point is minimizing the mass of the coldlink assembly as much as possible to reduce the structural loads that the supports must carry. This is where the use of lightweight conductor materials such as aluminum and sapphire pays off.

Maintaining a low emissivity (low IR absorptivity) on the coldlink surfaces is another priority so as to reduce radiation parasitics. Polishing and gold-plating coldlink surfaces, such as the gold-plated sapphire rod shown in Figure 6.31, is very useful. Flexible elements typically create a greater challenge, as their multilayer makeup typically gives them a naturally high IR absorptivity. These are best wrapped with low-emissivity single-layer insulation (SLI) such as gold or aluminum vapor deposited on Kapton or Mylar.

6.4.4.4 Providing for Easy Attachment and Removal of the Cold Load

With many cryogenic applications, gaining access to fasteners to allow installation and removal of the load from the cooler can be a difficult design issue. This is often resolved by providing a no-fastener blind mating between the cooler and the cold load by including a shrink-fit joint in the coldlink assembly. For example, the joint can be composed of an outer ring made of a high CTE material and a central low-CTE cylinder that is mounted to the mating half of the joint. The differential thermal expansion between the outer ring and the inner cylinder provides an easy slip fit between the two at room temperature, and then a low-ΔT rigid coupling at cryogenic temperatures. Work has been done over the years on pairs of materials that have the necessary high CTE difference and are resistant to cold welding between the mating elements. Classic material pairs include beryllium with copper and aluminum with molybdenum—as is used in the AIRS coupling shown in Figure 6.31 [Ross and Green, 1997]. To get repeatable performance requires careful selection of the gap between the two materials, and a high-quality surface finish on the mating surfaces.

A second area of cold load attachment often involves the use of a soft metal gasket such as indium to improve interfacial conduction in a bolted joint. After a period of time, such a joint can become diffusion bonded to some extent and very difficult to remove without exerting large forces. One means to allow large detachment forces to be applied in a way consistent with the fragility of many coldlink assemblies is to incorporate jack screws into the cold interface. Such screws allow one face to be carefully pried away from the other by tightening the jack screws.

6.4.4.5 *Providing for Active Temperature Control*

Many cryogenic applications require very tight regulation of the cold-load temperature, often down to the millikelvin level. To achieve very tight temperature regulation, one can couple the cold load to the cooler using a passive thermal filter involving combinations of thermal masses and thermal resistances. Another common approach is to provide active temperature control via a small heater on the cold load.

A third approach to cold-load temperature control is to use the variable cooling capability of the mechanical cryocooler to control the temperature via closed-loop control of the cryocooler's drive level. This latter approach has been used for many years in small tactical military Stirling coolers to provide rough (±1 K) control of cold-tip temperature. In the 1990s, this means of temperature control was expanded into space Stirling and PT coolers to provide precise millikelvin control for space-instrument focal plane temperatures [Clappier and Kline-Schoder, 1994]. This form of control does not add to the cold-end heat load, and is commonly available as a feature of many of today's Stirling and PT cryocoolers.

6.4.4.6 *Providing Measurements for Troubleshooting*

Quantifying cryogenic system cooling performance during system integration is an important aspect of system operation and troubleshooting. Invariably, during cooler operation, performance deviates from predictions due to a wide variety of reasons including manufacturing and thermal property variabilities and better definition of the actual application cryogenic loads.

To efficiently focus corrective actions, it is extremely useful to be able to quickly separate cryocooler performance issues from various load-related issues. One means of doing this is to provide a means of directly measuring the cryogenic load at the cooler attachment interface. This can often be done by adding a second temperature sensor to the opposite side of the flexlink element from the cooler cold tip. When combined with the cold-tip temperature sensor, this allows the differential temperature drop across the flexible link to be used as a measure of the heat conduction through the flexlink. The flexlink element can be easily calibrated by using a resistive heater on the cryogenic load side of the interface during cooler ground characterization testing. Because the thermal conductivity of the flexlink will vary with temperature, the calibration needs to be mapped over the temperature range of interest.

6.5 Cryocooler Application and Integration Considerations

In the previous section, generic topics of cryogenic system integration were discussed. However, when a mechanical cryocooler is used as the primary cooling instrument, a number of additional cryocooler-specific integration considerations become important. These include things to do to take the best advantage of cryocooler-unique capabilities, things to do to prolong the life and health of the cooler, and things to do to minimize the impact of less desirable cooler-generated environments.

Key cryocooler-specific integration topics include:

1. Providing for removal of the cryocooler's rejected heat with acceptable thermal gradients. This includes such things as understanding the amount and distribution of heat rejection between the compressor and expander heat rejection interfaces and understanding the dependence of cryocooler cooling performance on its heat sink temperature.

2. Providing structural support to hold and align the cryocooler in its intended application. The mounting structure must address both gravity mounting loads, differential expansion cooldown loads, and transportation loads if the application is to be transported to its final destination. The structural design must also limit the maximum static and dynamic displacements that will be transmitted into the cryocooler's cold-end flexbraid assembly and prevent warping loads that would distort or destroy the tight clearances and alignments of the cryocooler's sensitive piston and/or displacer mechanisms.

3. Managing cryocooler-generated vibration and minimizing any negative interactions between the cooler-generated vibration and the cryogenic application. This includes providing for the measurement of the cryocooler vibratory response through the use of either accelerometers or load cells; these measurements are often an integral part of a Stirling or PT cooler's closed-loop vibration suppression system.

4. Managing the electrical interfaces with the facility power source and managing the effects of any cryocooler-generated ripple currents on the input power bus or radiated EMI.

Each of these four topics is described in detail below. For more in-depth discussion of these and additional considerations of managing cryocooler system reliability through the use of redundancy, the reader is referred to Chapters 12 and 13 of the *Spacecraft Thermal Control Handbook, Vol. II: Cryogenics* [Ross, 2003e,f].

6.5.1 Cryocooler Thermal Interfaces and Heat Sinking Considerations

Integrating cryocoolers into a cryogenic system involves four key heat sinking/thermal issues plus some consideration of cryocooler internal parasitic loads and their sensitivity to gravity orientation.

6.5.1.1 Managing Compressor and Expander Heat Rejection

By their nature, cryocooler compressors draw and thus must dissipate a significant amount of power—typically >100 W for modest-size units and kilowatts for larger machines. Since the efficiency of a cryocooler strongly depends on its heat sink temperature, a fundamental design tradeoff involves increasing the thermal-system size, mass, and complexity to avoid large temperature gradients that increase heat sink temperature and degrade refrigeration performance. Overall, the performance sensitivity closely follows the generic Carnot equation and results in the cold end rising approximately 1 K in temperature for each 5 K rise in the cryocooler heat sink temperature [Ross and Johnson, 1998].

Most cryocoolers prefer to operate in the –20°C to +40°C temperature range, with non-operating temperatures in the range –40°C to 60°C. This temperature range is limited by constraints imposed by the cooler's precision mechanisms, electronics components, and internal construction materials. Often a heat sink temperature around 0°C is considered ideal, but such a low temperature is often not feasible from a system mass/temperature tradeoff point of view. In the end, most cryocooler systems end up requiring a very robust heat-rejection system to achieve the necessary temperature control and heat sink temperatures of around 20°C. As a result, a key lesson learned is to address the heat-rejection system early and thoroughly during the cryogenic system thermal design.

6.5.1.2 Spatial Distribution of Rejected Heat

Given the strong sensitivity to heat sink temperature, understanding where the heat is dissipated is critical to the heat sink design process. Generally, the primary source of the power that must be rejected by a cryocooler comes from the electrical power input to its compressor. In addition, there is a small amount of heat that is absorbed from the cryogenic load. However, for regenerative cryocoolers (e.g., PT and Stirling coolers), a significant fraction (25%–50%) of the power input to the compressor is rejected at the hot end of the expander rather than at the compressor. The division of heat between the compressor and the expander is also influenced by the relative temperatures of the two. If the compressor and expander are run at different temperatures, heat dissipation will shift toward the colder of the two [Kotsubo et al., 1992; Johnson and Ross, 1994].

GM, JT, and Brayton cryocoolers, on the other hand, tend to dissipate all their heat at the compressor. In summary, for sizing heat-rejection systems and calculating thermal gradients between the sensitive cooler stages and external heat sink temperatures it is critical to know how much heat is rejected and at what location.

6.5.1.3 Implications of Heat Sink Temperature on Cold-End Temperature Fluctuations

Because of the fundamental Carnot equation that governs the efficiency of cryocoolers, changing the heat sink temperature directly impacts the cooler thermal efficiency. The net result is that changes in cryocooler heat sink temperature due to any cause are likely to directly reflect into changes of the cooler's cold-end temperature, and thus into fluctuations in the cryogenic load temperature. Common causes of heat sink temperature variability include compressor cooling-loop control cycling, room-temperature variations, orbital heating variations in space, and changes in the power dissipated by the cryocooler. For a typical Stirling-cycle cryocooler, each 5 K change in the cryocooler heat sink temperature will result in the cold load changing approximately 1 K in temperature [Ross and Johnson, 1998].

The bottom line is that tightly controlling heat-sink temperature has a large benefit in helping to minimize cryocooler cold-tip temperature fluctuations and easing the problem of achieving precise control of the cold load temperature via closed-loop cryocooler stroke control or an independent heater on the cold load.

6.5.1.4 Implications of Cryocooler Temperature Control on Thermal Runaway

The strong dependency of cooler performance on heat-sink temperature also raises the possibility of thermal runaway under unfavorable heat sinking conditions. Common heat rejection temperature control modes include constant reject temperature (generally maintained via closed-loop temperature control), heat rejection temperature rising linearly with power dissipation (typical of conduction/convection to a constant-temperature heat sink), and heat rejection temperature dependent on the fourth root of power dissipation (typical for radiation to deep space). Any mode other than constant heat sink temperature can significantly alter the performance attributes of the cooler and raise the possibility of thermal runaway, whereby increased input power required at elevated heat sink temperatures further increases the heat sink temperature in an unstable spiral until maximum cryocooler stroke is exceeded. A good overview and analysis of the thermal stability of various heat rejection modes is described by Ross and Johnson [1998]. As described, operational stability limits are most in danger of being exceeded at high operating powers and the lowest operating temperatures for a given cooler.

In summary, to understand the system-level thermal implications of the cryocooler's operation it is necessary to analyze not just the cryocooler by itself, but the complete cryocooler system including its heat rejection system. In general, the safest cooling system is one that guarantees a roughly constant heat rejection temperature independent of compressor power dissipation.

6.5.1.5 Cold Finger Off-State Conduction

Parasitic conduction down the cold finger of Stirling, PT, and GM refrigerators can become an important design consideration in systems that incorporate nonoperating standby coolers of these types for added redundancy. This is because the parasitic heat load placed on operating coolers by a nonoperating cooler can be a substantial fraction (30%–50%) of the available cooling power of such coolers. Data on representative off-state conduction and means of measuring it are provided using both the transient warm-up rate method [Orlowska and Davey, 1987; Kotsubo et al., 1991] and measuring the actual conductance in a cryogenic conductance measurement facility [Kotsubo et al., 1991]. In general, the second technique is more accurate, as it assures true equilibrium thermal gradients within the off-cooler cold finger and regenerator.

6.5.1.6 PT Sensitivity to Gravity Orientation

For PT cryocoolers, the off-state conduction can additionally have a strong dependence on gravity orientation. This is caused by convection heat transfer occurring in the PT when the cold end of the PT becomes level with or higher than its hot end. Figure 6.32 shows example data for the AIRS PT measured at JPL [Ross, 2003b]. Note that the off-state conduction is a near-constant 0.5 W when the hot end of the pulse is pointed up, but abruptly increases to very high levels as a horizontal attitude is reached. When the hot end is pointed down, convection-induced loads increase to over 3 W.

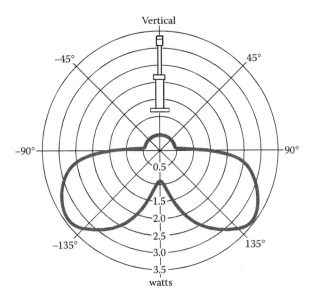

FIGURE 6.32
Thermal conductance of the AIRS PT as a function of angle with respect to gravity.

PT convection also affects the cooling performance of operational PT refrigerators by adding a gravity-dependent convection load onto the operating cooler. This convection loading tends to be quite dependent on the diameter/length aspect ratio of the PT itself and varies with angle similar to that shown in Figure 6.32. For very long and slim PTs the gravity effect is minimal, whereas short, squat PTs can have an appreciable gravity sensitivity. Even for sensitive PTs, the performance is minimally affected as long as the warm end of the PT is 10° or more above horizontal. Typical gravity-sensitivity data for operational PTs is provided by Ross, Johnson, and Rodriguez [2004].

6.5.2 Cryocooler Structural Mounting Considerations

From an engineering perspective, most cryocoolers are sensitive mechanical mechanisms that invariably demand a very well-engineered thermal–mechanical attachment if optimum system performance and long life are to be achieved. For cases where the coldhead is separate from the compressor, there will be a separate set of thermal–mechanical interfaces for the expander.

Key considerations that influence the cryocooler mechanical interface design include the following somewhat diverse items.

6.5.2.1 Surviving Transportation and Launch Acceleration Levels

Typical qualification-level random vibration environments for cryocoolers are around 0.1 g^2/Hz from 100 to 500 Hz, although for some applications, the level can be as high as 0.3 g^2/Hz. In general, cryocooler compressors and displacer "bodies" have little difficulty meeting such vibration environments.

However, a classic problem with cryocoolers is the fragility of the cold finger to vibration loads. This is because the structural robustness of the cold finger is in direct competition with minimizing thermal parasitic loading from conduction down the cold finger.

The result is a highly optimized cold finger structural design that may require some sort of auxiliary vibration restraint or added damping to survive high vibration levels. The resulting force levels are magnified by the attachment of a typical cold-load interface mass, which is often 50–100 g.

There are two common means of limiting vibration loads on the cold finger: cold finger bumper assemblies and add-on damper assemblies.

6.5.2.1.1 Cold Finger Bumper Assemblies

A cold finger bumper assembly, such as that shown in Figure 6.33a, is a separate structural support designed to limit the maximum dynamic deflection of the cold finger during high vibration levels. To avoid imparting a static deflection or parasitic thermal conduction path to the cold finger, the redundant structure provides bumpers that are separated away from the cold finger by a very small gap, typically a few thousandths of an inch. The gap is sized by the maximum deflection that can be withstood by the particular cold finger without risking damage during exposure to high vibration levels. Because the bumper assembly has to be in close proximity to the cold finger, it invariably must become an integral part of the coldlink and cryogenic thermal insulation implementation.

6.5.2.1.2 Cold Finger Damper Assemblies

As an alternative to the cold finger bumper assembly, one can limit the dynamic response of the cold finger to high vibration inputs by adding damping to its motion. One significant source of damping is just the flexbraid or S-link assembly itself. These assemblies provide a modest degree of damping ($Q \approx 20$) due to the internal rubbing that occurs between the assembly's many wires and foils. Often this is enough to allow the cold finger to survive the required vibration levels. However, development testing should be conducted with the selected flex-element to confirm the design's robustness prior to committing to the final hardware implementation. A second means of adding damping is to attach a separate particle damper, such as that shown in Figure 6.33b, to the cold finger's cold stage [Fowler et al., 2001]. Such dampers have the advantage of being tolerant to cryogenic temperatures and imparting no significant thermal loads on the system—just a minimal increase in cold surface area.

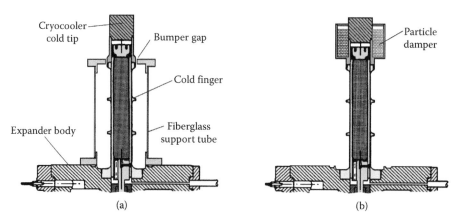

FIGURE 6.33
Example means of adding robustness to vibration loading: (a) cold finger bumper assembly, (b) cold finger damper assembly.

6.5.2.2 *Piston Sensitivity to Low-Frequency Excitation*

Most application structures have their resonant frequencies above 100 Hz as part of achieving sufficient stiffness to satisfy typical handling and alignment demands. In sharp contrast, most Stirling and PT cryocoolers have their fundamental drive frequency tuned in the range of 30–70 Hz, and commonly have highly resonant piston and balancer vibration modes as low as 20–30 Hz when not operating. A particularly sensitive vibration mode is the in-phase piston response of a dual-piston linear compressor when it is unpowered. In this mode, the two pistons travel in the same direction at the same time and do no gas compression. This mode has a strong coupling to low-frequency excitation and has a high amplification factor ($Q \approx 30$). The key issue is whether the excited motions during low-frequency excitation will cause the cooler piston, displacer, or balance motor assemblies to hit their end stops and possibly damage the internal parts or knock the cooler out of alignment.

During the cryosystem design process with such hardware, it is important that these cryocooler low-frequency vibration modes be specifically addressed to insure that the cooler hardware will safely withstand the vibration and transportation environments of the intended application. For cryocoolers intended to be launched into space, it is common to introduce special launch-vibration latches for cryocooler internal drive assemblies. For assemblies with motor drives, a favored means of introducing launch restraint is to short the drive motor coils during launch. Alternatively, some space systems power the coolers during launch and use closed-loop piston servo control to maintain the pistons in a centered position.

In circumstances where motor shorting is not an option, such as with Stirling coolers with unpowered mechanical displacers or balancers, the cooler system design needs to be thoroughly qualified for the low-frequency vibration environment anticipated. Chapter 11, Section 5 of the *Spacecraft Thermal Control Handbook, Vol. II: Cryogenics* presents a detailed discussion of Stirling cryocooler low-frequency vibration modes and means of predicting the expected damping level, with or without motor shorting [Ross, 2003d].

6.5.2.3 *Minimizing Cryocooler Warping Loads*

One of the greatest challenges facing the designers of cryocoolers is achieving long life. A key issue is the possibility of internal gaseous contamination from lubricants and wear products associated with bearings or rubbing surfaces. To address the related issues, there has been an unwritten rule that a multiyear-life cryocooler must avoid rubbing surfaces and maintain a tolerance to small particulate contamination as well. The flexure bearings and piston clearance seals incorporated into the Oxford-style Stirling cryocooler design concept are examples of the application of this rule. To maintain their tight noncontacting piston seals, linear compressors of this type face the challenge of maintaining tight manufacturing and assembly tolerances and a high degree of cooler dimensional stability in all operational environments.

Another important lesson learned in early cooler applications is how difficult it can be to structurally attach to a cryocooler's housing and heat-transfer interface without warping the cooler sufficiently to violate its tight internal running clearances. The act of just bolting a cooler into its support structure can be sufficient to cause rubbing of pistons and displacers in an early Oxford-style Stirling cooler. The culprit is redundant load paths or nonflat mounting surfaces that apply static warping loads into the cooler body. The radial running clearances of the compressor piston and moving displacer of the typical space cryocooler are around 0.0003″. Thus, only a very small permanent deflection of the compressor or displacer structure may be required to cause rubbing and accelerated wear of the compressor piston and/or displacer. With such coolers, one needs to be particularly sensitive about these issues and check at the time of system integration that running clearances have not been violated by warpage associated with the cooler's structural/thermal attachment, or by excessive cold finger side loads.

6.5.2.4 Managing Cryocooler-Generated Vibration

As shown in Figure 6.34, cryocoolers that incorporate low-frequency compressors, such as Stirling, PT, and GM coolers, can generate significant vibratory forces at their drive frequency and at every multiple of that frequency. The key sensitivity is the extent to which the cooler vibration excites mechanical system resonances that degrade application performance with respect to parameters such as optical resolution, pointing accuracy, or electronic noise. The lowest vibration levels are achieved with cryocoolers, such as turbo-Brayton or sorption JT, which do not incorporate low-frequency compressors.

In characterizing cooler-generated vibration, a useful parameter is the peak vibratory force imparted by the cooler into its supports when rigidly mounted. This force is the reaction force to moving masses within the cooler that undergo peak accelerations during various phases of the cooler's operational cycle. The accelerations can be from controlled motion such as the reciprocating motion of the compressor pistons, or natural vibratory resonances of the cooler's elastic structural elements. The example vibration spectrum shown in Figure 6.34 is for a representative Oxford-style linear compressor—both from a single compressor and from a balanced pair of compressors. Notice that the force levels from a balanced pair are relatively constant over the first several harmonics. Similar force levels are found for a single compressor that is balanced using an active or passive counter-balancer.

Although the level of vibratory force that is acceptable is a strong function of the specific application, a value on the order of 0.2 N (0.05 lb.) is fairly representative of that to be expected from typical coolers with two-piston, head-to-head designs. Although the levels shown in Figure 6.34 are for the piston-stroke axis, the measured levels in cross-axes can

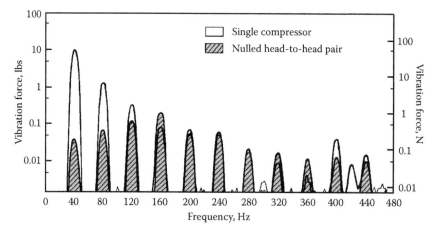

FIGURE 6.34
Example on-axis vibration force generated by an Oxford-style linear compressor operating at 40 Hz with and without using a second piston for head-to-head cancellation.

be very similar in magnitude due to internal resonances within the cooler. An extensive summary of cryocooler vibration characteristics and control methods is presented by Mon et al. [1995] and Ross [2003c].

To achieve further reductions in the on-axis vibration levels, higher-end Stirling and PT cryocoolers, such as those used for space applications, generally incorporate some level of active vibration suppression based on piston stroke control and closed-loop monitoring on the generated vibration. Such systems use active feedback to selectively null each of the first several harmonics in the drive axis by tailoring the electrical input to the individual linear drive motors.

A final category of cooler generated vibration is oscillatory motion of the coldblock itself caused by a combination of pressure-driven elongation of the cold finger at the drive frequency combined with some lateral vibration in response to higher-frequency cooler harmonics. These levels can be similar to the levels generated by a well-balanced compressor [Collins et al., 1995].

6.5.2.4.1 *Providing for Vibration Measurement*

Many space Stirling and PT cryocoolers utilize a closed-loop vibration suppression system to balance the vibratory forces generated by the opposing pistons. Because such a system requires a feedback signal proportional to the unbalanced vibration, either an accelerometer mounted on or near the cryocooler body, or a load cell mounted between the cryocooler and its support structure is required. Both types of feedback transducers have been used successfully. However, an accelerometer typically places fewer requirements on the cryocooler thermal/mechanical integration, and has been found to provide adequate sensitivity to allow effective nulling over the full range of cooler drive harmonics.

If a load-type vibration transducer is used, it must be integrated into the cooler's structural load path so that it shares a portion of the cooler generated vibration forces. At the same time it must be robust enough to survive launch loading levels, and must not lead to a significant thermal impedance in the cryocooler's heat-rejection path.

6.5.2.5 *Shielding Cryocooler Compressor Magnetic Fields*

In another area of integration sensitivity, Stirling and PT cryocoolers invariably result in the compressors being mounted in close proximity to the cold load. This raises the possibility of negative interactions with the compressor's electromagnetic fields, which are maximum at the cooler's 30–70 Hz drive frequency. Substantial data have been gathered over the years on the EMI signatures of space-rated Stirling and PT cryocoolers with respect to their ability to meet the ac magnetic field emissions requirements of the MILSTD-461C RE01 test specification (see, e.g., Johnson et al., 1995, 1999).

Although the cooler magnetic fields have been found to cause no negative interactions with most applications, means have also been developed to greatly suppress the fields using mu-metal shielding attached to the cooler body. Figure 6.35 shows the magnetic field reductions achieved by Johnson et al. using add-on mu-metal shields on the AIRS cryocoolers [Johnson et al., 1999]. A finding from these efforts is that the shields work well if they fully surround the compressor motors, but they need to be separated just enough so that they don't saturate and lose effectiveness. Their close proximity generally causes them to interface directly with the cryocooler support structure and thus become an integral part of the overall cooler structural integration task.

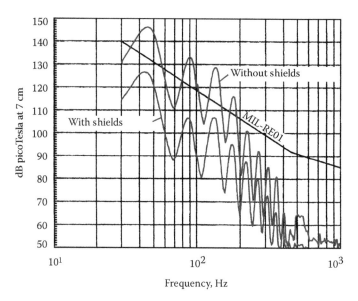

FIGURE 6.35
AC magnetic field emissions measured for the AIRS mechanical cooler with and without the addition of the flight mu-metal shields (versus MIL-STD 461C RE01 requirements).

The key point is for the system integrator to be aware of the sensitivities in this area and diligent in terms of exploring compatibilities and corrective actions prior to finalizing the application's structural/thermal design.

6.5.2.6 Providing for an External Vacuum Environment

Although the cryogenic application itself invariably requires a vacuum environment to be fully operational, an issue that often arises during system integration and testing is the desire to run key portions of the integrated cryocooler plus payload outside of its formal application vacuum chamber where access for measurements and troubleshooting is greatly improved. Providing for this external operational capability is often a useful part of the overall system design effort and may involve such things as the design of temporary vacuum bonnets or purge gas enclosures that allow the cooler to be turned on and run during various short-term tests.

Although a separate temporary vacuum enclosure is generally preferred if the payload design allows it, the purge solution is a viable alternative. However, a purge solution generally does not allow normal operation due to the heavy convection loading caused by the purge gas—typically a zero-moisture purge source such as LN2 boil-off.

6.5.3 Managing Cryocooler Electrical Functions and Interfaces

In addition to providing power to drive the cryocooler, cryocooler electronics are often tasked to provide a number of related functional capabilities such as closed-loop temperature control of the cryogenic stage, closed-loop vibration control, and acquisition and conversion of cooler performance data into digital form with engineering units. Incorporating these diverse functions into the electronics can result in embedded processors, digital logic, sensitive analog circuits, and digital communication interfaces in addition to the needed power-drive circuitry. See, for example, Harvey et al. [2004].

In addition, the cooler drive motors and electronics themselves generally generate electromagnetic signatures or EMI that can adversely interact with the cryogenic application's performance. Understanding and accommodating these diverse functions and their interfaces is important to successfully integrating an active cryocooler into a cryogenic system design.

6.5.3.1 Power Subsystem Considerations

The fundamental task of the cooler electronics power subsystem is to provide and control the power required by the cooler's drive motors. This includes accepting power from the facility power source and converting it into the form (voltage, current, frequency, phasing, etc.) needed to drive the cooler's motors and control their operation. The power source may be a dc source, such as a 28 V spacecraft, or an ac source (single- or three-phase) such as a commercial laboratory or production facility. Adapting the source power to the cooler's needs often involves such things as pulse width modulated (PWM) power amplifier stages, dc/dc converter circuitry, and inverters to generate ac drive waveforms at the cryocooler's drive frequency.

Additional features of the power drive circuitry can include such things as providing full voltage isolation from the power source to eliminate ground loops and noise, and providing high levels of suppression of conducted and radiated EMI from the power amplifier stages. The level of incorporation of EMI suppression and bus isolation are topics that need to be carefully examined and thoughtfully specified by the cooler integrator.

6.5.3.1.1 Power System EMI Considerations

Of the many issues involved in the design of cryocooler electronics, the power subsystem interface is one area in particular that has been a challenging area for many Stirling-cooler systems. The issue here is meeting power system requirements on allowable conducted emissions, particularly low-frequency ripple currents and maximum allowable inrush currents at turn-on [Johnson et al., 1995]. Low-frequency linear drive compressors, such as those used with Stirling and PT cryocoolers, introduce large pulsating currents onto the power bus at twice the cooler operating frequency. Unfortunately, filtering these currents can require massive filters, and not filtering them can result in unacceptable impacts on power system regulation. This has resulted in Stirling coolers connected to dc power systems having to add expensive auxiliary ripple-current filters [Johnson et al., 1999].

A number of the latest high-end Stirling and PT cryocooler electronics have introduced active ripple-current filtering as a cooler-electronics function [Harvey et al., 2004]. This solves the problem with minimal mass and efficiency impact. However, excessive ripple currents remain an issue for many Stirling and PT cooler electronics and must be carefully addressed by the cooler integrator.

Other power-system EMI issues that have been somewhat difficult to meet, but are more traditional in scope, include application requirements on high-frequency conducted and radiated emissions from electronics enclosures and cabling. These can have negative interactions with such things as wireless communication applications. The key point is to be aware of the sensitivities in this area and diligent in terms of exploring compatibilities and corrective actions prior to finalizing the application's system design. Example high-frequency EMI measurements are available in the cooler literature, for example Johnson et al. [1995, 1999].

6.5.3.1.2 Incorporation of Caging Relays

Sometimes embedded within the power drive electronics are caging relays meant to short the drive coils of linear motors to prevent excessive motion or possible damage during such things as peak spacecraft launch loads [Harvey et al., 2004]. Embedding shorting relays within the drive electronics can simplify the overall cryocooler system design by adding this functionality within the existing electronic packaging, cable routing, and communication interfaces, thus eliminating the need for the development of separate hardware to carry out this function.

6.5.3.2 Digital Control and Communication Considerations

In addition to powering the cryocooler's drive motors, the cooler electronics are commonly tasked with a number of operational control functions that are generally managed by an integrated digital controller or microprocessor. Such functions include controlling compressor drive power or piston-stroke level, providing closed-loop temperature control, providing closed-loop vibration control, and providing measurement of cooler control parameters such as cold-tip temperature, compressor piston stroke level, and measured vibration level [Harvey et al., 2004]. Additional functions include monitoring cooler health and safety parameters such as cooler maximum temperatures, excessive current levels, and excessive piston stroke levels.

6.5.3.2.1 Communication Interfaces

The electrical interface to the electronics from the host application generally consists of a serial command and telemetry link. Commands are used to set operating modes and such things as compressor stroke or speed, desired cold-tip temperature, and so on. Commands also request telemetry data, and, if required, diagnostic fault conditions or alarms set by the drive electronics. Communication interfaces and protocol need to be carefully examined and thoughtfully specified by the cooler integrator.

6.5.3.3 Electronics Environmental and EEE Parts Considerations

For Stirling and PT coolers, military and space applications have dominated much of the market focus and led to an emphasis on high-reliability electronics, with broad operational temperature ranges, broad input voltage ranges, and use of high-reliability electronics parts with conservative parts stress derating criteria. Many have also been designed and qualified for relatively severe particle radiation and launch vibration environments. However, the reliability and cost of cryocooler electronics are invariably an issue, and the drive electronics often cost two-thirds of the total cryocooler cost.

The key point is to be aware of the cost and reliability sensitivities in this area and diligent in terms of exploring electronics options and maturity prior to finalizing the cryocooler selection and the application's system design. Even with proven electronics designs, electronic parts selection and acquisition seems to always be an issue because of parts obsolescence issues and application-specific requirements.

6.5.4 Measuring Cryocooler Refrigeration Performance

The ability to cool a refrigeration load to a particular temperature is the most fundamental performance attribute of a cryocooler. Depending on the type of cryocooler, its performance generally depends on a wide variety of performance variables such as cold-load temperature, expander heat-sink temperature, compressor heat-sink temperature, compressor input power, compressor stroke, compressor drive frequency or speed (rpm), expander stroke and phase, and working-fluid fill pressure. As a result, understanding and predicting cryocooler performance in a given application can require an extensive set of performance data, together with knowledge of how to interpolate and extrapolate the data trends for the type of cooler being examined.

Over the years, Oxford-type Stirling and PT coolers have received great attention, and means of generating and displaying thermal performance data for this class of coolers have been highly refined. For example, an entire chapter is devoted to this topic in the *Spacecraft Thermal Control Handbook, Vol. II: Cryogenics* [Ross, 2003d].

However, many times, the performance of a specific cooler model is only specified in the product literature for a single operating point (e.g., 1 W at 80 K with a 20°C heat-sink temperature) even though the same cooler will perform usefully over a broad range of refrigeration temperatures, cooling loads, and operating conditions. The goal of refrigeration performance characterization is to map out this complete operating space of a cryocooler.

6.5.4.1 Measurement Procedures and Equipment

Accurately measuring the performance of a cryocooler requires careful control of the key performance-determining parameters, particularly the heat-sink temperature. For space applications, where cooling by convective heat transfer is generally not feasible, duplicating the expected conductive heat transfer paths and their associated thermal gradients is critically important. During measurements, it is necessary to insure that equilibrium conditions have been reached for a given cooling load prior to acquiring the cooler electrical input power data and the achieved cold-end temperature. Also, it is necessary to protect the cold end from parasitic radiation or conduction loads. This is achieved by shielding the cold end with MLI and by using low conductivity wiring for the cold-end thermometry and heater leads.

To prevent significant gaseous conduction loads, tests need to be conducted in a vacuum that is kept well below 10^{-4} torr. Figure 6.28, in Section 6.4.3, shows a handy means of roughly estimating gaseous conduction loads as a function of vacuum level and hardware surface area. However, much lower vacuum levels are required to prevent long-term buildup of contaminants that degrade the emittance of low-e surfaces and MLI. Minimizing parasitic load buildup during multi-month life tests can easily demand vacuum levels of 10^{-8} torr. Detailed calculational procedures for estimating contaminant buildup versus residual gas pressure levels and the corresponding emittance changes that will occur are presented by Ross [2003a].

An effective means of achieving the low-contaminant environments needed for long-term tests is to envelope the cryogenic hardware under test with a separate cryogenically cooled shield that is maintained well below the dew point of gases that could condense on the test hardware's surfaces. A common means of doing this is to use a separate GM cryocooler that is running at temperatures near 10 K to cool the shields. This prevents degrading contaminants from building up on the cryogenic surfaces of the test hardware.

6.5.4.2 Data Presentation Formats and Examples

A common means of displaying the refrigeration performance of regenerative cryocoolers such as Stirling, PT, and GM coolers, is a load-line plot. Such plots display the cryogenic load-carrying capability of a cryocooler as a function of its refrigeration temperature. The plot may include a second parameter such as stroke, input power, or heat-sink temperature. Since the heat-sink temperature has a strong effect, it is important to indicate on the plot the value of the heat-sink temperature used for the measurements. Figure 6.36 shows an example load-line plot for a Stirling cooler with a heat-sink temperature of 20°C and with the compressor piston stroke amplitude as a parameter.

As shown in Figure 6.36, cooling load versus temperature lines are generally quite linear, and this allows easy interpolation and extrapolation to additional conditions of interest from only a few measurements. A key attribute of each of the load lines is their slope and the temperature corresponding to zero refrigeration load (the "no-load" temperature). However, a weakness of this type of plot is that it does not provide information to conveniently understand the dependence of the measured cooling performance on input power or information on thermodynamic efficiency.

6.5.4.3 Multiparameter Plots

To expand the ability to view a cryocooler's efficiency performance dependencies simultaneously with its cooling capacity, refrigeration performance can be presented as input power versus cooling power, as shown in Figure 6.37. Radial lines from the origin are then lines of constant specific power, that is, watts of cooling power generated per watt of input power—a measure of efficiency. Refrigeration temperature is then displayed as a family of isotherms to allow easy determination of information over a broad range of temperatures. Since the input voltage and compressor stroke generally have a smooth dependency on the input power, these curves are often added, as shown in Figure 6.37.

The multiparameter-plot format can easily display five parameters simultaneously on a single two-dimensional plot; this greatly aids in understanding the strong interrelationships among many cryocooler performance variables. Such plots also allow documenting cryocooler performance dependence on other key operational variables, such as drive frequency, heat-sink temperature, and fill pressure [Johnson et al., 1997]. Even though it requires a bit more effort to generate such plots, building on the well-behaved load-line

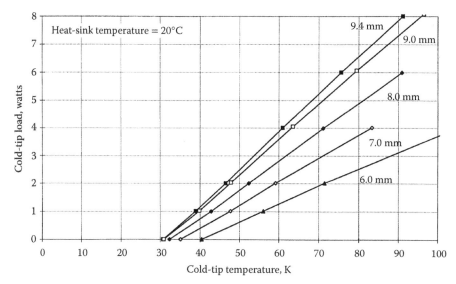

FIGURE 6.36
Example Stirling cooler load-line plot with piston stroke as a parameter.

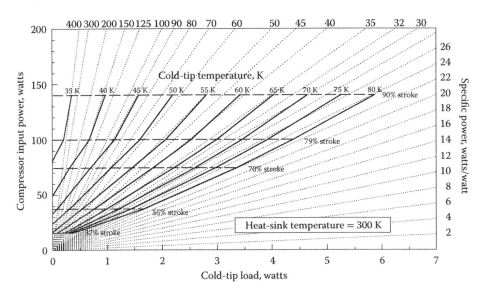

FIGURE 6.37
Multiparameter plot for the AIRS PT cryocooler with piston stroke as a parameter.

characteristics, only three to four measurement points are needed per stroke load line. Means of efficiently generating such plots are presented in Appendix A of the *Spacecraft Thermal Control Handbook, Vol. II: Cryogenics* [Ross, 2003g].

6.5.4.3.1 Displaying the Effect of Heat-Sink Temperature

Drawing from the Carnot relationship presented earlier as Equation 6.2 in Section 6.3.1.2, heat-sink temperature has a large influence on cryocooler performance. Using the multiparameter plot, one can plot this dependence for a measured cryocooler and develop a useful algorithm for computing the refrigeration performance dependence on heat-sink temperature.

Figure 6.38 shows the heat sink temperature dependency for a representative PT cryocooler by displaying the shift in the cooler performance caused by lowering the heat-sink temperature from 20°C to 0°C. Note that the new isotherms (lines of constant refrigeration temperature) for a 0°C heat-sink temperature are positioned on top of similar 20°C-isotherms corresponding to refrigeration temperatures approximately 3 K warmer; that is, the performance at 55 K with a 0°C heat-sink temperature is the same as the performance at 58 K with a 20°C heat-sink temperature. Note that this shift is also quite constant over the complete range of refrigeration temperatures plotted.

For this particular cooler, the proportionality constant ($\Re$) between heat-sink temperature and cold-end temperature is 20°C/3 K ≈ 7. This fixed relationship between cold-tip temperature and change in heat-sink temperature has been found to be approximately true for a wide variety of cryocoolers, not just the PT cooler shown here. However, the actual value of the proportionality ($\Re$) is found to vary somewhat between cooler models from a low value of around 2°C/K to a high value of around 10°C/K [Ross and Johnson, 1998]. If measured values do not exist for a refrigerator of interest, a good mean value for estimating purposes is around 5°C/K.

Based on this empirically derived finding between cold-tip temperature (Θ) and heat-sink temperature (T), one can derive the cooling power $P(\Theta_A, T_A)$ at cold-end temperature

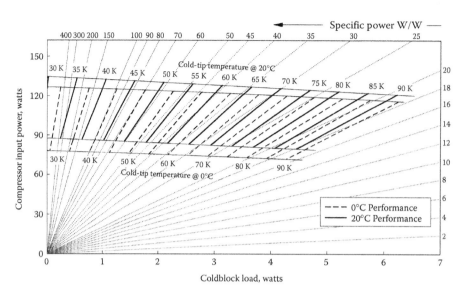

FIGURE 6.38
Multiparameter plot generated with data taken at two heat-sink temperatures.

Θ_A and heat-sink temperature T_A as equal to the cooling power $P(\Theta_B, T_0)$ at cold-end temperature Θ_B at the baseline heat-sink temperature T_0, that is

$$P(\Theta_A, T_A) = P(\Theta_B, T_0); \quad \text{where } \Theta_B = \Theta_A - (T_A - T_0)/\mathfrak{R} \qquad (6.8)$$

where $\mathfrak{R}$ is the measured change in heat-sink temperature required to shift the cold-end performance by 1 K. Equation 6.8 allows a performance plot such as that shown in Figure 6.37 to be used to determine refrigeration performance for a broad range of heat-sink temperatures.

6.6 Summary

This chapter has provided an overview of the common means of achieving cryogenic temperatures, including both passive systems involving the use of liquid and frozen cryogens, as well as active cryocoolers. In addition, the critical aspects of cryogenic cooling system design and sizing—including load estimation and margin management—and cryocooler application and integration considerations have been reviewed. Because this field is very extensive, a single chapter precludes delving into specific details and must by necessity be limited to providing an introductory description of the available technologies and to summarizing the key decision factors and engineering considerations in the acquisition and use of cryogenic cooling systems.

For the process of pursuing this topic in more detail, the reader is directed to the many references in the References section of this chapter, but even more important, to the prominent journals and proceedings that they are published in. Within the cryogenics community, there is a relatively finite number of prominent conferences and journals covering the field. Within the United States, there are two primary conferences held every 2 years on staggered years: the Cryogenic Engineering Conference/International Cryogenic Materials Conference (CEC/ICMC), which publishes the *Advances in Cryogenic Engineering* proceedings, and the International Cryocooler Conference, which publishes the *Cryocoolers X* proceedings. Both of these conferences have been active for more than 30 years and contain a wealth of knowledge in their past proceedings. Complementing these conference proceedings is the monthly journal *Cryogenics*, which publishes a wealth of cryogenics-related articles, including the proceedings of the Space Cryogenics Workshop, which is held every 2 years with a specific focus on cryogenics applications in space.

Also available, but somewhat less accessible, are the proceedings of the International Cryogenic Engineering Conference/International Cryogenic Materials Conference (ICEC/ICMC) that is held every 2 years, typically in Europe or Asia.

In terms of texts, the *Spacecraft Thermal Control Handbook, Vol II: Cryogenics* provides an extensive review of cryogenics technologies associated with space missions [Donabedian, 2003a]. The key drawback here is the absence of data on ground-based technologies such as GM cryocoolers.

Acknowledgments

Some of the research reported in this chapter was conducted at the Jet Propulsion Laboratory (JPL), California Institute of Technology—under a contract with the National Aeronautics and Space Administration. The author would like to express his appreciation of the many scientists and engineers who contributed to the reported technologies, and would like to express a special thanks to Dr. Dean Johnson, Jet Propulsion Laboratory, Saul Miller, The Aerospace Corp., Jeff Raab, Northrup Grumman AS, and Dr. Chao Wang, Cryomech, for reviewing this chapter and providing valuable technical comments and suggestions.

References

Arentz, R.F., et al., Design and verification of Stirling cooler interfaces suitable for long-lifetime, space-borne sensor systems, *Cryocoolers 8*, Plenum Publishing Corp, New York, NY (1995), pp. 855–867.

Arkhipov, V.T., et al., Multicomponent gas mixtures for J-T cryocoolers, *Cryocoolers 10*, R. Ross Jr., ed., Kluwer Academic/Plenum Publishers, New York, NY (1999), pp. 487–495.

Boiarski, M.J., Brodianski, V.M. and Longsworth, R.C., Retrospective of mixed refrigerant technology and modern status of cryocoolers based on one-stage, oil-lubricated compressors, *Advances in Cryogenic Engineering*, Vol. 43, Plenum Press, New York, NY (1998), pp. 1701–1708.

Bonney, G.E. and Longsworth, R.C., Considerations in using Joule-Thomson coolers, *Proceedings of the Sixth International Cryocoolers Conference*, Vol. 1, David Taylor Research Center, Bethesda, MD (1990), pp. 231–244.

Bradley, P., Radebaugh, R., Huber, M., Lin, M. and Lee, Y., Development of a mixed-refrigerant Joule-Thomson microcryocooler, *Cryocoolers 15*, ICC Press, Boulder, CO (2009), pp. 425–432.

Bradshaw, T.W., Delderfield, J., Werrett, S.T. and Davey, G., Performance of the Oxford miniature Stirling cycle refrigerator, *Advances in Cryogenic Engineering*, Vol. 31, Plenum Press, New York, NY (1986), pp. 801–809.

Bradshaw, T.W. and Orlowska, A.H., A closed-cycle 4K mechanical cooler for space applications, *Proc. of 9th European Symposium on Space Environmental Control Systems*, Florence, Italy (1991).

Bradshaw, T.W., Orlowska, A.H. and Jewell, C., Life test and performance testing of a 4K cooler for space applications, *Cryocoolers 10*, Kluwer Academic, New York, NY (1999), pp. 521–528.

Clappier, R.R. and Kline-Schoder, R.J., Precision temperature control of Stirling-cycle cryocoolers, *Advances in Cryogenic Engineering*, Vol. 39, Plenum Press, New York, NY (1994), pp. 1177–1184.

Collins, S.A., Johnson, D.L., Smedley, G.T. and Ross, R.G., Jr., Performance characterization of the TRW 35K pulse tube cooler, *Advances in Cryogenic Engineering*, Vol. 41B, Plenum Publishing Corp., New York, NY (1995), pp. 1471–1478.

Crawford, L., Radiant coolers, *Spacecraft Thermal Control Handbook, Vol. II: Cryogenics*, M. Donabedian, ed., The Aerospace Press, El Segundo, CA (2003), pp. 55–90.

Dobak, J., Yu, X. and Ghaerzadeh, K., A novel closed-loop cryosurgical device, *Advances in Cryogenic Engineering*, Vol. 43, Plenum Press, New York, NY (1998), pp. 897–902.

Dolan, E.X., Swift, W.L, Tomliison, B.J., Gilbert, A. and Bruning, J., A single stage reverse Brayton cryocooler: Performance and endurance tests on the engineering model, *Cryocoolers 9*, R. Ross Jr., ed., Plenum Press, New York, NY (1997), pp. 465–474.

Donabedian, M., Chapter 15: Cooling systems, *The Infrared Handbook*, revised edition, IRIA Series in Infrared & Electro-Optics, George J. Zissis and William L. Wolfe, eds., IRIA Center, Washington, DC (1993), pp. 15–1 to 15–85.

Donabedian, M., *Spacecraft Thermal Control Handbook, Vol. II: Cryogenics*, The Aerospace Press, El Segundo, CA (2003a).

Donabedian, M., Chapter 2: Cryogenic fluid storage, *Spacecraft Thermal Control Handbook, Vol. II: Cryogenics*, The Aerospace Press, El Segundo, CA (2003b), pp. 13–29.

Donabedian, M., Chapter 6: Cryogenic radiator design and comparative performance, *Spacecraft Thermal Control Handbook, Vol. II: Cryogenics*, M. Donabedian, ed., The Aerospace Press, El Segundo, CA (2003c), pp. 91–117.

Fowler, B.L., Flinta, E.M., Olson, S.E., *Design Methodology for Particle Damping*, Proc. SPIE 4331, Smart Structures and Materials 2001: Damping and Isolation, 186, July (2001).

Gifford, W.E. and Longsworth, R.C., Pulse tube refrigeration progress, *Advances in Cryogenic Engineering*, Vol. 10B, Plenum Press, New York, NY (1965), pp. 69–79.

Harris, R., Chenoweth, J., and White, R., *Cryocooler Development for Space Flight Applications, SPIE Technical Symposium Paper 280-10*, Washington, DC, April (1981).

Harvey, D., et al., Advanced cryocooler electronics for space, *Cryogenics*, Vol. 44, Issues 6–8, June–August (2004), pp. 589–593.

Johnson, D.L. and Ross, R.G., Jr., Spacecraft cryocooler thermal integration, *Proceedings of the Spacecraft Thermal Control Symposium*, Albuquerque, NM, November (1994).

Johnson, D.L., Smedley, G.T., Mon, G.R., Ross, R.G., Jr. and Narvaez, P., Cryocooler electromagnetic compatibility, *Cryocoolers 8*, Plenum Publishing Corp., New York, NY (1995), pp. 209–220.

Johnson, D.L., Collins, S.A., Heun, M.K. and Ross, R.G., Jr., Performance characterization of the TRW 3503 and 6020 Pulse Tube Coolers, *Cryocoolers 9*, Plenum Publishing Corp., New York, NY (1997), pp. 183–193.

Johnson, D.L., Collins, S.A. and Ross, R.G., Jr., EMI performance of the AIRS cooler and electronics, *Cryocoolers 10*, Plenum Publishing Corp., New York, NY (1999), pp. 771–780.

Johnson, W.R., *Final Report, Thermal Performance of Multilayer Insulations*, NASA CR-134477, LMSC-D349866, prepared for NASA LeRC by Lockheed Missiles and Space Co., 5 April (1974).

Kawecki, T., High Temperature Superconducting Space Experiment II (HTSSE II) overview and preliminary cryocooler integration experience, *Cryocoolers 8*, Plenum Publishing Corp., New York, NY (1995), pp. 893–900.

Konkel, C. and Bradley, W., Design and qualification of flight electronics for the HST NICMOS reverse Brayton cryocooler, *Cryocoolers 10*, R.G. Ross Jr., ed., Kluwer Academic Plenum Publishers, New York, NY (1999), pp. 439–448.

Kotsubo, V., Johnson, D.L. and Ross, R.G., Jr., Cold-tip off-state conduction loss of miniature Stirling cycle cryocoolers, *Advances in Cryogenic Engineering*, Vol. 37B, Plenum Press, New York, NY (1991), pp. 1037–1043.

Kotsubo, V.Y., Johnson, D.L., and Ross, R.G., Jr., Calorimetric thermal-vacuum performance characterization of the BAe 80 K space cryocooler, SAE Paper No. 929037, *Proceedings of the 27th Intersociety Energy Conversion Engineering Conference*, P-259, Vol. 5, San Diego, CA, August 3–7, (1992), pp. 5.101–5.107.

Kutzner, K., Schmidt, F. and Wietzke, Radiative and conductive heat transmission through superinsulations—Experimental results for aluminum coated plastic films, *Cryogenics*, July (1973), pp. 396–404.

Lin, E.I., Stultz, J.W., and Reeve, R.T., Test-derived effective emittance for Cassini MLI blankets and heat loss characteristics in the vicinity of seams, AIAA paper 95-2015, 30th AIAA Thermophysics Conference, San Diego, CA, June 19–22 (1995).

Longsworth, R.C. and Steyert, W.A., JT refrigerators for fast cooldown to < 100 K and < 80 K, *Proceedings of the Third Interagency Meeting on Cryocoolers*, David Taylor Research Center, Bethesda, MD (1988), pp. 133–148.

Longsworth, R.C., Boiarski, M.J., and Klusmier, L.A., Closed cycle throttle refrigerator, *Cryocoolers 8*, Plenum, New York, NY (1995), pp. 537–541.

Longsworth, R.C., 80 K throttle-cycle refrigerator cost reduction, *Cryocoolers 9*, Plenum, New York, NY (1997), pp. 521–528.

Longsworth, R.C., Considerations in applying open cycle JT cryostats to cryosurgery, *Cryocoolers 11*, Kluwer Academic Plenum Press, New York, NY (2001), pp. 783–792.

Marquardt, E.D., Radebaugh, R. and Dobak, J., A cryogenic catheter for treating heart arrhythmia, *Advances in Cryogenic Engineering*, Vol. 43, Plenum Press, New York, NY (1998), pp. 903–910.

Maytal, B.-Z. and Pfotenhauer, J.M., *Miniature Joule-Thomson Cryocooling: Principles and Practice*, Springer (2012).

Miller, C.D., Development of the long-lifetime solid nitrogen dewar for NICMOS, *Advances in Cryogenic Engineering*, Vol. 43, New York, NY: Plenum Press (1998a), pp. 927–933.

Miller, C.D., Pre- and post-launch performance of the NICMOS dewar, *Advances in Cryogenic Engineering*, Vol. 43, Plenum Press, New York, NY (1998b), pp. 935–940.

Mon, G.R., Smedley, G.T., Johnson, D.L. and Ross, R.G., Jr., Vibration characteristics of Stirling-cycle cryocoolers for space application, *Cryocoolers 8*, Plenum Publishing Corp., New York, NY (1995), pp. 197–208.

Nast, T.C. and Murray, D.O., Orbital cryogenic cooling of sensor systems, AIAA Paper 76-979, *Proceedings of the Systems Designs Driven by Sensors, AIAA Technical Specialists Conference*, Pasadena, CA, October 19-20 (1976).

Nast, T.C., A review of multilayer insulation theory, calorimeter measurements, and applications, *Recent Advances in Cryogenic Engineering - 1993*, ASME HTD-Vol. 267, J.P. Kelley and J. Goodman, eds., American Society of Mechanical Engineers, New York, NY (1993), pp. 29–43.

Orlowska, A.H. and Davey, G., Measurement of losses in a Stirling-cycle cooler, *Cryogenics*, vol. 27 (1987), pp. 645–651.

Petach, M. and Michaelian, M., Mid infrared instrument (MIRI) cooler cold head assembly acceptance testing and characterization, *Cryocoolers 18*, ICC Press, Boulder, CO (2014), pp. 11–17.

Raab, J. and Tward, M., Northrop Grumman aerospace systems cryocooler overview, *Cryogenics*, Vol. 50, Issue 9, September (2010), pp. 572–581.

Radebaugh, R., Zimmerman, J. and Smith, D.R., A comparison of three types of pulse tube refrigerators: New methods for reaching 60 K, *Advances in Cryogenic Engineering*, Vol. 31, Plenum Press, New York, NY (1986), pp. 779–789.

Radebaugh, R. Refrigeration for superconductors, *Proc. IEEE, Special Issue on Applications of Superconductivity*, vol. 92, October (2004), pp. 1719–1734.

Ravex, A., Trollier, T., Sentis, L., Durand, F., Crespi, P., Cryocoolers development and integration for space applications at air liquide, *Proceedings of the Twentieth International Cryogenic Engineering Conference (ICEC 20)*, Elsevier (2005), pp. 427–436.

Ross, R.G., Jr. and Green, K.E., AIRS cryocooler system design and development, *Cryocoolers 9*, Plenum Publishing Corp., New York, NY (1997), pp. 885–894.

Ross, R.G., Jr. and Johnson, D.L., Effect of heat rejection conditions on cryocooler operational stability, *Advances in Cryogenic Engineering*, Vol. 43B (1998), pp. 1745–1752.

Ross, R.G., Jr., Cryocooler load increase due to external contamination of low-ϵ cryogenic surfaces, *Cryocoolers 12*, Kluwer Academic/Plenum Publishers, New York, NY (2003a), pp. 727–736.

Ross, R.G., Jr., AIRS pulse tube cooler system level performance and in-space performance comparison, *Cryocoolers 12*, Kluwer Academic/Plenum Publishers, New York, NY (2003b), pp. 747–754.

Ross, R.G., Jr., Vibration suppression of advanced space cryocoolers—An overview, *Proceedings of the International Society of Optical Engineering (SPIE) Conference*, San Diego, CA, March 2-6, (2003c).

Ross, R.G., Jr., Chapter 11: Cryocooler performance characterization, *Spacecraft Thermal Control Handbook, Vol. II: Cryogenics*, The Aerospace Press, El Segundo, CA (2003d), pp. 217–261.

Ross, R.G., Jr., Chapter 12: Cryocooler reliability and redundancy considerations, *Spacecraft Thermal Control Handbook, Vol. II: Cryogenics*, The Aerospace Press, El Segundo, CA (2003e), pp. 263–284.

Ross, R.G., Jr., Chapter 13: Cryocooler integration considerations, *Spacecraft Thermal Control Handbook, Vol. II: Cryogenics*, The Aerospace Press, El Segundo, CA (2003f), pp. 285–324.

Ross, R.G., Jr., Appendiex A: Constructing a cryocooler multiparameter plot, *Spacecraft Thermal Control Handbook, Vol. II: Cryogenics*, The Aerospace Press, El Segundo, CA (2003g), pp. 605–608.

Ross, R.G., Jr., Estimation of thermal conduction loads for structural supports of cryogenic spacecraft assemblies, *Cryogenics*, Vol. 44, Issue: 6–8, June–August, (2004), pp. 421–424.

Ross, R.G., Jr., Johnson, D.L. and Rodriguez, J.I., Effect of gravity orientation on the thermal performance of Stirling-type pulse tube cryocoolers, *Cryogenics*, Vol. 44, Issue: 6–8, June–August (2004), pp. 403–408.

Ross, R.G., Jr., Chapter 11: Aerospace coolers: A 50-year quest for long-life cryogenic cooling in space, *Cryogenic Engineering: Fifty Years of Progress*, K. Timmerhaus and R. Reed, eds., Springer Publishers, New York, NY (2007), pp. 225–284.

Ross, R.G., Jr., Johnson, D.L., Elliott, D., AIRS pulse tube coolers performance update–Twelve years in space, *Cryocoolers 18*, ICC Press, Boulder, CO (2014), pp. 87–95.

Saes, *St 171 and St 172 - Sintered Porous Getters*, SAES Getters Group (2007).

Scott, Russell B., *Cryogenic Engineering*, Met-Chem Research, Inc., Boulder, CO (1988), pp. 146–147.

Selzer, P.M., Fairbank, W.M. and Everitt, C.W.F., A superfluid plug for space, *Advances in Cryogenic Engineering*, Vol. 16, Plenum Press, New York, NY (1971), pp. 277–281.

Sherman, A., History, status and future applications of spaceborne cryogenic systems, *Advances in Cryogenic Engineering*, Vol. 27, Plenum Press, New York, NY (1982), pp. 1007–1029.

Sixsmith, H., Miniature cryogenic expansion turbines—A review, *Advances in Cryogenic Engineering*, Vol. 29, Plenum Press, New York, NY (1984), pp. 511–523.

Sugimura, R.S., Russo, S.C. and Gilman, D.C., Lessons learned during the integration phase of the NASA IN-STEP Cryo system experiment, *Cryocoolers 8*, Plenum Publishing Corp., New York, NY (1995), pp. 869–882.

Swift, W. and Sixsmith, H. Performance of a long life reverse Brayton cryocooler, *7th International Cryocooler Conference Proceedings*, Air Force Phillips Laboratory Report PL-CP—93-1001, Kirtland Air Force Base, Albuquerque, NM, April (1993), pp. 84–97.

Swift, W.L., Dolan, F.X. and Zagarola, M.V., The NICMOS Cooling System—5 years of successful on-orbit operation, *Advances in Cryogenic Engineering*, Vol. 53, Amer. Institute of Physics, Melville, NY (2008), pp. 799–806.

ter Brake, H.J.M., et al., 14.5 K hydrogen sorption cooler: Design and breadboard tests, *Cryocoolers 16*, ICC Press, Boulder, CO (2011), pp. 445–454.

Viehmann, W. and Eubanks, A.G., *Effects of Surface Contamination on the Infrared Emissivity and Visible-Light Scattering of Highly Reflective Surfaces at Cryogenic Temperatures*, NASA Technical Note TN D6585, NASA Goddard Space Flight Center, February (1972).

Wade, L.A., et al., Hydrogen sorption cryocoolers for the Planck mission, *Advances in Cryogenic Engineering*, Vol. 45, Plenum Press, New York, NY (2000), pp. 499–506.

Wang, C. and Gifford, P.E., Two-stage pulse tube cryocoolers for 4 K and 10 K operation, *Cryocoolers 12*, Kluwer Academic/Plenum Press, New York, NY (2003), pp. 293–300.

Wang, C., Characteristics of 4 K pulse tube cryocoolers in applications, *Proceedings of the Twentieth International Cryogenic Engineering Conference* (ICEC 20), Beijing, China: Elsevier, Ltd. (2005), pp. 265–268.

Werrett, S.T., Peskett, G.D., Davey, G., Bradshaw, T.W. and Delderfield, J., Development of a small Stirling cycle cooler for space applications, *Advances in Cryogenic Engineering*, Vol. 31 (1986), pp. 791–799.

Williams, B., Jensen, S., and Batty, J.C., An advanced solderless flexible thermal link, *Cryocoolers 9*, Plenum Publishing Corp., New York, NY (1997), pp. 807–812.

Xu, M.Y., Yan, P.D., Koyama, T., Ogura, T., and Li, R., Development of a 4 K two-stage pulse tube cryocooler, *Cryocoolers 12*, Kluwer Academic/Plenum Press, New York, NY (2003), pp. 301–307.

Zagarola, M., McCormick, J., Cragin, K., Demonstration of an ultra-miniature turboalternator for space-borne turbo-Brayton cryocooler, *Cryocoolers 17*, ICC Press, Boulder, CO (2012), pp. 453–460.

7

Cryogenic Piezoelectric Materials for Transducer Applications

Shujun Zhang, Hyeong Jae Lee, Xiaoning Jiang,
Wenbin Huang, Stewart Sherrit, and Matthew M. Kropf

CONTENTS

7.1 Introduction

In recent years, there is growing scientific research interest and industrial demands relating to sensors, actuators, motors, and other transducer devices operable at the cryogenic temperature range, which is commonly defined as below –100°C [Barriere et al., 2013; Kogbara et al., 2014; Weisensel et al., 1998]. In particular, industrial, aerospace, and medical cryogenic systems require innovative transduction devices to meet their performance and affordability goals. The coldest spot on Earth has been reported to be near Dome Argus in East Antarctica with a temperature of –93.2°C, which set a record for natural cold in August 2010 [Scambos, 2013]. Even though cryogenic environments do not occur naturally on Earth, there are a variety of existing terrestrial applications for cryogenic technology including superconducting magnets in nuclear magnetic resonance spectroscopy systems [Smith and Blandford, 1995] and magnetic resonance imaging systems [Bushong, 2003], sensitive infrared detectors [Druart et al., 2008], cryomilling [Witkin and Lavernia, 2006], material processing, hardening [Kalia, 2010; Zurecki, 2005], and quenching [Barron, 1982; Zurecki, 2005]. Many of these systems require actuators to control structure shapes, to operate local valves to control cryofluid flows, or to position structures during operation.

In the aerospace area, there are a variety of potential mission targets identified in the Decadal Survey that have cryogenic environments [Squyres, 2011]. For example, unlike Earth, other bodies in the solar system do reach the cryogenic temperature range, as shown in Figure 7.1. The outer planets including the dwarf planet Pluto and the moons of Saturn and Neptune all reach temperatures below –150°C. Even though the average temperatures on the Moon and Mars are within the range of Earth temperatures, the lack of substantial atmosphere on these bodies produces extreme temperature ranges with the poles of Mars reaching –140°C and the nighttime temperature of the Moon reaching –170°C. Although Mercury is the closest planet to the Sun, it has temperature extremes that reach –173°C. The need for reliable surface sampling and handling technology has been identified as a challenge that must be met if sampling missions to these destinations are to move past the planning stage. As well as the standard actuators required for exploration systems [Sherrit, 2005], novel actuators and sensors are required for valves [Sherrit et al., 2014], vibrators [Sherrit et al., 2009], drills [Bar-Cohen and Zacny, 2009], penetrators [Bao et al., 2006], sonar [Towner et al., 2006], microphones [Fulchignoni and Ferri, 2006], and other transduction applications.

Another space application for cryogenic actuators is in infrared telescopes that may be ground-based, airborne, or space telescopes. The common component of each of these telescopes is the infrared solid-state detector that must be cooled to cryogenic temperatures. In addition, infrared space telescopes require cryo-cooling systems to maintain sensitivity, which requires the use of valves. A variety of missions have been proposed

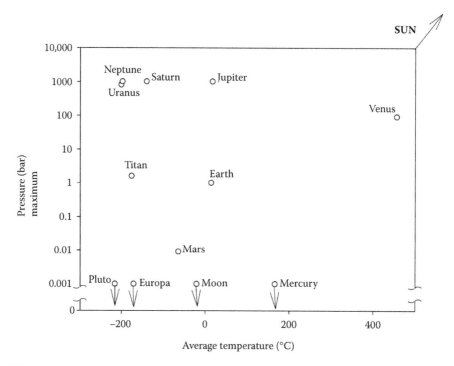

FIGURE 7.1

Estimated pressure and temperature for a selection of proposed mission destinations. (From Sherrit, S., *Conference on Smart Structures and Materials*, International Society for Optics and Photonics, 2005, pp. 335–346.)

using infrared cameras or detectors including SIRTF/SAFIR, the mid infrared instrument (MIRI) of the James Webb Space Telescope (JWST) [Stockman, 1997]. JWST, previously known as Next Generation Space Telescope (NGST), was designed as a large space infrared telescope to achieve a 1000 times greater sensitivity than any currently existing or planned facilities [Stockman, 1997]. The design requirements include a large primary mirror (6–8 m diameter), orbit at the second Lagrange point (L2), and an operation temperature between 35 and 65 K. Due to the long distance (1.5 million kilometers) between the telescope and the space shuttle, JWST's orbit will be unserviceable, rendering the active optics a necessity. Cryogenic actuators are used to position the mirror segments by image plane wavefront sensing. Occasional refocus is needed every few days after the initial configuration. Based on the three-mirror anastigmatic design, the telescope has secondary and tertiary mirrors to deliver images that are free of optical aberrations, which could be achieved using adaptive optics consisting of cryogenic deformable mirrors. Cryogenic actuator arrays with large stroke, high force, high resolution and low power consumption become very critical for the development of the deformable mirrors. On the other side, for infrared detection, a cryogenic environment could greatly diminish the noise caused by the Stefan–Boltzmann law, which provides a huge advantage in the sensitivity over those detectors that are not cooled. Cryogenic sensors, actuators, and motors are hence essential for proper operation of these space instruments.

The Spitzer Space Telescope [Finley et al., 2004], shown in Figure 7.2, has detectors that must be cooled to only about 5° above absolute zero to ensure that the telescopes internal heating does not interfere with its observations of cold cosmic objects. To accomplish this, the telescope has a cryostat of liquid helium that requires low-power cryogenic actuators to open venting valves.

Some scientific instruments in which cryogenic devices are becoming more common include probes, stages, and scanners of the scanning tunneling microscope (STM) and the atomic force microscope (AFM) for surface science study [Saitoh et al., 2009], shown in Figure 7.3. The precise tip-surface positioning in a cryogenic environment is possible because of very low thermal drift. On the other hand, some special phenomena, for example, charge density wave (CDW), superconductivity, and structural phase transition, are only observable at a cryogenic state.

Other commercial and space applications of cryogenic transducers include the transfer and transportation of liquefied gases [Park et al., 2008]. The processes of separating air into its components and the resulting production of liquid oxygen, liquid nitrogen, liquid helium, and liquid argon are some of the primary commercial cryogenic applications. Cryogenic fluid devices including valves are needed to control the flow of these liquid gases in various scenarios, for example, liquid oxygen and liquid hydrogen for space shuttle propulsion, liquid nitrogen for food freezing, cooling of chamber systems for high-vacuum state, cooling of infrared detector and medical applications, liquid argon for plasma technologies, etc.

In addition to these cryogenic actuator applications, transducer materials have many sensing applications at low temperatures. One typical application is the monitoring of the integrity of structures working in cryogenic surroundings. For instance, pressure vessels in liquid rocket engines can fail due to many causes including corrosion, pitting, stress corrosion cracks, seam weld cracks, and dents due to internal or external impacts [Qing et al., 2008]. The detection of defects and monitoring of their growth are important for ensuring the safety and reliability of advanced space exploration vehicles/propulsion systems. Also, the sensing of cryogenic liquid level, flow pressure, and velocity is important for many cryogenic fluid handling applications [Fisher and Malocha, 2007].

FIGURE 7.2
Spitzer Space Telescope during assembly. (Courtesy of JPL/Caltech/NASA.)

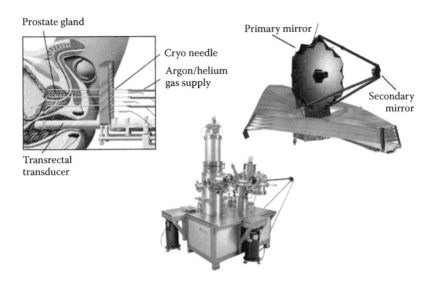

FIGURE 7.3
Common applications of cryogenic transducers (left: cryogenic surgery (http://www.prostatecancercentre.com/treatments/cryo.html, courtesy of Prof. Christopher Eden); center: cryogenic STM (http://www.omicron.de/en/products/low-temperature-spm/instrument-concept, © Oxford Instruments 2013); right: James Webb Space Telescope (http://james-webb-space-telescope.com/, © 2015 James Webb Space Telescope, designed by Themes & Co.)

Various transduction apporaches for actuation and sensing have been developed to meet such operational requirements, and these include electromagnetic, magnetostrictive, electrostrictive, and piezoelectric actuations. A magnetostrictive material referred to as Terfenol D is used. The advantages of magnetostrictive materials appear to be (1) higher stiffness, (2) larger strains, (3) higher load capacity, (4) increased performance at cryogenic temperatures, and (5) lower amplifier volt-amp requirement. Disadvantages appear to be higher mass and the generation of a magnetic field [Wada, 1993]. On the other hand, piezoelectric actuation/sensing is promising for cryogenic applications due to its high-precision displacement control, high force output, quick response, and low power consumption.

In this chapter, typical cryogenic transduction techniques will be surveyed by introducing each technique with pros and cons. The general transduction materials, including magnetostrictive, electrostrictive, and piezoelectric/ferroelectric materials, will be reviewed, with emphasis on the development milestones of ferroelectric materials. The electromechanical properties of lead zirconium titanate (PZT) ceramics and relaxor-$PbTiO_3$ crystals at cryogenic temperatures are summarized. The remainder of the chapter addresses recent developments on cryogenic actuator and sensors.

7.2 Advantages and Drawbacks of Existing Electromechanical Conversion Mechanisms

Conventional actuators/motors are based on the electromagnetic effect because of the low material cost and good compatibility with the common electric power source. However, resistive losses associated with windings induce Joule heating, which prevents the conventional electromagnetic transducers from the applications where low temperature environment needs to be maintained [Weisensel et al., 1998]. An approach to circumvent this problem is to use superconductive coil, whose electrical resistance drops abruptly to zero when the coil temperature is cooled below a characteristic critical temperature. Theoretically, an electric current flowing through a loop of superconductive wire can persist indefinitely with no power source, thus ruling out the generation of heat and the change in the environment temperature.

In addition to the electromagnetic effect, several electromechanical or magnetomechanical coupling effects have been employed for developing cryogenic devices, including magnetostriction, electrostriction, and piezoelectricity. Materials exhibiting these effects belong to a group of special materials named smart materials. These designed materials have one or more properties that can be significantly changed in a controlled fashion by external stimuli such as stress, or electric or magnetic fields.

Magnetostriction is a magnetomechanical coupling effect in ferromagnetic materials that causes them to change their dimensions during the process of magnetization. It arises from a reorientation of the atomic magnetic moments aligned to the applied magnetic field. When the atomic magnetic moments are completely aligned, saturation occurs and increasing the applied magnetic field can produce no further magnetostriction. The saturated magnetization is the most fundamental measure of a magnetostrictive material. High magnetostriction is usually possessed by iron alloys with rare earth elements such as Tb and Dy [Ma et al., 2005]. For instance, polycrystalline $Tb_{0.60}Dy_{0.40}$ has shown magnetostrictions at 77 K of 3000 microstrains for an applied field of 4.5 kOe [Dooley et al., 1999].

Giant magnetostrictive materials offer superior strain up to 1% over the temperature range of 1–150 K. In stark contrast to piezoelectric materials, the magnetostrictive effect becomes more significant at low temperature [Henderson, 1997; Henderson et al., 1998]. A main challenge in the application of magnetostrictive transducers such as actuators is the generation of magnetic field for exciting the magnetostrictive materials, which require a relatively modest magnetic field of 1000 Oe to saturate. Despite this low field requirement, heating of a nonsuperconducting solenoid coil remains a problem for any applications sensitive to thermal injection. Therefore, superconducting coil is needed for the operation of magnetostrictive devices at low temperature. Traditional superconducting materials (metallic) have a transition temperature below 30 K, which is not appropriate for higher-temperature operation. High-temperature superconducting ceramic materials have been reported with transition temperatures as high as 138 K. However, a complex fabrication process, high material cost, and bulky size inhibit the magnetostrictive transducers from being mature devices.

Electrostriction describes the shape change of all dielectric materials under the application of an electric field. It arises from a slight displacement of ions in the crystal lattice due to the Coulomb force upon being exposed to an external electric field. Among various materials, high electrostrictive strain (~1%–5%) has been observed in poly(vinylidene fluoride-trifluoroethylene) [P(VDF-TrFE)] copolymers at room temperature, making them promising for many applications. However, the high strain level decreases quickly with decreasing temperature, and below 160 K, only a small strain level (~0.1%) sustains [Ang et al., 2003], although it is still comparable with ceramics. Electroceramics, such as doped $SrTiO_3$ and $Pb(Mg_{1/3}Nb_{2/3})O_3$ (PMN), have also been used for developing electrostrictive cryogenic actuators [Mulvihill et al., 2002]; multilayer PMN electrostrictive actuators were used on Hubble telescope mirror, providing superior adjustment of the telescope image because of negligible strain hysteresis [Wada, 1993]. There exist a broad range of available materials for electrostrictive actuators; however, extremely high electric field (~1 MV/m) is commonly required for activating the electrostriction, which may prohibit a compact transducer system.

TABLE 7.1

Comparison of Common Cryogenic Transduction Techniques

Cryogenic Techniques	Typical Strain Level	Advantages	Disadvantages
Magnetostriction [Dooley et al., 1999]	1%	High strain level; better performance at low temperature	Need for superconducting coil; high material cost; bulky system
Electrostriction [Mulvihill et al., 2002]	0.01%–0.1%	Broad materials choice; low power consumption; immunity to magnetic field	Low strain level; high electric voltage
Piezoelectricity [Jiang et al., 2009]	0.2%	Immunity to magnetic field; low power consumption; high efficiency	Degraded performance at low temperature

Piezoelectricity, which changes the shape of certain materials when exposed to external electric field, has found numerous application areas including sensors, actuators, and transducers under different temperatures and environments [Zhang and Shrout, 2010; Zhang et al., 2015]. Immunity to magnetic field, low power consumption, efficient conversion of electricity into mechanical energy or vice versa, and low heat generation make piezoelectric devices superior compared with other techniques in many applications. Conventional piezoelectric ceramic materials lose 75% strain at a temperature of 40 K compared with room-temperature strain. Single crystal piezoelectrics exhibit significantly higher piezoelectric performance at both room temperature and cryogenic temperatures [Park and Shrout, 1997], for example, d_{33} of single-crystal piezoelectrics (PMN-PT or PZN-PT) at 30 K is about equal to the d_{33} of PZT-5A at room temperature, indicating promising cryogenic actuation using single-crystal piezoelectrics [Martin et al., 2012]. Relaxor ferroelectric single crystals of $Pb(Zn_{1/3}Nb_{2/3})O_3$ (PZN), $Pb(Mg_{1/3}Nb_{2/3})O_3$, and their solid solutions with the normal ferroelectric $PbTiO_3$ (PT) demonstrated that a consistent high strain (~0.2%) could be achieved from <001> oriented rhombohedral crystals from room temperature to 150 K, with minimal strain hysteresis [Liu et al., 2002]. Common cryogenic transduction techniques are summarized in Table 7.1 [Dooley et al., 1999; Mulvihill et al., 2002; Li et al., 2013, 2014]. Among these various transduction techniques, piezoelectric transducers are attractive because of the aforementioned features. In the next section, representative piezoelectric transducers and their applications will be presented in detail.

7.3 Figure of Merit of Piezoelectric Materials

The piezoelectric mechanism has been extensively employed for transducer applications because of its high precision displacement control, high force output, and low power consumption. The interrelationships of a piezoelectric transducer among the electric displacement D, electric field E, mechanical stress T, and elastic strain S can be described with the following linear constitutive equations:

$$S, D \xrightarrow[s^E, \varepsilon^T]{d} (T, E) \begin{cases} S = s^E T + \mathbf{d}^t E \\ D = \mathbf{d}.T + \varepsilon^T E \end{cases} \tag{7.1}$$

where ε^T is a 3×3 matrix of permittivity coefficients under constant stress and $\mathbf{d}$ is a 3×6 matrix of piezoelectric coefficients, s^E is the 6×6 matrix elastic compliance coefficient $(1/c^E)$, and $\mathbf{d}^t$ is the transpose of $\mathbf{d}$ (6×3 matrix).

Actuator applications require a large displacement, and piezoelectric materials offer such a displacement, which is proportional to piezoelectric strain coefficient d_{eff}, according to the linear constitutive Equation 7.1. Note that piezoelctric actuators can be classified into two categories based on the type of drive field: unidirectional displacement devices with a dc field, (i.e., nonresonant devices), and (Equation 7.2) resonant displacement devices with an ac field at their resonant frequnecies. For the case of nonresonant devices, the figure of merit (FOM) is piezoelectric strain coefficient (d_{eff}) as the strain is directly proportional to d_{eff}; however, the strains of resonant devices are dependent on not only the piezoelectric strain coefficient, but also the mechanical quality factor (i.e., $S = d_{\text{eff}}Q_m E$); thus, the product of piezoelectric strain coefficient and mechanical quality factor, Q_m, are generally considered as the FOM for resonance-based devices. For the resonant devices, the selection of primary vibration modes is also of importance, as it produces a profound impact on the device performance. The vibration modes show the dependence on the elastic boundary conditions of the materials, as listed in Table 7.2, and the electromechanical coupling, defined as the square root of mechanical energy over the total (mechanical and electrical) energy of the oscillating transducer at the resonance frequency. Schematic representations of different mode vibrations, such as radial, extensional, transverse, and shear modes of piezoelectric materials, are shown in Figure 7.4 with equations to determine the electromechanical coupling of each mode [Zhang et al., 2005].

Lead zirconate titanate (PZT) ceramics are the most widely used piezoelectric materials owing to their high piezoelectric and electromechanical properties near the morphotropic phase boundary (MPB) [Berlincourt, 1971; Jaffe and Berlincourt, 1965]. Examples of the piezoelectric properties based on crystal symmetry can be found in Table 7.3. Note that although Rochelle salt crystals have relatively high piezoelectric coefficients, d_{14}, compared with other piezoelectric materials, their use is limited as the crystal is water soluble, and the ferroelectric phase only exists between −18 and 24°C [Valasek, 1921]; thus, they have been replaced by other piezoelectric materials, such as polycrystalline ferroelectric ceramics [Tichỳ et al., 2010].

Owing to their high piezoelectric activities, ferroelectric materials have gained extensive attention, exhibiting a spontaneous polarization (P_s) in the absence of an electric field. Figure 7.5 shows the electromechanical coupling of various piezoelectric materials as a function of relative permittivity. Note that although the piezoelectric coefficients of nonferroelectric materials, such as quartz, show excellent stability under harsh environments, they suffer

TABLE 7.2

Electromechanical Coupling Factors of Piezoelectric Materials with Different Vibration Modes

Mode	Boundary Conditions	Coupling Factor
Length extensional mode, k_{33}	$T_1 = T_2 = 0, T_3 \neq 0$	$d_{33}/\sqrt{s_{33}^E \varepsilon_{33}^T}$
Thickness extensional mode, k_t	$S_1 = S_2 = 0, S_3 \neq 0$	$e_{33}/\sqrt{c_{33}^D \varepsilon_{33}^S}$
Thickness shear mode, k_{15}	$S_1 = S_2 = S_3 = 0, S_5 \neq 0$	$e_{15}/\sqrt{c_{44}^D \varepsilon_{11}^S}$
Transverse mode, k_{31}	$T_2 = T_3 = 0, T_1 \neq 0$	$d_{31}/\sqrt{s_{11}^E \varepsilon_{33}^T}$
Radial mode, k_p	$T_3 = 0, T_1 = T_2 \neq 0$	$k_{31}\sqrt{2/(1-\sigma)}, \sigma = -s_{12}^E/s_{11}^E$

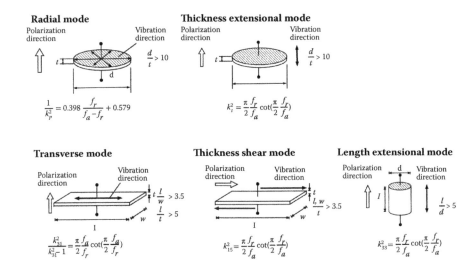

FIGURE 7.4
Schematic of various resonant modes and equations to determine the electromechanical coupling coefficients of piezoelectric materials.

TABLE 7.3

Piezoelectric Strain Coefficients, d_{ij} (pC/N) of Various Single Crystals and Polycrystalline Ceramics

	Symmetry	d_{31}	d_{32}	d_{33}	d_{36}	d_{11}	d_{14}	d_{15}	d_{22}	d_{24}	d_{25}
Tourmaline	3m	0.34		1.83				3.63	−0.33		
LiNbO$_3$	3m	−1		6				68	21		
LiTaO$_3$	3m	−3		9.2				26	8.5		
α-Quartz	32					2.3	−0.67				
KDP	$\bar{4}$2m				69.6		4.2				
ADP	$\bar{4}$2m				−145		5.2				
Rochelle salt	222				12		2300				−56
BaTiO$_3$*	∞m	−78		190				260			
PZT-5H*	∞m	−274		593				741			

Source: Newnham, R.E., *Properties of Materials: Anisotropy, Symmetry, Structure: Anisotropy, Symmetry, Structure,* Oxford University Press, Oxford, 2005.
Note: Ferroelectric polycrystalline ceramics must be poled to be piezoelectically active.
*Polycrystalline ferroelectric ceramics.

from low piezoelectric activities, as shown in Figure 7.5, and may not generate sufficient displacement in an actuator at cryogenic temperatures. In addition, lead-free piezoelectrics based on the perovskite families of KNN (K,Na-Niobates) and NBT (Na,Bi-titanate) have been developed and subsequently characterized in relation to their PZT counterparts, due to the environmental concerns. As the environmental regulations become more and more strict, research on lead-free piezoelectric materials is one of the popular topics in the materials field. As shown in Figure 7.5, KN lead-free crystals were reported to possess high thickness shear coupling of 0.8 [Xu, 1991], while KNN-LT crystals exhibited high coupling of 0.9, with very low permittivity [Huo et al., 2014]. Relaxor-PT piezoelectric materials, such as Pb(Mg$_{1/3}$Nb$_{2/3}$)O$_3$-PbTiO$_3$ (PMN-PT), have received attention because they can provide relatively high relative permittivities (~6000), nearly twice that of conventional PZT ceramics,

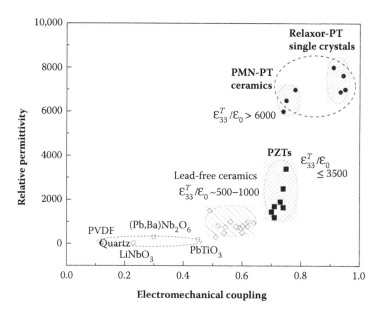

FIGURE 7.5
Relative permittivity for various piezoelectric materials as a function of electromechanical coupling factor.

while possessing comparable electromechanical coupling factors (k_{33} = ~0.78). Of particular significance is that some of the relaxor-PT systems can be grown into single-crystal form, pushing the coupling up to 0.9–0.95 [Zhang and Li, 2012].

7.4 Materials Properties versus Temperature

7.4.1 Cryogenic PZT Ceramics

The thermal stability of dielectric and piezoelectric properties is very important for most electromechanical applications [Zhang and Li, 2012]. The main limitation for the implementation of piezoelectric materials at cryogenic temperatures is their low strain due to the clamping effects of ferroelectric domains. For ferroelectric PZT ceramics, ferroelectric domains and hysteresis are important characteristics as the piezoelectric and electromechanical properties are strongly influenced by the ferroelectric domain wall movements. The contributions related to domain wall movement to the piezoelectric activities are referred to as extrinsic mechanisms that are thermally activated processes, and more than 50% of the net piezoelectric responses of PZT ceramics arise from these extrinsic contributions. Therefore, when PZT materials are used at cryogenic temperatures, most of the extrinsic contributions are frozen out; consequently, the materials lose their piezoelectric performance. For example, to facilitate domain wall motion, donor dopants, that is, higher-valency ions than those of host atoms (e.g., La^{3+} for Pb^{2+} or Nb^{5+} for Ti^{4+}/Zr^{4+}), are typically added to pure PZT composition, which leads to enhanced dielectric, piezoelectric, and electromechanical properties. These materials are so-called "soft" PZTs (e.g., PZT5H and PZT5A), and are typically used for ultrasonic medical imaging, pressure sensors, and actuators operating at low input powers in the order of micro to milliwatt range.

In contrast, the use of acceptor dopants, which creates acceptor-oxygen vacancy defect dipoles, that is, an internal bias field, stabilizes the domain wall motion, effectively increasing coercive field (E_C) while decreasing the dissipation factors, both dielectric and mechanical losses [Carl and Hardtl, 1978; Haerdtl, 1982; Okazaki and Maiwa, 1992; Takahashi, 1982]. This effect is called the "hardening" effect. PZT ceramics modified with acceptor dopants, such as Mn^{3+} and Fe^{3+}, increase the mechanical quality factor, and these materials are classified as "hard" PZT ceramics (e.g., PZT4 and PZT8), and they are typically used for high-power ultrasonic transducers, such as high-intensity focused ultrasound, ultrasonic drilling, and underwater sonar projectors, where low losses are critical to reduce the heat generation.

Figure 7.6 shows examples of temperature-dependent longitudinal (d_{33}) and transverse (d_{31}) piezoelectric properties of soft PZT ceramics (PZT5H and PZT5A), compared with hard PZT (PZT4) counterparts. As shown, the PZT5H ceramics exhibited the highest piezoelectric properties at room temperatures; however, the piezoelectric strain coefficient (d_{33}) decreases to less than 40% of their room temperature values at low temperatures, and the piezoelectric responses of all PZT ceramics converge when the operating temperature is decreased to $-150°C$, indicating that the extrinsic contributions are frozen out below this temperature. It is interesting to note that the piezoelectric activity of hard PZT4 has a relatively small drop, from 240 pm/V of d_{33} at room temperature to 230 pm/V of d_{33} at $-150°C$ because of the small domain wall (extrinsic) contributions at room temperature [Zhang et al., 1994].

Piezoelectric co-fired stacks or multilayers are the most commonly used electromechanical actuator configurations, where a number of thin poled piezoelectric layers are connected mechanically in series and electrically in parallel. In comparison with a single piece of thick piezoelectric element, the effective piezoelectric strain coefficient (d) and capacitance (C) of stacks are proportional to the number of the piezoelectric layers (n) (i.e., $d^*_{33} = n \times d_{33}$; $C^* = n \times C$), which increases the dynamic strain of piezoelectrics for a given ac electric field. The performance of a piezoelectric stack actuator, such as the displacement (ΔL) and the maximum force (F) of the piezo stack, can be calculated according to Equation 7.2, where ΔL and F are proportional to the number of rings (n) and the surface area, respectively:

$$\Delta L = nd_{33}U$$

$$F_b = \frac{d_{33}A}{s^E_{33}(h)}(nU) \tag{7.2}$$

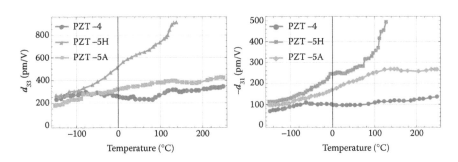

FIGURE 7.6
Piezoelectric strain coefficient of various PZT ceramics as a function of temperature. (Data from Hooker, M.W., NASA, Report No. NASA-CR-1998-208708, 1998.)

where ΔL and F_b are the displacement and block force, respectively, A is the surface area, h ($=n \times t$) is the total thickness, and U is voltage.

To examine the effect of cryogenic temperature on transducer performance requires careful electromechanical characterization techniques, as the piezoelectric materials and devices display a variety of nonlinearities associated with thermal effects and dependencies on boundary conditions. The details regarding the characterization techniques are discussed elsewhere [Sherrit and Mukherjee, 2007], and the general outline for transducer modeling based on the determination of the material constants is shown in Figure 7.7. One popular method to model a piezoelectric transducer is to use the Butterworth–Van Dyke (BVD) model and determine the circuit parameters, since this method is relatively simple and an effective tool in designing actuators around resonance. These techniques can also be used to characterize a piezoelectric material when it is cut to a specific geometry with a proper aspect ratio to isolate a specific resonance mode. In general, however, the BVD circuit model can be used to extract the resonance and antiresonance frequencies along with the effective coupling and mechanical Q of the transducer. Figure 7.8 shows the schematic diagram of the equivalent circuit, with circuit parameters R_1, C_1, L_1, known as the motional resistance, capacitance, and inductance, while C_o is known as the static capacitance. From the BVD circuit parameters, the change in frequency of the fundamental resonance mode, piezoelectric strain coefficient, electromechancial coupling factor, and mechanical quality factor can be derived according to the equations shown in Figure 7.8.

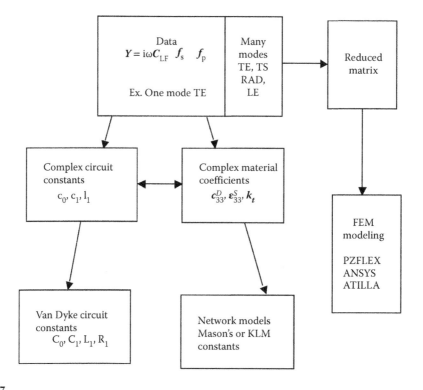

FIGURE 7.7
A schematic diagram that outlines the methods to model piezoelectric transducers. The transducer's effective properties can be calculated from the material coefficients by using the material coefficients in finite element modeling or in one-dimensional network models. (From Sherrit, S., and Mukherjee, B.K., Characterization of piezoelectric materials for transducers, Arxiv, http://arxiv. org/ftp/arxiv/papers/0711/0711.2657.pdf, 2007.)

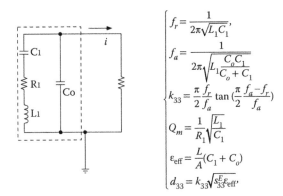

$$
\begin{cases}
f_r = \dfrac{1}{2\pi\sqrt{L_1 C_1}}, \\[2ex]
f_a = \dfrac{1}{2\pi\sqrt{L_1 \dfrac{C_o C_1}{C_o + C_1}}} \\[2ex]
k_{33} = \dfrac{\pi}{2}\dfrac{f_r}{f_a} \tan\left(\dfrac{\pi}{2}\dfrac{f_a - f_r}{f_a}\right) \\[2ex]
Q_m = \dfrac{1}{R_1}\sqrt{\dfrac{L_1}{C_1}} \\[2ex]
\varepsilon_{\text{eff}} = \dfrac{L}{A}(C_1 + C_o) \\[2ex]
d_{33} = k_{33}\sqrt{s^E_{33}\varepsilon_{\text{eff}}},
\end{cases}
$$

FIGURE 7.8
Schematic diagram of the equivalent circuit of the transducer around resonance. The components noted by subscript 1 represent the motional branch of the transducer. ε_{eff}, d_{33}, k_{33}, and Q_m are effective dielectric permittivity, piezoelectric strain coefficient, electromechanical coupling factor, and mechanical quality factor, respectively. L and A are length and area of the piezoelectric stack, respectively.

Figure 7.9 shows temperature effects on electromechancial properties for piezoelectric stack actuators made from PZT5H materials (PI Ceramic—30 × P-885.51), which are determined using BVD parameters of piezoelectric stacks, showing typical behaviors of PZT ceramics, where the effective coupling factor exhibits a decreasing trend with decreasing temperature, with the values being on the order of 0.65 and 0.58 at room temperature and –160°C, respectively. In contrast, the mechanical quality factor (mechanical energy stored divided by the energy loss in each cycle) of the PZT stack increased with decreasing temperature from 20 to 55 when the temperature was changed from 20 to –160°C, which is a consequence of the clamping of domain wall motion as temperature is decreased. The temperature dependence of capacitance and piezoelectric longitudinal coefficient (d_{33}) can be seen in Figure 7.10, where the properties gradually decrease as the temperature decreases; at –160°C, they are approximately half of the room-temperature values. At cryogenic temepratures, the reduction of piezoelectric coefficients of PZT ceramics is due to the following mechanisms: (1) the intrinsic piezoelectric response of PZT in a single domain will reduce on decreasing the temperature, based on thermodynamic analysis [Haun et al., 1989]); (2) the extrinsic piezoelectric response (contribution of domain wall motion [Carl and Hardtl, 1978; Härdtl, 1982]) will also decrease at cryogenic temperature because the activation energy required to make domains change from one minima state to another is increased with decreasing temperature.

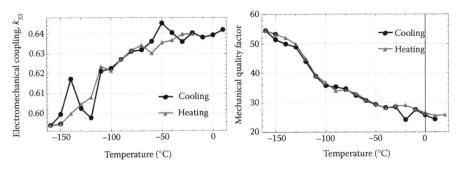

FIGURE 7.9
The effective electromechanical coupling factor and mechanical Q of PZT-based stacks (PI PICMA 885.51) as a function of temperature.

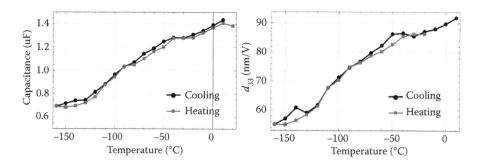

FIGURE 7.10
The capacitance and piezoelectric strain coefficient of PZT-based stacks (PI PICMA 885.51) as a function of temperature.

7.4.2 Cryogenic Relaxor-PT Crystals

$A(B_1,B_2)O_3$-type perovskites, referred to as relaxors or relaxor-ferroelectrics, are another very important class of oxides, and are distinguished from normal ABO_3-type ferroelectric materials due to different characteristics from normal ferroelectrics. For instance, as can be seen in the chemical formula $A(B_1,B_2)O_3$, the B site of relaxors is occupied by a mixture of cations, such as a low-valence cation (Mg^{2+}, Fe^{3+}, Sc^{3+}, or Zn^{2+}...) and a high-valence cation (Ta^{5+}, Nb^{5+}, or W^{6+}...). These random occupations of the B-site in the same crystallographic sites lead to the onset of relaxor behaviors, such as a diffuse phase transition over a Curie temperature range, and strong frequency-dependent dielectric permittivity [Cross, 1987; Smolenskii, 1970].

The incorporation of $PbTiO_3$(PT) into a PMN or PZN system leads to morphotropic phase boundary (MPB) solid solutions, subsequently exhibiting enhanced dielectric and piezoelectric properties analogous to a PZT system. The advantages of relaxor-ferroelectric solid solutions over normal ferroelectric materials can be attributed to their higher dielectric permittivity and comparable piezoelectric properties to a PZT system, as shown in Figure 7.5. Most importantly, some of the relaxor-ferroelectrics can be grown into large single crystals, which can offer a significantly higher electromechanical coupling factor, exceeding 90% of k_{33}, resulting in a large improvement in the overall performance of piezoelectric devices [Zhang et al., 2015].

The origin of such high electromechanical properties of relaxor-PT single crystals is mainly due to the polarization rotation effect [Damjanovic et al., 2003], and the extrinsic contributions of domain engineered single crystals such as <001> orientation are significantly lower than those of polycrystalline ceramics due to the enhanced stability of the domain-engineered 4R structures [Park and Shrout, 1997], as shown in Figure 7.11. It was explained that on the microscopic scale, poling along <001> in rhombohedral relaxor-PT crystals creates four degenerate domain variants, and the dipoles in domains can be aligned along any one of the four equivalent <111> directions. This leads to stable 4R multidomain configurations; therefore, poling with small external electric field has little effect on domain reorientation, resulting in low strain hysteresis. According to the studies [Li et al., 2010], the extrinsic contribution of relaxor-PT crystals was found to be less than 10% of the total piezoelectric response across the compositional range of 0.25 < × < 0.35 in $(1 - x)$PMN − xPT, where MPB in PMN-PT is situated at approximately 33% PT.

The temperature-dependent longitudinal-mode properties for domain engineered relaxor-PT crystals are shown in Figure 7.12. Similar to the polycrystalline PZT ceramics,

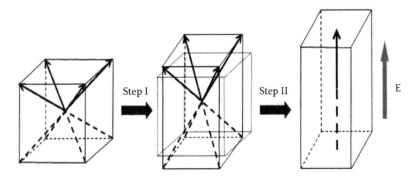

FIGURE 7.11
Illustrations of four-domain state with electric field in <001> direction. (Reprinted with permission from Zhang, S. and Li, F., *J. Appl. Phys.*, 111, 031301, 1–50, 2012. © 2012, AIP Publishing LLC).

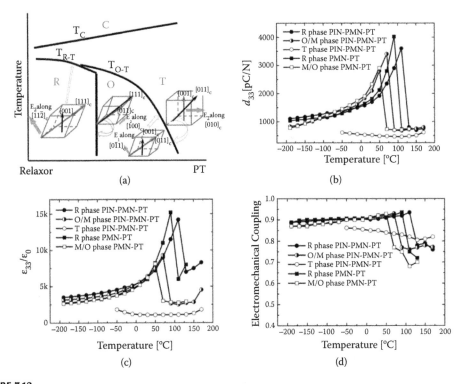

FIGURE 7.12
Schematic phase diagram and the polarization rotations for relaxor-PT-based crystals (a). Temperature dependence of longitudinal-mode piezoelectric coefficient (b), relative permittivity (c), and electromechanical coupling factors (d) for relaxor-PT crystals (Reprinted with permission from Zhang, S. and Li, F., *J. Appl. Phys.*, 111, 031301, 1–50, 2012; Li, F. et al., *Appl. Phys. Lett.*, 96, 192903, 1–3, 2010. © 2012, AIP Publishing LLC.)

both piezoelectric and dielectric properties of relaxor-PT crystals exhibit a strong temperature dependence, where the values of piezoelectric and dielectric properties are approximately half of the room-temperature values at –200°C. The decreasing trends of the dielectric constant and piezoelectric coefficient observed in rhombohedral (R) and orthorhombic/monoclinic (O/M) crystals with decreasing temperature were thought to inherently associate with the temperatures being deviated from R–T and O/M-T phase transition points, respectively. For tetragonal (T) PIN-PMN-PT crystals, on the other hand, the dielectric and piezoelectric coefficients are quite stable over the temperature range of –50 to 180°C, due to the fact that there is no phase transition in this temperature range. Of particular significance is that the high electromechanical coupling factor (~0.9) was found to maintain similar values from room temperature down to –200°C, due to the fact that both dielectric and piezoelectric variations are similar, which is promising for cryogenic transducer applications.

Figures 7.13 and 7.14 show a comparison between PZT5H ceramics and PZN-8%PT single crystals in terms of the piezoelectric coefficients as a function of temperature, suggesting that both polycrystalline PZT ceramics and single-crystal PZN-PT materials showed decreased piezoelectric coefficients with decreasing temperature. However, PZN-PT crystals were found to still possess high piezoelectric d_{31}, on the order of –240 pm/V at –250°C, compared with the room-temperature value of PZT ceramics. In addition, the effective longitudinal piezoelectric d_{33} was found to exhibit the room-temperature value of soft PZT materials at –200°C, while below –200°C, the performance

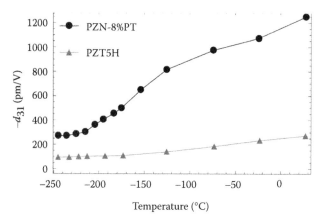

FIGURE 7.13
Temperature-dependent piezoelectric transverse coefficients (d_{31}) for PZN-8%PT single crystals compared with those of PZT5H ceramics. (Data from Jiang, X., and Rehrig, P.W., *Advanced Piezoelectric Single Crystal Based Actuators*, presented at the Proceedings of SPIE, pp. 253–262, 2005.)

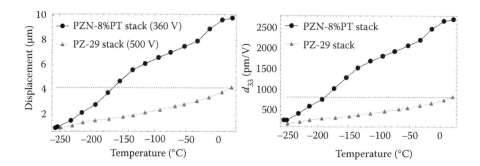

FIGURE 7.14
Temperature-dependent displacements and calculated piezoelectric strain coefficients (d_{33}) for stacked PZN-8%PT compared with those of stacked PZ-29 (PZT5H). For displacement measurements, the input voltages of PZN-8%PT crystals and PZ-29 ceramic stacks are 360 and 500 V, respectively. (Data from Fasik, J.C., *An Inchworm Actuator for the Next Generation Space Telescope*, Burleigh Instruments, Fishers, NY, 1998.)

of both materials converges to the low level of piezoelectric coefficients [Jiang and Rehrig, 2005]. Table 7.4 summarizes the temperature-dependent longitudinal piezoelectric coefficients of selected piezoelectric materials.

The strain versus electric field (S-E) loops of PIN-PMN-PT crystals and PMN-PT ceramics were measured at temperatures from 30°C to –150°C, as shown in Figure 7.15 [Li et al., 2010]. It can be seen that both the amplitude and hysteresis of strain reduce on decreasing the temperature for PIN-PMN-PT crystals and PMN-PT ceramics, indicating that both the intrinsic piezoelectricity and the extrinsic piezoelectricity decrease with decreasing temperature. Based on the thermodynamic analysis, the shear piezoelectric response of a single domain state will decrease as temperature moves away from the polymorphic phase transition (PPT) temperature. The reduction of shear piezoelectric response corresponds to an effective "hardening" of the polarization rotation, leading to the decreased intrinsic piezoelectricity. The extrinsic contribution in domain engineering crystals is believed to relate to an electric field induced phase transition (phase boundary motion) or polar nanoregions (PNR) reorientation. Therefore, as the temperature shifts downward from PPT, the motion of the phase boundary or PNR becomes "frozen," leading to a decrease in the extrinsic parameter. For PMN-PT ceramics, on the other hand, the drastic reduction of extrinsic contribution (almost no hysteresis at –150°C) is mainly due to the temperature clamping effect of domain wall motion [Li et al., 2010].

TABLE 7.4

Temperature-Dependent Longitudinal Piezoelectric Coefficients (d_{33}, pm/V) of Various Piezoelectric Materials

Temp. (°C)	PZT4	PZT5-A	PZT5H	Stacked PZ-29 (PZT5H)	Stacked PZT-8%PT Single Crystal
25	250	300	600	840	2720
–50	290	280	385	560	2080
–150	230	175	270	360	1300
–200	–	–	–	270	600
–250	–	–	–	185	265

Source: Jiang, X. and P. W. Rehrig, *Advanced Piezoelectric Single Crystal Based Actuators,* presented at the Proceedings of SPIE, 253–262, 2005; Fasik, J. C., *An Inchworm Actuator for the Next Generation Space Telescope,* Burleigh Instruments, Fishers, New York, 1998.

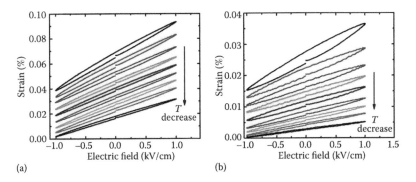

(a)   (b)

FIGURE 7.15
Electric-field-induced strain of (a) PMN-PT ceramics, (b) PIN-PMN-PT crystal. (Reprinted with permission from Li, F., et al., *Appl. Phys. Lett.,* 96, 192903, 1–3, 2010. © 2010, AIP Publishing LLC.)

7.5 Recent Development of Cryogenic Actuators and Sensors

7.5.1 Multilayer Stack Actuators

A single piezoelectric plate has limited deformation due to the finite allowable strain. To enlarge the stroke for cryogenic applications, which require a stroke of ~100 μm at temperatures <20–300 K, multilayer single-crystal stack actuators can be employed. Piezoelectric plates can be shunt connected using the copper shim embedded between the plates. Actuator stroke output of a stack actuator is given by the following formula:

$$\Delta L = d_{33} \frac{V}{t} L \tag{7.3}$$

where ΔL is the stroke, V is the driving voltage, t is the layer thickness, and L is the active length of the stack (the sum of all active layer thickness). Each layer will make equal contributions and act together to yield the total stroke. Given the stack length and driving voltage, a higher d_{33} and thinner layer will result in a larger stroke output. In addition to the large stroke, stack actuators could offer high blocking force and thus large load-carrying capacity. Jiang et al. developed a PMN-PT stack with 36-layer 5 mm × 5 mm × 0.5 mm plates, and it could yield over 50 μm stroke at a driving voltage of 700 V with the resonance frequency at 37 kHz [Jiang and Rehrig, 2005]. They further demonstrated the ability of a piezoelectric stack actuator array for correcting the low-order aberrations of a deformable mirror with peak to valley errors as large as 40 μm [Jiang et al., 2004b].

Strain from the active materials can be amplified through specially designed passive structures to obtain a larger stroke. The blocking force will be sacrificed to maintain constant power consumption. Among various devices, the flextensional actuator (cymbal type) has been regarded as a popular architecture because of the large amplification ratio and simple configuration. Two different piezoelectric drivers can be used for flextensional actuators, as shown in Figure 7.16. One is pizeo-plate driven, called "31" mode. Electric field is applied to the plate in the Z direction; the contraction of the plate in X direction is then amplified into a large displacement in the Z direction. This "31" mode flextensional actuator is known for a large stroke and low profile, but the blocking force is usually low. The other flextensional actuator utilizes a stack actuator as a driver, and is also called a "33" mode flextensional actuator. The force output from "33" mode flextensional actuators is known to be much higher than that of "31" mode.

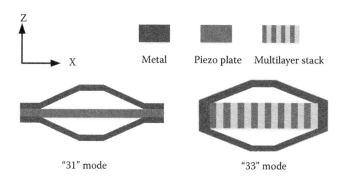

FIGURE 7.16
"31" and "33" mode flextensional actuators.

Jiang et al. [2009] developed a "33" mode flextensional actuator using 300 pieces of PMN-PT single-crystal plates with dimensions of 10 mm × 10 mm × 0.14 mm. A stroke of 76 μm was achieved with 150 V applied voltage at room temperature. About 50% strain was retained at 77 K with lower hysteresis compared with room-temperature stroke. Multiple actuators can be assembled in series to further enlarge the stroke. The stroke of the actuator stack comprising six flextensional elements was measured to be >130 μm at 77 K under the driving of 150 V. Xu et al. [2013] studied a similar actuator for obtaining large displacement (96.5 μm), high load capability (5 kg force), and broad bandwidth (7.3 kHz resonance). The actuator maintained 66% of its room-temperature displacement at 77 K.

7.5.2 Piezoelectric Motors

Although the combination of multilayer stack actuators and amplification schemes could generate 100 μm stroke at cryogenic temperature, they do not meet the requirements for operation of each primary mirror segment and the secondary mirror of the deformable mirror, at 30 K with 20 nm resolution, 6 mm stroke, and position set-hold with power-off. There is some recent research work reported on magnetostrictive cryogenic actuation with superconducting coils by Energen [Joshi, 2000]. However, superconducting technology is not yet allowing mature magnetostrictive cryogenic actuation, and the fabrication process, cost, and volume of actuators are not comparable with piezoelectric counterparts. Compared with piezoelectric actuators, piezomotors including inchworm motors, picomotors, and ultrasonic motors hold advantages such as large stroke in two directions and position set-hold in power-off states.

Inchworm motors typically use three piezo-actuators, as shown in Figure 7.17: one lateral piezo-actuator with a fixed center part and two clutch piezo-actuators mounted on the two sides of the lateral one. Extension and contraction of the lateral piezo-actuator are transferred into the linear motion of a shaft through the frictional force from the clutch piezo-actuators. The working steps can be broken down as follows. When the lateral piezo-actuator extends, the front clutch actuator is in extension status to grab the shaft while the rear clutch is in relaxation; thus, the shaft moves forward. The front clutch will then relax and the rear clutch will extend. The contraction of the lateral actuator will thereby drag the shaft forward again.

Compared with inchworm motors, relatively high-speed motion due to the high frequency is an attractive feature of ultrasonic motors. Ultrasonic motors are classified into two main categories based on the function mechanism: standing wave and traveling wave vibrations. The ultrasonic vibration induced in the stator is used to impart the motion to the rotor through dry friction. Recall the knowledge of wave propagation. The standing wave is expressed by

$$u_s(x,t) = A \cos kx \cos \omega t \tag{7.4}$$

while the traveling wave is expressed as

$$u_t(x,t) = A \cos(kx - \omega t) \tag{7.5}$$

where u is the displacement, A is the amplitude, k is the wave number, x is the coordinate, ω is the angular frequency, and t is time. Using a trigonometric relation, the traveling wave can be transformed as

$$u_t(x,t) = A \cos kx \cos \omega t + A \cos(kx - \pi/2)\cos(\omega t - \pi/2) \tag{7.6}$$

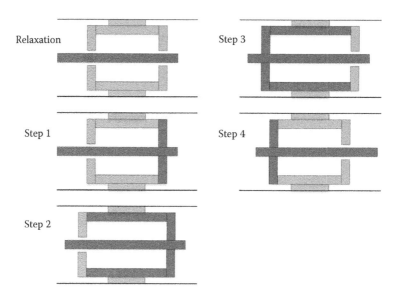

FIGURE 7.17
Schematic of inchworm motors. Step 1: The front clutch actuator is extended to grab the shaft. Step 2: The lateral piezo-actuator is driven to move the shaft forward. Step 3: The rear clutch is actuated, while the front clutch is released from the shaft. Step 4: The lateral piezo-actuator is deactivated, and the shaft moves forward due to the contraction (With kind permission from Springer Science+Business Media: *Piezoelectric Actuators and Ultrasonic Motors*, Vol. 1, 1997, Uchino, K.)

This leads to an important result, that is, a traveling wave can be generated by superimposing two standing waves whose phases differ by 90° from each other both in time and in space. More generally, the phase difference can be chosen arbitrarily except for 0, $\pm\pi$ as long as it is identical in space and time.

A typical standing-wave-based ultrasonic motor relies on the wobbling motion. A wobbling motion can be generated by activating two orthogonal bending motions of a bar structure simultaneouly but with a 90° phase difference [Dong et al., 2002], which can be attained using two piezoelectric actuators bonded orthogonally on the bar structure. Jiang et al. developed a single-crystal linear piezomotor based on the wobbling mode, as shown in Figure 7.18 [Dong et al., 2005; Jiang et al., 2004a, 2006]. The stator comprises two single-crystal ring stacks and a stainless steel bar. Each stack was coated with a two-segment electrode under which PMN-PT crystal segments were poled in reverse directions with respect to each other. Upon application of drive voltage to a stack, one segment will expand and the other will contract, leading to a bending vibration. Two such ring stacks are assembled orthogonally in the stator. When a pair of voltage signals with a 90° phase difference are applied to these two ring stacks, a wobbling motion at the center part can be excited with high efficiency. Through the frictional block, the wobbling motion of the stator can be transformed into the linear motion of either a slider or a shaft using a screw. A stroke >10 mm, step resolution ~20 nm, and response time ~2 ms under the driving voltage of 60 $V_{\text{p-p}}$ were obtained with the prototypes at 77 K.

Recently, Li et al. [2014] investigated a millimeter-size piezoelectric ultrasonic liner micromotor using a single $Pb(In_{1/2}Nb_{1/2})O_3–Pb(Mg_{1/3}Nb_{2/3})–PbTiO_3$ crystal. To activate the first-bending wobbling mode of the piezoelectric bar stator, electric driving voltages with 90° phase difference were applied to two pairs of diagonal electrodes simultaneously. The maximum linear speeds obtained are 88 mm/s at 173 K and 45 mm/s at 98 K.

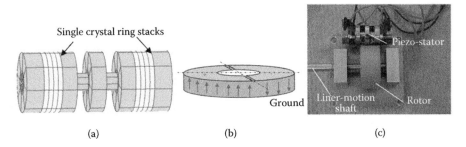

FIGURE 7.18

(a) Construction of piezoelectric composite stator operating on center wobbling mode. (b) Schematic view of a single-crystal ring stack for the piezomotor. (c) Photograph of the assembled motor with screw. (From Jiang, X., et al., *Ultrasonics Symposium, 2004 IEEE*, pp. 1314–1317, © 2004a, IEEE.)

Dong et al. fabricated a piezoelectric single-crystal PMN-PT/Ti-alloy traveling wave ultrasonic motor [Dong et al., 2008]. Nine PMN-PT were bonded on a Ti-alloy ring to form a uniform structure and activate a nine wavelength flexure traveling wave. The maximum acheviable torque output was >1.5 kg cm, the minimum step resolution was <0.072°/step, and the power consumption at a 25% duty cycle was ~2 W under 77 K.

Daisuke developed a bolt-clamped Langevin-type transducer using a PMN-PT single crystal [Daisuke et al., 2012]. Two crystal rings, two ring electrodes, and four quartered electrodes were clamped together by the transducer body and a nut. Two flexural-mode vibrations are generated in a perpendicular direction to each other, and a traveling wave is generated at the tip of the transducer when sinusoidal voltages are applied to the quartered electrodes. The sinusoidal voltages have a phase difference of 90°. The speed could reach 144 rpm at an ultra-low temperature of 4.4 K.

In brief, piezoelectric single crystals could offer large strain output under moderate electrical voltage at cryogenic temperature. A high electromechanical coupling coefficient allows the low profile and low power consumption of the piezoelectric devices. A variety of strain amplification mechanisms can be adopted for magnifing the stroke of actuators. Attributed to the prosperous development of tranditional ultrasonic motors, cryogenic motors can be designed and fabricated. Performance of typical actuators and motors are summarized in Table 7.5. Overall, high stroke, wide functional temperature range, and large bandwidth enable piezoelectric actuators and motors to satisfy the requirements for most practical applications.

TABLE 7.5

Summary of Cryogenic Piezoelectric Actuators and Motors

Type	Motion (Stroke/Speed)	Resolution	Response Time
Multistack actuators [Jiang et al., 2004b]	76 μm at 77 K	20 nm	
Flextensional stack actuators [Jiang et al., 2009]	250 μm at 77 K	20 nm	
Wobbling mode motor [Jiang et al., 2009]	50–100 mm/s at 77 K	78 nm	2 ms
9λ mode ultrasonic motor [Dong et al., 2008]	60 rpm at 77 K	0.1 μm	3.6 ms
Langevin-type ultrasonic motor [Daisuke et al., 2012]	144 rpm at 4.4 K		

7.5.3 Active Tuning in Cryogenic Superconducting Radio Frequency (SRF) Cavities

Another application utilizing piezoelectric actuators at cryogenic temperatures to enable modern physics experiments is in SRF cavities. SRF cavities provide key capabilities enabling advances in particle and nuclear physics experiments. Advances in the efficiency of SRF cavities are today being pushed toward the theoretical limits, which provide the potential of utilizing the technology for applications ranging from the generation of radio-active beams for nuclear astrophysics to creating new methods of light-emitting sources [Padamsee, 2001]. One of the techniques that is facilitating improvements in the efficiency of SRF cavities is active tuning utilizing piezoelectric actuators.

SRF cavities operating in pulsed mode at cryogenic temperatures exhibit high-magnitude electromagnetic fields, which lead to mechanical deformations of the cavity wall on the order of micrometers. This deformation is of sufficient scale to negatively alter the cavity's resonant bandwidth, and thus requires additional RF power consumption, decreasing the system efficiency. This process is referred to as Lorentz detuning. One technique to actively combat this detuning effect is the deployment of piezoelectric actuators. In order for the active tuning approach to be successful, the piezoelectric transducers must operate with extensions of several micrometers at temperatures as low as 1.8 K.

Research quantifying the low temperature performance of piezoelectrics, predominantly PZT in multilayer stacks, has been conducted down to 1.8 K in free and variably preloaded states [Martinet, 2006; Fouaidy, 2007]. Furthermore, the active detuning devices have been tested for resilience to fast neutrons and the dielectric properties measured as a function of temperature [Fouaidy, 2005]. This research has resulted in the determination that, despite significant drop in performance at decreasing temperatures, these devices could provide the necessary extension to be used as active tuning mechanisms in SRF cavities. In the process, piezoelectric material properties and device performance at cryogenic temperatures have been systematically studied; relaxor-PT single crystals are expected to show an especially large advantage over PZT ceramics.

7.5.4 Structural Health Monitoring

Structural health monitoring (SHM) approaches can be broadly classified into two categories: passive SHM and active SHM. Passive SHM methods, such as monitoring strain, acceleration, and acoustic emission, are relatively mature methods but have limited utility because they can only infer the presence of damage from the passive measurement but cannot directly interrogate the structure to detect the damage. In contrast, active SHM methods could initiatively detect damage presence, location, and intensity. Piezoelectric transducers are the most widely used devices for active SHM; they can be bonded to a structure for both transmitting and receiving ultrasonic elastic waves to achieve damage detection.

When a piezoelectric transducer is attached to a mechanical structure with the application of an oscillating electric signal, the mechanical structure will vibrate at the excitation frequency. Given the constant excitation voltage, the vibration amplitude of the substrate varies at different excitation frequencies and reaches extreme values near its mechanical resonances, resulting in enlarged or suppressed deformation of the piezoelectric transducer due to the strain transfer. This alters the electromechanical impedance spectrum of the piezoelectric transducer as a reflection of the property of the substrate structure. As damage occurs in the structure, the vibration modes differ from the pristine state and can be monitored by the piezoelectric transducer. On the other hand, for the pitch-catch method (shown in Figure 7.19), the vibration of the piezoelectric transmitter couples with the plate substrate

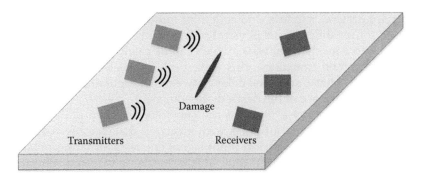

FIGURE 7.19
Piezoelectric-transducer-based structural health monitoring by wave propagation.

to excite an acoustic wave. A Lamb wave is formed after a certain distance and travels along the plate structure, which acts as a wave guide. Another piezoelectric transducer is attached at a distant place to capture the traveling wave information. Any damage along the wave path would affect the properties of the Lamb wave including frequency components, magnitude, and traveling time. Through the analysis of the received signal and the comparison with the excitation signal, the location and severity of the damage can be interpreted.

Santoni-Bottai and Giurgiutiu have successfully used piezoelectric wafer active sensor–based electromechanical impedance and the pitch-catch methods for damage detection of complex composite specimens dipped in liquid nitrogen [Giurgiutiu et al., 2011; Santoni-Bottai and Giurgiutiu, 2012]. In their experiment, defects including cracks, corrosions, disbands, and cracks under bolts were introduced into a metallic spacecraft specimen, which was constructed from an aluminum skin bonded with stiffeners using an adhesive. A 100-kHz central frequency three-count Hanning window tone-burst signal was applied to the transmitter, at which the nondispersive fast axial wave can be separated from the dispersive low-speed flexural wave. The location of disbands and cracks can be detected precisely. For electromechanical impedance measurement, the frequency spectrum covers 1–700 kHz. A drastic change of the readout (i.e., different resonances and impedance magnitude) was observed from the transducer attached at the disbanded area, confirming the validity of this method.

Qing et al. developed a piezoelectric-sensor-based SMART Tape system using both PZT (APC 850) and PMN-PT single crystals for the SHM of liquid rocket engines [Qing et al., 2008]. They demonstrated that the developed system can withstand operational levels of vibration and shock energy on a representative rocket engine duct assembly and is functional under the combined cryogenic temperature and vibration environment.

7.5.5 Cryogenic Liquid Sensing

Sensing at cryogenic temperatures is required for many critical applications, for example, monitoring of fuel tanks of aerospace vehicles. There are many challenges for cryogenic sensors, including the extreme cold, which makes many sensors inoperable due to freeze-out of conduction carriers, mechanical stress and strain that impacts reliability, undesirable device heat generation in the vessel, and so on. In principle, acoustic devices can successfully operate at cryogenic temperature without any serious performance degradation. In particular, surface acoustic wave (SAW) devices operate as sensors, and certain embodiments are passive, wireless, and coded for multisensor applications. SAW sensors are widely used for sensing various physical phenomena. The sensor transduces an input electrical signal into a mechanical wave, which can be easily influenced by physical phenomena. The device then transduces this wave back to an electrical signal. Changes in frequency, amplitude, phase, and time delay between the input and output signals can be used to measure the presence of the desired phenomena.

The basic device consists of a piezoelectric substrate and two pairs of interdigitated electrodes as the input and output ports with a certain distance between them, shown in Figure 7.20. The sinusoidal electrical input signal creates alternating polarity between the adjacent sets of fingers of the IDT; thus, an alternating tensile and compressive strain between fingers of the electrode could be generated by the piezoelectricity, producing a mechanical wave at the surface. The wavelength of the mechanical wave is identical to the distance between the fingers with the same polarity and thus the same mechanical status. Any change of stress, strain, or mass could alter either the pitch between the fingers or the distance between two sets of electrodes, and will be reflected in the output electric signal.

Fisher and Malocha used the lithium niobate ($LiNbO_3$) SAW sensor to detect the level of cryogenic liquid nitrogen [Fisher and Malocha, 2007]. Wave energy damping and phase shift were measured as the main indicators. Aoki et al. [2004] employed both Rayleigh-SAW and shear horizontal (SH) SAW for the measurement of liquid helium. Attenuation coefficients of Rayleigh-SAW and SH-SAW are related with the acoustic impedance of liquid helium, which depends on the density, in other words, pressure of the liquid under a certain temperature. The attenuation coefficient can be calculated from the measurements of single-transit and triple-transit signals. For instance, the attenuation coefficient of a 30 MHz SAW was measured to increase from 0.1 Np/cm at 1 bar to 0.3 Np/cm at 20 bar at 4.2 K.

In brief, piezoelectricity-enabled wave-guide techniques including Lamb waves in a thin plate structure and SAW can be employed for various sensing applications. In addition, the mechanical property of the structure can be coupled with the electrical response of a piezoelectric transducer, whose electromechanical impedance spectrum could thus indicate the structure integrity. These sensing techniques are summarized in Table 7.6.

7.5.6 MEMS and NEMS Devices at Cryogenic Temperature

Piezoelectric materials are also utilized in MEMS devices at cryogenic temperatures, such as radio frequency (RF) MEMS switches. The performance of these devices at cryogenic temperatures is becoming more important as they are integrated into microwave

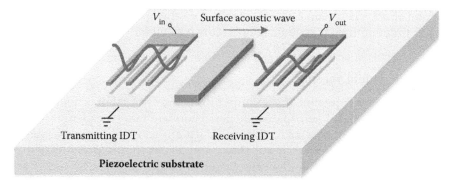

FIGURE 7.20
Schematic of surface acoustic wave sensor.

TABLE 7.6

Summary of Cryogenic Piezoelectric Sensing Techniques

Type	Measurements	Applications
Electromechanical impedance [Giurgiutiu et al., 2011]	Resonance frequencies, phases, and impedance	Damage detection for cryogenic structures
Lamb wave [Qing et al., 2008]	Traveling time, frequency components, and magnitude	Damage detection for cryogenic structures
Surface acoustic wave [Aoki et al., 2004]	Frequency shift, phase, magnitude	Monitoring of liquid level, biological matter, humidity, etc.

systems used in satellite communications and spacecraft diagnostic monitoring systems [Brown et al., 2008].

RF MEMS switches have been used together with high-T_C superconductors (HTS). Such devices have utilized piezoelectrics in RF MEMS switches to achieve switched HTS microstrip resonators and HTS tunable filters [Noel et al., 2003; Prophet et al., 2005]. While the application of RF MEMS HTS devices was shown to be possible, serious questions as to the performance and lifespan of these devices at cryogenic temperatures had to be addressed [Su, 2006; Brown, 2008]. Based on the interdependent nature of MEMS devices and the materials that comprise them, the basis for performance loss and failure mechanisms can be difficult to pinpoint. This difficulty compounds the effort to improve cryogenic temperature performance. To this end, research has attempted to isolate specific aspects of performance and device mortality to individual experiments and simulations.

One performance-degrading effect of piezoelectric RF MEMS devices results in the mismatch of thermal expansion coefficients between the metallic strips and the piezoelectric substrate, most commonly quartz. This mismatch in thermal expansion coefficients gives rise to stress concentrators over temperature variations, which the substrate, as actuator, must overcome. This phenomenon had been observed [Goldsmith and Forehand, 2005] and modeled [Pacheco et al., 2000] in RF MEMS switches. Later, this performance mechanism was quantified in terms of both increasing actuation voltage, to overcome stress created by thermal expansion mismatches, and device mortality [Su et al., 2006].

Finally, cryogenic temperature capable MEMS and NEMS are now being designed and used to study quantum fluids and superfluidity. Most often these micro- and nanostructures have been deployed in liquid helium, such as comb-drive oscillators, to study the fluid properties of superfluids on new length scales [Gonzalez et al., 2013].

The deployment of micro- and nanoscale piezoelectric devices in the cryogenic environment has immediate applications improving satellite communication and aerospace instrumentation but also serves to facilitate a finer resolution in experiments probing toward a better understanding of the physical world.

7.6 Summary

Cryogenic transducers are gaining increasing interest and playing a bigger role in commercial, medical, and space applications. The future innovation of cryogenic actuators and motors could arise from, but is not limited to, smart materials and structures with excellent cryogenic properties, for example, wider functional temperature range with more stable properties, higher energy conversion coefficients, immunity to the effects from other physical fields, smart structures to amplify the strain from the materials, new strategies to transfer the mechanical energy from the material to the structure, etc. Higher sensitivity, better survivability and reliability with relation to thermal and vibrational shocks, wireless sensor networks, and multifunctionality are the pursuits of the future cryogenic transducers. Lower cost, ease of fabrication, and convenient integration could further assist cryogenic devices in finding broader application areas and becoming more competitive in future markets.

Acknowledgments

Some of the research reported in this chapter was conducted at the Jet Propulsion Laboratory (JPL), California Institute of Technology, under a contract with the National Aeronautics and Space Administration (NASA). Some of the reported research has been funded by the Robotics Collaborative Technology Alliance (RCTA), PA No. W911NF-08-R-0012, United States Army Research Laboratory (ARL).

The authors would like to thank Zhenxiang Cheng, Institute for Superconducting and Electronic Materials, University of Wollongong, Australia; Laurent Lebrun, Laboratoire de Génie Electrique et Ferroélectricité (LGEF), Institut National Des Sciences Appliquees, Lyon, France; and Dragan Damjanovic, Ecole polytechnique fédérale de Lausanne (EPFL), Switzerland, for reviewing this chapter and providing valuable technical comments and suggestions.

References

Ang, C., Z. Yu, and L. E. Cross, Electrostrictive and dielectric properties of stretched poly(vinylidene fluoride–trifluoroethylene) copolymers at cryogenic temperatures, *Applied Physics Letters*, vol. 83, pp. 1821–1823, 2003.

Aoki, Y., Y. Wada, Y. Sekimoto, W. Yamaguchi, A. Ogino, M. Saitoh, et al., Application of surface acoustic wave sensors for liquid helium-4 and helium-3, *Journal of Low Temperature Physics*, vol. 134, pp. 945–958, 2004.

Bao, X., Y. Bar-Cohen, Z. Chang, S. Sherrit, M. Badescu, S. Du, et al., USDC-based rapid penetrator of packed soil, in *Smart Structures and Materials*, 617108, pp. 1–6, 2006.

Bar-Cohen, Y. and K. Zacny, *Drilling in Extreme Environments: Penetration and Sampling on Earth and Other Planets*. Hoboken, NJ: John Wiley & Sons, 2009.

Barriere, J.-C., M. Berthé, M. Carty, B. Duboué, J. Fontignie, D. Leboeuf, et al., Cryomechanism: A cryogenic rotating actuator, presented at the *SPIE Optical Engineering Applications*, 2013.

Barron, R., Cryogenic treatment of metals to improve wear resistance, *Cryogenics*, vol. 22, pp. 409–413, 1982.

Berlincourt, D., Piezoelectric crystals and ceramics, in *Ultrasonic Transducer Materials*, O. E Mattiat, ed. New York, NY: Plenum, pp. 63–124, 1971.

Brown, C., A. S. Morris, III, A. I. Kingon, and J. Krim. Cryogenic performance of RF MEMS switch contacts. *Journal of Microelectromechanical Systems.*, vol. 17, 6, p. 1460, 2008.

Bushong, S. C., *Magnetic Resonance Imaging*, St Louis, MO: Elsevier Health Sciences, 2003.

Carl, K. and K. Hardtl, Electrical after-effects in Pb(Ti, Zr)O$_3$ ceramics, *Ferroelectrics*, vol. 17, pp. 473–486, 1978.

Cross, L. E., Relaxor ferroelectrics, *Ferroelectrics*, vol. 76, pp. 241–267, 1987.

Daisuke, Y., K. Takefumi, S. Koichi, K. Masataka, and T. Dai, An ultrasonic motor for use at ultralow temperature using lead magnesium niobate–lead titanate single crystal, *Japanese Journal of Applied Physics*, vol. 51, p. 07GE09, 2012.

Damjanovic, D., M. Budirmir, M. Davis and N. Setter, Monodomain versus polydomain piezoelectric response of 0.67PMN-0.33PT single crystals along nonpolar directions, *Applied Physics Letters*, vol. 83, pp. 527–529, 2003.

Dong, S., S. Cagatay, K. Uchino, and D. Viehland, A 'center-wobbling' ultrasonic rotary motor using a metal tube-piezoelectric plate composite stator, *Journal of Intelligent Material Systems and Structures*, vol. 13, pp. 749–755, 2002.

Dong, S., L. Yan, D. Viehland, X. Jiang, and W. S. Hackenberger, A piezoelectric single crystal traveling wave step motor for low-temperature application, *Applied Physics Letters*, vol. 92, 153504, pp. 1–3, 2008.

Dong, S., L. Yan, N. Wang, D. Viehland, X. Jiang, P. Rehrig, et al., A small, linear, piezoelectric ultrasonic cryomotor, *Applied Physics Letters*, vol. 86, p. 053501, 2005.

Dooley, J. A., C. A. Lindensmith, R. G. Chave, N. Good, J. Graetz, and B. Fultz, Magnetostriction of single crystal and polycrystalline Tb$_{0.60}$Dy$_{0.40}$ at cryogenic temperatures, *Journal of Applied Physics*, vol. 85, pp. 6256–6258, 1999.

Druart, G., N. Guérineau, R. Haïdar, E. Lambert, M. Tauvy, S. Thétas, et al., MULTICAM: A miniature cryogenic camera for infrared detection, in *Photonics Europe*, pp. 69920G, pp. 1–10, 2008.

Fasik, J. C., *An Inchworm Actuator for the Next Generation Space Telescope*, Fishers, NY: Burleigh Instruments, 1998.

Finley, P. T., R. A. Hopkins, and R. B. Schweickart, Flight cooling performance of the Spitzer Space Telescope cryogenic telescope assembly, in *Astronomical Telescopes and Instrumentation*, pp. 26–37, 2004.

Fisher, B. H. and D. C. Malocha, Cryogenic liquid sensing using SAW devices, in *Frequency Control Symposium, 2007 Joint with the 21st European Frequency and Time Forum. IEEE International*, pp. 505–510, 2007.

Fulchignoni, M. and F. Ferri, Recent results on Titan from the HASI instrument, in *Geophysical Research Abstracts*, p. 10178, 2006.

Giurgiutiu, V., B. Lin, G. Santoni-Bottai, and A. Cuc, Space application of piezoelectric wafer active sensors for structural health monitoring, *Journal of Intelligent Material Systems and Structures*, vol. 22, pp. 1359–1370, 2011.

Goldsmith, C. L. and D. I. Forehand. Temperature variation of actuation voltage in capacitive MEMS switches. IEEE Microwave and Wireless Components Letters, vol. 15, 10, p. 718, 2005.

Gonz´alez, M., P. Zheng, E. Garcell, and Y. Lee, H. B. Chan. "Comb-drive MEMS Oscillators for Low Temperature Experiments". *Review of Scientific Instruments*, 84, 025003, 2013.

Härdtl, K., Electrical and mechanical losses in ferroelectric ceramics, *Ceramics International*, vol. 8, pp. 121–127, 1982.

Haun, M. J., E. Furman, Z. Zhuang, S. Jang, and L. Cross, Thermodynamic theory of the lead zirconate-titanate solid solution system, *Ferroelectrics*, vol. 94, pp. 313–313, 1989.

Henderson, D., Building a cryogenic inchworm motor: technical challenges and proposed solutions, *NGST Annual Technical Challenge Review*, NASA/Goddard Space Flight Center, Greenbelt, MD, 1997.

Henderson, D., and J. C. Fasik, Inchworm motor developments for the Next-Generation Space Telescope (NGST), *Proceedings of SPIE*, pp. 252–256, 1998.

Hooker, M. W., Properties of PZT-based piezoelectric ceramics between 150 and 250°C. NASA Report No. NASA-CR-1998-208708, 1998.

Huo, X. Q., L. Zheng, S. Zhang, R. Zhang, G. Liu, R. Wang, B. Yang, W. Cao, and T. Shrout, Growth and properties of Li, Ta modified KNN lead free piezoelectric single crystals, *Phys. Status Solidi-RRL*, vol. 8, pp. 86–90, 2014.

Jaffe, H. and D. Berlincourt, Piezoelectric transducer materials, *Proceedings of the IEEE*, vol. 53(10), pp. 1372–1386, 1965.

James Webb Space Telescope [Online]. Available at: http://james-webb-space-telescope.com/.

Jiang, X., W. B. Cook, and W. S. Hackenberger, Cryogenic piezoelectric actuators, in *SPIE Optical Engineering+ Applications*, 74390Z, pp. 1–8, 2009.

Jiang, X., P. Rehrig, W. Hackenberger, S. Dong, and D. Viehland, Single crystal ultrasonic motor for cryogenic actuations, in *Ultrasonics Symposium, 2004 IEEE*, pp. 1314–1317, 2004a.

Jiang, X., P. Rehrig, W. Hackenberger, J. Moore, S. Chodimella, and B. Patrick, Single crystal piezoelectric actuators for advanced deformable mirrors, in *ASME 2004 International Mechanical Engineering Congress and Exposition*, pp. 37–42, 2004b.

Jiang, X. and P. W. Rehrig, *Advanced Piezoelectric Single Crystal Based Actuators*, presented at the Proceedings of SPIE, pp. 253–262, 2005.

Jiang, X., P. W. Rehrig, W. S. Hackenberger, and T. R. Shrout, Cryogenic actuators and motors using single crystal piezoelectrics, in *Advances in Cryogenic Engineering: Transactions of the Cryogenic Engineering Conference-CEC*, pp. 1783–1789, 2006.

Joshi, C. H., Cryogenic magnetostrictive actuators and stepper motors, pp. 240–249, 2000.

Kalia, S., Cryogenic processing: a study of materials at low temperatures, *Journal of Low Temperature Physics*, vol. 158, pp. 934–945, 2010.

Kogbara, R. B., B. Parsaei, S. R. Iyengar, Z. C. Grasley, E. A. Masad, and D. G. Zollinger, Evaluating damage potential of cryogenic concrete using acoustic emission sensors and permeability testing, presented at the *SPIE Smart Structures and Materials+ Nondestructive Evaluation and Health Monitoring*, 2014.

Li, F., L. Jin, Z. Xu, D. Wang and S. Zhang, Electrostrictive effect in PMN-PT crystals, *Applied Physics Letters*, vol. 102, 152910, pp. 1–5, 2013.

Li, F., L. Jin, Z. Xu and S. Zhang, Electrostrictive effect in ferroelectrics: An alternative approach to improve piezoelectricity, *Applied Physics Reviews-Focused Reviews.*, vol. 1, 011103, pp. 1–21, 2014.

Li, F., S. Zhang, Z. Xu, X. Wei, J. Luo, and T. R. Shrout, Piezoelectric activity of relaxor-$PbTiO_3$ based single crystals and polycrystalline ceramics at cryogenic temperatures: Intrinsic and extrinsic contributions, *Applied physics letters,* vol. 96, 192903, pp. 1–3, 2010.

Li, X., Y. Wu, Z. Chen, X. Wei, H. Luo, and S. Dong, Cryogenic motion performances of a piezoelectric single crystal micromotor, *Journal of Applied Physics*, vol. 115, p. 144103, 2014.

Liu, S.-F., S.-E. Park, L. E. Cross, and T. R. Shrout, Temperature dependence of electrostriction in rhombohedral $Pb(Zn_{1/3}Nb_{2/3})O_3–PbTiO_3$ single crystals, *Journal of Applied Physics*, vol. 92, pp. 461–467, 2002.

Low Temperature UHV STM and AFM Technology [Online]. Available at: http://www.omicron.de/en/products/low-temperature-spm/instrument-concept.

Ma, T., C. Jiang, and H. Xu, Magnetostriction in <110> and <112> oriented crystals $Tb_{0.36}Dy_{0.64}(Fe_{0.85}Co_{0.15})_2$, *Applied Physics Letters*, vol. 86, 162505, pp. 1–3, 2005.

Martin F., H. Brake, L. Lebrun, S. Zhang and T. Shrout, Dielectric and piezoelectric activities in (1-x) PMN-xPT single crystals from 5K to 300K, *Journal of Applied Physics*, vol. 111, 104108, pp. 1–5, 2012.

Mulvihill, M., R. Shawgo, R. Bagwell, and M. Ealey, Cryogenic Cofired Multilayer Actuator Development for a Deformable Mirror in the Next Generation Space Telescope, *Journal of Electroceramics*, vol. 8, pp. 121–128, 2002.

Newnham, R. E., *Properties of Materials: Anisotropy, Symmetry, Structure: Anisotropy, Symmetry, Structure*. Oxford: Oxford University Press, 2005.

Noel, J., Y. Hijazi, J. Martinez, Y. A. Vlasov, and G. L. Larkins, Jr. A switched high-Tc superconductor microstrip resonator using a MEMS switch. *Superconductor Science and Technology,* vol. 16, 12, p. 1438, 2003.

Okazaki, K. and H. Maiwa, Space charge effects on ferroelectric ceramic particle surfaces, *Japanese Journal of Applied Physics*, vol. 31, pp. 3113–3116, 1992.

Pacheco, S.P., L.P.B., Katehi, and C.T.-C., Nguyen. Design of low-actuation voltage RF MEMS switch. *IEEE MTT-S International Microwave Symposium Digest*, p. 165, 2000.

Park, J. M., R. P. Taylor, A. T. Evans, T. R. Brosten, G. F. Nellis, S. A. Klein, et al., A piezoelectric microvalve for cryogenic applications, *Journal of Micromechanics and Microengineering*, vol. 18, p. 015023, 2008.

Park, S. and T. Shrout, Ultrahigh strain and piezoelectric behavior in relaxor based ferroelectric single crystals, *J. Appl. Phys.*, vol. 82, pp. 1804–1811, 1997.

Prophet, E. M., J. Musolf, B. F. Zuck, S. Jimenez, K. E. Kihlstrom, and B. A. Willemsen. Highly-selective electronically-tunable cryogenic filters using monolithic, discretely-switchable MEMS capacitor arrays. *IEEE Transactions on Applied Superconductivity*, vol. 15, no. 2, p. 956, 2005.

Qing, X. P., S. J. Beard, A. Kumar, K. Sullivan, R. Aguilar, M. Merchant, et al., The performance of a piezoelectric-sensor-based SHM system under a combined cryogenic temperature and vibration environment, *Smart Materials and Structures*, vol. 17, p. 055010, 2008.

Saitoh, K., K. Hayashi, Y. Shibayama, and K. Shirahama, A low temperature scanning probe microscope using a quartz tuning fork, *Journal of Physics: Conference Series*, vol. 150, p. 012039, 2009.

Santoni-Bottai, G. and V. Giurgiutiu, Damage detection at cryogenic temperatures in composites using piezoelectric wafer active sensors, *Structural Health Monitoring*, vol. 11, pp. 510–525, 2012.

Scambos, T., A. P., G. Campbell, T. Haran, M. Lazzara, The Coldest Place on Earth: −90°C and below from Landsat 8 andother satellite thermal sensors, in *AGU Fall Meeting*, San Francisco, 2013.

Sherrit, S. Smart material/actuator needs in extreme environments in space. *Proceedings of the SPIE Smart Structures and Materials Symposium*, Vol. 5761,1, pp. 335-346, May 2005.

Sherrit, S., K. Frankovich, X. Bao, and C. Tucker, Miniature piezoelectric shaker mechanism for autonomous distribution of unconsolidated sample to instrument cells, in *SPIE Smart Structures and Materials+ Nondestructive Evaluation and Health Monitoring*, 72900H, pp. 1–9, 2009.

Sherrit, S. and B. K. Mukherjee, Characterization of piezoelectric materials for transducers, *Arxiv*, http://arxiv.org/ftp/arxiv/papers/0711/0711.2657. pdf, 2007.

Sherrit, S., W. Zimmerman, N. Takano, and L. Avellar, Miniature cryogenic valves for a Titan Lake sampling system, in *SPIE Smart Structures and Materials+ Nondestructive Evaluation and Health Monitoring*, 90613J, pp. 1–10, 2014.

Smith, I. C. and D. E. Blandford, Nuclear magnetic resonance spectroscopy, *Analytical Chemistry*, vol. 67, pp. 509R–518R, 1995.

Smolenskii, G., Proceeding of 2nd Meeting on Ferroelectricity, Kyoto, 1969, *Journal of Physical Society of Japan*, vol. 28, p. 26, 1970.

Squyres, S., Vision and voyages for planetary science in the decade 2013–2022, *National Research Council Publications*, 2011.

Stockman, H. S., The Next Generation Space Telescope. Visiting a time when galaxies were young, HS. Space Telescope Science Institute, Baltimore, MD (USA), The Association of Universities for Research in Astronomy, Washington, DC (USA), Jun 1997, XIX+ 163 p., vol. 1, 1997.

Su, H.T., I. Llamas-Garro, M.J. Lancaster, M. Prest, J.-H. Park, J.-M. Kim, C.-W. Baek and Y.-K. Kim. Performance of RF MEMS switches at low temperatures. *Electronics Letters*, vol. 42, 21, 2006.

Takahashi, S., Effects of impurity doping in lead zirconate-titanate ceramics, *Ferroelectrics*, vol. 41, pp. 143–156, 1982.

Tichý, J., J. Erhart, E. Kittinger, and J. Prívratská, *Fundamentals of Piezoelectric Sensorics: Mechanical, Dielectric, and Thermodynamical Properties of Piezoelectric Materials*. Berlin, Germany: Springer Verlag, 2010.

Towner, M., J. Garry, R. Lorenz, A. Hagermann, B. Hathi, H. Svedhem, et al., Physical properties of Titan's surface at the Huygens landing site from the Surface Science Package Acoustic Properties sensor (API-S), *Icarus*, vol. 185, pp. 457–465, 2006.

Uchino, K., *Piezoelectric Actuators and Ultrasonic Motors*, vol. 1. New York, NY: Springer Science & Business Media, 1997.

Valasek, J., Piezoelectric and allied phenomena in Rochelle salt, *Physical Review*, vol. 17, p. 475, 1921.

Wada, K. B., Summary of precision actuators for space structure, *JPL Internal Document, D-10659*, Pasadena, CA, March 1993.

Weisensel, G. N., O. D. McMasters, and R. G. Chave, Cryogenic magnetostrictive transducers and devices for commercial, military, and space applications, *Proceedings of SPIE*, vol. 3326, pp. 459–470, 1998.

Witkin, D. and E. J. Lavernia, Synthesis and mechanical behavior of nanostructured materials via cryomilling, *Progress in Materials Science*, vol. 51, pp. 1–60, 2006.

Xu, T.-B., L. Tolliver, X. Jiang, and J. Su, A single crystal lead magnesium niobate-lead titanate multi-layer-stacked cryogenic flextensional actuator, *Applied Physics Letters*, vol. 102, p. 042906, 2013.

Xu, Y. H., *Ferroelectric Materials and their Applications*. New York, NY: North-Holland, 1991.

Zhang, Q., H. Wang, N. Kim, and L. Cross, Direct evaluation of domain-wall and intrinsic contributions to the dielectric and piezoelectric response and their temperature dependence on lead zirconate-titanate ceramics, *Journal of Applied Physics*, vol. 75, pp. 454–459, 1994.

Zhang, S., E. Alberta, R. Eitel, C. Randall and T. Shrout, Elastic, piezoelectric and dielectric characterization of modified BS-PT ceramics, *IEEE Transactions Ultrasonics Ferroelectrics Frequency Control*, vol. 52, pp. 2131–2139, 2005.

Zhang, S. and F. Li, High performance ferroelectric relaxor-PbTiO$_3$ single crystals: Status and perspective, *Journal of Applied Physics*, vol. 111, 031301, pp. 1–50, 2012.

Zhang, S., F. Li, X. Jiang, J. Kim, J. Luo and X. Geng, Advantages and challenges of relaxor-PT ferroelectric crystals for electroacoustic transducers—A review, *Progress in Materials Science*, vol. 68, pp. 1–66, 2015.

Zhang, S., and T. Shrout, Relaxor-PT single crystals: observations and developments, *IEEE Transactions Ultrasonics Ferroelectrics Frequency Control*, vol. 57, pp. 2138–2146, 2010.

Zheludev, I. S. and A. Tybulewicz, *Physics of Crystalline Dielectrics*, vol. 2. New York, NY: Plenum Press, 1971.

Zurecki, Z., Cryogenic quenching of steel revisited, http://www.airproducts.com/~/media/Files/PDF/industries/metals-cryogenic-quenching-of-steel-revisited-33005019GLB.pdf, 2005.

8

Low Temperature Motors and Actuators: Design Considerations and a Case Study

Joe Flynn and Yoseph Bar-Cohen

CONTENTS

8.1 Introduction

The general distinction between a motor and an actuator is that a motor is the element that makes the motion (mostly rotation of an electromagnetic driver) while an actuator is a complete activation device with a gear that enables the movement of a mechanism. Mobility, manipulation, and articulation of mechanisms require the use of actuators and, in some applications, they enable extremely high precision displacement (e.g., optical devices). Each degree of freedom (DoF) of a mechanism typically requires a dedicated actuator and, with the rise in the number of DoF, the complexity significantly increases. In such systems as robots, the mass and volume of the actuators take up a significant percentage of the total system. Actuators function as the equivalent of muscles and they are operated by a source of energy (typically electric current, hydraulic, heat, or pneumatic pressure) that is converted into motion. The control system that drives an actuator can be a simple fixed mechanical or electronic system, computer driven or operated by a human user. The general types of actuators that are used include electric (such as ac, dc, brushed, and brushless motors), pneumatic, hydraulic, piezoelectric, and shape-memory alloys. The operation of actuators is significantly different from the operation of natural muscles, which are both compliant and linear in behavior [Full and Meijer, 2004].

There are many low temperature applications where a motor is needed to perform mechanical functions. For applications on Earth, 180 K (−135.8°F) is the lowest temperature that has ever been recorded and this was made by the NASA's Aqua Satellite in 2010 [science.nasa.gov]. As far as the temperatures that are involved with potential planetary applications, measurements made by NASA's Lunar Reconnaissance Orbiter (LRO) suggest that the permanently shadowed craters at the Moon's South Pole may be even cooler than Pluto. In portions of these craters, Diviner has recorded less than 35 K (–397°F) as the minimum daytime temperature [space.com NASA/UCLA]. Thus, the temperature range between 35 and 180 K represents a fairly reasonable minimum goal for having motors that would support Earth and extraterrestrial low temperature applications.

The number of applications for low temperature motors and actuators is quite large but, more importantly, many of these applications are mission critical. An example is a system of nozzles controlled by valves operated by low temperature actuators that release propellant to maintain the position of a satellite in its orbital slot. If any of the actuators fail, the satellite would be rendered useless. Specific planetary applications include driving astronomical instruments with infrared detectors (e.g., deformable mirrors for space-based astronomical imaging systems), as well as cryogenic focusing and active optics control mechanisms. Other low temperature applications include valves, pumps, manipulators and activation mechanisms, latches, robotic actuation, solar panel positioners, and drive actuators of rovers.

The leading actuation mechanisms that are potentially operational at low temperatures include electromagnetic, piezoelectric, as well as magneto- and electro-strictive based mechanisms. The use of piezoelectric transducers to drive a motor is described and discussed in great detail in Chapter 7. This chapter is focused on electromagnetic motors and actuators. For planetary applications, actuators are used to operate robotic devices as well as space mechanisms and instruments. These include rovers, release and deployment mechanisms, antennas, positioning devices, aperture opening and closing devices, etc. Increasingly, actuators need to have minimal mass, volume, and power consumption, as well as the capability of operating at extremely cold environments. The miniaturization of conventional electromagnetic motors (ac and dc) is limited by practical manufacturing difficulties. These electromagnetic motors, which are the most widely used in compact mechanisms, employ speed-reducing gears, which can reduce the speed from many thousands of rpm and thus reach higher torque. The use of gear trains adds mass, volume, and complexity as well as reduces the system reliability due to the increase in the number of system components. The miniaturization of conventional electromagnetic motors is limited by manufacturing constraints and loss of efficiency. Furthermore, conventional actuators are backdrivable making high precision control difficult to obtain or requiring additional braking mechanisms.

Commercial and NASA-led development efforts have resulted in several motors for operation at cryogenic temperatures. The motors include the ones developed by Ball Aerospace [Slusher, 1999; Streetman, 2002] and the one developed by Dynamic Structures and Materials (DSM). The latter was the result of a Small Business Innovative Research (SBIR) contract producing a cryogenic piezoelectric valve [Park, 2009] where a piezoelectric stack operating as a switch is used to move a diaphragm. Another reported cryogenic motor is the Nanometer Resolution Linear Actuator that was developed for spacecraft optical instruments [Nalbandian and Hatheway, 2000]. This motor was developed jointly by Alson E. Hatheway, Inc. (AEH) and Moog, Schaeffer Magnetics Division (SMD) to demonstrate a low-mass, high-precision actuator. The importance of low temperature actuators and the limits of the state of the art were well recognized by the engineers at the Jet Propulsion Lab who developed the Curiosity Rover for the

Mars Science Lab (MSL) mission (landed on Mars in August 2012). The motors were of the brushed dc type and made by Maxon, Switzerland; they were designed to sustain the temperature changes on the surface of Mars, which can range from around –120°C to 25°C, and the atmospheric pressure of about 6 torr. The mission had to be postponed 2 years due to the unavailability of the low temperature motors at the scheduled original launch date in 2010.

The requirements and specifications of low temperature motors depend on their application. For planetary ones, particularly those that are operated in the vicinity of infrared systems, the requirements include being small and low mass, with exceptionally low power dissipation, minimal dry lubrication, while exhibiting low or zero lost motion. Designing a high-reliability, low temperature motor or actuator to withstand the effects of temperature necessitates knowing what the effects entail. When choosing components that experience wear (such as bearings), in addition to having a high reliability it is essential to know the motor or actuator's expected lifetime. As an example, the useful lifetime of geosynchronous orbit satellites averages about 15 years and the limit is primarily imposed by the exhaustion of propellant aboard [Kurtin, 2013]. Other deep-space-related spacecraft systems, especially those that are intended to orbit other planets, may require much longer expected lifetimes. In any case, it is desirable to design low temperature motors or actuators that have a life expectancy longer than any already known system/component lifetime.

In this chapter, design aspects that govern the development of low temperature motors and actuators are described and discussed. The description follows a case study that is intended to help understand the components function and the requirements for making a low temperature motor. A key parameter is the coefficient of thermal expansion (CTE), which dictates the selection of the materials that are used to produce low temperature motors or actuators [Wigley, 1971; Timmerhaus et al., 2013]. In addition, it is critical to dissipate the generated heat away from the motor. Besides the selection of construction materials, one needs to select materials to operate as effective lubricant for the moving parts, adhesive for the motor construction, and electrical potting materials for the motor packaging.

8.2 CTE and Material Selection Considerations

The CTE (α, or α_1) is a material property that represents the coefficient to be used to determine the extent to which a material expands upon heating. It is the measure of the change in length with temperature for a solid material and it is expressed as

$$\frac{(l_f - l_o)}{l_o} = \alpha_1 (T_f - T_o)$$

where l_o and l_f are, respectively, the original and final lengths with the temperature change from T_o to T_f. The CTE or α_1 has units of reciprocal temperature (K^{-1}) such as (μm/m)/°C.

Generally, CTE is not a concern in motor and actuator design, since the temperature range at which it is intended to operate does not vary enough to cause dimensional changes that are sufficiently significant to interfere with its operation. However, if a motor is cooled, for example (as in the case study that is discussed later in this chapter) from room temperature

~293.15 K down to 40 K, the through bolts used to hold the motor together will contract lengthwise. The significant dimensional change would either stretch or break the bolts and/or damage or distort the motor end bells or case. In a perfect scenario, all of the motor and actuator's materials would have identical CTEs and, therefore, would expand and contract at the same rate. Unfortunately, this is not possible, since both motors and actuators require for efficient operation, at least two materials that have largely different CTEs. The first material is silicon steel, which is necessary to carry the flux in the motor's magnetic circuit, that is, stator, rotor, etc. The second material is copper, which is wound on the silicon steel stator (the motor's electrical phase windings) to create magnetic fields. If the copper is wrapped around the silicon steel stator too tightly, the difference in the CTEs creates a situation where if the motor or actuator is cooled to a low enough temperature, the copper would break due to the fact that the two materials are contracting at different rates.

Both motors and actuators have relatively small air gaps between stationary and moving parts (in the case study below, these air gaps are typically in the order of 0.127–0.508 mm). Often times, certain parts used in motors and actuators are held in place by a press fit, that is, the stator of a motor is often pressed into a cylindrical tube case and a moving rotor resides in the stator. When the operating temperature of the motor transitions from the hottest to the coldest end of the temperature range and the dimension variation is large enough, it would result in the stator moving around in the case. This may have a catastrophic effect where the stator would crash into the rotor.

8.3 Fabricating Low Temperature Motors and Actuators

The details of fabricating motors vary among manufacturers. To understand the process, a case study example is given in this section and it is based on the lead author's study at QM Power Inc. This study was performed under NASA SBIR contract number NNX11CB84C, titled "Lightweight High Efficiency Electric Motors and Actuators for Low Temperature Mobility and Robotics Applications" for which Boeing performed low temperature tests in their helium cryostat. This case study included the process used to test new materials and new motor constructions designed to avoid negative effects caused by differing CTEs. The design and material integrity for this low temperature motor product was specified to operate at temperatures from 403 K down to 40 K.

Low temperature motors and actuators are in low demand, which makes them expensive to manufacture. However, new manufacturing methods and materials could help lower their total cost. The new material that was tested in this study is Bluestone™, which allows some of the motor case parts to be manufactured using stereolithography (SLA). The ability to use an SLA process for low-quantity manufacturing of low temperature motor case parts substantially lowers the cost of the motor. The Bluestone™ case was made in two parts where the first part houses all of the motor components and the second part is an end cap or shield (see Figure 8.1).

Since the motor contains materials with different CTE, a study of the effects of expansion and contraction was completed for all of the motor parts. The stainless steel through bolt was determined to shrink 0.4 mm when cooled from 293 K down to 40 K, placing excessive strain on both the bolt itself and the case ends. To allow for the 0.4-mm shrinkage, a Belleville spring washer (Figure 8.2) or conical washer was added on the through bolt to apply a preload to the bolted section.

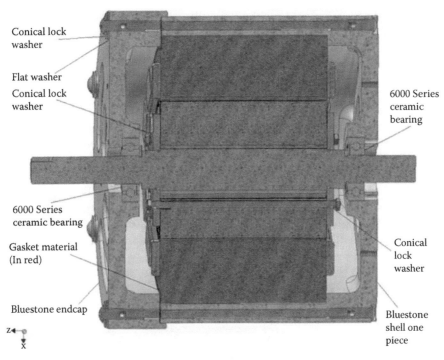

FIGURE 8.1
The low temperature motor with the Bluestone™ components.

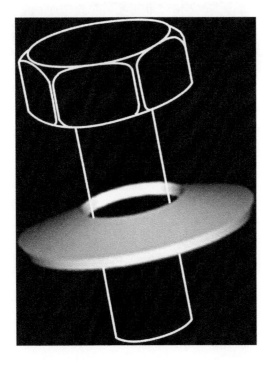

FIGURE 8.2
Belleville spring washer.

Next, the torque was calculated for the Belleville washer when compressed about 70%, which leaves 0.23-mm travel (compression) remaining for each end. This gives a total of 0.229 mm × 2 = 0.46 mm total available compression for shrinkage of the 6–32 through bolt. The 6–32 through bolt is estimated to shrink: 0.000009 mm/mm/K × 253 K × 101.6 mm = 0.229 mm. Thus, there is 0.23 mm remaining at 40 K during the reduction in temperature from 293 K down to 40 K (253 K total temperature delta from standard room temperature of 293 K).

The torque value to compress the Belleville spring washers to a height of 100/70 = 0.03 (max compress travel)/n; n = 0.5 mm; height = 1.6 mm (washer free height) –0.5 mm = 1.07 mm total height of washer was then calculated. The Belleville washer (cup washer) is shown in Figure 8.3 on the through bolt in the "material test article."

Among other issues to be considered is the shrinkage of the silicon steel lamination stack captured inside the Bluestone™ housing, which would allow movement of the lamination stack, which might result in the stator contacting the rotor. This issue is resolved by placing a silicone sealant between the Bluestone™ housing and the stator lamination stack (Figure 8.4).

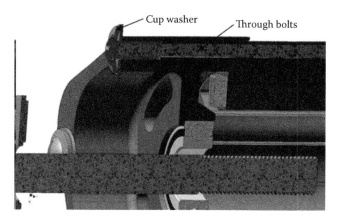

FIGURE 8.3
The Belleville (cup) washer on the through bolt in the "material test article" being tested in the cryostat.

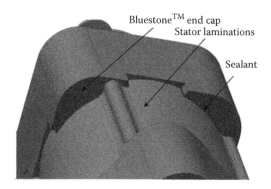

FIGURE 8.4
Silicon sealant placed between the Bluestone™ housing and the stator lamination stack.

The next issue that needed to be resolved was the motor shaft shrinkage at 40 K, which would place a stress on the hybrid ceramic bearings. This issue was resolved by placing beryllium copper wavy washers between the shaft and the bearings (Figure 8.5). The ceramic bearings selected for the low temperature motor are cryogenic certified, silicon nitride (Si_3N_4), and were chosen since they are one-third the weight of their stainless steel counterparts and do not require lubrication.

In addition to testing the SLA Bluestone™ material's ability to maintain its integrity when cooled from 293 K down to 40 K, the following have also been observed or monitored:

- When the copper windings used in the motor phases were excited, they produced fields that were magnetically opposing, creating forces that mechanically "flexed" the copper conductors forming the winding. The rate of this mechanical flexing could be quite high for a motor operating at 1800 rpm or 30 revolutions per second. If the copper conductors forming a winding were not bonded to prevent movement, they would eventually break due to fatigue. In Earth ambient motors, the windings are typically bonded using shellac but this would not be suitable for cryogenic use, since shellac is already brittle at 293 K. Two cryogenic certified epoxies produced by Master Bond, EP37-3FLFAO and EP29LPSP, were applied to a winding, which was used as part of the cryogenic testing.

- As pointed out already, the through bolts used to hold the cryogenic motor together contracted in length up to 0.483 mm when the temperature was reduced from 293 K down to 40 K. To address this issue, Bellville spring washers were used to absorb the high pressures that the through bolts created due to contraction. If it was not absorbed, it could result in the breakage of the through bolt. The selection of the proper Bellville spring washer and its initial tension and flexibility were an integral part of the final assembly. These Bellville spring washers, adjusted to a given initial torque, were a part of the cryogenic testing.

- Cryogenically certified Si_3N_4 ceramic bearings were used in the construction of the motor. These bearings had to be able to withstand the same pressure as the compression pressure of the Bellville spring washers and were also used as part of the cryogenic testing.

- As the temperature was reduced from 293 K down to 40 K, the stator, comprising silicon steel laminations, contracted and could become loose within the motor's outer case and contact the rotor. To prevent any stator movement in relation to the rotor, the stator was compressed between the two end shields using "wavy" washers. This assembly was also used as part of the cryogenic testing.

FIGURE 8.5
Beryllium copper wavy washers placed between the shaft and the bearings of the motor.

8.4 Testing the Motors and Actuators and Analysis of the Test Results

Multiple configurations were incorporated in the same test article that was cooled from 293 K down to 40 K. The following is a list of the configurations and what was measured or observed during or after the cooling:

1. Belleville (conical) washers (McMaster Carr part number 91235A109, 17-7 PH Stainless Steel) were installed at each end cap T-bolt. Each of the four all-thread attachments had different torque values: 68 (normal), 102, 136, and 169 N-cm. After cooling from 293 K down to 40 K and returning to room temperature, the Bellville spring washers were examined for sustaining the preset torque with material deformation (flattening). The comparison of the "flattening" of the four torque values for the Bellville washer in relation to the Bluestone™ end bells and laminate material shrinkage during the cooling from 293 K down to 40 K was observed. These data helped in determining the optimal torque settings for the Bellville spring washers at 293 K.

2. Large washers with RTV Silicone applied to one side were observed after cooling from 293 K down to 40 K. One washer had Permatex Red High-Temp RTV Silicone applied that was 0.25 mm thick, and the second washer had Permatex Ultra-Grey Rigid RTV Silicone applied that had the same thickness. The two types of silicone were visually examined for any material deformation or separation from the washers.

3. Bluestone™ material for each end cap was tested for durability during the cooling from 293 K down to 40 K. A 9.5-mm stainless steel bolt was installed through the center finger tight, with the large washers containing the RTV Silicone on each end. Shrinkage of the 9.5-mm stainless steel bolt was tested for the strain on the Bluestone™ end cap structure and the RTV associated with the large washers.

4. A laminate stack was wound with 20-gauge copper magnet wire, nine turns, on two poles. Master Bond Epoxy EP29LPSP was applied to one of the poles and EP37-3FLFAO to the other. After cooling from 293 K down to 40 K and returning to room temperature, the epoxy on each pole was visually examined for material deformation and the ability to allow the copper to contract and expand. X-ray and continuity tests were used to determine if any copper wire breakage occurred.

5. Permatex Red High-Temp RTV Silicone applied on the gap between the laminate and one end cap was examined after exposure to the temperature drop from 293 K down to 40 K. Evidence of RTV Silicone material separation from either the Bluestone™ or laminate contact surface was studied.

6. The enclosed cured epoxy samples were cold soaked at 40 K and observed for any abnormalities.

8.4.1 Cryogenic Test Results

The SLA Bluestone™ material used in the test motor's end bells experienced cracking with some sections breaking completely off (Figures 8.6 and 8.7). X-rays of the materials used in the test article after being exposed to a temperature of 40 K are shown in Figure 8.8.

FIGURE 8.6
The motor Bluestone™ end shield with broken tab.

FIGURE 8.7
The motor Bluestone™ end shield showing cracking.

FIGURE 8.8
Two x-rays of test article after being exposed to a temperature of 40 K.

The x-rays did not yield much information about the fissures and cracks in the Bluestone™ material, since it appeared to be fairly transparent in these images. The other materials that were tested did not show any defects in the x-ray images.

The following is the analysis of the configurations and parts after testing:

1. The Belleville (conical) spring washer: There was a slight reduction in torque on all four T-nuts as follows: 67.8 N-cm reduced to 56.5 N-cm, 101.7 N-cm reduced to 67.8 N-cm, 135.6 N-cm reduced to 90.4 N-cm, and 169.5 N-cm reduced to 101.7 N-cm. Also, a flattening of the Belleville spring washers was observed.

2. The two types of RTV Silicone: Both RTV Silicone types lost adhesion to the washer and both experienced material cracking, separating into pieces. The Permatex Red High-Temp Silicone experienced more cracking and separations than the Grey.

3. Two end caps made from a SLA Bluestone™ material: The Bluestone™ material experienced failure, cracking at several areas. There was evidence that the Bluestone™ had shrinkage that exceeded the laminate steel material, as one end cap had its centering lugs break off. This only occurred on the one end where the Red RTV Silicone filled the gap at the four lugs to simulate laminate centering. The Silicone turned solid at cryogenic temperatures, not allowing for the contraction of the Bluestone™, and thus the ears broke off.

4. The Scotchcast laminate stack with the wounds: Both cured epoxy samples showed no evidence of material failure, unlike the RTV Silicones in item (2). The wound epoxy coils showed no evidence of copper wire breaks or shorting to the armature. The ohm meter connected to the wire ends showed a closed circuit, indicating no copper wire breakage. It also showed an open circuit between wire ends and the laminate, indicating no grounded copper wire. The Scotchcast material showed no evidence of flaking or separation from the laminates.

8.4.2 Summary of the Test Conclusions

1. The Bluestone™ material cannot be used at temperatures requiring "cryo" temperature conditions.

2. The Silicone RTV cannot be used at temperatures requiring "cryo" temperature conditions.

3. The Master Bond Epoxy EP29LPSP and EP-37-3FLFAO can be used at temperatures requiring "cryo" temperature conditions.

4. The Belleville (conical) washer, McMaster Carr part number 91235A109, can be used at temperatures requiring "cryo" temperature conditions.

5. The 3M Scotchcast can be used to bond motor laminates at cryogenic temperatures.

8.5 Materials Selection for the Motor End Bell and Case

Certain materials have already been either used or tested in cryogenic applications such as stainless steel and titanium and could be easily used in any low temperature motor or actuator application. One problem is the high cost per kg to place objects into space: ~$4540 per kg or ~$10,000 per lb. [NASA: Advanced Space Transportation Program: Paving the Highway to Space]. Stainless steel is heavy, having a density of ~7913 kg/m^3, and while titanium is a little more than half this weight at ~4533 kg/m^3, it is very difficult to machine and is several times the cost of stainless steel. A dissembled low temperature stainless steel motor is shown in Figure 8.9.

Other materials that could be suitable for parts used in low temperature motors and actuators are as follows: N610 Glass filled Carbon Fiber Composite; Acculam® Epoxyglas G10, which is a halogen-free epoxy resin binder, with a woven fiberglass substrate, and is used in structural and cryogenic applications; and AlBeMet®. AlBeMet® is the trade name

FIGURE 8.9
Stainless steel case cryogenic motor parts.

for a beryllium and aluminum metal matrix composite material derived by a powder metallurgy process.

Of the aforementioned materials that would be suitable for low temperature motors and actuators, AlBeMet® would certainly top the list. When used in satellite structures, AlBeMet® provides properties that are roughly halfway between aluminum and beryllium, at the same time reducing the cost of finished components. Reducing the amount of the raw material that is used and the related part fabrication cost attain these savings. Aluminum beryllium's performance-to-cost relationship has allowed Materion's customers to successfully use AlBeMet® as a replacement for traditional aluminum. This possibility is rarely justified when comparing beryllium to aluminum [Materion Corporation, www.materion.com]. Table 8.1 shows the physical properties of AlBeMet® 162 and 140 compared with common aluminum properties, and Figure 8.10 shows a graph of the CTE over a temperature range. Table 8.1 shows that the specific stiffness of AlBeMet® is more than twice as high as the stiffness of aluminum and it has lower to equal CTE.

TABLE 8.1

Physical Properties of AlBeMet® 162 and 140 Compared with Common Aluminum Properties [Based on AlBeMet® DataSheet]

Material	2024T6(AL)	6061T6(AL)	AM162	AM140
Density, g/cm³ (lb./in³)	2.77 (0.100)	2.70 (0.100)	2.10 (0.076)	2.28 (0.082)
Modulus, GPa (Msi)	72 (10.5)	69 (10.0)	193 (28)	150 (22)
Poisson's ratio	0.024	0.024	0.17	0.23
CTE @ 25°C ppm/°C (ppm/°F)	22.9 (12.7)	23.6 (13.1)	13.9 (7.7)	16.5 (9.1)
Thermal conductivity @ W/mK (BTU/h Ft °F)	151	180	210	204
Specific heat @ 20°C J/kgK (BTU/lb °F)	875	896	1506	
Electrical conductivity @ 20°C, % IACS (International Annealed Copper Standard)	38	43	49	
Damping capacity 25°C, 500 Hz	1.05×10^{-2}	1.05×10^{-2}	1.5×10^{-3}	
Fracture toughness K_{1c} Ksi $\sqrt{in}$ (MPa·$\sqrt{m}$)	23 (25)	23 (25)	10–21 (121–23)	

Source: Courtesy of Materion Brush Inc.

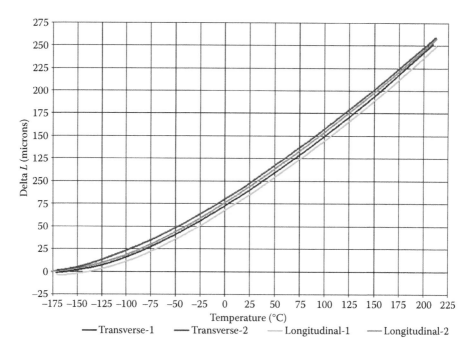

FIGURE 8.10
Delta L data for thermal expansion of AlBeMet® 162 over –175°C to 210°C temperature range.

Of the materials used in low temperature motors and actuators, the ones that have the fewest selection options are those used for the magnetic steel, typically silicon steel, and those used in the conductors, typically copper; the conductors could also use gold, silver, or aluminum. None of the aforementioned metals would be candidates for magnetic steels or conductors but would be suitable for the end bells, case, shaft, and other structural components used in low temperature motors and actuators.

8.6 Heat Transfer in a Low Temperature Motor or Actuator

Motors and actuators produce heat while they are operating, and this is equal to the system's losses. The primary source of the losses or production of heat is the electrical current (amperes) flowing through a conductor's resistance (ohms). This loss is expressed in watts as a function of the current squared times the resistance through which it flows, or:

$$I^2R = W$$

where I is the current in amperes, R is the resistance in ohms, and W is the watts.

A second primary source of heat-producing loss is magnetic core loss where a magnetic core, typically silicon steel, is the magnetic material used to confine and guide magnetic fields in electromechanical devices such as motors and actuators. Core losses in a motor or actuator consist of the sum of two types of losses: hysteresis and eddy current. Both types of losses share the same physical mechanism that creates them or the changing magnetic field induced into the core from the current flowing through the motor conductors in varying magnitude and the frequency at which the current changes directions. Eddy current losses are minimized by using laminations to build the motor or actuator's core as opposed to just using a solid block of steel. By laminating the core, the resistance increases, resulting in a decrease in eddy currents. Hysteresis losses are minimized by using high-grade silicon steel.

In very general terms, eddy currents are currents that are induced in the core by the changing magnetic field and flowing through the resistance of the core material. Hysteresis losses are equal to the energy required to reverse a magnetic domain or the coercive force that is generally presented in a BH curve, as shown in Figure 8.11, for the material being used or considered. As can be seen in the magnetic hysteresis loop [Basic Electronics Tutorials Site by Wayne Storr] in Figure 8.11, the lower the coercive force, the less the hysteresis losses.

All electromechanical devices including low temperature motors and actuators produce heat when operating, with the two major sources of heat production being core losses and I^2R losses in the motor conductors, where the quantity of the heat produced is based upon the resistance of a material. The resistance of materials increases with temperature, and therefore if the heat produced during operation is not transferred away from the system, the resistance of the materials increases, which also increases losses, resulting in the materials retaining more heat.

There are four main mechanisms of heat transfer: radiant, advection, conduction, and convection. They involve transfer of heat from objects to an adjoining substance; therefore, transferring heat away from a motor or actuator is not normally a concern when operated in Earth atmosphere.

It seems counterintuitive that if a motor or actuator or some other electromechanical device is operating at 80 K in outer space, there would be any problem with transferring heat away from the device. Generally, efforts are made to increase the area from which heat

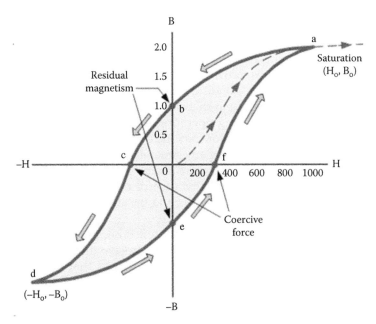

FIGURE 8.11
Magnetic hysteresis loop.

can be radiated into space and to increase the conductive area to surrounding hardware. However, when operating in outer space, where most low temperature motors and actuators are operated, they are likely being operated in vacuum. In vacuum, the main mechanism of heat transfer is radiation, which depends on large temperature differences to be effective. Even when operating in vacuum, heat can still be transferred by conduction to an object(s) to which the motor or actuator is attached. In outer space applications, the heat can be transferred to an adjoining object such as a part of a frame upon which the object is mounted, thereby increasing the total effective area available to dissipate heat into space. The use of low temperature motors and generators in outer space applications requires an entire system approach. Using a drive motor on a lunar rover as an example, the motor is attached to the rover's frame, which has multiple members and other attachments.

The development of a motor involves a whole system solution in which the low temperature motor or actuator is being used, and since no two systems would be the same, a general design solution does not exist. Nonetheless, heat transfer when implementing low temperature motors and actuators cannot be ignored or addressed lightly. Not properly addressing heat transfer could result in motor or actuator losses and overload on the electrical power supply or ultimately in the destruction of the motor or actuator. This is a good example of the many disciplines, thermodynamics in this case, besides those required to design a motor, that are required when designing low temperature systems.

8.7 Summary

The understanding of how motors and actuators operate involves a mix of disciplines. The focus of this chapter has been on design solutions related to low temperature motors and actuators; they were discussed in this chapter though a description of a case study. The design considerations affecting motor performance were reviewed in relation to low temperatures ranging from 40 to 403 K. Generally, any low temperature motor or actuator that is purchased or designed should use fully space/cryogenic qualified lubrication (if any), bearings, and cryogenic materials in the construction, including a titanium or invar case, wiring, laminates, and lamination bonding alternatives. Every attempt should be made to match the CTE of all components, so that the motor or actuator will maintain operating tolerances over the widely varying operating temperatures. Thermal characteristics of the low temperature motor or actuator must be designed to operate in a hard vacuum with partial gravity, especially displacing any generated heat that would have any adverse effects such as lowering the low temperature motor or actuator's efficiency. Additional design requirements include high reliability, ease of maintenance, low system volume, low mass, high efficiency, low operating power, and the ability to sustain launch loads and vibrations.

Although not directly discussed in this chapter, other considerations may require attention too. These include the ability to handle abrasive dust, solar and cosmic radiation exposure, and low out-gassing. Other design issues that may require attention include modular design characteristics, fail-safe operation, and reliability when designing pumps of fluids and slurries or mechanisms of moving biomass, particulates, and solids.

Acknowledgments

A portion of the data and descriptions presented in this chapter were performed under NASA SBIR contract number NNX11CB84C, titled "Lightweight High Efficiency Electric Motors and Actuators for Low Temperature Mobility and Robotics Applications" awarded to QM Power Inc. where Boeing was a subcontractor on the award to perform low temperature tests in their helium cryostat.

Some of the research reported in this chapter was conducted at the Jet Propulsion Laboratory (JPL), California Institute of Technology, (Caltech) under a contract with the National Aeronautics and Space Administration (NASA). The authors would like to thank Ted Iskenderian and Mircea Badescu, JPL/Caltech, Pasadena, CA, for reviewing this chapter and for providing valuable technical comments and suggestions.

References

Full, R. J., and K. Meijir, Metrics of natural muscle function, Chapter 3, in: Bar-Cohen, Y. (ed.), *Electroactive Polymer (EAP) Actuators as Artificial Muscles—Reality, Potential and Challenges*, 2nd Edition, ISBN 0-8194-5297-1, Bellingham, WA: SPIE Press, Vol. PM136 (March 2004), pp. 73–89.

Kurtin, O. D., *Satellite Life Extension: Reaching for the Holy Grail*, Via Satellite Publications (March 1, 2013), http://www.satellitetoday.com/publications/2013/03/01/satellite-life-extension-reaching-for-the-holy-grail/

Nalbandian, R., and A. E. Hatheway, Development of a cryogenic nanometer-class repeatability linear actuator, *Aerospace Mechanisms Symposium 2000*, New York, NY: AIAA (2000).

Park, J. M., *A Piezoelectrically Actuated Cryogenic Microvalve With Integrated Sensors*, PhD dissertation. University of Michigan (2009), pp. 1–103.

Slusher, R., *Motion Reducing Flexure Structure*, U.S. Patent Number 5,969,892, Oct. 19, 1999.

Streetman, S., L. Kingsbury, *Cryo Micropositioner*, U.S. Patent Number 6,478,434, Nov. 12, 2002.

Timmerhaus, K. D., R. W. Fast, R. Reed (eds.), *Advances in Cryogenic Engineering*, ISBN-10: 1461398762; ISBN-13: 978-1461398769, New York, NY, Springer (2013), p. 1178.

Wigley, D., *Mechanical Properties of Materials at Low Temperatures*, The International Cryogenics Monograph Series, ISBN-10: 0306305143; ISBN-13: 978-0306305146, New York, NY, Springer (1971), p. 326.

Internet References

Materion Corporation, www.materion.com

NASA: Advanced Space Transportation Program: Paving the Highway to Space, http://www.nasa.gov/centers/marshall/news/background/facts/astp.html

NASA's Aqua Satellite, http://www.science.nasa.gov

NASA/UCLA, http://www.space.com

Wayne Storr: Basic Electronics Tutorials Site, http://www.electronics-tutorials.ws/electromagnetism/magnetic-hysteresis.html

9

Sample Handling and Instruments for the In Situ Exploration of Ice-Rich Planets

Julie C. Castillo, Yoseph Bar-Cohen, Steve Vance, Mathieu Choukroun, Hyeong Jae Lee, Xiaoqi Bao, Mircea Badescu, Stewart Sherrit, Melissa G. Trainer, and Stephanie A. Getty

CONTENTS

9.1 Introduction

NASA's key science goals for the exploration of the solar system seek a better understanding of the formation and evolutionary processes that have shaped planetary bodies and emphasize the search for habitable environments. Efforts are also made to detect and quantify resources that could be used for the support of human exploration. These themes call for chemistry and physical property observations that may be best approached by *in situ* measurements. NASA's planetary missions have progressively evolved from remote reconnaissance to *in situ* exploration with the ultimate goal to return samples. This chapter focuses on the techniques, available or in development, for advanced geophysical and chemical characterization of icy bodies, especially Mars's polar areas, Enceladus, Titan, Europa, and Ceres. These astrobiological targets are the objects of recent or ongoing exploration whose findings are driving the formulation of new missions that involve *in situ* exploration.

After reviewing the overall objectives of icy body exploration (Section 9.1), we describe key techniques used for addressing these objectives from surface platforms via geophysical observations (Section 9.2) and chemical measurements (Section 9.3).

9.1.1 Science Rationales for the Exploration of Mars's Polar Regions

Mars is the fourth planet from the sun, the second smallest planet in the solar system, after Mercury, and a neighboring body to Earth. It was named after the Roman god of war, and it is also known as the "Red Planet" because of its reddish appearance resulting from the iron oxide that is prevalent on its surface. Polar temperature ranges from –143°C to –125°C, and operating on its surface requires the use of mechanisms that can tolerate low temperatures. Its surface contains a thin atmosphere with a pressure of about one-hundredth of Earth's atmosphere, and it is filled with impact craters, volcanoes, valleys, deserts, and polar ice caps. Its rotational period and seasonal cycles are somewhat similar to those of Earth, as is the axial tilt that produces the seasons. For its potential of harboring life in some form including extinct, Mars has been one of the most attractive bodies for planetary exploration. This includes the five orbiting spacecraft—2001 Mars Odyssey, Mars Express, Mars Reconnaissance Orbiter, MAVEN, and Mars Orbiter Mission—and the landed missions—Mars Exploration Rovers Spirit and Opportunity, the Phoenix lander, and the Mars Science Laboratory Curiosity. In particular, the Phoenix mission deployed a lander at high latitude where ice was found as well as perchlorates [Hecht et al., 2009]. Several missions [e.g., Ice Breaker by McKay et al., 2013; Cryobot by Aharonson et al., 2009; http://web.gps.caltech.edu/~oa/simulations.shtml] have been proposed to gain further insight into the structure of Mars's polar caps and their isotopic makeup. The latter can help constrain the origin of Mars's ice, which bears implications to constrain the origin of Earth's water and organics.

9.1.2 Enceladus Exploration

Saturn's moon, Enceladus, was discovered in 1789 by William Herschel. Enceladus is only about 500 km (310 mi) in diameter and has been extensively studied, first by the two Voyager spacecraft that passed nearby in the early 1980s, and more recently, since 2004, by the Cassini-Huygens Mission. Enceladus' surface reflects almost all the sunlight that strikes it, and it is mostly covered by clean ice with a surface temperature as low as –198°C.

In 2005, the Cassini orbiter discovered a water-rich plume venting from Enceladus' South Polar Region. Large rifts located around the South Pole, called "Tiger Stripes," shoot geyser-like jets of water vapor, other volatiles, and solid material, including sodium chloride crystals and ice particles, into space, at approximately 200 kg (440 lb) per second [Hansen et al., 2006]. More than 100 geysers have been identified [Porco et al., 2014] whose activity is driven by the tidal stress exerted by Saturn. Some of the water vapor falls back as "snow," whereas the rest escapes, and may be the source of most of the material making up Saturn's E ring [Kempf et al., 2008]. The heat production inferred from surface temperature measurements at the Tiger Stripes is about 15 GW [Spencer and Nimmo, 2013] and has not been explained yet. Further, in 2014, Cassini data provided evidence for a large south polar subsurface ocean of liquid water within Enceladus with a thickness of around 10 km [Iess et al., 2014]. These observations make Enceladus a major target for *in situ* exploration with focus on the South Polar Region and sampling of the plumes by geochemistry instruments during high-velocity flybys [Lunine et al., 2015]. Landers for Enceladus exploration have been proposed [e.g., Konstantinidis et al., 2015] that could help further the assessment of Enceladus' habitability potential.

9.1.3 Titan's Surface Exploration

Saturn's largest satellite Titan was discovered by Christiaan Huygens in 1655, and the fact that it had an atmosphere was established by Kuiper in 1944, who detected absorption bands due to methane at infrared wavelengths from ground-based observations. Sixty years later, the Cassini-Huygens mission unveiled Titan's complex world, which is very much akin to Earth in many regards but shaped by the cycle of hydrocarbons, especially methane.

Decades of modeling efforts, validated by ground-based and spacecraft observations (*Voyager I*, 1980, and *Cassini-Huygens*, since 2004), elucidated the complex chain of UV-induced photochemical reactions that take place in the atmosphere and the rates at which they proceed [e.g., Yung et al., 1984; Coustenis, 2005; Coustenis et al., 1991, 2003; Wilson and Atreya, 2004; Lavvas et al., 2008]. Using these constraints, Lavvas et al. [2008] established a refined photochemistry model of the atmosphere, and Cordier et al. [2009] used the outputs to estimate the fluxes of photochemical products to the surface. This information establishes two essential starting points about Titan: (1) the dominant products of the photochemistry are ethane C_2H_6 and propane C_3H_8, which are in liquid state at Titan's surface conditions (1.5 bar of N_2, 92–94 K); (2) the higher-molecular-weight compounds would be in the solid state on the surface, where they may form the observed dunes [e.g., Lorenz et al., 2008), and/or contribute to evaporitic materials tentatively detected by the visual and infrared mapping spectrometer (VIMS) [e.g., Barnes et al., 2011]. These solid organic compounds are thus expected to form a blanket cover on the presumed water ice bedrock. A recent study discusses the possibility of the emergence of a form of life not based on liquid water in Titan's hydrocarbon lakes [Stevenson et al., 2015].

Titan, and particularly its lakes, has been identified as a high-priority target by the NRC Planetary Science Decadal Survey "Visions and Voyages" for 2013–2022. Several mission concepts have been suggested for the follow-on exploration of Titan with focus on the habitability potential of its surface and subsurface. The Titan and Saturn System Mission concept developed in 2008 [Coustenis et al., 2009] introduced an architecture involving an orbiter, balloon, and surface element. More recent concepts have focused on landers, especially lake landers [e.g., Stofan et al., 2013; Mitri et al., 2014], and even submarines [http://www.gizmag.com/nasa-titan-submarine-concept/35960/].

9.1.4 Europa Exploration

Europa is the smallest of Jupiter's Galilean satellites and the sixth largest moon in the solar system. Its average surface temperature is about −171°C. It was discovered by Galileo Galilei in 1610. Ground-based observations (e.g., radar) and flybys by the Voyager spacecrafts have revealed a complex surface dominated by ice and salt compounds associated with tectonic features (e.g., ridges). The Galileo mission, launched in 1989, provided the majority of the current data on Europa, revealing a geologically young surface, as indicated by the variety of tectonic features and the scarcity of impact craters. Galileo also uncovered the presence of a deep ocean from the detection of an induced (time-variable) magnetic field. That same technique also led to the discovery of deep oceans in the other Galilean moons Ganymede and Callisto. Because of its active geology and expected high tidal heat input, Europa's ocean is believed to be relatively close to the surface (<25 km) [Pappalardo, 2010] and in contact with a rocky core, while the thick hydrospheres of Ganymede (~800 km) [Vance et al., 2014] and Callisto (~200–400 km) [Schubert et al., 2004] suggest the presence of high-pressure ice phases at the interface with the rocky core. Ongoing analyses of Galileo datasets over the past 20 years continue to reveal surprises at Europa, including evidence for subduction of the brittle upper crust of Europa's ice, a key feature of plate tectonics on Earth [Kattenhorn and Prockter, 2014]. Also recently, Hubble Space Telescope observations of Jupiter's magnetosphere in the vicinity of Europa have revealed evidence of transient water vapor plumes [Roth et al., 2014]; these have yet to be confirmed by subsequent observations.

Because of the strong likelihood of having a global ocean in contact with a rocky mantle, Europa is a primary target for future exploration, as illustrated by the numerous mission concepts developed since even before the end of the Galileo mission (Figure 9.1). These missions all focus on getting a better understanding of Europa's internal structure and

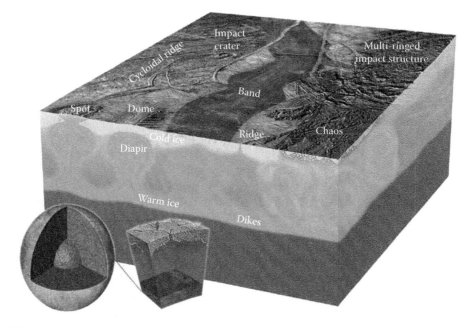

FIGURE 9.1

Europa's hydrosphere structure based on the current state of knowledge inferred from observations by the Galileo mission. (Courtesy of NASA/JPL-Caltech.)

surface composition. Several concepts have considered some form of *in situ* platform, such as penetrators [Gowen et al., 2011] and legged landers [Pappalardo et al., 2013].

9.1.5 Ceres Exploration

Ceres is the largest body in the asteroid belt. It was also the first asteroid to be discovered when it was observed by Giuseppe Piazzi in 1801. Ceres's density suggests that it contains about 50% of ice in volume [McCord et al., 2011]. Most of our knowledge of Ceres to date comes from ground-based and Earth-orbiting telescopes, especially NASA's Hubble Space Telescope [Thomas et al., 2005]. An extended vapor cloud at Ceres was detected with the Herschel Space Telescope [Küppers et al., 2014]. This is evidence for the presence of ice in Ceres's subsurface that was predicted by Thomas et al. [2005]. Ground-based infrared observations enabled the discovery of carbonates and possibly of brucite [Rivkin et al., 2006], which are signatures of formation in a hydrothermal environment. The Dawn mission achieved rendezvous at Ceres in early March 2015 and will carry out in-depth investigations of Ceres's chemistry and geology that will help answer questions on Ceres's origin and habitability potential. The prospect of shallow ice tables or even cold traps in shadowed craters has already prompted interest for surface exploration as a possible precursor to a Europa lander [Poncy et al., 2009]. While the Dawn mission is in its early stage at Ceres it is expected that it will obtain critical data needed to formulate a follow-on *in situ* mission.

9.2 Geophysical Exploration Techniques

This section reviews key geophysical techniques that have been developed with the objective to characterize the physical properties of outer planet satellites with *in situ* platforms. Such techniques complement global-scale observations of the gravity and magnetic fields. These techniques are as follows: acoustic radar with application to the characterization of Titan's lake; seismometry with application to Europa's deep interior; and ground-penetrator radar with application to Enceladus in particular. Other techniques are notable, such as electric fields, induction, etc.

9.2.1 Sonar: Acoustic Radar

Conceived at a time when it was widely thought that Titan had a global ocean, the Surface Science Package (SSP) carried by the Huygens probe, and deployed by the Cassini Orbiter in January 2004, had the ability to measure sound velocity [Zarnecki et al., 2002]. This measurement could have been used to determine the relative amounts of methane and ethane in the ocean. It also had a limited ability to do sonar measurements.

The key challenge to using sonar, also known as acoustic radar, is to employ piezoelectric transducers that are operational at the extremely low temperatures (~90 K), which is specifically applicable to Titan's lakes. The issues involve a very large thermal expansion mismatch that may be associated with the construction materials and the composition of the liquids (methane/ethane) in these lakes. To address the challenges, one needs to take advantage of the piezoelectric materials that can potentially be used to serve as transmitters and receivers of acoustic waves. The required transducer needs to perform optimally at the temperature of 90 K. Also, operation in liquid methane/ethane requires addressing the related physics that affects the response and performance of the sonar as an analyzer. Combining the use of a transducer array [Sherman and Butler, 2007] and phase control, one may be able to scan the terrain without physically moving the transducer. Combining the use of an array and phase control as well as vehicle movement, 3D mapping can be accomplished, which is scientifically desirable.

Sonar can be used to perform ultrasonic analysis for investigating Titan's wet subsurface beneath and around its hydrocarbon lakes [e.g., Stofan et al., 2007; Hayes et al., 2008]. Such a scientific instrument, based on the transmission and reception of acoustic waves by one or more piezoelectric transducers, can be installed on an *in situ* platform (lander, floater, submarine, rover, etc.) and measure the bathymetry of the lakes. This ultrasonic analyzer may be essential to measure directly the structure of the subsurface, and its interactions with the lakes: depth of liquid percolation, stratigraphy (porosity gradients and/or discontinuities between subsurface materials), existence and thickness of the organic deposits blanket, location of bedrock beneath, connectivity of the lakes, and existence and extent of a deep-seated aquifer.

One sonar system for Titan exploration has been reported in the literature. It was part of the Meteorology and Physical Properties Package (MP3), conceived at John Hopkins University/Applied Physics Laboratory [Lorenz et al., 2012], and proposed as part of the scientific payload for the Titan Mare Explorer (TiME), a Discovery 12 Step 2 mission concept. The sonar of the MP3 package was solely dedicated to measuring bathymetry of Ligeia Mare, using standard (low-efficiency) PZT transducers. Alternatively proposed by JPL, and described herein, is the use of recent developments in high-efficiency piezoelectric transducer technology to achieve similar capabilities with less power. Such an instrument would enhance the scientific return of an ultrasonic analyzer by enabling the capability of sounding the subsurface beneath the hydrocarbon lakes (Figure 9.2). The terrain beneath the piezoelectric transducer can be scanned without physically moving the transducer. This can be accomplished by the use of a transducer array and phase control. The image of the lake bottom and the subsurface can be made in 3D by combining the operation of the sounding device with the vehicle movement.

Sonar as an analytical instrument can be used to emit and receive sound waves, where the reflected or backscattered echoes from acoustic interfaces are used to measure the distance of objects by analyzing the travel time between the transducer and the objects/layers. This principle is also used in Navy sonar systems, but the focus in this application is beyond mapping the lake bottom surface and immersed objects. It is interesting to note that the same principle is also widely used in diagnostic medical imaging and nondestructive testing. The main differences between medical imaging and underwater sonar are the operating frequency and the power level. In general, megahertz ultrasound (1–10 MHz) is used for nondestructive testing and medical imaging for high resolution, and this type of ultrasonic test does not require high power level (typically much less than 1 W). In contrast, for great distance range finding and imaging, sonar transducers use higher power with low to moderate frequencies in the Hz–kHz range. This is also a requirement for an ultrasonic instrument that may be used as an analyzer for the subsurface of Titan's lakes, but the frequency needs to be focused on the ~1–100 kHz range.

9.2.1.1 Transducer Composition and Behavior at Cryogenic Conditions

Piezoelectric transducers convert applied electrical signals into acoustic radiation and are designated as projectors. Furthermore, transducers that convert received acoustic radiation into electrical signals are designated as hydrophones. The performance criteria for projectors and hydrophones are quite different; for example, the major concern of projectors is high power output, while that of hydrophones is high sensitivity [signal-to-noise ratio (SNR)].

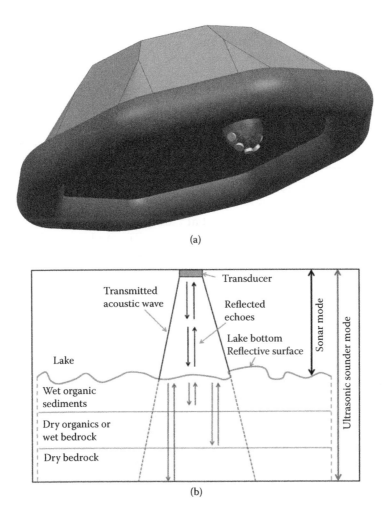

(a)

(b)

FIGURE 9.2
(a) The envisioned marine lander and the ultrasonic analyzer for the exploration of Titan. (b) Illustration of instrument operation compared with conventional sonar systems.

Piezoelectric transducers can serve as both projectors and hydrophones, in which the performance is greatly dependent on the properties of the piezoelectric materials such as mechanical Q_m (inverse of mechanical loss) and electromechanical k coupling. High values of these coefficients allow generating broad bandwidth signals and provide high sensitivity and increased power efficiency. Currently, the majority of piezoelectric materials for such transducers are ferroelectric materials due to their high electromechanical properties, which arise from the two types of contributions: the intrinsic (lattice effects) and extrinsic (the motion of ferroelectric–ferroelastic domain walls) contributions in ferroelectric materials. One of the most important characteristics of this kind of material is the morphotropic phase boundary (MPB), which refers to the boundary between two compositions where the two phases are present in equivalent energy states. MPB is an important concept for ferroelectric materials as MPB compositions offer enormously high dielectric and piezoelectric properties as a result of enhanced intrinsic contributions. Lead zirconate titanate (PZT) is one of the most widely used piezoelectric materials because of its MPB characteristics [Jaffe et al., 1971].

The piezoelectric response contains not only the intrinsic contribution, but also an extrinsic contribution caused by the movement of non-180° domain walls, which is strongly temperature dependent. MPB-based PZTs are generally tailored with dopant ions, which impede or facilitate domain wall movements. Importantly, in PZT ceramics, more than 50% of the net piezoelectric responses arise from these extrinsic contributions; therefore, when PZT materials are used at cryogenic temperatures, most of the extrinsic contributions are frozen out. Consequently, the materials lose their piezoelectric performance; for example, the piezoelectric d coefficient was reported to decrease from 760 to 220 pC/N when the operating temperature was decreased from 300 to 30 K [Park and Shrout, 1997; Hackenberger et al., 2008]. This indicates the necessity for appropriate piezoelectric materials to be used to make an ultrasonic analyzer for operation at cryogenic temperatures. It is interesting to note that the transducer of the JHU/APL sonar system is made of PZT-5A [Lorenz et al., 2012].

Recently, the domain engineered <001> relaxor-PT single-crystal family, such as PZN-PT, PMN-PT, and PIN-PMN-PT, has been studied extensively due to their extremely high piezoelectric responses, strain over 1.7%, piezoelectric constant d_{33} over 2000 pC/N, and electromechanical coupling factor k_{33} over 90%, with almost nonhysteretic strain-field behavior [Park and Shrout, 1997]. Of particular significance is that, in contrast to PZT ceramics, the mechanical Q_m values can be tailored by the crystallographic orientation, being on the order of 200 and >800 for <001> and <011> oriented PMN-PT crystals, respectively, without sacrificing the electromechanical k coupling. Since the origin of such high electromechanical properties of relaxor-PT single crystals is the polarization rotation effect (i.e., intrinsic contributions), the property degradation at cryogenic temperatures is much lower than in PZT ceramics, making them promising candidates for cryogenic transducers from the perspective of bandwidth and power efficiency of transducers [Fu and Cohen, 2000]. The relaxor-PT single-crystal transducers, specifically <110> oriented binary PMN-PT or ternary PIN-PMN-PT, can be used to produce a probe that can potentially sustain in the very cold conditions, is inert to potential chemical reactions, and is constructed of materials with minimal thermal mismatch.

9.2.1.2 Estimates for the Operation of Sonar on Titan

The estimation of detection range for a sonar transducer for a given input power is one of the most important considerations for the success of research related to potential missions to Titan. In order to accurately estimate the detection range for mapping the topography of Titan's lake, the elastic properties of propagating media need to be known. Table 9.1 shows reference properties of various media from several sources, which allow for the estimation of the transmission range and/or transmission loss of acoustic waves from sonar using a given input power.

For the detection of acoustic wave from a sonar system, the SNR should be higher than the detection threshold (DT), that is, SNR > DT, where SNR can be written as follows:

$$\text{SNR} = L_S - L_N = (\text{SL} - 2\text{TL}) - (\text{NL} - \text{DI}) \tag{9.1}$$

where SL is source level, DI is the directivity index (in the case of omni-directional, DI = 1), TL is the one-way transmission loss, and NL is the background noise level. The value of the detection threshold of a sonar transducer is dependent on the transducer performance and signal processing method.

TABLE 9.1

Material Properties Used for Assessing the Ability of the Proposed Ultrasonic Analyzer to Detect Subsurface Interfaces

Material	Vp (km/s)	Vs (km/s)	Elastic modulus (GPa)	Attenuation
Methane (94 K)	1.520 94 K [Singer, 1969] 1.490 96 K [Straty, 1974]	–	0.9–1.9 GPa [Marx and Simmons, 1984; Shimizu et al., 1996])	Alpha/$f^2 \times 10^{17}$: 5.6–6.2 cm^{-1}/s^2 [Singer, 1969; Straty, 1974]
Ethane (95 K)	1.974 [Tsumura and Straty, 1977]	–		
H$_2$O ice (90 K)	4.2 [Proctor, 1966]	1.98 [Proctor, 1966]	5–9 GPa Young modulus for polycrystalline ice	~1 db/100 m at –25°C for 35 and 60 MHz [Johari and Charrette, 1975]
Benzene (solid, 170 K)	~2.8 [Heseltine et al., 1964]	–	5.72 GPa Linear extrapolation yields 6.5 at 90 K [Heseltine et al., 1964]	~2–4 db/cm at 255 K. [Heseltine et al., 1964]; 3.1 × 10^5 cm/s at 273 K [Liebermann, 1959]

The source level (SL) is defined as the intensity of the radiated acoustic wave relative to the intensity in medium referenced to 1 µPa at 1 m, given in Equation 9.1:

$$SL = 10\log\left(\frac{I_S}{I_0}\right) = 10\log(W_a) + 168.9 \text{ dB re: } 1 \text{ µPa at } 1 \text{ m} \tag{9.2}$$

where W_a is the acoustic output power and I_0 is the reference sound intensity of methane, referenced to 1 µPa at 1 m.

The transmission loss (TL) includes all the effects of the energy losses, such as geometric losses (spreading, spherical, or cylindrical), and attenuation due to scattering, viscosity, and adsorption. The main source of attenuation is generally associated with absorption, where acoustic energy is converted into heat energy. In this estimation, only spreading and absorption losses are considered and these are the main causes of transmission loss of acoustic waves. Transmission loss due to the effects of spreading and absorption can be expressed as follows:

$$TL = 20\log(x) + \alpha x \tag{9.3}$$

where x is the distance and α the is attenuation coefficient. Note that the attenuation coefficient has a strong frequency dependence, that is, there are much greater losses at a higher frequency. In addition, the attenuation is also temperature dependent, generally increasing with decreasing temperature. Unfortunately, the attenuation coefficients of most materials at 90 K are not available; thus, from this estimation, we used the attenuation values shown in Figure 9.3.

The materials presented in Table 9.1 are used as a reference to estimate one-way sound pressure level (SPL) at distance through the structural model presented in Figure 9.4. Using a given 10 W input power, the transmission loss and pressure level of the acoustic wave (predicted by Equations 9.2 and 9.3, respectively) can be estimated as a function of distance with different frequencies (see Figure 9.2). From the figure, it can be seen that the most

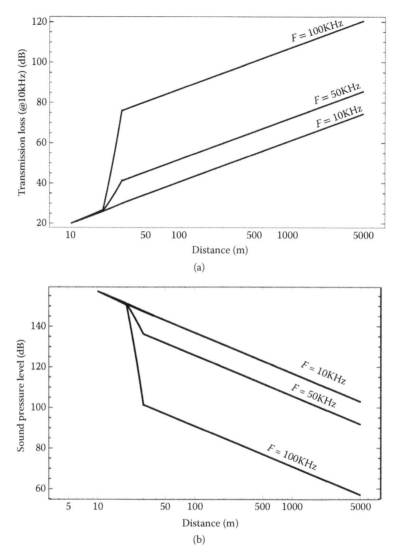

FIGURE 9.3

Transmission loss (a) and sound pressure level (b) of sound wave through our nominal subsurface model for an ultrasonic analyzer using 10 W input power.

dominant factor for transmission loss is spreading loss for sonar range (<10 kHz); however, with increasing frequency (ultrasound >20 kHz), the attenuation coefficient factor becomes a significant factor in figuring the transmission range. Although this is a rough estimation, it is promising that if we design the sonar below 20 kHz, it is possible for sounding the subsurface beneath Titan's lakes with 10 W.

If we assume that the transducer can detect the sound when DT >0, we can also estimate the detection range with Equation 9.1 for a given input power, assuming that NL − DI is 0, which is shown in Figure 9.4. From the figure, it can be seen that the sonar transducer can detect sound up to 50 kHz for a given 10 W. Therefore, using a frequency of a few tens of kHz, the sonar system as an ultrasonic analyzer can enable the transmission of sound

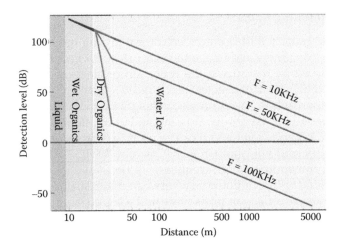

FIGURE 9.4
Influence of the operating frequency on the threshold detection level, using a nominal 10 W input power, for our nominal subsurface structural model. This assumes no significant suspended elements from wind or wave action.

through the various layers of the Titan subsurface, down to several hundreds of meters, providing detailed information on its structure.

The use of acoustic radar to perform analysis on Titan may enable detection of various anomalies and discontinuities as well as the characterization of material properties. Generally, detection requires an acoustic property mismatch between the propagating medium and the presence of other materials such as cryo liquids, sediments, run-off, and so forth.

9.2.1.3 Pulse-Echo versus Phase-Shift Techniques

Successful mapping of targets in Titan's lake requires effective signal processing and analysis. The most widely used methods of transmitting and receiving ultrasonic signals are the pulse-echo and the phase-shift techniques. For the case of the pulse-echo technique, the same transducer is used for both a projector and a hydrophone, where the accuracy depends on the accurate measurements of the time-of-flight (ToF) value between two successive echoes; the distance of the object is estimated from the product of the ToF and the sound velocity when the path is homogeneous. Although it is proven to be simple and inexpensive, difficulties arise from the fact that the received signals may contain significant noise and additional reflections from various sources. Potential noise sources include ambient lake noise, internal probe noise, preamplifier noise, and the reverberation from the lake surface. These unwanted echoes make the measurement of the ToF difficult.

In a phase-shift method, the projector generates a continuous wave, whose echo is detected by a separate hydrophone. In this case, the distance information is determined by comparing the phase shift between the transmitted and the reflected waves. Although better performance than with pulse-echo can be obtained as the transducer can be optimized for only projecting or receiving acoustic signals, complex hardware is required to measure phase shift. To effectively extract the distance and directional information from the received signals in real time, various signal processing methods and algorithms need to be investigated including cross-correlation analysis and deconvolution for time-frequency analysis.

9.2.2 Ground-Penetrating Radar

Ground-penetrating radar (GPR) is a nondestructive testing method that is based on the use of radar pulses to examine and image the subsurface for geophysical characterization. The technique is based on measuring the reflection of electromagnetic radiation in the microwave band (UHF/VHF frequencies) of the radio spectrum.

Given that this approach is based on electromagnetic radiation, the method is operational at low temperatures. Generally, this method is applied for use in various media including rock, soil, ice, and water and it can be used to detect subsurface objects, changes in material properties, and voids and cracks [Conyers, 2004; Daniels, 2004; Jol, 2009]. The technique involves transmitting high-frequency radio waves in the range of 10 MHz to 1 GHz into the ground. When a boundary between materials having different dielectric constants is encountered, reflection takes place and a signal is returned to the surface. An antenna is used to receive the returned signals, and the response provides information about the ground content. The operating principle is somewhat similar to seismic and ground sonar tests, except that electromagnetic energy is used instead of acoustic energy. Like seismometry, described later, the timing of the reflected signal yields information on the geometry (thickness, slope) of stratifications. The signal amplitude, compared to the amplitude of the source, carries information on the properties of the probed material. The material response to electromagnetic radiation is generally referred to in terms of "transparency." A pure water ice layer is transparent to the signal while a material rich in impurities tends to scatter and attenuate the signal, so that its recorded amplitude may be a small fraction of the source intensity. The effective depth of penetration of GPR is limited by the transmitted center frequency, the electrical conductivity of the ground, and the radiated power. The higher the frequency, the lower the penetration depth, but the higher the resolution that can be obtained, and in choosing the operating frequency, one needs to do a trade-off between the resolution and the depth of penetration. In addition, higher electrical conductivity of the subsurface is associated with higher attenuation and therefore decreased depth of penetration. Generally, using low frequencies provides depth of subsurface penetration that can be achieved with GPR testing and can be as much as several kilometers for pure water ice. When testing dry soils and hard rocks such as granite and limestone, which have relatively low conductivity, the depth of penetration is limited to several meters. On the other hand, moist soils having high electrical conductivity lead to penetration that is in the range of a few centimeters.

Ground-penetrating radar measurements have been used on Earth for characterizing the thickness and structure of ice sheets [e.g., Legarsky et al., 2001]. This technique has also been applied to Mars during two ongoing missions: the MARSIS experiment on ESA's Mars Express mission and the SHAllow RADar investigation on the Mars Reconnaissance Orbiter [Seu et al., 2004]. While its penetration into Titan's surface is limited to a few centimeters, the Cassini Orbiter Ku-Band radar returned significant information on the surface properties of this singular world: radiometry (for temperature), altimetry, synthetic aperture radar imaging, and dielectric properties [e.g., Elachi et al. 2006; Hayes et al. 2010; Le Gall, et al. 2011; Lorenz et al. 2014]. The Cassini Orbiter also performed bistatic radio science experiments (Cassini-Titan-Earth geometry) providing information on the volumetric properties of Titan's surface [Marouf et al., 2014].

While the latter are remote observations, more recently GPR was used as part of the bistatic radar system COmet Nucleus Sounding Experiment by Radiowave Transmission (CONSERT) of the Rosetta orbiter and its lander Philae. CONSERT did not meet its primary objectives following the half-successful landing of Philae. However, measurements acquired during Philae's descent returned partial permittivity mapping of comet 67P [Plettemeier et al., 2015].

In situ ground-penetrating radar experiments have been suggested for several Mars missions, including the WISDOM radar [Ciarletti et al., 2011] proposed to the ESA's Exo-Mars mission that was eventually descoped. The Mars 2020 payload includes a radar system called RIMFAX (Radar Imager for Mars SubsurFAce eXperiment) (Figure 9.5) that will be used to search for buried ice on Mars as part of the mission thrust for characterizing Mars's habitability and identifying potential samples of interest for return to Earth. *In situ* GPR was also suggested for the exploration of Enceladus [e.g., Konstantinidis et al., 2015] for characterization of the ice shell. It is remarkable that the most recent generation of that type of instrument is less than 1 kg and requires less than 1 W of power [Kim et al., 2012]. While still under development, the projected penetration depth for that class of radar is of the order of a few hundred meters for pure ice down to a few meters if ice is mixed with rocks.

9.2.2.1 Dielectric Properties of Planetary Surfaces

The interaction of material with electromagnetic waves is described via two properties: (1) the electrical permittivity $\varepsilon = \varepsilon' - i\varepsilon''$ (or dielectric constant when measured relative to the value for air) and (2) the magnetic permeability $\mu = \mu' - i\mu''$, where prime and second refer to the real and imaginary parts of these complex parameters, respectively. The real part is a measure of how much energy is stored and radiated by the material, while the imaginary part is a measure of the amount of energy that is lost in the material as a consequence of energy attenuation effects, which is generally a function of temperature, physical and chemical properties, and impurities.

When a radar pulse reaches the boundary between two materials of differing dielectric properties, a portion of the incident energy is reflected backwards while the remainder is transmitted forward. As successive dielectric interfaces are encountered, the signal experiences additional reflections and losses by absorption (affecting both the transmitted and reflected portions of the signal) until its strength is so attenuated that the reflected signal can no longer be detected above the ambient noise.

The penetration depth is primarily a function of the nature and temperature of the sounded terrains. Penetration depth can be estimated from observations at terrestrial

FIGURE 9.5
Illustration of the observations to be obtained by the RIMFAX radar planned for NASA's Mars 2020. (From Hamran, S.E. et al., *AGU Fall Meeting*, 2014, http://adsabs.harvard.edu/abs/2014AGUFM.P11A3746H.)

analogs and experience gained from measurements at Mars [e.g., Phillips et al., 2008]. Mars Advanced Radar for Subsurface and Ionospheric Sounding (MARSIS) aboard the European Mars Express spacecraft has demonstrated very low attenuation within the Mars polar ice, although Mars polar ice is believed to contain up to 20% nonice impurities. The primary reason for low bulk attenuation is the cold temperature of the ice.

Both the Antarctica and Mars polar caps are relevant analogs to Titan's and Enceladus' subsurface because they associate water ice and clathrate hydrates (gas molecules encaged by water molecules, e.g., Choukroun et al., 2013), which are predicted to be major constituents of both Titan's and Enceladus' icy shells [e.g., Tobie et al., 2006; Kieffer et al., 2006; Fortes, 2007]. On the other hand, limited knowledge of Titan's surface composition and structure, for example, porosity at various scales, precludes a firm prediction of the penetration depth that would be achieved. Ground-penetrating radar investigations at terrestrial ice sheets for the same range of frequencies considered for these instruments (mean frequency and bandwidth of a few tens of MHz) have reached depths of up to 3 km [e.g., Studinger et al., 2003; Holt et al., 2006]. SHARAD's observations have achieved penetration depths of 0.1 km (in porous regolith) to 1.5 km at Mars's north polar layered terrains. The very low temperatures (i.e., < 200 K) at the Saturnian satellites would enhance the transparency of the material and the radar signal penetration. Based on these results, the expected cold ice temperature expected at both Titan and Enceladus would make it possible for radar to detect horizons as deep as many tens of kilometers.

The SHARAD instrument (SHAllow RADar on Mars Reconnaissance Orbiter) was used as a reference for assessing the science return expected at Enceladus from *in situ* radar. The instantaneous bandwidth equal to or greater than 10 MHz would yield a vertical altimetry resolution of 10 m and would allow the resolution of horizons Enceladus' crust with the same level of accuracy. A preliminary study developed for the Titan and Saturn System Mission [Coustenis et al., 2009] suggests that a mean frequency of 20 MHz is representative of the range of frequencies that can achieve the required spatial resolution of a few kilometers as well as subsurface sounding a few kilometers deep in an icy body like Enceladus.

9.2.3 Seismology

Seismic investigations offer the most comprehensive view into the deep interiors of planetary bodies. Developing missions (InSight and Europa Lander) have identified seismometry as a critical measurement to constrain the interior structure and thermal state of astrobiological targets. By pinpointing the radial depth of compositional interfaces, seismic investigations can complement otherwise nonunique composition and density structures inferred from gravity and magnetometry studies, such as those planned for NASA's Europa mission and ESA's JUICE mission. Seismic investigations also offer information about fluid motions within or beneath ice, which complement magnetic studies, and they can record the dynamics of the shell, which would shed new light on crack formation and propagation. Characterizing the internal workings of oceanic icy moons will figure prominently in the next Planetary Science Decadal Survey, and seismic investigations offer the best view into their deep internal structure—the compositions and radial extent of their icy layers, rocky mantles, and metallic cores, if present. Such information is needed if NASA is to seriously consider putting a submarine into a deep ocean. The evidence for salty oceans within Europa, Ganymede, Callisto, Enceladus, and Titan has produced intense public and scientific interest in the possibility of life in extraterrestrial oceans. Seismometry's utility for addressing these problems of intense public interest thus makes it a key tool for future icy moon exploration.

The highest priority goals of both NASA's Europa mission and ESA's JUICE mission are to "characterize the ice shell and any subsurface water, including their heterogeneity, and the nature of surface-ice-ocean exchange," and to "characterize the extent of the ocean and its relation to the deeper interior," respectively. The studied Europa Lander [Pappalardo et al., 2013] would also address these goals, using a multiband seismometer (MBS: 0.1–50 Hz and 125–250 Hz) based on the Exomars and Insight instruments. The architecture for doing this called for three-axis seismometers in each of the lander's six legs, with required spacing between them of several meters or more.

9.2.3.1 Identification and Modeling of Possible Seismic Sources

Tides may be a major source of seismicity in icy moons, as they are in Earth's moon [Bulow et al., 2007], where almost all deep-focus lunar seismic events are clearly related to tidal stress. The locations are identical within the ability of the seismic network to locate them, and the waveforms of the repeat events are almost identical. It thus appears that there are some "special locations" where the tidal stress triggers slip on preexisting fault planes. Tides also trigger a small part of Earth's seismic activity, such as in the well-studied Cascadia subduction zone [Royer et al., 2015]. Europa's ice may see more tidally generated activity; its ice is likely more brittle than the lunar mantle rock [Nixon and Schulson, 1987; Schulson, 2006], but the driving tides, while stronger, have a much longer period. Forced librations in longitude—periodic variations in rotation rate, driven by the torques acting on a triaxial body—very likely create fluid motions in subsurface oceans in librating icy satellites [Le Bars et al., 2015]. In Europa and many other icy moons, gravitationally forced librations will create volume-filling turbulent flow, a possible seismic source similar to that seen from turbulent flow in terrestrial rivers [Tsai et al., 2012; Gimbert et al., 2014]. Riverine flows and flows through glacial channels [Tsai and Rice, 2012] may also be useful models for the downflow of dense brines from chaos regions on Europa into its underlying ocean [Sotin et al., 2002]. Turbulent flows in Europa's ocean may result from natural currents as well [Soderlund et al., 2014], providing a potential source of seismic information.

In the case of Europa, a major source of seismic waves comes from the tidal opening and closing of fractures [Lee et al., 2003], or slip along subduction-like features [Kattenhorn and Prockter, 2014]. Estimated signal strengths up to 100 Hz for impact and fracture sources span a range of ±80 dB relative to 1 J/Hz [Lee et al., 2003], and are detectable using available technology [e.g., Kovach and Chyba, 2001; Pappalardo et al., 2013].

Previous analyses of seismology on Europa [Cammarano et al., 2006; Panning et al., 2006] have concentrated on low-frequency signals (0.001–0.1 Hz) for which a normal-mode analysis is relevant. They concluded that events with magnitude 5 or larger could occur, and that mm accuracy displacement measurements would suffice to characterize the thickness of the ice shell and the depth of the ocean. They note that penetration of seismic waves originating near the surface into the suboceanic silicate region would be limited, but seismic sources within the silicate region would allow determination of the deeper structure, as has been done for the Moon [Goins et al., 1981; Garcia et al., 2011]. Seismic techniques to determine the internal structure of a body in the past required three or more seismic stations in order to locate the event (e.g., Mars and Lunar Network missions [Mocquet et al., 1998]). The InSight mission to Mars has determined a way to avoid this by measuring the travel times of surface waves that travel more than one time around the planet, or possibly detecting the location of impacting meteoroids that produce seismic events with orbital imaging. Active seismology using an artificial seismic source has been proposed [e.g., Scheeres, 2011], although it is difficult to implement. Ambient noise seismology is the new frontier and enables investigations of structures without quakes or active sources but requires sophisticated processing. Jennifer Jackson and Zhongwen Zhan at campus have been doing experiments on terrestrial ice flows that are relevant to Europa and Titan. A previously unconsidered aspect is generation of acoustic signals in the ocean itself. Earth's oceans generate a weak seismic hum [Kedar, 2011; Ardhuin, 2015]. Tidal flexure and librational response can be modeled to quantify forced motion in Europa's ocean, as well as seismic sources associated with collisions.

The architecture study, commissioned by NASA (Science Definition Team 2012 Study Report; Pappalardo et al., 2013) assumed three-axis seismometers on each of the lander's six legs. The investigation would probe the detailed structure of Europa's ice and ocean [Kovach and Chyba, 2001; Lee et al., 2003; Cammarano et al., 2006; Panning et al., 2006] during a nominal mission lasting 9 eurosols (32 days). A multiband seismometer model payload instrument, with capabilities based on Exomars, has also been adopted for Insight: low-pass 0.1–75 Hz and high-pass 125–250 Hz. The Europa Lander architecture is achievable in spite of Europa's intense radiation, but its seismic investigation warrants further analysis. The study referred to implementing such a multi-seismometer approach, but did not assess the improvement in accuracy of having at least three of the lander's feet firmly coupled to the underlying ice. Seismic investigations of Antarctic ice shelves as well as ice sheets with subglacial lakes provide the best analog for Europa, as noted in the Europa Lander study [Robin, 1958; Bentley, 1964; Roethlisberger, 1972]. Observations from Antarctic ice shelves suggest that certain phase amplitudes are very different than expected, for example with some phases being more strongly attenuated than might initially be assumed, and other phases being enhanced by resonance effects.

In summary, the state of the literature so far suggests the suitability of probing seismo-acoustic signals at frequencies from mHz to 100 kHz. Frequencies lower than 0.1 Hz could probe very long wavelength modes correlated global geophysical processes. A sufficiently accurate low-frequency seismometer would also measure the tidal changes in local gravity [Kawamura et al., 2015]. Measurements of the tidal Love numbers [Latychev et al., 2009] would provide additional information about internal structure and a check on gravity science measurements.

9.2.3.2 Broadband Seismometers: Status Quo

A seismometer can be thought of as mass suspended on a spring with a natural frequency of

$$f_0 = \frac{1}{2\pi}\sqrt{\frac{k}{m}}$$

where k is the spring constant and m is the suspended mass. When the ground moves at a frequency much higher than f_0, the suspended mass remains stationary, and the relative displacement between the mass and the ground is measured as the seismic signal. When the frequency of ground motion is much lower than f_0, the suspended mass moves with the ground and the seismometer loses its sensitivity. Therefore a figure of merit for a broadband seismometer is f_0: the lower its value, the better.

Two classes are relevant to icy satellite applications: very broadband (VBB) high-dynamic-range sensors and micro-electro-mechanical systems (MEMS) sensors. Each class of sensors has its own set of challenges.

VBBs: The larger, typically more sensitive, yet more complex VBBs are similar to top-of-the line terrestrial VBBs (Figure 9.6), typically comprising a large spring-mass system

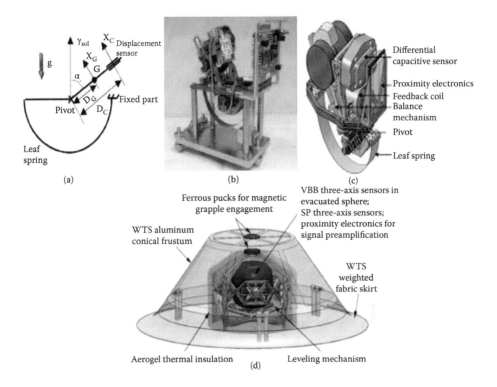

FIGURE 9.6
Insight's seismic experiment for interior structure (SEIS) inverted-pendulum seismometer. (a) Principle, (b) single-axis prototype, (c) block diagram, and (d) shown within the whole seismometer and wind/thermal shield (WTS) system.

and highly sensitive feedback electronics. The InSight VBB's seismometer to be launched in March 2016 was developed by France's CNES (Centre National d'etudes spatiales) and IPGP (Institut de Physique du Globe de Paris) with input from JPL. InSight's VBB has been tested extensively for Mars conditions down to $-65°C$. However, the seismometer is not designed to operate in the icy moon temperature ($\sim-160°C$) and radiation environment. Its projected sensitivity is $10^{-10}g/\sqrt{Hz}$ at 0.01–1 Hz.

MEMS seismometers: MEMS accelerometers are widely used for both industrial and scientific applications. Nevertheless, only a handful of candidate MEMS technologies have the potential of meeting the projected sensitivity performance of highly sensitive broadband seismometers. As an example, the InSight SEIS Short Period (SP) Seismometer is etched out of a single silicon wafer and employs a double cantilever spring support of a proof mass whose motion is recorded by a capacitive transducer. Its projected sensitivity is $10^{-9}g/\sqrt{Hz}$ at 0.04–0.11 Hz.

9.3 Chemical Measurements

Here we address instrumentation that is optimized for *in situ* chemical analyses of icy planetary environments. This includes standoff and contact techniques to elucidate the composition of planetary targets such as Titan, Europa, Enceladus, Ceres, and small icy bodies. Not included in this discussion are remote sensing techniques, which typically are not required to meet specific operational constraints as implied by cryogenic conditions.

The operating conditions for the majority of *in situ* flight instruments are generally controlled by the spacecraft to be within the range of −40 to 50°C, with survival conditions as low as −60°C. It is assumed in this section that the adaptation of these instruments to operate at cryogenic conditions is not the most efficient path toward technology development, given the success and robustness of current instrument lines and the other technical challenges present. Rather, we focus here on the interface through which the environment is interrogated and the sample is acquired and processed for delivery to the instrument, which is maintained in the spacecraft within the normal range of operating temperature.

9.3.1 Airless Bodies

Enceladus, Europa, Ceres, and other small icy bodies differ in important ways that can give us clues to the formation, differentiation, and evolution of our solar system. These environments provide a few key similarities, however, that enable the use of instrument techniques and technologies that are compatible with cryogenic temperatures and high vacuum. Generally, opportunities for *in situ* investigations of the chemistry of these bodies are sufficiently sparse or infrequent to merit a comprehensive analysis of their chemistry at every opportunity. Future mission architecture studies have included flybys, orbiters, surface landers, and even ice penetrators. As a result, development efforts for instrumentation designed to analyze the composition of these planetary environments need to consider various candidates for sample presentation. A flyby or orbiting instrument may encounter, for example, rarified neutrals and ions originating from the planetary surface, dust or ice particles originating from the subsurface due to plume activity, or ejecta generated by a secondary impactor. In contrast, a lander may encounter loose regolith, consolidated bedrock, ice formations, or plume ejecta on the surface of a planetary body. For those outer planet satellites that are thought to possess salty oceans, a deep ice probe (e.g., Philberth probe or Cryobot) could encounter a high-pressure, saline, aqueous sample. Two limiting cases that are most likely to be encountered in the next 10–20 years are (1) neutrals, ions, dust, and ice from orbit; and (2) loose regolith and condensed ice and rock on the surface.

9.3.1.1 Orbital Instruments: Neutrals, Ions, Dust, and Ice

The quantitative abundances of these species in the vicinity of a planetary body will of course depend on the specific characteristics of the target body. In general, however, low number densities of these species, and an altitude-dependent density gradient, are expected for airless bodies. Successful analyses of exospheric species have been performed in past missions to the Moon [Mahaffy et al., 2014a], Venus [Niemann et al., 1980], Enceladus [Waite et al., 2009], and comets [e.g., Nordholt et al., 2003]. The design of these experiments are not explicitly driven by the cryogenic conditions of the surface, so further discussion of these measurements is outside of the scope of this volume and will not be included here.

Lofted surface dust and ice grains present an opportunity to characterize surface and subsurface materials from orbit, but *in situ* analysis implies the need to capture and ingest the dust particle into an instrument, such as a mass spectrometer. The scientific return on dust composition measurements is considerable: these dust and ice grains may be more representative of surface composition than sputtered neutrals, and in addition, the mineral or ice matrix of the dust particle could serve to protect encapsulated organics from damaging ultraviolet or energetic particle radiation that may be characteristic of the orbital environment. Along with these compelling reasons to measure composition of dust and ice grains from orbit, there are challenges in these measurements as well. In general, any flyby or orbiter will encounter the planet with a relative speed of several kilometers per second. This can be detrimental to the spacecraft structure itself, but of direct consequence to *in situ* analyses; impact speeds of this magnitude can pose challenges to the ingestion or acquisition of exospheric constituents by analytical instrumentation.

Experiments supporting and subsequent to the Stardust sample return mission to comet Wild 2 [Brownlee, 2014] have shown that high impact speeds can alter or volatilize constituents of the dust particles [Burchell et al., 2012]. Even with this limitation, key measurements on Earth of the returned samples detected a variety of organic species [Sandford et al., 2006]. For certain *in situ* analyses, high impact speed is an advantage. For example, several dust analyzers have used the high impact speed to promote dissociation of the dust particles for elemental analysis [Srama et al., 2004; Horyani et al., 2008, 2014; Colangeli et al., 2007]. For those investigations, however, that aim to understand the molecular composition of indigenous organics or mineral composition, a softer collection or ingestion approach is preferred. Where possible within mission constraints, a combined strategy is adopted of minimizing orbital speed, where possible, and employing capture surfaces that impart low momentum to the dust particle during capture.

9.3.1.2 Landed Instruments: Loose Regolith, Condensed Ices, and Consolidated Rock

A landed mission offers a rich regime for conducting *in situ* investigations of composition of a region of a planetary surface. The surface is naturally stationary with respect to the payload, which provides a key advantage for instrumentation. Standoff techniques can be employed for survey analyses, and precise sample selection and acquisition for detailed chemical and mineralogical analyses become possible. Several flight instruments serving as examples of the following *in situ* techniques are currently in use on Mars, as part of the Mars Science Laboratory [Grotzinger et al., 2012], and at Comet 67/P Churyumov-Gerasimenko, as part of the orbiter and lander architecture of the Rosetta mission [Glassmeier et al., 2007].

9.3.1.2.1 Standoff Measurements

Standoff techniques that do not fundamentally require the instrument to contact the surface to be studied include x-ray fluorescence and spectroscopy, reflectance imaging and spectroscopy, Raman spectroscopy, laser-induced breakdown spectroscopy, and laser ablation mass spectrometry.

Reflectance imaging and spectroscopy, even for an active instrument, only requires the interaction of a photon source with the surface under study. Photon response of the surface is measured at a single-pixel or array detector plane and can indicate composition by probing electronic transitions, surface refraction, or rotational-vibrational resonances of elements, minerals, and molecules [Rieder et al., 2003; Christensen et al., 2003; Morris et al., 2004; Edwards and Christensen, 2013]. X-ray spectroscopy is a particular class of instrument that uses an illumination source to generate characteristic x-rays that indicate the elemental composition of a solid sample. Traditionally, high-Z, rock-forming elements are detectable using a Cm-244 source, but new techniques are enabling the ability to detect elements as low as carbon using low-energy electron sources, often coupled to electron microscopy capabilities [Feldman et al., 2003]. New instruments now in development are allowing for unprecedented spatial resolution for application on planetary surfaces, such as for PIXL (Planetary Instrument for X-RAY Lithochemistry) on the Mars 2020 Rover

[Allwood et al., 2015]. Particle detection, in general, can also be similarly employed to assign elemental composition to characteristic gamma rays and neutrons generated either passively or actively from surface and subsurface layers.

Minimal heat is imparted to the sample by these instruments, and often only a fixed working distance or range of working distances is required for these standoff measurements. In some cases, the electrical potential of the planetary surface would need to be defined as ground, but that electrical contact can be established outside of the analytical field of view. Under some conditions, the illumination of a cryogenic sample with intense infrared emission from an illumination source may volatilize some of the most labile constituents of an icy sample, and this effect would require environmental testing of the instrument during development. An environment rich in aromatic organics is a natural target for ultraviolet fluorescence detection [Bhartia et al., 2008].

A special category of reflectance spectroscopy is Raman spectroscopy, in which the use of a continuous-wave or pulsed laser allows very precise excitation of individual phonon or vibrational modes of a sample constituent. This produces extremely narrow spectroscopic features that are diagnostic of minerals and molecules with greater specificity than with traditional light spectroscopy and has been of interest for *in situ* planetary surface investigations for several decades [Wang et al., 1998; Marshall and Marshall, 2014]. The Raman instrument traditionally has required close placement of its optics to the surface under study, but no ingestion of the sample is required. Recent advancements have relaxed the proximity requirement, and in fact, remote Raman spectroscopy has been demonstrated in planetary analog surfaces in the laboratory and in a field-portable configuration [Sharma et al., 2003]. A recent field study of water ice crystalline structure in terrestrial icebergs has demonstrated the operation of a Raman spectrometer using visible illumination at 532 nm at working distances up to 120 m [Rull et al., 2011]. For cryogenic environments that are expected to be particularly rich in aromatic organics or certain sedimentary minerals, special measures may be required, such as time-resolved gating techniques [Blacksberg et al., 2010], to mitigate effects of fluorescence that can diminish the sensitivity of the instrument.

Also benefiting from recent advances in *in situ* laser technology is the implementation of laser-induced breakdown spectroscopy [Wiens et al., 2012]. This standoff technique is capable of inducing photon emission from an ablated plume with characteristic energies to allow identification of primarily inorganic elemental composition of a rock surface at a distance of several meters. Such a technique is, in principle, compatible with deployment in a cryogenic environment, with provisions for temperature control of the laser and spectrometer subsystems. For a planetary surface dominated by water ice, the need for localized sample volatilization can lead to a temperature dependence in analyte yield. Reduction in signal is also expected to accompany operation on an airless body, due to lower plasma temperature and more rapid expansion of the plume into a vacuum environment. For highly pure water ice containing only trace minerals or salts, the effective transparency of the target deserves consideration in selecting laser wavelength and assessing expected instrument performance and effective penetration depth. For example, the laser wavelength used on Mars by ChemCamis, the fundamental frequency of a pulsed Nd:KGW solid-state laser, and the use of similar near-infrared wavelengths have been demonstrated on icy sample compositions relevant to the Mars polar regions [Arp et al., 2004] and Europa [Pavlov et al., 2011].

Laser mass spectrometry has also been considered in a standoff configuration and shown to be viable in laboratory testing and simulation [Brinckerhoff et al., 2000; Li et al., 2012]. In such an instrument, a laser is focused beyond the dimensions of the instrument body to promote ablation and ionization of elemental species from a planetary surface. On an airless body, ions can be generated at distances up to 100 cm between the planetary surface and the inlet of the mass spectrometer and guided into a time-of-flight mass analyzer for elemental identification.

9.3.1.2.2 *Measurements Employing Sample Ingestion*

A subclass of instruments can enable detailed, highly precise compositional analyses of ingested solid sample. These instrument types and techniques include gas-phase mass spectrometry, evolved gas analysis, tunable laser spectroscopy and cavity ring-down spectroscopy, gas chromatography, liquid chromatography and capillary electrophoresis, laser desorption/ionization mass spectrometry, and x-ray diffraction.

Mass spectrometry has been the gold standard for *in situ* analysis of the chemical composition of planetary environments since the dawn of solar system exploration. Mass spectrometers provide a robust tool and nearly universal detection of a variety of environmental constituents across a wide range of concentrations that can approach eight orders of magnitude. Mass spectrometers have been dispatched to a variety of exotic extraterrestrial environments: to discover argon and helium on the lunar surface, measure noble gases in Jupiter's high pressure atmosphere, probe the depths of the Venus pressure cooker, and explore the organics at the top of the Titan atmosphere, to name but a few examples. An assortment of mass spectrometer types have been utilized in planetary exploration, including magnetic sector [Biemann et al., 1976], quadrupole [most recently, Mahaffy et al., 2012, 2014a, 2014b], time-of-flight [Balsiger et al., 2007], and ion traps [Brinckerhoff et al., 2013].

A method that has found wide use in the generation of gas-phase samples from a solid powder is pyrolysis. Typically, a powdered sample is generated by a robotic crusher and delivered to an oven. When connected to an analytical instrument, evolved gas analysis can be used to elucidate classes of minerals that may be present, the presence and nature of bound water, and surface-bound or preserved organics that may have been encapsulated in solid surfaces. The challenge of working on a cryogenic surface is principally in the acquisition and processing of a sample without imparting excess heat to the sample. However, surface ices that feature a highly volatile component may offer elegant hardware solutions to conducting evolved gas analysis, in that lower maximum temperatures of the heating elements may be required to analyze the condensed species on a cold surface. This could allow for new implementations in heating elements that interact directly with the surface, via a heated probe for example, to produce gas-phase analyte for subsequent ingestion and analysis.

The options for gas-phase analysis of evolved volatiles are numerous. Several types of gas-phase analyzer are discussed briefly here, but this list should not be construed to be comprehensive. An example of a gas-phase analyzer that can provide high sensitivity to certain key compounds is tunable laser spectroscopy (TLS). In this technique, a sample cavity with opposing mirrors on each end, known as a Herriott cell, is filled with evolved gas. A laser pulse of known wavelength is reflected between the mirrors for a large number of passes, and for a molecule that absorbs resonantly at the laser wavelength, the abundance of the targeted species can be quantified very precisely [Tarsitano and Webster, 2007; Mahaffy et al., 2012].

Alternatively, the gaseous products can be directed to a gas chromatograph. This technique is widely used to study labile organic composition in a variety of terrestrial samples. With sufficient temperature control of the analytical column, gas chromatography (GC) provides separation of a complex organic gas-phase mixture by compound via interactions with a stationary resin coating the interior of the GC column. Chiral separation of amino acids has also been designed into *in situ* instrumentation [Mahaffy et al., 2012; Goesmann et al., 2007].

When starting with an icy planetary surface, an alternative to complete volatilization to the gas phase can be considered. Melting the ice into the liquid state offers new opportunities in instrumentation. Liquid-phase analysis is typically more sensitive for the study and quantification of certain classes of compounds that require further processing to be analyzed by GC, such as amino acids and nucleobases. These prebiotic compounds typically possess functional groups that often preclude intact volatilization for gas-phase analysis without the use of a derivatizing agent. Derivatization can be done *in situ* [Mahaffy et al., 2012; Goesmann et al., 2007] but this step adds complexity and risk for a mission in which the existence and abundance of organic species are poorly constrained. Alternatively, analysis of the liquid by capillary electrophoresis or liquid chromatography can be considered as a more optimized approach for the *in situ* search and identification of prebiotic organic species. The challenges in implementing these liquid techniques lie in the requirements to maintain stable optimal operating temperatures and the transport of solvents during the mission cruise phase without loss to leakage or alteration of the solvent or buffer chemistry after launch.

A class of instruments employing a laser to directly investigate the composition of a solid planetary sample offers an alternative to these highly capable but involved analytical techniques. Laser desorption/ionization is under development for the Mars Organic Molecule Analyzer (MOMA) mass spectrometer 2018 ESA/Roscosmos ExoMars mission [Brinckerhoff et al., 2013]. A laser desorption/ionization mass spectrometer (LDMS) offers some particular advantages for use on an airless body. For example, a mass spectrometer that benefits from high vacuum for sensitivity reasons, such as a quadrupole mass filter or time-of-flight mass analyzer, can operate at an airless body without the need for a supplementary vacuum pump or the need to transport a sample into a dedicated vacuum enclosure. An unprepared solid powder, chip, or core can be presented at the focal plane of the instrument for direct interrogation of the sample composition. A pulsed laser, particularly with a wavelength in the ultraviolet range, can desorb and ionize inorganic and organic species directly from a solid sample [Brinckerhoff, 2005; Wurz et al., 2012]. Additional features, such as extraction, chromatographic separation, and laser post-ionization, can be employed in advanced instrument concepts for higher performance as required by the mission.

9.3.1.2.3 Challenges to Instruments Requiring Sample Ingestion

For instruments that require ingestion of a sample, there are a few key common challenges presented by handling regolith in icy environments. The regolith will be composed of particulate and adsorbed species with varying degrees of volatility; for comprehensive, quantitative regolith analysis, care must be taken to avoid fractionation of the sample during acquisition and processing. In addition, for some planetary targets, the acquisition mechanism will need to operate in a microgravity environment. Heritage mechanisms for sample crushing and portioning, such as the ones that have been developed for Mars Science Laboratory, are not suitable for a microgravity environment. Missions in operations and in development have devised approaches to address these challenges, but more work is required.

A range of sample handling requirements is represented by the aforementioned instruments, but the degree of regolith handling can be divided into the following categories: scooping and portioning, melting, and pyrolyzing. For all mechanisms, the icy or cryogenic regolith will contact the hardware, and the extent to which that hardware is maintained at ambient temperature will determine the degree of loss of volatile species during handling. For instruments optimized for liquid analysis, such as liquid chromatography, ion chromatography, and capillary electrophoresis, melting is a key approach to introducing a sample into a fluidic subsystem. Airless bodies present a particular constraint on this process, in that sublimation heat loss can dominate the heating process at low pressures. Therefore, melting must be done in a hermetically sealed container to prevent loss of key constituents into the gas phase. Furthermore, even for a sealed system, a fraction of condensed analyte will be lost to produce the head pressure during the initial melting process. A gas pressurization subsystem may then be considered to counteract quantitative losses due to the combined effects of change of phase and dead volumes of the hardware.

Consolidated icy surfaces present a particular challenge to sampling hardware, not previously encountered in predecessor missions to the Moon or Mars: water ice is extremely hard, especially at cryogenic temperatures. This aspect will be discussed in detail in Chapter 10. Once the mechanical requirements of the surface drill are addressed, additional challenges arise due to sample volatilization and differential adhesion or cohesion properties of the acquired materials. In the following section, we treat the specific case of surface solids and liquids on Titan to allow the presentation of surface material to an *in situ* analyzer.

9.3.2 Titan Surface and Liquids

Here we explore some of the challenges and possible solutions for probing the composition of the surface environment of Titan, which provides an array of sampling reservoirs such as cryogenic fluids and frozen surface materials, and in which a quantitative investigation of the chemical inventory is paramount. Regardless of the "flavor" of instrumentation deployed, for many instruments, there is a requirement that the components for analysis be gas-phase ions prior to introduction into the analyzer. For atmospheric investigations this entails careful design of the inlet, dictated by the sampling environment and ambient pressure. Sampling of solid and liquid phases necessitates an additional step to get analytes into the gas phase, with optional complementary separation approaches prior to introduction into the mass spectrometer. The sample capture and preparation approach needs to be tailored to both the conditions particular to the exploration environment as well as the instrument requirements.

9.3.2.1 Sampling Lakes and Seas on Titan

Exploration of Titan's lakes and seas, thought to be composed primarily of methane and ethane, represents a primary science goal for Titan and Outer Planet science. Titan lake and sea landers have been studied as a potential element in the Titan Saturn System Mission [ESA, 2009], as a stand-alone mission concept [Vision and Voyages, 2011], the Discovery 12 Step 2 mission concept Titan Mare Explorer (TiME) [Stofan et al., 2013], as well as exploration concepts described earlier. Primary science goals for measurements of the Titan seas include the determination of the chemical make-up of the liquid with the isotopic compositions of major elements, the inventory of dissolved gases, and the identification of trace molecules and suspended solids that result from the active CH_4 photochemical cycle. Furthermore, the lakes in the South Pole in particular may exhibit seasonality [Hayes et al., 2011], and so a temporal investigation of the composition over the long (~7.5 yr/season) seasonal cycle on Titan would be desirable.

Mass spectrometry, as a "universal" detector, is likely to play a major role in future *in situ* investigations of the Titan lakes. We therefore frame this discussion using mass spectrometry as a strawman analyzer to facilitate the following discussion of sampling from a cryogenic liquid. The primary challenges for sampling the lakes for compositional analysis are as follows: (1) the sampling mechanism must function in the cryogenic fluid at approximately 90 K; and (2) the acquired liquid sample must be converted into gas and transferred to the ion source of the mass spectrometer, with the conversion of the sample to the gas phase preserving the integrity of the sample to ensure proper quantitative analysis. This must be accomplished despite a wide array of unknowns, such as the bulk composition, which drives the range of boiling points and specific heats, as well as the particulate load and sizes and the liquid viscosity (Lorenz et al., 2010).

The challenge of developing a sampling system that is compatible with the cryogenic fluids may be the most straightforward to address. The propulsion, medicine, food, and aerospace industries have developed cryogenic fluid management technologies that operate well within the conditions of the Titan seas. Valves that function at the cryogenic temperatures are commonly designed to operate with LN_2 (77 K), LO_2 (55 K), or LHe (4 K). NASA has led efforts to advance cryocooler technologies using LHe in support of Earth and Space Science Missions, with a recent push for the James Webb Space Telescope instrumentation. Given the mass, power, and volume constraint of planetary instruments, miniaturized automated valves designed for other spaceflight applications will be necessary components for *in situ* fluid sampling on Titan. Materials typically used in cryogenic valves, metals (e.g., stainless steel, copper, brass, aluminum and its alloys) and plastics (Teflon®, Vespel®, Torlon®), should be robust in the hydrocarbon liquids, which are not particularly corrosive. However, any proposed valve system would need to be tested against relevant mixtures for possible degradation of mechanisms or valve seats.

Cryogenic seals for the valve and inlet ports that may protrude from the spacecraft for sampling are well understood, and have been used in every single flight project and mission that has involved liquid cryogens. The most common method of creating a cryogenic seal uses indium wire that is compressed into a groove. Indeed indium uniquely preserves its malleability to very cold temperatures, so the sealing properties are retained. Indium seals have proven effective in past missions such as COBE, SHOOT, ASTRO-E I, and ASTRO-E II. The testing programs on these past projects, on the ground and in flight, have proven the versatility, repeatability, and maturity of designs that use indium to create cryogenic seals. Conflat flanges can also be used to achieve a cryogenic seal, but these types of seals require far more mass and have higher He leak rates, although that may not be an issue in the liquid hydrocarbons.

9.3.2.1.1 Mass Spectrometer Gas Analysis and Sample Vaporization

The conversion of the acquired liquid sample into the gas phase for measurement, while preserving the sample fidelity, is a much greater challenge to the proper investigation of the sea composition. There are a variety of thermal complications that arise when sampling from a multicomponent cryogenic lake, including the provision of proper insulation of the spacecraft to minimize heat flow through the spacecraft skin to prevent local vaporization and thus compositional changes in the sea [Lorenz, 2015]. Then, provided that a sampling port on the exterior of the spacecraft has access to unaltered fluid, the sampling apparatus must be maintained at the sea temperature during capture. A warm enough sampling reservoir that drives the fluid to the onset of boiling will alter the sample composition and may lead to rapid gas expansion that drives fluid out of the open port. This requires proper thermal coupling to the cryogenic sea, with limited thermal coupling to the interior of the spacecraft during sampling. Furthermore, during sample capture, either there should be no heat generated by the mechanism itself or such heat should be redirected and dissipated away from the sampling region.

This is particularly the case if the quantification of the bulk composition of the lake is a required science measurement. The phase diagram in Figure 9.7 shows the limited tolerance for temperature increases under the conditions of the Titan Sea before the sample begins to fractionate due to vaporization. Here, the bubble point is defined as the temperature (at a given pressure) where the first bubble of vapor is formed. The vapor will have a different composition than the parent liquid. Correspondingly, the dew point is the temperature (at a given pressure) at which a vapor will begin to condense into liquid. For a pure compound, the bubble point and dew point are equivalent. This simple approximation based on Raoult's Law for mixtures indicates that if the temperature of the sample is raised to 40 K, the CH_4 will begin to preferentially enter the vapor phase enough to change the sample composition by nearly 50%. With a slightly more complicated model for the sea composition such as that proposed by Cordier et al. [2009] and Tan et al. [2013], temperature changes of even a few K could lead to significant loss of the N_2 and dissolved noble gases in the system (see Table 9.2). An update to the thermodynamic analysis suggests that the ethane component may be slightly higher, which will only serve to increase the viscosity [Cordier et al., 2013a].

Temperature increases above the bubble point shown in Figure 9.7 will lead to rapid fractionation and alteration of the liquid sample. Above the dew point (boiling point), the composition of the vapor phase will reflect the composition of the original liquid sample. Below the bubble point, temperature increases will lead to preferential evaporation of the more volatile component (Figure 9.8).

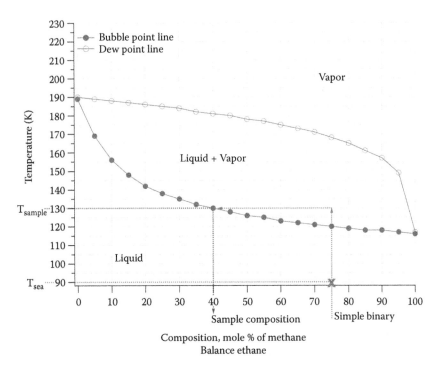

FIGURE 9.7
Phase diagram for CH_4 and C_2H_6 mixtures at 1500 mbar pressure shows the tolerance of the cryogenic hydrocarbons against warming by the sampling mechanism. An example is illustrated with the "Simplified Ligeia Binary" mixture of 25% C_2H_6 and 75% CH_4 (Table 9.1).

TABLE 9.2

Candidate Lake Compositions and Thermochemical Properties

	Pure CH₄	Pure C₂H₆	Simplified Ligeia[a] Binary	Cordier et al. (2009)
X [CH₄]	1		0.75	0.10
X [C₂H₆]		1	0.25	0.74
X [N₂]				0.005
X [C₃H₈]				w0.07
X [C₄H₁₀]				0.085
Bubble point (K)[b]	116	189	120	150
Specific heat C_p (kJ/kg K)	3.29	2.27	3.04	2.4
Viscosity at 94 K (µPa s)	208	1141	305	1423
Expansion ratio at 250 K, 1.5 bar	388:1	295:1	344:1	282:1

Source: Adapted from Lorenz, R.D. et al., *Icarus,* 207, 932–937, 2010.

   X denotes the mole fraction of each component.
[a] From Mastrogiuseppe et al. [2014] and Malaska et al. [2014].
[b] Assumes ideal mixture, P_{surf} = 1500 mbar.

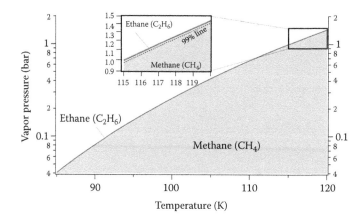

FIGURE 9.8
The vapor pressure curve for the "Simplified Ligeia Binary" mixture of 75% methane and 25% ethane. The headspace vapor of the liquid is dominated by the more volatile component, methane, which leads to sample fractionation even at temperatures below the bubble point. The dashed line in the inset indicates that > 99% of the vapor phase is composed of CH₄ even just up to the bubble point.

One approach is to design a valve that also serves as the sample chamber and vaporization region, minimizing the transfer of cryogenic fluids into the warm spacecraft and thus mitigating issues with sample fractionation. A concept for this type of all-in-one cryogenic inlet was developed for coupling to a neutral mass spectrometer on a Titan lake lander [Trainer et al., 2012]. For the NMS inlet concept, a custom valve was designed at the bottom of a sampling chamber that was completely submerged in the sea and thus thermally coupled to the sampling region. The valve seal utilizes geometries discussed earlier for cryovalves, but the actuation mechanism and other temperature-sensitive components are housed internally to avoid large temperature fluctuations. This design thus has the benefit of maximizing the use of high heritage components. The concept of operations shown in Figure 9.9 demonstrates this type of sampling approach. With the chamber surrounded

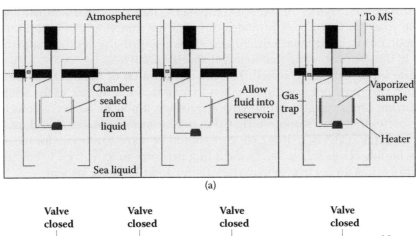

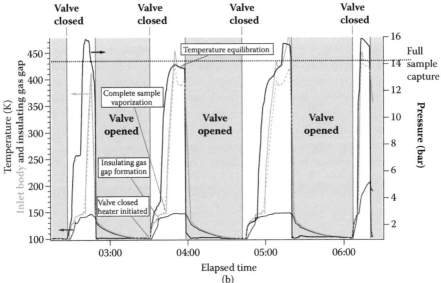

FIGURE 9.9
(a) A cryogenic fluid inlet concept for quantitative sampling of the bulk composition of a Titan sea with a mass spectrometer or other gas analyzer. (b) Testing at NASA Goddard Space Flight Center in cryogenic fluid demonstrated the duty cycle of this approach.

by fluid during sample acquisition, it is easily held at the sea temperature for capture of a high-fidelity, nonfractionated sample. The valve then seals the chamber from the sea to perform sample vaporization. The temperature of the captured fluid is not raised until the seal from the environment is made, preventing premature boiling and escape. The small amount of cryogenic sea that is heated by the inlet chamber (right panel) serves as an insulation pocket to minimize the power applied during heating. An inlet modeled after this concept has been designed and tested to flight-readiness [Trainer et al., 2012]. Repeated sampling capability and robust leak-tight seals in a model cryogenic fluid were demonstrated in a laboratory system (Figure 9.9).

This inlet allowed for rapid (< 1 h) recooling due to excellent thermal connectivity to the sampling fluid. The cryogenic valve maintained the required seal through repeated sealing/heating cycles with no loss of sample gas at high pressures. The volume of sample captured is driven by the chamber dimensions and tailored to the needs of the sampling instrumentation, with excess gas diverted through an exhaust line. The rapid vaporization approach drives the majority of the sea components into the vapor and through heated lines in the spacecraft and thus provides quantitative sampling of the fluid. Following sample vaporization, additional enrichment steps using flight heritage approaches can take place in the gas processing line to target specific analytes.

For science investigations in which the bulk composition of the sea fluid is not a high priority, such care does not need to be taken to maintain the sample at the ambient temperature during capture. Rather, following acquisition it may be advantageous to boil off the major species in the sea sample (CH_4, C_2H_6) to concentrate the less abundant, less volatile compounds for analysis. In this case, care should be taken to account for the expansion ratio of the liquids (Table 9.2), possibly including pressure relief valves, to ensure that the generated pressures do not overwhelm the valving and plumbing system.

9.3.2.1.2 *Liquid Sample Distribution and Analysis*

Other approaches for chemical analysis may require maintenance of the sea fluids in the liquid state for analysis, and thus could utilize cryogenic valves that are maintained at cold temperatures without concerns for resistance to heating. The use of piezoelectric valves and liquid distribution systems for Titan sea fluids has been explored [Sherrit et al., 2014]. The advantages of utilizing piezoelectric actuation in a cryogenic fluid distribution system is that this approach allows proportional flow control and is capable of generating high force with low power consumption, on the µW to mW scale [Park et al., 2008]. A major disadvantage of piezoelectric actuation is its small stroke, even for large voltages. At cryogenic temperatures, the displacement is further reduced due to degradation of the piezoelectric coefficient. One proposed solution to overcome this challenge is the fabrication of a large effective perimeter in the valve plate of the MEMS device [Brosten et al., 2007]. The use of a relaxor-PT single crystal, discussed in Section 9.2.1.1, also shows promise to make the use of piezoelectric microvalves feasible for fluid management.

A liquid distribution system would rely on a pumping mechanism to draw in and move fluid through a controlled system. For very small volumes, the development of cryogenic micropumps for cooling systems on satellites, detector arrays, and superconducting magnets may be leveraged for liquid sampling on Titan. Micropump designs that utilize electrohydrodynamic pumping, in which the throughput and power requirements are proportional to the dielectric properties of the fluid, may have a range of performance dependent on the liquid characteristics (e.g., composition, density, temperature, and viscosity) [Darabi and Wang, 2005]. A mechanical pump developed for high-energy physics applications could provide higher throughput, and may be more robust against a variety of fluid properties [Grohmann et al., 2005]. These and other proposed pumping solutions need to be vetted and tested for function in simulated Titan sea conditions.

Liquid distributions systems could be designed to interface with a variety of analyzers. Some that could provide valuable scientific investigations in the Titan seas are nuclear magnetic resonance (NMR) spectroscopy, capillary electrophoresis, or even liquid chromatography or supercritical-fluid chromatography coupled to MS, common in the petroleum industry [Barman et al., 2000]. For instance, solution-state NMR spectroscopy can provide high-resolution structural analysis of complex mixtures and large compounds, but can also provide quantitative analysis in a nondestructive manner. This approach has been used to analyze the chemical structure of analogs of Titan photochemical aerosol [He and Smith, 2014]. Microfluidic capillary electrophoresis has been demonstrated as a viable technique for separating and identifying nitrogen functional groups in complex Titan-like organics [Cable et al., 2014]. However, there are no flight-ready cryogenic instruments for liquid analysis currently available.

In addition, as discussed earlier, if quantitative analysis of the bulk composition were not a priority measurement, the fluid could also be passed through this type of distribution system and heated for introduction into a gas analyzer. Moreover, if the sample is distilled as described above, the fluidic analysis can be performed at accordingly higher temperatures, relaxing the cryogenic temperature requirements on the instrumentation.

9.3.2.1.3 Challenges to In Situ Liquid Sampling

The composition of the Titan lakes and seas is unknown, since there are many environmental components that could be present in the liquid or solid phases in these reservoirs. A significant fraction is likely made of methane, ethane, propane, and even nitrogen, although the surface is not cold enough for the formation of liquid nitrogen; as the dominant atmospheric component (~94% at the surface [Niemann et al., 2010]) this gas will be incorporated into precipitable fluids and surface liquids. There are additional trace compounds, such as benzene, hydrogen cyanide, and acetonitrile, which may also be present as liquids or solutes. Competing models suggest a variety of compositional mixtures in equilibrium with the atmosphere at the surface [Cordier et al., 2009, 2013a; Tan et al., 2013]. Recent Cassini observations have indicated that the northern sea Ligeia Mare is primarily composed of methane [Mastrogiuseppe et al., 2014]. However, the liquid compositions of Titan's surface liquid reservoirs may vary significantly and even differ from one lake to the next. Further, density and other fluid variations could lead to stratification within the lakes and seas, rather than a homogeneous mixture. Thus, the possible exploration targets for analysis of the cryogenic liquids on Titan could vary spatially, vertically, and temporally. The compositional variation alone presents a significant challenge in that the properties of the fluid and constraints placed on sampling and analysis may not be tightly bound, and developed sampling mechanisms and instrumentation will need to function in a wide array of conditions (e.g., Table 9.1). As discussed above, the integrity of captured and processed samples may have tight operating margins, particularly for the more volatile fluid components. Moreover, the uncertainty in the measured properties of candidate liquids at Titan-relevant conditions is high [Cordier et al., 2013b].

Particulates could be composed of deposited haze aerosols [Tomasko and West, 2009] or precipitated organic solutes [e.g., Malaska and Hodyss, 2014]. Evidence of "evaporite" deposits in dry lake beds suggests that, at least in evaporating bodies, suspended matter is likely [Barnes et al., 2011; Cordier et al., 2013b]. These suspended particulates provide challenges to the integrity and efficacy of sampling systems. Particulate matter is a particular hazard to valve seals in all conditions, and may prove more damaging at cryogenic temperatures where materials are more hardened and brittle. For microfluidic sampling of the fluid, particulate matter could also lead to clogging, as particle diameters of hundreds of nm to several μm are on the order of the fluid channel or valve conductance dimensions. Typically particle filters are used to protect valve seats and other small channels from clogging. However, this approach may be limiting for scientific investigations in which the suspended material may itself be the target analyte. In addition, the filter lifetime will be a limiting factor and must be considered depending on the required number of samples and mission duration.

A heavy particulate load may also have implications for the fluid viscosity, and therefore the sampling and fluid handling mechanisms. Although viscosity is expected to vary substantially as a function of composition, and thus may vary from sea to sea, or at different times of year, the calculated dynamic viscosities for different candidate sea compositions range from 150 to ~2000 µPa s [Lorenz et al., 2010]. This is a manageable range given that it does not greatly exceed that of water (894 µPa s) or blood (3000–4000 µPa s) at 25°C, two common fluids used in lab-on-a-chip microfluidics. However, high concentrations of suspended particulates may greatly increase the viscosity, to the point of producing a mud-like slurry of fluid. The range of conditions on target lakes or seas and the effect on cryogenically adapted pumping and fluid management systems are not well characterized. Fluid viscosity is not normally a consideration for cryogenics where viscosities are quite low (much less than water) and particulates are filtered out, and will need to be taken into consideration for any sampling system developed for Titan.

9.3.2.2 Ice and Organic Surfaces on Titan

The solid surface of Titan is also a high-priority exploration target, with a large variety of geologic units including alluvial systems, dunes, mountains, and cryovolcanic flows [Aharonson et al., 2013]. The general surface composition has been proposed to be water-ice bedrock covered in hydrocarbon and organic deposits, such as the aforementioned atmospheric aerosol and lake evaporites [Clark et al., 2010]. The only *in situ* explorer sent to Titan's surface to date, the Huygens Probe, landed in what was later determined to be damp ground [Lorenz et al., 2006]. The Huygens GCMS detected a release in volatiles upon landing and subsequent warming of the surface, suggesting a surface saturated in atmospheric constituents such as ethane, acetylene, and carbon dioxide [Niemann et al., 2010].

The discussion of potential stand-off and *in situ* compositional measurements for airless bodies in Section 9.3.2 covers the types of instrumentation appropriate for exploration of the Titan surface. The environmental challenge of operating such instrumentation on Titan's surface as compared to an airless body is the high atmospheric pressure, ~1.5 bar. The dense atmosphere provides challenges for instrumentation that requires low pressure or vacuum conditions. Mass spectrometers, for example, would require high-conductance pumps to maintain low pressures inside the mass analyzers. This has been accomplished previously for short time frames on Titan using chemical getter and ion pumps [Niemann et al., 2002]. For extended missions requiring instrument evacuation on Mars, hybrid turbomolecular pumps have been used to provide the necessary vacuum conditions [Mahaffy et al., 2012]. A laser desorption/ionization mass spectrometer, such as the MOMA instrument under development for ExoMars, would be well-suited for the analysis of Titan's organic surface materials. MOMA-MS utilizes a fast-actuating solenoid aperture valve to enable LDMS operations at Mars ambient pressures (4–8 mbar) without requiring a vacuum seal to the sample [Arevalo et al., 2015]. Both of these components are designed to work at Mars ambient pressures and spacecraft-controlled temperatures (−45°C for the aperture valve). Although the pressure conditions at Titan are several orders of magnitude higher, similar technologies may be adapted through the use of differential pumping with secondary pumps.

9.3.2.3 Application Example: Tunable Laser Spectroscopy for Low Temperature Atmosphere on Titan

Tunable laser spectroscopy has developed over the last three decades as a powerful way to understand planetary climate cycles. TLS *in situ* instruments have participated in measurement campaigns using airplanes, helicopters, balloons, ground vehicles, and on foot. Most recently a tunable laser spectrometer visited the surface of Mars aboard the Curiosity mission as part of the Sample Analysis at Mars suite. TLS-SAM has begun addressing questions about the loss of water from the atmosphere [Webster et al., 2013a] and the possibility for present-day release and uptake of methane [Mumma et al., 2009; Webster et al., 2013b].

In the case of Titan, rapid, direct, and sensitive hydrocarbon characterization is needed for future aerial exploration of the atmospheres of the outer planets, or *in situ* exploration of Titan. A planetary alkane TLS (PA-TLS) could investigate Titan's complex interplay of geology, hydrology, and meteorology by assessing isotopic and compositional fractionation associated with exchanges across the multiple hydrocarbon reservoirs (Figure 9.10). Such an instrument would be a natural outgrowth from the successful Mars TLS, but it would require attention to sample handling to characterize the range of relevant conditions found on Titan, as discussed earlier. For example, isotopic characterization of methane is optimal for mbar pressures found on Mars, where line broadening is minimized. A cryogenic vacuum would thus need to be included, which could also prove helpful for analyzing liquids sampled from Titan's lakes.

Evaporation and precipitation of liquids and vapors also fractionate the isotopic composition of atmospheric gases due to their different diffusion rates. Craig [1961] describes a "meteoric line" for water in Earth's atmosphere: the linear relation between D/H and $^{18}O/^{16}O$. They couple this information with studies of evaporation of water in the laboratory similar to what we propose to do here. A PA-TLS can assess the relation between $^{12}C/^{13}C$ and D/H in methane (as CH_3D) and also for ethane with cryogenic hydrocarbon liquids.

The isotopic tracer concept applies across the series of hydrocarbons with increasing chain length. Telling et al. [2009] pointed out that the trend in $^{12}C/^{13}C$ and D/H in molecules in terrestrial hydrocarbon deposits indicates chemical processes occurring at shallow crustal depths [Clark et al., 2010]. In general, heavy-isotope enrichment increases with the size of hydrocarbon molecules. Measuring the trend of the $^{12}C/^{13}C$ or D/H ratio can reveal the chemical processes that generate the molecules. Processes occurring at high temperature and pressure, and in low-pressure gas-phase conditions, produce molecules where the

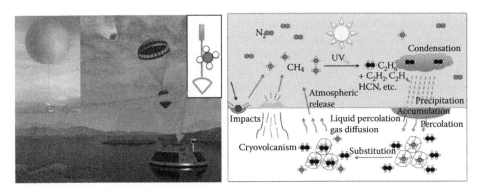

FIGURE 9.10
In Titan's Earth-like hydrocarbon cycle, alkanes move between gas, liquid, and solid reservoirs on Titan [right: adapted from Choukroun and Sotin, 2012]. Rapid, direct, and sensitive measurements of these and other molecules in Titan's atmosphere using tunable diode laser spectroscopy (middle schematic) could track Titan's hydrocarbon cycle, with high spatial and temporal fidelity needed to distinguish among the various processes. This approach would fit well on a probe, balloon, or surface explorer (left image).

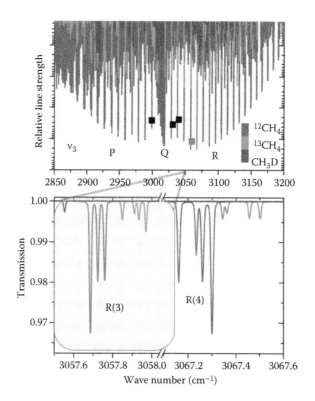

FIGURE 9.11
Methane absorption in the region of interest for PA-TLS, based on the development of the Mars TLS instrument.

heavier isotope is depleted as molecular sizes increase. The decreasing ratio is consistent with a free-radical formation mechanism.

Equations of state for fluid–gas systems under planetary conditions [Tan et al., 2013; Glein and Shock, 2013] provide the tools for modeling transport and exchange of hydrocarbons on other planets.

Figure 9.11 shows methane absorption in the region of interest for PA-TLS, based on the development of the Mars TLS instrument [Webster and Mahaffy, 2011]. Methane absorbs more strongly in this region than ethane or propane, and has well-separated absorption lines for both ^{13}C and D isotopologues, providing unambiguous isotope ratios. A PA-TLS instrument would incorporate a 3057 cm^{-1} laser for measuring methane under conditions obtained in planetary atmospheres. The strongest lines at 3057+{0.58,0.69,0.98} cm^{-1} allow sensitive measurements of absolute abundances of {CH$_3$D,^{12}CH$_4$,^{13}CH$_4$}.

The well-characterized 3057 cm^{-1} region serves as a reference for determining ethane compositions in the 2979 cm^{-1} and 2968 cm^{-1} regions, by subtracting methane absorbance effects [Harrison et al., 2010]

9.4 Summary/Conclusions

This chapter has highlighted some of the many techniques that are available for the *in situ* exploration of icy bodies, driven by discoveries achieved by recent missions. Many of the geophysical techniques are similar to those used on Earth for ice sheet exploration and provide powerful approaches for characterizing the thermophysical properties of planetary bodies. High-resolution chemistry measurements have to handle the challenges that come with sample handling and processing but big science questions at Titan and other icy bodies are motivating the development of innovative technologies. The past 10 years have seen an increased number of concepts concerning the observation of icy bodies from *in situ* platforms as the natural next step in our exploration of these objects, driven by the decadal science objectives to improve our understanding of their origin, evolution, and potential habitability for life.

Acknowledgments

Some of the research reported in this chapter was conducted at the NASA Goddard Space Flight Center (GSFC). Further, some of the research reported in this chapter was conducted at the Jet Propulsion Laboratory (JPL), California Institute of Technology (Caltech), under a contract with the National Aeronautics and Space Administration (NASA). The authors would like to thank Robert Hodyss and James Cutts, JPL/Caltech, Pasadena, CA; and Tibor Kremic, NASA's Glenn Research Center, Cleveland, OH, for reviewing this chapter and providing valuable technical comments and suggestions.

References

Aharonson, O., A.G. Hayes, J.L. Lunine, et al. (2009). An asymmetric distribution of lakes on Titan as a possible consequence of orbital forcing. *Nature Geoscience* **2**, 851–854.

Aharonson, O., A.G. Hayes, P.O. Hayne, R.M. Lopes, A. Lucas, and J.T. Perron (2013). Titan's surface geology. In: *Titan: Surface, Atmosphere, and Magnetosphere*. I.C.F. Mueller-Wodarg, C.A. Griffith, E. Lellouch and T.E. Cravens, Eds. Cambridge, UK: Cambridge University Press, 63–101.

Allwood, A., B. Clark, D. Flannery, et al. (2015). Texture-specific elemental analysis of rocks and soils with PIXL: The planetary instrument for x-ray lithochemistry on Mars 2020. *2015 IEEE Aerospace Conference Proceedings*, pp. 1–13, doi: 10.1109/AERO.2015.7119099.

Ardhuin F., L. Gualtieri, and E. Stutzmann (2015). How ocean waves rock the Earth: Two mechanisms explain microseisms with periods 3 to 300 s, Geophysical Research Letters, *AGU Journal*, doi: 10.1002/2014GL062782, http://onlinelibrary.wiley.com/doi/10.1002/2014GL062782/abstract.

Arevalo Jr., R., W.B. Brinckerhoff, F. Van Amerom, et al. (2015). Design and demonstration of the Mars Organic Molecule Analyzer (MOMA) on the ExoMars 2018 Rover. Proceedings of the *IEEE Aerospace Conference, Big Sky, MT* Print ISBN: 978-1-4799-5379-0; DOI: 10.1109/AERO.2015.7119073, 1–11.

Arp Z., D. Cremers, R.C. Wiens, et al. (2004). Analysis of water ice and water ice/soil mixtures using laser-induced breakdown spectroscopy: Application to Mars polar exploration. *Applied Spectroscopy* **58**, 220A–244A.

Balsiger H., K. Altwegg, P. Bochsler, et al. (2007). Rosina–Rosetta orbiter spectrometer for ion and neutral analysis. *Space Science Reviews* **128**, 745–801.

Barman, B.N., V.L. Cebolla, and L. Membrado (2000). Chromatographic techniques for petroleum and related products. *Critical Reviews in Analytical Chemistry* **30**, 75–120.

Barnes J.W., J. Bow, J. Schwartz, et al. (2011). Organic sedimentary deposits in Titan's dry lakebeds: Probable evaporite. *Icarus* **216**, 136–140.

Bentley, C.R (1964). The structure of Antarctica and its ice cover. In: *Research in Geophysics. Vol. 2, Solid Earth and Interface Phenomena.* H. Odishaw., Ed., Cambridge, MA: Massachusetts Institute of Technology Press, 335–389.

Bhartia, R., W.F. Hug, E.C. Salas, et al. (2008). Classification of organic and biological materials with deep ultraviolet excitation. *Applied Spectroscopy* **62**, 1070–1077.

Biemann, K., J. Oro, P. Toulmin, et al. (1976). Search for organic and volatile inorganic compounds in two surface samples from the chryse planitia region of Mars. *Science* **194**, 72–76.

Blacksberg, J., G.R. Rossman, and A. Gleckler (2010). Time-resolved Raman spectroscopy for *in situ* planetary mineralogy. *Applied Optics* **49**, 4951–4962.

Brinckerhoff, W.B. (2005). On the possible in situ elemental analysis of small bodies with laser abla-tion TOF-MS. *Planetary and Space Science* **53**, 8, 817–838, doi:10.1016/j.pss.2005.04.005.

Brinckerhoff, W.B., G.G. Managadze, R.W. McEntire, A.F. Cheng, and W.J. Green (2000). Laser time-of-flight mass spectrometry for space. *Review of Scientific Instruments* **71**, 536.

Brinckerhoff, W.B., V.T. Pinnick, F.H.W. van Amerom, et al. (2013). Mars organic molecule ana-lyzer (MOMA) mass spectrometer for ExoMars 2018 and beyond, *IEEE Aerospace Conference Proceedings*.

Brosten, T.R., J.M. Park, A.T. Evans, et al. (2007). A numerical flow model and experimental results of a cryogenic micro-valve for distributed cooling applications. *Cryogenics* **47**, 501–509.

Brownlee, D. (2014). The stardust mission: Analyzing samples from the edge of the solar system. *Annual Reviews of Earth and Planetary Sciences* **42**, 179–205.

Burchell, M.J., M.J. Cole, M.C. Price, and A.T. Kearsley (2012). Experimental investigation of impacts by solar cell secondary ejecta on silica aerogel and aluminum foil: Implications for the Stardust Interstellar Dust Collector. *Meteoritics & Planetary Science* **47**, 671–683.

Bulow, R.C., C.L. Johnson, B.G. Bills, and P.M. Shearer (2007). Temporal and spatial properties of some deep moonquake clusters. *Journal of Geophysical Research: Planets (1991–2012)* **112**(E9).

Cable, M.L., S.M. Hörst, C. He, et al. (2014). Identification of primary amines in Titan tholins using microchip nonaqueous capillary electrophoresis. *Earth and Planetary Science* **403**, 99–107, doi: 10.1016/j.epsl.2014.06.028.

Cammarano, F., V. Lekic, M. Manga, M. Panning, and B. Romanowicz (2006). Long-period seismol-ogy on Europa: 1. Physically consistent interior models. *J. Geophys. Research: Planets, AGU Journal*, doi:10.1029/2006JE002710, http://onlinelibrary.wiley.com/doi/10.1029/2006JE002710/abstract.

Choukroun, M., and C. Sotin (2012). Is Titan's shape caused by its meteorology and carbon cycle? *Geophysical Research Letters* **39**, L04201.

Choukroun, M., S. Kieffer, X. Lu, and G. Tobie (2013). Clathrate hydrates: Implications for exchange processes in the outer solar system. In: *The Science of Solar System Ices.* Gudipati, M.S., Castillo-Rogez, J., Eds., New York, NY: Springer.

Christensen, P.R., G. Mehall, S.H. Silverman, et al. (2003). Miniature thermal emission spectrometer for the mars exploration rovers. *Journal of Geophysical Research: Planets* **108**, 8064.

Ciarletti, V., C. Corbel, D. Plettemeier, P. Cais, S.M. Clifford, and S.E. Hamran (2011). WISDOM GPR designed for shallow and high-resolution sounding of the Martian subsurface. *Proceedings of the IEEE* **99**(5), 824–836.

Clark, R.N., J.M. Curchin, J.W. Barnes, et al. (2010). Detection and mapping of hydrocarbon deposits on Titan. *Journal of Geophysical Research: Planets* **115**, doi:10.1029/2009JE003369.

Colangeli, L., J.J. Lopez Moreno, P. Palumbo, et al. (2007). GIADA: The grain impact analyser and dust accumulator for the Rosetta space mission. *Advances in Space Research* **39**, 446–450.

Conyers, L.B. (2004). *Ground-Penetrating Radar for Archaeology*, Walnut Creek, CA: AltaMira Press Ltd.

Cordier, D., O. Mousis, J.I. Lunine, P. Lavvas, and V. Vuitton (2009). An estimate of the composition of Titan's lakes. *Astrophysical Letters* **707**, L128.

Cordier, D., O. Mousis, J.I. Lunine, P. Lavvas, and V. Vuitton (2013a). ERRATUM: An esti-mate of the chemical composition of Titan's lakes. *The Astrophysical Journal* **768**, L23–26, doi:10.1088/2041-8205/768/1/L23.

Cordier, D., J.W. Barnes, and A.G. Ferreira (2013b). On the chemical composition of Titan's dry lakebed evaporites. *Icarus* **226**, 1431–1437.

Coustenis, A (2005). Formation and evolution of Titan's atmosphere. *Space Science Reviews* **116**, 171–184, doi:10.1007/s11214-005-1954-2.

Coustenis, A., A. Salama, B. Schulz, et al. (2003). Titan's atmosphere from ISO mid-infrared spectroscopy. *Icarus* **161**, 383–403.

Coustenis, A., B. Bézard, D. Gautier, A. Marten, R. Samuelson (1991). Titan's atmosphere from Voyager infrared observations: III. Vertical distributions of hydrocarbons and nitriles near Titan's North Pole. *Icarus* **89**, 152–167.

Coustenis, A., J. Lunine, D.L. Matson, C. Hansen, K. Reh, P. Beauchamp, J.-P. Lebreton, and C. Erd (2009). *The Joint NASA-ESA Titan Saturn System Mission (TSSM) Study*. 40th Lunar and Planetary Science Conference, (Lunar and Planetary Science XL), held March 23–27, 2009 in The Woodlands, Texas, id.1060.

Craig, H (1961). Isotopic variations in meteoric waters. *Science, American Association for the Advancement of Science* **133**, 3465, 1702–1703.

Daniels, D.J., (Ed.) (2004). Ground Penetrating Radar, 2nd Edition, *Institution of Engineering and Technology*, UK, ISBN-10: 0863413609, ISBN-13: 978-0863413605, 752.

Darabi, J., and H. Wang (2005). Development of an electrohydrodynamic injection micropump and its potential application in pumping fluids in cryogenic cooling systems. *Journal of Microelectromechanical Systems* **14**(4), 747–755, doi:10.1109/JMEMS.2005.845413.

Edwards, C.S., and P.R. Christensen (2013). Microscopic emission and reflectance thermal infrared spectroscopy: Instrumentation for ouantitative *in situ* mineralogy of complex planetary surfaces. *Applied Optics* **52**, 2200–2217.

Elachi, C., S. Wall, M. Janssen, et al. (2006). Titan radar mapper observations from Cassini's T_3 fly-by. *Nature* **441**, 709–713, doi:10.1038/nature04786.

ESA (2009). TSSM in-situ elements. Assessment Study Report, ESA-SRE(2008)4, European Space Agency, 12 February 2009.

Feldman, J.E, J.Z. Wilcox, T. George, D.N. Barsic, and A. Scherer (2003). Elemental Surface Analysis at ambient pressure by electron-induced x-ray fluorescence. *Review of Scientific Instruments* **74**, 1251.

Fortes, A.D (2007). Metasomatic clathrate xenoliths as a possible source for the south polar plumes of Enceladus. *Icarus* **191**(2), 743–748.

Fu, H., and R.E. Cohen (2000). Polarization rotation mechanism for ultrahigh electromechanical response in single-crystal piezoelectrics. *Nature* **403**(6767), 281–283.

Garcia, R.F., J. Gagnepain-Beyneix, S. Chevrot, P. Lognonne (2011). Very preliminary reference Moon model. *Physics of Earth and Planetary Interior* **188**, 96–113.

Gimbert, F., V.C. Tsai, and M.P. Lamb (2014). A physical model for seismic noise generation by turbulent flow in rivers. *Journal of Geophysics Research: Earth Surface* **119**, 2209–2238, doi:10.1002/2014JF003201, http://web.gps.caltech.edu/~tsai/files/Gimbert_etal_JGR2014.pdf.

Glassmeier, K-H., H. Boehnhardt, D. Koschny, E. Kuhrt, and I. Richter (2007). The Rosetta mission: Flying towards the origin of the solar system. *Space Science Reviews* **128**, 1–21.

Goesmann, F., H. Rosenbauer, R. Roll, et al. (2007). Cosac, the cometary sampling and composition experiment on Philae. *Space Science Reviews* **128**, 257–280 (2007).

Goins, N.R., A.M. Dainty, and M.N. Toksöz (1981). Seismic energy release of the Moon. *Journal of Geophysical Research: Solid Earth, AGU Journal*, doi:10.1029/JB086iB01p00378, http://onlinelibrary.wiley.com/doi/10.1029/JB086iB01p00378/abstract.

Gowen R.A., A. Smith, A.D. Fortes, et al. (2011). Penetrators for in situ subsurface investigations of Europa. *Advances in Space Research*, Science Direct, Elsevier, Melbourne, Australia, 48, 725–742.

Grohmann, S., R. Herzog, T.O. Niinikoski, et al. (2005). Development of a miniature cryogenic fluid circuit and a cryogenic micropump. *Cryogenics* **45**, 432–438, doi:10.1016/j.cryogenics.2005.03.004.

Grotzinger, J.P., J. Crisp, A.R. Vasavada, et al. (2012). Mars science laboratory mission and science investigation. *Space Science Reviews* **170**, 5–56.

Hackenberger, W., J. Luo, X.N. Jiang, K.A. Snook, P.W. Rehrig, S.J. Zhang, and T.R. Shrout (2008). Recent developments and applications of piezoelectric crystals. In: *Handbook of Advanced Dielectric, Piezoelectric and Ferroelectric Materials–Synthesis, Characterization and Applications*, Z. Ye (Ed.), Melbourne, Australia: Elsevier Science, 73–100.

Hamran, S.E., H.E.F. Amundsen, L.M. Carter, R.R. Ghent, J. Kohler, M.T. Mellon, and D.A. Paige (2014). The RIMFAX Ground Penetrating Radar on the Mars 2020 Rover, AGU Fall Meeting. http://adsabs.harvard.edu/abs/2014AGUFM.P11A3746H.

Hansen, C.J., L. Esposito, A.I.F. Stewart, J. Colwell, A.P. Hendrix, W.D. Shemansky, and R. West (2006). Enceladus' water vapor plume. *Science* 311(5766), 1422–1425, doi:10.1126/science.1121254.

Harrison J. J., N,D.C. Allen, and P. F. Bernath (2010). Infrared absorption cross sections for ethane (C_2H_6) in the 3μ region. *Journal of Quantitative Spectroscopy & Radiative Transfer* 111, 357–363.

Hayes, A., O. Aharonson, P. Callahan, et al. (2008). Hydrocarbon lakes on Titan: Distribution and interaction with a porous regolith. *Geophysical Research Letters* 35, L09204, doi:10.1029/2008GL033409.

Hayes, A.G., A.S. Wolf, O. Aharonson, et al. (2010). Bathymetry and Absorptivity of Titan's Ontario Lacus. *Journal of Geophysical Research: Planets* 115, E09009.

Hayes, A.G., O. Aharonson, J.I. Lunine, et al. (2011). Transient surface liquid in Titan's polar regions from Cassini. *Icarus* 211(1), 655–671.

He, C., and M.A. Smith (2014). A comprehensive NMR structural study of Titan aerosol analogs: Implications for Titan's atmospheric chemistry. *Icarus* 243, 31–38.

Hecht, M.H., S.P. Kounaves, R.C. Quinn, et al. (2009). Detection of perchlorate and the soluble chemistry of Martian soil at the Phoenix lander site. *Science* 325(5936), 64–67, doi:10.1126/science.1172466.

Heseltine, J.W., D.W. Elliott, and O.B. Wilson (1964). Elastic constants of single crystal benzene. *The Journal of Chemical Physics* 40, 2584, doi:10.1063/1.1725566.

Holt, J.W., M.E. Peters, S.D. Kempf, D.L. Morse, and D.D. Blankenship (2006). Echo source discrimination in single-pass airborne radar sounding data from the Dry Valleys, Antarctica: Implications for orbital sounding of Mars. *Journal of Geophysical Research* 111, E06S24, doi:10.1029/2005JE002525.

Horyani, M., Z. Sternovsky, M. Lankton, et al. (2014). The Lunar Dust Experiment (LDEX) Onboard the Lunar Atmosphere and Dust Environment Explorer (LADEE) Mission. *Space Science Reviews* 185, 93–113.

Horyani, M., V. Hoxie, D. James, et al. (2008). The student dust counter on the New Horizons mission. *Space Science Reviews* 140, 387–402.

Iess, L., N.J. Rappaport, R.A. Jacobson, et al. (2010). Gravity field, shape, and moment of inertia of Titan. *Science* 327, 1367–1369.

Jaffe, B., W.R. Cook, and H. Jaffe (1971). *Piezoelectric Ceramics*. New York, NY: Academic.

Johari, G.P., and P.A. Charrette (1975). The permittivity and attenuation in polycrystalline and single-crystal ice Ih at 35 and 460 MHz. *Journal of Glaciology* 14, 293–303.

Jol H.M., (Ed.) (2009). *Ground Penetrating Radar Theory and Applications*, Melbourne, Australia: Elsevier Science, ISBN-10: 0444533486, ISBN-13: 978-0444533487, 544.

Kattenhorn, S.A., and L.M. Prockter (2014). Evidence for subduction in the ice shell of Europa. *Nature Geoscience* 7, 762–767, doi:10.1038/ngeo2245. http://www.nature.com/ngeo/journal/v7/n10/full/ngeo2245.html.

Kawamura, T., N. Kobayashi, S. Tanaka, and P. Lognonné (2015). Lunar surface gravimeter as a lunar seismometer: Investigation of a new source of seismic information on the Moon. *Journal of Geophysical Research–Planets, AGU Journal*, doi:10.1002/2014JE004724. http://onlinelibrary.wiley.com/doi/10.1002/2014JE004724/abstract.

Kedar, S (2011). Source distribution of ocean microseisms and implications for time dependent noise tomography, *Comptes Rendus Geoscience* 343, 548–557, doi:10.1016/j.crte.2011.04.005.

Kempf, S., U. Beckmann, G. Moragas-Klostermeyer, et al. 2008. The E ring in the vicinity of Enceladus: I. Spatial distribution and properties of the ring particles. *Icarus* 193(2), 420–437.

Kieffer, S.W., X. Lu, C.M. Bethke, J.R. Spencer, S. Marshak, and A. Navrotsky (2006). A clathrate reservoir hypothesis for Enceladus' south polar plume. *Science* 314(5806), 1764–1766.

Kim, S.-S., S.R. Carnes, and C.T. Ulmer (2012). Miniature Ground Penetrating Radar (GPR) for Martian Exploration: Interrogating the shallow subsurface of Mars from the surface, Concepts and Approaches for Mars Exploration, Workshop held at Houston, TX, June 12–14 (2012), http://www.lpi.usra.edu/meetings/marsconcepts2012/pdf/4094.pdf

Konstantinidis, K., C.L.F. Martinez, B. Dachwald, et al. (2015). A lander mission to probe subglacial water on Saturn's moon Enceladus for life. *Acta Astronautica* **106**, 63–89.

Kovach, R.L., and C.F. Chyba (2001). Seismic detectability of a subsurface ocean on Europa. *Icarus* **150**, 279–287.

Kuiper, G.P (1944). Titan: A satellite with an atmosphere. *The Astrophysical Journal* **100**, 378–388.

Küppers, M., L. O'Rourke, D. Bockelée-Morvan, et al. (2014). Localized sources of water vapour on the dwarf planet (1) [thinsp] Ceres. *Nature* **505**(7484), 525–527.

Latychev K., J.X. Mitrovica, M, Ishii, N.-H. Chan, and J.L. Davis (2009). Body tides on a 3-D elastic earth: Toward a tidal tomography. *Earth and Planetary Science Letters* **277**, 86–90, doi:10.1016/j.epsl.2008.10.008.

Lavvas, P.P., A. Coustenis, and I.M. Vardavas (2008). Coupling photochemistry with haze formation in Titan's atmosphere, Part II: Results and validation with Cassini/Huygens data. *Planetary and Space Science* **56**, 67–99.

Le Bars, M., D. Cébron, and P. Le Gal (2015). Flows driven by libration, precession, and tides. *Annual Review Fluid Mechanics* **47**, 163–193.

Lee, W.H.K., J.C. Lahr, and C.M. Valdes (2003). The HYPO71 earthquake location program. In: *International Handbook of Earthquake and Engineering Seismology, Part B*. W.H.K. Lee, H. Kanamori, P.C. Jennings, and C. Kisslinger. Eds., Amsterdam: Academic Press, 1641–1642.

Le Gall, A., M.A. Janssen, L.C. Wye, et al. (2011). Cassini SAR, radiometry, scatterometry and altimetry observations of Titan's dune fields. *Icarus* **213**, 608–624.

Legarsky, J.J., S.P. Gogineni, and T.L. Akins (2001). Focused synthetic aperture radar processing of ice-sounder data collected over the Greenland ice sheet. *Geoscience and Remote Sensing, IEEE Transactions on* **39**(10), 2109–2117.

Li, X, W.B. Brinckerhoff, G.G. Managadze, et al. (2012). Laser ablation mass spectrometer (LAMS) as a standoff analyzer in space missions for airless bodies. *International Journal of Mass Spectrometry* **323–324**, 63–67.

Liebermann, L. (1959). Resonance absorption and molecular crystals. II. Benzene. *The Journal of Acoustical Society of America* **31**, 1073–1075.

Lorenz, R.D., E. Stofan, J.I. Lunine, et al. (2012). MP3–A meteorology and physical properties package to explore air-sea interaction on Titan. In: *43rd Lunar and Planetary Science Conference*, Houston, Texas. 19–23.

Lorenz, R.D., K.L. Mitchell, R.L. Kirk, et al. (2008). Titan's inventory of organic surface materials. *Geophysical Research Letters* **35**, 02206.

Lorenz, R.D., H.B. Niemann, D.N. Harpold, S.H. Way and J.C. Zarnecki (2006). Titan's damp ground: Constraints on Titan surface thermal properties from the temperature evolution of the Huygens GCMS inlet. *Meteoritics & Planetary Science* **41**(11), 1705–1714.

Lorenz, R.D., C. Newman, and J.I. Lunine (2010). Threshold of wave generation on Titan's lakes and seas: Effect of viscosity and implications for Cassini observations. *Icarus* **207**, 932–937.

Lorenz, R.D., R.L. Kirk, A.G. Hayes, et al. (2014). A radar map of Titan Seas: Tidal dissipation and ocean mixing through the throat of Kraken. *Icarus* **237**, 9–15.

Lorenz, R.D (2015). Heat rejection in the Titan surface environment: Potential impact on science investigations. *Journal of Thermophysics and Heat Transfer,* doi:10.2514/1.T4608.

Lunine, J.I., H. Waite, F. Postberg, L. Spilker, and K. Clark (2015). Enceladus life finder: The search for life in a habitable moon. In: *Lunar and Planetary Science Conference* (Vol. 46, p. 1525).

Mahaffy, P.R., C.R. Webster, M. Cabane, et al. (2012). The sample analysis at mars investigation and instrument suite. *Space Science Reviews* **170**, 401–478.

Mahaffy, P.R., R.R. Hodges, M. Benna, et al. (2014a). The neutral mass spectrometer on the lunar atmosphere and dust environment explorer mission. *Space Science Reviews* **185**, 27–61.

Mahaffy, P.R., M. Benna, T. King, et al. (2014b). The neutral gas and ion mass spectrometer on the Mars atmosphere and volatile evolution mission. *Space Science Reviews* **195**(1), 49–73.

Malaska, M.J., K. Mitchell, M. Choukroun, et al. (2014). *Where is Titan's ethane?* Laurel, MD: Titan Through Time Workshop.

Malaska, M.J., and R. Hodyss (2014). Dissolution of benzene, naphthalene, and biphenyl in a simulated Titan lake. *Icarus* **242**, 74–81.

Marouf, E.A., A.J. Kliore, N.J. Rappaport, et al. (2014). First Cassini Radio Science Bistatic Scattering Observation of Titan's Northern Seas, American Geophysical Union, Fall Meeting 2014, abstract #P22A-04.

Marshall, C.P., and A.O. Marshall (2014). Raman spectroscopy as a screening tool for ancient life detection on Mars. *Philosophical Transactions A* **372**, 20140195.

Marx, S.V., and R.O. Simmons (1984). Ultrasonic sound velocities and elastic constants of liquid and crystalline CD4 and C6H12. *Journal of ChemicalPhysics* **81**, 944, doi:10.1063/1.447695.

Mastrogiuseppe, M., V. Poggiali, A. Hayes, et al. (2014). The bathymetry of a Titan sea. *Geophysical Research Letters* **41**(5), 1432–1437.

McCord, T.B., J. Castillo-Rogez, and A. Rivkin (2011). Ceres: Its origin, evolution and structure and Dawn's potential contribution. In: *The Dawn Mission to Minor Planets 4 Vesta and 1 Ceres*. New York, NY: Springer, *Space Science Reviews Journal* **163**(1), 63–76, doi:10.1007/s11214-010-9729-9.

McKay, C.P., C.R. Stoker, B.J. Glass, et al. (2013). The icebreaker life mission to Mars: A search for biomolecular evidence for life. *Astrobiology* **13**(4), 334–353. http://adsabs.harvard.edu/abs/2013AsBio..13..334M

Mitri G., A. Coustenis, G. Fanchini, A. Gerard Hayes, L. Iess, and K. Khurana (2014). The exploration of Titan with an orbiter and a lake probe, *Planetary and Space Science, Elesvier* **104**. doi:10.1016/j.pss.2014.07.009 (08/2014), http://dx.doi.org/10.1016/j.pss.2014.07.009

Mocquet, A. (1998). A search for the minimum number of stations needed for seismic networking on Mars. 1999. *Planetary Space Science* **47**, 397–409.

Morris, RV., G. Klingelhofer, B. Bernhardt, et al. (2004). Mineralogy at Gusev Crater from the Moss-Bauer spectrometer on the Spirit Rover. *Science* **305**, 833–836.

Nixon, W.A. and E.M. Schulson (1987). A micromechanical view of the fracture toughness of ice. *Journal Physique* **48**, 313–319.

Niemann, H.B., J.R. Booth, J.E. Cooley, et al. (1980). Pioneer Venus orbiter neutral gas mass spectrometer experiment. *IEEE Transactions on Geoscience and Remote Sensing* **GE-18**, 60.

Niemann, H.B., S.K. Atreya, S.J. Bauer, et al. (2002). The gas chromatograph mass spectrometer for the Huygens probe. *Space Science Reviews* **104**(1–2), 553–591.

Niemann, H.B., S.K. Atreya, J.E. Demick, et al. (2010). Composition of Titan's lower atmosphere and simple surface volatiles as measured by the Cassini-Huygens probe gas chromatograph mass spectrometer experiment. *Journal of Geophysical Research* **115**(E12), doi:10.1029/2010JE003659.

Nordholt, J.E., D.B. Reisenfeld, R.C. Wiens, et al. (2003). Deep space 1 encounter with comet 19P/Borrelly: Ion composition measurements by the PEPE mass spectrometer. *Geophysical Research Letters* **30**, 1465.

Panning M., V. Lekic, M. Manga, F. Cammarano, and B. Romanowicz (2006). Long-period seismology on Europa: 2. Predicted seismic response. *Journal of Geophysical Research* **111**(E12), doi:10.1029/2006JE002712, http://onlinelibrary.wiley.com/doi/10.1029/2006JE002712/abstract.

Pappalardo, R.T (2010). *Seeking Europa's Ocean, Proceedings International Astronomical Union (IAU) Symposium No.269*, C. Barbieri, S. Chakrabarti, M. Coradini and M. Lazzarin, Eds., 101–114, doi:10.1017/S1743921310007325.

Pappalardo R.T., S. Vance, F. Bagenal, et al. (2013). Science potential from a Europa lander. *Astrobiology Journal* **13**(8), 740–773, doi:10.1089/ast.2013.1003, http://www.ncbi.nlm.nih.gov/pubmed/23924246.

Park, S.E., and T.R. Shrout (1997). Ultrahigh strain and piezoelectric behavior in relaxor based ferroelectric single crystals. *Journal of Applied Physics* **82**, 1804–1811.

Park, J.M., R.P. Taylor, A.T. Evans, et al. (2008). A piezoelectric microvalve for cryogenic applications. *Journal of Micromechanics Microengineering* **18**, doi:10.1088/0960-1317/18/1/015023.

Pavlov, S.G, E.K. Jessberger, H-W. Hubers, et al. (2011). Minaturized laser-induced plasma spectrometry for planetary in situ analysis—The case for Jupiter's moon Europa. *Advances in Space Research* **48**, 764–778.

Phillips, R.J., M.T. Zuber, S.E. Smrekar, et al. (2008). Mars north polar deposits: Stratigraphy, age, and geodynamical response. *Science* **320**, 1182–1185, doi:10.1126/science.1157546.

Poncy, J., O. Grasset, V. Martinot, and G. Tobie (2009). Exploring medium gravity icy planetary bodies: An opportunity in the inner system by landing at ceres high latitudes, EGU General Assembly 2009, held 19–24 April, 2009 in Vienna, Austria http://meetings.copernicus.org/egu2009, 6968.

Porco, C., D. DiNino, and F. Nimmo (2014). How the geysers, tidal stresses, and thermal emission across the south polar terrain of Enceladus are related. *The Astronomical Journal* **148**(3), 45.

Plettemeier, D., C. Statz, J. Abraham, et al. (2015). Insights gained from data measured by the CONSERT instrument during Philae's descent onto 67P/C-G's surface. *Geophysical Research* **17**, EGU2015–6693, http://meetingorganizer.copernicus.org/EGU2015/EGU2015-6693.pdf.

Proctor, T.M (1966). Low Temperature speed of sound in single-crystal ice. *Journal of Acoustical Society of America* **39**, 972–977.

Rieder, R., R. Gellert, J. Bruckner, et al. (2003). The new Athena alpha particle x-ray spectrometer for the Mars exploration rovers. *Journal of Geophysical Research: Planets* **108**, 8066.

Rivkin, A.S., E.L. Volquardsen, and B.E. Clark (2006). The surface composition of Ceres: Discovery of carbonates and iron-rich clays. *Icarus* **185**(2), 563–567.

Robin, G. de Q (1958). Glaciology III. Seismic shooting and related investigations Norwegian-British-Swedish Antarctic Expedition, 1949–1952. *Scientific Results V. Norsk Polarinstitutt, Oslo*, 134

Roethlisberger, H (1972). Seismic Exploration in Cold Regions, Cold Regions Science and Engineering Monograph 11-A2a, Hanover, NH.

Roth, L., J. Saur, K.D. Retherford, D.F. Strobel,, P.D. Feldman, M.A. McGrath, and F. Nimmo (2014). Transient water vapor at Europa's south pole. *Science* **343**(6167), 171–174, doi:10.1126/science.1247051, https://www.sciencemag.org/content/343/6167/171.abstract.

Royer, A.A., A.M. Thomas, and M.G. Bostock (2015). Tidal modulation and triggering of low-frequency earthquakes in northern Cascadia. *Journal of Geophysical Research* **120**, 384–405.

Rull F, A. Vegas, A. Sansano, and P. Sobron (2011). Analysis of arctic ices by remote Raman spectroscopy. *Spectrochimica Acta Part A: Molecular and Biomolecular Spectroscopy* **80**, 148–155.

Sandford, S.A., J. Aléon, C.M. Alexander, et al. (2006). Organics captures from comet 81P/Wild 2 by the Stardust spacecraft. *Science* **314**, 1720–1724.

Scheeres, D.J (2011). Mathematics in Earth Orbit: The Dynamics of Earth's Artificial Orbital Population, invited presentation at the International Conference on Mathematical Modeling in Industry, São Paolo, Brazil, December 2011.

Schulson, E.M., (2006). The fracture of water ice Ih: A short overview. *Meteoritics & Planetary Science* **41**(10), 1497–1508, http://onlinelibrary.wiley.com/doi/10.1111/j.1945-5100.2006.tb00432.x/pdf.

Schubert, G., J.D. Anderson, T. Spohn, and W.B. McKinnon (2004). Interior composition, structure and dynamics of the Galilean satellites. In: *Jupiter: The Planet, Satellites and Magnetosphere*. F. Bagenal, T.E. Dowling, W.B. McKinnon. Eds., Cambridge, UK: Cambridge University Press, 281–306.

Seu, R., D. Biccari, R. Orosei, et al. (2004). SHARAD: The MRO 2005 shallow radar. *Planetary and Space Science* **52**(1), 157–166.

Sharma, S.K, P.G. Lucey, M. Ghosh, H.W. Hubble, and K.A. Horton (2003). Stand-off Raman spectroscopic detection of minerals on planetary surfaces. *Spectrochimica Acta Part A: Molecular and Biomolecular Spectroscopy* **59**, 2391–2407.

Sherman, C.H., and J.L. Butler (2007). *Transducers and Arrays for Underwater Sound*, Vol. 124. Office of Naval Research.

Sherrit, S., W. Zimmerman, N. Takano, and L. Avellar (2014). Miniature cryogenic valves for a Titan lake sampling system. *Proceedings SPIE 9061, Sensors and Smart Structures Technologies for Civil, Mechanical, and Aerospace Systems* **90613J**, doi:10.1117/12.2045185.

Shimizu, H., N. Nakashima, and S. Sasaki (1996). High-pressure brillouin scattering and elastic properties of liquid and solid methane. *Physical Review B* **53**, 111–115.

Singer, J.R (1969). Excess ultrasonic attenuation and volume viscosity in liquid methane. *Journal of Chemical Physics* **51**, 4729, doi:10.1063/1.1671860.

Soderlund, K.M., B.E. Schmidt, J. Wicht, and D.D. Blankenship (2014). Ocean-driven heating of Europa's icy shell at low latitudes. *Nature Geoscience Letter* **7**, 16–19, doi:10.1038/ngeo2021, http://www.nature.com/ngeo/journal/v7/n1/full/ngeo2021.html.

Sotin, C., J.W. Head III, G. Tobie (2002). Europa: Tidal heating of upwelling thermal plumes and the origin of lenticulae and chaos melting. *Geophysical Research Letters* **29**, 74-1–74-4.

Spencer J. R., and F. Nimmo (2013). Enceladus: An active ice world in the saturn system. *Annual Review of Earth and Planetary Sciences*, **41**, 693–717.

Srama, R., T.J. Ahrens, N. Altobelli, et al. (2004). The cassini cosmic dust analyzer. *Space Science Reviews* **114**, 465–518 (2004).

Stevenson, J.M., W.A. Fouad, D. Shalloway, D. Usher, J. Lunine, W.G. Chapman, and P. Clancy (2015). Solvation of nitrogen compounds in Titan's seas, precipitates, and atmosphere. *Icarus* **256**, 1–12.

Stofan, E., R. Lorenz, J. Lunine, E. Bierhaus, B. Clark, P. Mahaffy, and M. Ravine (2013). TiME—The Titan Mare Explorer. In: Proceedings of the IEEE Aerospace Conference, Big Sky, MT, March, paper #2434.

Stofan, E.R., C. Elachi, J.I. Lunine, et al. (2007). The lakes of Titan. *Nature* **445**, 61–64.

Straty, G.C (1974). Velocity of sound in dense fluid methane. *Cryogenics* **14**, 367–370.

Studinger, M., et al. (2003). Ice cover, landscape setting, and geological framework of Lake Vostok, East Antarctica. *Earth Planetary Science Letters* **205**, 195–210.

Tan, S.P., S.J. Kargel, and G. Marion (2013). Titan's atmosphere and surface liquid: New calculation using Statistical Associating Fluid Theory. *Icarus* **222**, 53–72.

Tarsitano, C.G., and C.R. Webster (2007). Multilaser Herriot cell for planetary tunable laser spectrometers. *Applied Optics* **46**, 6923–6935.

Thomas, P.C., J.W. Parker, L.A. McFadden, C.T. Russell, S.A. Stern, M.V. Sykes, and E.F. Young (2005). Differentiation of the asteroid Ceres as revealed by its shape. *Nature Letters* **437**(7056), 224–226, doi:10.1038/nature03938m, http://hubblesite.org/pubinfo/pdf/2005/27/pdf.pdf.

Tobie, G., J.I. Lunine, and C. Sotin (2006). Episodic outgassing as the origin of atmospheric methane on Titan. *Nature* **440**(7080), 61–64.

Tomasko, M.G., and R.A. West (2009). Aerosols in Titan's atmosphere. In: *Titan from Cassini-Huygens*, R. Brown, J.P. Lebreton and J.H. Waite, Eds., New York, NY: Springer, 297–321.

Trainer, M.G., P.R. Mahaffy, E.R. Stofan, J.I. Lunine, and R.D. Lorenz (2012). Measuring the composition of a cryogenic sea. International Workshop on Instrumentation for Planetary Missions, Greenbelt, MD, Abstract #1033.

Tsai, V.C., and J.R. Rice (2012). Modeling turbulent hydraulic fracture near a free surface, *Journal of Applied Mechanics*. **79**(3), 31,003, doi:10.1115/1.4005879.

Tsai, V.C., B. Minchew, M.P. Lamb, and J.-P. Ampuero (2012). A physical model for seismic noise generation from sediment transport in rivers. *Geophysical Research Letteers* **39**, L02404, doi:10.1029/2011GL050255.

Tsumura, R., and G.C. Straty (1977). Speed of sounds in saturated and compressed fluid ethane. *Cryogenics* **17**, 195–200.

Vance S., M. Bouffard, M. Choukroun, and C. Sotin (2014). Ganymede's internal structure including thermodynamics of magnesium sulfate oceans in contact with ice. *Planetary and Space Science Journal, Elsevier* **96**, 62–70, doi:10.1016/j.pss.2014.03.011, http://www.sciencedirect.com/science/article/pii/S0032063314000695.

Vision and Voyages (2011). *Vision and Voyages for Planetary Science in the Decade 2013–2022.* Washington DC: National Academies Press.

Waite, J.H., W.S. Lewis, B.A. Magee, et al. (2009). Liquid water on Enceladus from observations of ammonia and Ar-40 in the plume. *Nature* **460**(7254), 487–490.

Wang, A., L.A. Haskin, and E. Cortez (1998). Prototype Raman spectroscopic sensor for *in situ* mineral characterization on planetary surfaces. *Applied Spectroscopy* **52** 477–487.

Wiens, R.C., S. Maurice, B. Barraclough, et al. (2012). The ChemCam instrument suite on the Mars Science Laboratory (MSL) Rover: Body unity and combined systems tests. *Space Science Reviews* **170**, 167–227.

Wilson, E.H., and S.K. Atreya (2004). Current state of modeling the photochemistry of Titan's mutually dependent atmosphere and ionosphere. *Journal of Geophysical Research* **109**, E06002.

Wurz P., D. Abplanalp, M. Tulej, M. Iakovleva, V. A. Fernandes, A. Chumikov, and G. G. Managadze (2012). Mass spectrometric analysis in planetary science: Investigation of the surface and the atmosphere. *Solar System Research* **46**(6), 408–422.

Yung, Y.L., M. Allen, and J.P. Pinto (1984). Photochemistry of the atmosphere of Titan—Comparison between mode and observations. *Astrophysiscal Journal Supplement Series* **55**, 465–506.

Zarnecki J.C., M.R. Leese, J.R.C. Garry, N. Ghafoor, and B. Hathi(2002). Huygens' surface science package. *Space Science Reviews* **104**, 593–611.

Internet References

http://www.gizmag.com/nasa-titan-submarine-concept/35960/
http://web.gps.caltech.edu/~oa/simulations.shtml

10

Drilling and Breaking Ice

Kris Zacny, Gale Paulsen, Yoseph Bar-Cohen, Xiaoqi Bao, Mircea Badescu, Hyeong Jae Lee, Stewart Sherrit, Victor Zagorodnov, Lonnie Thompson, and Pavel Talalay

CONTENTS

10.1 Ice Drilling Methods

More than 170 years ago, in 1841, Louis Agassiz, one of the creators of glacial theory, made his first attempt to drill to the bed of Unteraargletscher, Alpines [Clarke, 1987]. Since that time, various systems for ice drilling have been designed and tested—many of them were quite effective, but some proved temperamental.

In general, penetration of ice is required mainly for scientific or exploration purposes. As such, different scientific goals drive different measurement and sampling requirements. To accomplish some measurements, scientists may need a continuous ice or permafrost core of certain diameter. Other requirements may be satisfied by just analyzing cuttings or fluids. If instruments can be packaged into a small-diameter tube, downhole measurements can be performed by lowering such an instrument into an existing hole.

The aforementioned scientific goals, however, need to be traded against engineering and logistics complexity as well as cost. Engineering mainly refers to the ease of developing a certain drilling approach. For example, a hot water drill where a borehole (BH) is created using a hot water jet is easier and faster than that using a wireline system [Just, 1963]. Logistics plays a role if the system needs to be transported over a long distance and into a remote location. Sometimes equipment needs to be air lifted, and in turn, an entire drill system has to be broken down into manageable pieces and then reassembled. Probably the ultimate challenge is to transport the drilling system to another planet: this has been done before at a higher cost and system complexity. Cost is a major driver in all endeavors. Very often a compromise has to be reached between science output and engineering complexity due to lower budgets.

In general, drilling a hole and analyzing cuttings is the simplest approach. A hole needs to be drilled of any diameter with the fastest method possible, and there is no need to keep the BH open. However, not all science questions can be addressed by examination and analysis of cuttings. In addition, cuttings are mixed up and also mixed with the drilling fluid, which might affect measurements. The depth from which cuttings come is not precise but rather represents a specific range. If these issues do not affect science output, this approach would be optimal. Drilling a hole and lowering a probe solve some of the problems related to knowing where a sample came from. However, a noncontact sonde is limited to certain types of measurements and also has limited sensitivity. If measurement with a sonde is an option and can be performed relatively fast, there is no need to keep a BH open and any type of hole drilling method that is fast and inexpensive can be used. There is no doubt that the best sample is an ice core. A core can preserve gas, dust, and rock fragments that are extremely useful for investigating past history. However, capturing of an ice core and retrieving it is a much more complex and slow operation [Koshima et al., 2002]. Normally a core is captured in sections, and hence the BH has to be kept open to prevent shrinkage by using a low freezing temperature and environmentally friendly fluid. The drill itself is complex, since it needs a core chamber, cutting chamber, and various subsystems to perform drilling, bottomhole cleaning, drill stabilizing, and so forth. After the core is out, it also needs to be cleaned and stored at below freezing temperature for the entire time period.

Depending on the nature of the ice destruction at the bottom of the hole, all developed methods of ice drilling can be divided into mechanical, thermal, and thermo-mechanical (Figure 10.1). The last method has never been used in the practice of drilling operations on the glaciers, although there are a few articles relating to the development of thermo-mechanical drills for making holes in ice [Koci, 1994].

The special thermal drills proved an attractive alternative for ice drilling. There are different drilling tools that use heat to make holes in ice. These include electric thermal coring drills, hot points, hot-water drilling systems, steam drills, and others [Hodge, 1971; Ignatov, 1965; Howorka, 1965; Smith et al., 2006]. The specific energy required for thermal drills in the range of 590–680 MJ m^{-3} is two orders of magnitude higher than the energy required for mechanical systems, which need 1.9–4.8 MJ m^{-3} [Koci and Sonderup, 1990]. Mechanical drilling tools can use percussion, but most commonly, they use rotary drilling with sharp cutters [Talalay, 2003].

The main feature of the electromechanical cable-suspended drills is that an armored cable with a winch is used instead of a pipe string to provide power to the downhole motor system and retrieve the downhole unit. The use of armored cable allows a significant reduction in power and material consumption, a decrease in the time of round-trip operations, and a simplification in the cleaning of the hole from the cuttings.

The material removal function is critically important to all drilling systems, as the presence of excessive material at the bottom of the BH leads to decreasing penetration rates and even to loss of the drill.

According to Mellor and Sellmann [1976], material removal systems can be grouped into the following categories: (1) direct lifting; (2) lateral displacement of material (e.g., in compressible snow-firn layers); (3) lifting of cuttings by circulation medium (air or liquids); and (4) dissolving. The first two methods and also air circulation are used for so-called "dry"

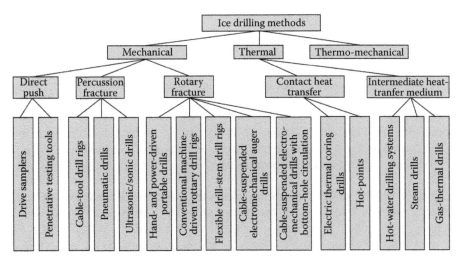

FIGURE 10.1
Classification of ice drilling methods.

drilling when the BH shaft is filled with air, but the experience of deep drilling shows significant closure in an open hole. The deepest "dry" BH were 411 m drilled by the mechanical method at Site-2, Greenland, 1956–1957 [Lange, 1973] and 952 m drilled by a thermal coring drill at Vostok Station, Central Antarctica, 1972 [Korotkevich and Kudryashov, 1976].

For drilling at a greater depth, it is necessary to prevent BH closure through visco-plastic deformation of ice by filling it with a fluid [Talalay and Hooke, 2007; Talalay et al., 2014b]. Various low temperature drilling fluids have been proposed for drilling in ice. These are broken down into three main groups: (1) two-component kerosene-based fluids with density additives, (2) alcohol compounds, and (3) ester compounds [Talalay and Gundestrup, 2002; Talalay et al., 2014a].

Ice drilling methods, depending on depth and application, can be broken into four groups:

1. *Near-surface shallow drilling* up to 50 m for ablation stakes installation, temperature measurements at the bottom of the active layer, revealing of anthropogenic pollution, and so forth

2. *Shallow drilling* up to 400 m for structure study of the snow and firn zones and for research of current climate variability, such as interhemispheric relationships and the timing and expression of the Little Ice Age

3. *Intermediate drilling* up to 1500 m for the study of marginal parts of the Antarctic and Greenland sheets, for accessing holes through ice shelves, and for understanding the behavior and timing of the last deglaciation

4. *Deep drilling* up to the depth of 4000 m for fundamental studies of ice-sheet dynamics and thermal regime, for access to subglacial environment, and for recovering as much old ice core as possible to reconstruct the climate of Earth for at least 1.4 million years and beyond.

10.2 Mechanical Drilling in Ice

10.2.1 Direct-Push Drilling and Penetrating

Direct-push drilling technology includes methods that advance a drill by pushing, hammering, or vibrating. While this does not meet the proper definition of drilling, it does achieve the same result—a BH and, if applicable, a core. Direct-push drilling tools do not remove cuttings from the hole, and deepening is achieved by compression of formation. Direct-push drilling cannot be used in solid ice but it can be considered for investigations of the compressible snow-firn layers. Two methods of direct-push snow-firn drilling include (1) drive sampling and (2) penetrative testing.

The drive sampling is the primary method of field measuring snow depth, depth-integrated density, and snow water equivalent, as it is considerably less destructive to the snowpack and faster to use than traditional snow pit techniques. A *snow sampler* (also referred to as "snow tube") commonly consists of a metal or plastic tube (sometimes in sections for portability) with a cutter head fixed at its lower end and driving wrench for operating the sampler. Drilling with snow samplers is done by pushing and, if necessary, rotating the sampler down through the snowpack and subsequently extracting a core (Figure 10.2).

FIGURE 10.2
Surveyor takes snow core sample and reads the snow depth using the gauge on the tube (From R. Abramovich, USDA NRCS Snow Survey, http://blogs.usda.gov/2011/03/02/forecasting-western-waters-%E2%80%93-carrying-on-a-tradition/.)

In most snow samplers, the cutter head has wedge-shaped teeth and is able to penetrate various types of snow, through crusted and icy layers, and in some cases, through solid ice layers of appreciable thickness that may form near the surface. The cutter head should not compact the snow so that an excessive amount of snow is accepted by the interior of the cutter [Guide to Hydrological Practices, 2008]. As snow samplers do not include a core catcher, the cutter head must seize the core base with sufficient adhesion to prevent the snow core from falling out when the sampler is withdrawn from the snow. Small-diameter cutter heads retain the sample much better than large cutter heads, but larger samples increase the accuracy in measurements.

There are many snow samplers in use with varying diameters and cutter arrangements (Mt. Rose Sampler, Federal Snow Sampler, Rosen Sampler, Adirondack Sampler, and others). Typically, snow tube samplers measure snow density within 5–10% of snow pit estimates.

In addition to collecting core samples, numerous penetrative instruments have been used to measure mechanical properties (shear strength, hardness, inherent cohesion, etc.) of snow and firn. One of the simplest methods of snow avalanche observation is estimation of the spatial extent of distinct weak snow layers or significant changes in layer hardness using the ski pole–like penetrometer. The ram penetrometer (also called "Swiss ramsonde") has, for many years, been used to test snow hardness [McClung and Schaerer,

2006]. The standard ram penetrometer consists of a 1-m lead section tube with a 40-mm-diameter conical tip and an apex angle of 60°, which is driven into the snow by means of a weight dropped onto the anvil on the tube top.

The ram penetrometer, developed at the end of the 1930s by R. Haefeli, is still used today in an almost unaltered form. However, snow is often so thinly stratified that the Rammsonde can detect only major layers. Further development of snow penetrometers was focused on recording the penetration resistance continuously using digital sensors (Digital Thermo-Resistograph, SnowMicroPen, SABRE probe).

McCallum [2014] adapted standard cone penetrometer testing (CPT) to allow penetrative testing in hard polar firn to depths of 10 m. To use it in Antarctica, a special container was manufactured in which the CPT and ancillary equipment are stored and transported, and in which a CPT operator can stand. Also, 35.6-mm-diameter cones having a corresponding cross-sectional area of 10 cm^2 and sleeve area of 150 cm^2 were specially produced to measure resistance on the cone tip and friction on the cone sleeve (Figure 10.3). They also have the capacity to measure pore pressure. Standard steel alloy rods, each of length 1 m, are used to transfer the penetrative force from the hydraulically driven rams to the cone at depth. Field tests proved that the CPT can be used efficiently in polar environments to potentially provide estimates of physical parameters in hard firn to substantial depth.

10.2.2 Percussion Drills

"Percussion ice drilling" implies a process of drilling BH by impact disintegration of ice. Percussion drills can be classified as (1) cable-tool drills, (2) pneumatic drills for blast holes, (3) rotary-percussion drills that combine both the rotary and percussive action, and (4) sonic drills that employ the use of high-frequency, resonant energy to advance a drill bit into subsurface formations.

The cable tool is the oldest drilling method and originated over 4000 years ago in China. It involves the use of percussion by repeatedly lifting and dropping a heavy string of drilling tools or weight into a BH. The material is pulverized at the bottom of the hole forming granulate, which is then removed with a bailer when the penetration rate becomes unacceptable. The first ever "scientific" drilling in ice was carried out by Agassiz using a cable-tool rig. In 1841, he drilled a 50-m-deep hole in the Alps. The following year after 6 weeks of tedious effort, 16-, 32.5-, and 60-m-deep holes had been drilled (Figure 10.4). For the first 3 days of drilling of each hole, the typical drill rate was 13 m/day (with four men); this decreased to 3–4 m/day (with eight men) as the hole depth increased. Discouraged, Agassiz made no further attempts to drill through glaciers [Clarke, 1987].

Although some attempt to develop reliable cable-tool rigs was made by the Institute of Geography, USSR Academy of Sciences, in the 1960s [Tsykin, 1966], nowadays cable-tool

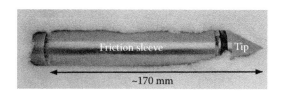

FIGURE 10.3
Cone incorporating 35.6-mm-diameter tip and 135-mm-long cylindrical friction sleeve. (From McCallum, A., *J. Glaciol.*, 60, 2014; reprinted with permission of the International Glaciological Society.)

FIGURE 10.4
Cable tool used by L. Agassiz in 1842. (From Clarke, G.K.C., *J. Glaciol. Spec.*, 1987. With permission of the International Glaciological Society.)

drilling is not used because the drilling equipment is rather heavy and energy-demanding, the drilling depths are shallow, and the quality of core is poor.

Handheld pneumatic drills or jackhammers are powered by compressed air, which accelerates a free mass toward the steel drill pipe. Upon impact, the free mass bounces back while the drill pipe penetrates into the subsurface. Conventional pneumatic drills with 1.2-m and 2.4-m-long steel augers were quite effective to drill blast holes during the experimental tunneling at the base of an ice cliff near Camp Tuto, Northwest Greenland [Rausch, 1958]. Holes 44.5 mm in diameter were drilled at a rate of 1.5 m min^{-1} for the 2.4-m deep holes.

The rotary-percussion drilling combines both the rotary and percussive action. A hole is formed when energy is transmitted through the drill rod to the drill bit, prompting it to thrust into the formation using a repeated hammering motion. Ice fragmentation occurs at highly pressurized contact zones between the bit and the ice. Bit rotation creates new impact positions for the bit, and new ice is fragmented advancing the bit. The behavior of rotary percussion drills in silty ice was tested in conjunction with a tunneling project in Northwest Greenland [McAnerney, 1970]. Rotary-percussion drills performed poorly in vertical holes and showed better results horizontally. The average rate of penetration was rather slow, near 0.2 m min^{-1}, presumably because of the poor cuttings transportation.

10.2.3 Hand- and Power-Driven Portable Ice Drills

These drills are the small systems that can drill holes to maximum depths of ~50 m. Depending on the tasks, portable drills could be either noncore devices or capable of recovering cores. They are relatively lightweight and do not require a drilling fluid. Portable ice drills can be subdivided into (1) noncore augers, (2) noncore "piston" drills, (3) core augers, (4) core drills with tooth and annular bits, and (5) mini-drills.

Hand and mechanically driven noncore augers are used to drill holes through sea/lake/river ice for winter fishing, ice measurements, and hydrological research. Common augers include a rotating helical screw blade called a "flighting" to act as a screw conveyor to remove cuttings from the bottom of the hole. The rotation of the blade causes cuttings to move up the hole being drilled. The auger is driven from the surface directly or by a series of extensions that are added as drilling proceeds into the ice. The weight of a noncore portable ice auger should be small enough to transport it by man or a small vehicle.

A good example of a noncore augering set is the "Ice Thickness Kit" produced by Kovacs Enterprises, Inc. The Kovacs ice auger is 50 mm in diameter, 1 m long, and joins together via a patented push-button connector, which allows for quick connection of one auger section to another (Figure 10.5). The same connection is used to attach a 51-mm-wide ice cutting bit. This method of assembly means that there are no pins or connector bolts that can be lost and no bolts on which clothing can snag. A few options to drive the auger are provided: hand brace, 0.5-inch electric drill, heavy-duty electric drill, and the engine drive. The heavy-duty electric drill and the engine drive turn the auger at the optimum speed of 550–650 rpm. Drilling rates of 4 m min^{-1} in ice are achievable with this power drive. The maximum achievable depth is 24 m through a multiyear pressure ridge and 23 m through a grounded ice island (tabular iceberg) using a 0.5-inch electric drill.

FIGURE 10.5
Shallow drilling on Shackleton Glacier, Canadian Rocky Mountains. (Courtesy of Kovacs Ice Drilling Equipment.)

Hand ice augers for winter fishing are produced in Northern Europe, Canada, USA, Russia, and China, although Rapala® VMC Corporation is the global leader in this category. The sizes of commercially produced augers are 4.5 inch, 6 inch, 8 inch, and 10 inch, although the most popular is an 8-inch system. The depth of the drilling is limited by the length of the auger and its extension, and in most cases, it is not more than 1.2 m. The drilling of a hole to a depth of 1 m by hand takes 5 min in dry ice. Drilling in the wet ice takes longer because it is necessary from time to time (every 20 cm or so) to retrieve the auger from the hole for cleaning.

Nowadays propane, gasoline, and electric-powered augers are becoming more efficient and, in turn, lightweight (Figure 10.6). There are several options on the market including brands such as Eskimo, Jiffy, and Strikemaster. Engine sizes may vary along with weights and performance. There are a variety of different power options: both two-stroke and four-stroke gasoline models are available. Propane-powered ice drills are much cleaner and can be operated even indoors. Battery-powered models require less maintenance, but charging the battery is an ongoing problem. Augers are coated with an anti-adhesion coating (permanent color or Teflon coating) to prevent ice buildup. In principle, the depth of portable ice auger drilling can be increased by adding auger extensions, but in practice, it is limited by human or engine capacity.

Noncore "piston" drills use the drill head as piston to recover cuttings (Figure 10.7). Cutters are associated with the slotted disk and fed ice chips during cutting on the upper surface of the disk. The disk connected with the brace handle chucks out cuttings when lifting. The efficiency of "piston" ice drills is lower than that of auger drills, and hence they are not often used nowadays.

FIGURE 10.6
Drilling with Jiffy gasoline-powered auger through the ice cover of Arctic Lake with ice thickness about 2 m. (Photograph by J. Briner, University of Buffalo [http://post.queensu.ca/~pearl/cf8/cf8pics.html].)

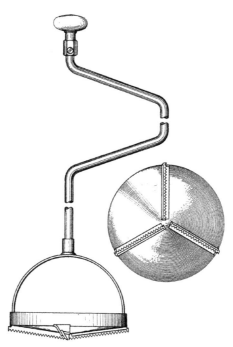

FIGURE 10.7
R. D. Green's "piston" ice drill (U.S. Pat. Des. 141,685, 1945).

The portable ice core augers use the same drilling principle as the noncoring augers, but instead of a full-diameter screw conveyor, a single-core barrel (on rare occasions, double-core barrel) with flights is used. The prototype of the ice core auger was designed by Brooks in 1932 (Figure 10.8). Cutters were located at the lower end of the flights. This ice auger looks very similar to modern designs that are widely used in shallow ice drilling.

Most hand augers take cores with a diameter in the range 75–100 mm and length less 1 m. Although designed primarily as handheld tools, with hand rotation, they later became powered augers, driven by a handheld electric motor or by a light gasoline engine. In the case of hand rotation, the auger is rotated manually using either a T-handle (Figure 10.9a) or for more speed but less torque, a brace handle [Bentley et al., 2009]. On the surface, the extension rods are added to reach greater depths; they are usually connected by a male–female joint and fixed with various types of retainers: some are quick-release push-button types that have a spring-loaded ball like Lockwell or Hartwell pins, while some have a pivoting arm or wire spring clip.

The maximum depth of drilling is limited by the weight of the auger and the drill rods, which have to be retrieved each time a core section is recovered. When the penetration depth exceeds 6 m, two or three people are needed to raise and lower the drill and to hold the string while adding or removing extension rods. A support clamp at the mouth of the hole is useful for supporting the weight of the string, and a tripod or gin pole helps in raising or lowering (Figure 10.9b). A block and tackle could be used to multiply the lifting force by a factor of four. The use of a tripod to assist in lifting the drill string increases the depth capacity. The deepest hole drilled using a hand-drilling method with a tripod was 55 m in Ward Hunt Ice Shelf, Ellesmere Island, in 1960 [Ragle et al., 1964]. Drilling beyond a depth of about 30 m is better achieved with an electromechanical drilling method due to the increasing weight and handling of the drill stem and the trip time in and out of the BH.

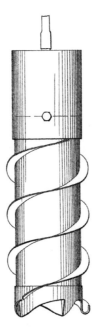

FIGURE 10.8
F. W. Brooks' ice core auger (U.S. Pat. 1,857,585, 1932).

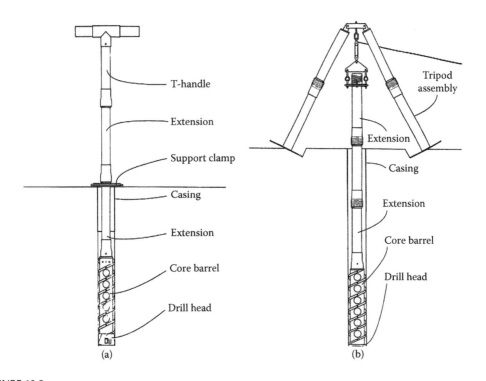

FIGURE 10.9
(a) Working position of hand ice core auger with T-handle. (b) Auger lifting with tripod. (From Koci, B.R., and Kuivinen, K.C., *J. Glaciol.*, 30, 105, 1984. With permission of the International Glaciological Society.)

During the last 60 years, many different types of ice core augers were designed and tested: CRREL, PICO, Kovacs, UCPH, IDDO, and so forth. All of them are similar, and only differ by material type, design of drill heads, and connections. Hereafter we describe the SIPRE auger that was developed in the early 1950s and is known worldwide [Rand and Mellor, 1985]. The classical SIPRE auger kit includes a core barrel with a drill head, a driving head, five extension rods with connector pins, a turning brace and T-handle, a starting mandrel, tools, and a box. The entire system weights ~40 kg (Figure 10.10). The overall length of the SIPRE auger core barrel is approximately 1 m when the drill head is fitted. On the outer surface of the core barrel, two flights are welded with a helix pitch of 200 mm. The outside helix angle is set at 30°. In some modifications, the core barrel has a dull sandblast finish, while some are chrome-plated, and others are Teflon-coated.

The drill head is equipped with two chisel-edge mild steel cutters having 30° rake angle and 20° clearance angle. The nominal internal diameter of the drill head is 3 inches (76.2 mm), and the outer diameter is 4 inches (110.9 mm). The drilling speed is set by elevating screws that can be shimmed with washers; this limits the angle of the helical penetration path. The drill head does not normally have a core-catcher (though it can), since the core is usually retained by cuttings jammed between the core and the barrel wall.

Hand rotation was sufficient to achieve penetration rates of up to 0.5 m min^{-1} in ice when the cutters are sharp, but rates in the range 0.15–0.35 m min^{-1} are more typical. The SIPRE auger can also be rotated by an electric motor or gasoline engine turning at about 200–500 rpm. With motor drive, penetration rates of up to 1.7 m min^{-1} have been achieved [Rand and Mellor, 1985].

In subsequent years, the SIPRE auger has been used to drill thousands of holes throughout the cold regions on Earth. Even though there have been many other improved versions of hand ice augers, the SIPRE auger is still popular in the ice-coring science community (Figure 10.11).

FIGURE 10.10
Three-inch SIPRE auger kit. (Courtesy of Byrd Polar Research Center, The Ohio State University [http://www.icedrill.org/graphics/pics/home_sipreauger_lg.jpg].)

FIGURE 10.11
Drilling through Lake Joyce, Dry Valleys, Antarctica ice with hand-powered SIPRE auger, 2003. (Courtesy of B. Hall, The University of Maine [http://climatechange.umaine.edu/Research/Expeditions/2003/MillenialLakes/Spire.html].)

A few versions of handheld ice core drills with tooth bits were designed in Canada and USSR in the 1950–1960s [Miller, 1954; Ward, 1954; Tsykin, 1962]. These drills did not use augers for chip transportation and had no means of removing the ice cuttings from the vicinity of the cutting blades. Since the cuttings were allowed to remain in the hole, the corer would soon become inefficient or ineffective.

Another ice core drill with an annular bit was developed by N. V. Cherepanov. The drill consisted of a massive metal ring with a square or rectangular cross section (Figure 10.12). To attach the cutter, a slot was made in the ring at an angle of 40°–45° and a width of 25–30 mm. This slot provided the cuttings with easy access to the upper flat surface of the ring. Cuttings were cleared by frequently (every 5–6 cm of penetration) lifting the ring to the surface.

Portable mini-drills are used for sampling firn or ice from trenches and pits to a depth of few dozen centimeters, for density and impurity measurements. A good example of such a drill is a portable Livingston Island mini-drill. It consists of three pieces: a drill head, a 0.46-m-long core barrel, and a drive nut [Casas et al., 1998]. The total weight of the engine and mini-drill is ~6 kg. The chipmunk drill is perhaps the smallest coring drill, driven by a commercial 0.5-inch electric drill. The drill collects 2-inch (50.8 mm) diameter cores in solid ice (Figure 10.13). It has two barrels, one 15 cm long and another 50 cm long [Bentley et al., 2009].

10.2.4 Conventional Machine-Driven Rotary Drill Rigs

The concept of conventional drilling with pipe string is used frequently in mineral exploration where BH may be from dozens to a few thousand of meters in depth. In this approach, the bit is attached to the drilling pipe string rotated at a surface using a

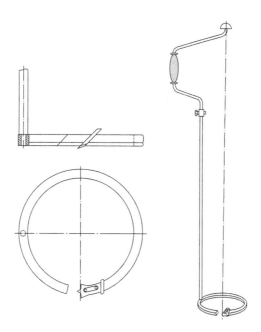

FIGURE 10.12
Ice core drill with annular bit (USSR Pat. 472,237, 1975).

FIGURE 10.13
Chipmunk in use in Greenland. (From Bentley et al., *Drilling in Extreme Environments. Penetration and Sampling on Earth and Other Planets,* 2009. Copyright Wiley-VCH Verlag GmbH & Co., KgaA, Weinheim. Reproduced with permission.)

rotary rig. Some kind of a fluid or air circulation is used for cleaning the hole of cuttings and cooling the bit. In some cases, augers could be used for lifting cuttings up the BH.

Cuttings removal systems of conventional machine-driven rotary drill rigs can be grouped into the four following categories: (1) dry drilling without a circulation/transportation system; (2) continuous or discontinuous screw transport by auger; (3) blow-out by air circulation; and (4) wash-out by fluid circulation. All these methods could be applied to noncore and core drilling. To remove the core from a conventional core barrel, the entire drill string has to be removed from the hole. This is a time-consuming approach, as each rod has to be removed one at a time. Wire-line drilling is considered a good method to decrease the time of tripping operations. With wire-line drilling, a core barrel can be removed from the bottom of the hole without removing the drill string. All these methods were applied at the early stage of drilling in glaciers in the 1950s–1970s [Miller, 1951; Ract-Madoux and Reynaud, 1951; Ward, 1952; Heuberger, 1954; Schytt, 1958; Kapitsa, 1958; Treshnikov, 1960; Bazanov, 1961; Tongiorgi et al., 1962; Lange, 1973; Rand, 1977].

Generally, drilling performance with conventional machine-driven rotary drill rigs was poor because of a number of problems. The deepest BH was drilled to a depth of 411 m using a Failing Model 314 rotary type drill rig with air circulation at the Site, Greenland, in 1957 [Lange, 1973]. Core drilling was done with the drill string held manually in a tension and at a rotational speed of 40–75 rpm. The rate of penetration ranged from 3.8 to 15.2 m h^{-1}. The core quality was generally fair with cracking or disking.

Renewed interest in conventional rotary drill rigs comes from geologists desiring subglacial bedrock samples to, for example, validate geotectonic and paleogeographic reconstructions. In 2014, the University of Minnesota began construction of the Rapid Access Ice Drill (RAID) that should be able to penetrate the Antarctic ice sheets to a depth of 3300 m in order to take cores of the deepest ice samples across the glacial bed, and continue coring into bedrock below [Goodge and Severinghaus, 2014]. RAID is based on a modification of an industry-standard diamond rock-coring system (Figure 10.14). The most critical area for RAID development is to find a way to efficiently separate the chips from the drilling fluid once the chips reach the surface. This process may ultimately limit how fast the system can penetrate ice. The separation process must be fast enough so that the drilling fluid can be pumped back down the hole.

It needs to be emphasized that in order to use commercial drill rigs, many of its subsystems such as hydraulics, the fluid processing system, and so on need to be redesigned to operate at low temperatures. In addition, commercial rigs are very heavy and large and often require numerous trucks for transportation. Hence, their use in Antarctica could be difficult.

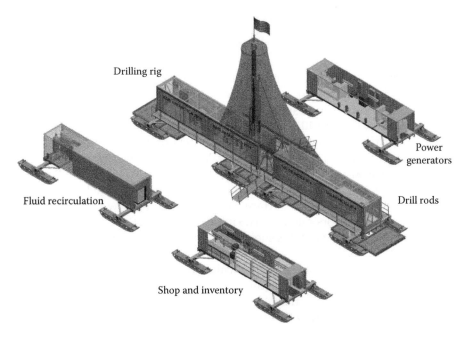

FIGURE 10.14
RAID operational layout. (From Goodge, J., and Severinghaus, J., *U.S. Ice Drilling Program*, 2014.)

10.2.5 Flexible Drill-Stem Drill Rigs

Flexible drill-stem technology utilizes bending conduit (coiled tubing or reinforced hose) to rapidly reel bits, motors, and other tools in and out of the BH. Such operations proceed quickly compared to using a jointed pipe conventional drilling rig because connection time is eliminated.

A combination of hot-water and mechanical drilling systems using high-pressure hose was suggested by Das et al. [1992] for quick access to interesting areas as well as the ability to obtain quality ice or rock core samples. When core samples are required, a downhole positive displacement motor or PDM and a core barrel replace the hot-water nozzle. The core diameter is expected to be between 200 and 300 mm. For ice drilling, hot water is used to melt chips created during the drilling process. The drill thus requires no chip storage area. Tests carried out at the University of Alaska demonstrated the ability to use warm water as a chip transport/melting medium (Koci, 1994). A 15-cm core over 1 m in length was retrieved successfully despite the presence of rocks, sand, leaves, and cracks in the ice.

The usage of a coiled tubing drill system for fast access to the subglacial environment and bedrock coring has been proposed by Clow and Koci [2002]. A coiled tubing drilling system, however, requires major developments to adapt these systems for exploring subglacial environments. The minimum operating temperature for commercial composite tubing is currently −40°C. At colder temperatures, the tubing liner becomes too brittle. The long-term performance of other subsurface components (e.g., mud motors, circulation subassemblies, and orienting tools) at very low temperatures and repeated drilling cycles should be evaluated, and component modifications are likely required. The coiled tubing itself is subject to fatigue every time it is coiled out. A total of 100–200 trips is the usual lifetime of a coil, depending upon the axial and pressure loading.

A similar system for rapid drilling of an access hole in ice sheets with minimal resources and logistics support, called RADIX, is being developed at the University of Bern [Schwander et al., 2014]. It is intended to drill a "pinhole" with a diameter of ~20 mm with small-diameter PDM. The prototype of the downhole unit with a 30-mm-diameter drill bit was tested on the Plaine Morte glacier, a temperate glacier in Switzerland, in 2013. The depth reached during the test was limited to 3.5 m due to clumping preventing chips from rising in the annulus. Therefore, the main problem of the system is ice cuttings transport in the small space between the hose and the wall of the BH.

The University of Wisconsin designed a sled-mounted flexible air drill rig called Rapid Air Movement (RAM) (Figure 10.15). The bottom-hole assembly comprises an aggressive 4-inch (101.6 mm) drill head with four tool steel cutters that is driven at a high rotational speed of up to 2500 rpm by a two conventional 0.7 kW air-tool turbines coupled together through a planetary gear reducer [Bentley et al., 2009]. The drill is suspended on a continuous air hose and weighs nearly 90 kg. Ice chips are blown out of the hole by the exhaust air from the high-flow-rate turbine. Surface components include two sled-mounted diesel-engine-driven air compressors rated for 11.3 m^3 min^{-1} at 1.4 MPa, a winch to hold 100 m of 1.5-inch (38.1 mm) air hose, a 6.5-kW generator, spares, and tool storage. The total shipping weight of the RAM drill is 10.3 t.

The RAM drill is the fastest of all mechanical drills, capable of drilling speeds of up to 3 m min^{-1}. It was successfully used to bore many hundreds of seismic holes with average depths of 46–74 m in West Antarctica. A maximal depth of 90 m was routinely attained at the lower Thwaites Glacier during 25-min runs. The RAM system has been tried out once at the South Pole with lesser success (it could not penetrate deeper than 63 m). Hence, the drill performance is very sensitive to the local characteristics of the firn.

FIGURE 10.15
RAM drill, West Antarctica. (From Bentley et al., *Drilling in Extreme Environments. Penetration and Sampling on Earth and Other Planets*, 2009. Copyright Wiley-VCH Verlag GmbH & Co., KgaA, Weinheim. Reproduced with permission.)

10.2.6 Cable-Suspended Electromechanical Auger Drills

The first concept of the cable-suspended electromechanical auger drill (or just "shallow drill") was based on the SIPRE hand auger driven by a suspended motor with an antitorque device [Suzuki, 1994]. Two such drills were built in Japan in the 1970s. At the same time, a novel shallow drill was built at the University of Iceland and used in 1972 to drill a 415-m-deep hole on Vatnajökull glacier, Iceland [Árnason et al., 1974]. Shortly afterwards, two other drills were designed at the University of Bern [Rufli et al., 1976] and at CRREL [Ueda and Garfield, 1968]. The last two drills became excellent prototypes for shallow drills development, and that is why this type of drill is often referred to as a "Rufli-Rand drill."

The upper part of the shallow drill includes (Figure 10.16): (1) cable termination to connect the drill with the armored cable; (2) in most of the drills, a slip-ring device to prevent cable damage when the antitorque fails, and a rotation sensor to detect an antitorque failure during penetration; (3) the antitorque system to prevent spinning of the nonrotated section; (4) in some drills, a hammer to ease core breaking and retrieve a sticking drill; and (5) not always, a pressure instrumentation chamber, containing the sensors and controls.

The lower part of the shallow drill includes: (1) a coring head equipped with cutters and core catchers; (2) a core barrel with spiral flights and a chip chamber; (3) in most drills, an outer barrel (jacket); and (4) a geared motor. Torque limiters (friction clutch) were included in some drill designs. Some methods to increase friction between cuttings and the inner surface of the outer barrel, such as ribs or grooves, are provided in most of the drills.

If an auger drill has a double-core barrel design, cuttings are either (a) dropped into the core barrel and stored in its upper part (U-type—unit storage for cuttings and core, Figure 10.17a), or (b) further pushed up and stored above the barrel (S-type—separate storage for cuttings and core, Figure 10.17b) [Suzuki and Shimbori, 1985]. In the U-type drills,

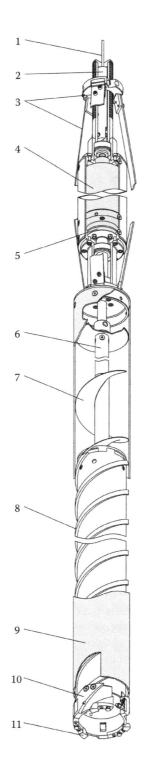

FIGURE 10.16
Basic structure of ice shallow drill: 1, cable; 2, cable termination; 3, antitorque leaf springs; 4, motor with reducer; 5, clutch; 6, shaft; 7, booster; 8, rotating core barrel; 9, jacket; 10, drill head; 11, cutters. (Reprinted from Hong, J. et al., *Ann. Glaciol.*, 55(68), 65–71(7), 2014. With permission of the International Glaciological Society.)

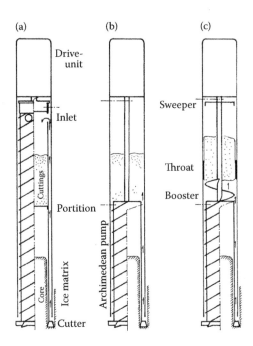

FIGURE 10.17

Schematic of double-core barrel auger drills. (a) U-type, (b) S-type, and (c) S-type with booster (throat was not yet tested). (From Suzuki, Y., and Shimbori, K., *Natl. Inst. Polar Res. Mem.*, 39, 1985.)

cuttings are packed only by gravity and the cuttings density is near 400 kg m^{-3}. The pushing action in the S-type increases the amount of cuttings. To increase density even more, the shaft has a booster: a 100-mm (one pitch) long, single, spiral, slightly inclined inward (Figure 10.17c), which gives additional upward force and compresses cuttings. Although the power consumption in the last case is slightly higher (~10% or so), cuttings density reaches 670 kg m^{-3} for warm ice and 600 kg m^{-3} for cold ice [Suzuki, 1984]. This means that S-type drill could be considerably shorter than a U-type drill while capturing the same length core.

During the last few decades, dozens of electromechanical auger drills have been designed. The details of the various internal components give each electromechanical auger drill its own unique operating capacities. A description of some of the shallow drills and drilling performance is given below.

10.2.6.1 Icelandic Drill

The Icelandic Drill with double S-type core barrel and 15-cm-high booster welded to the axle just above the core barrel was built by the University of Iceland [Árnason et al., 1974]. The steel armor of the cable is fastened to the cable termination cylinder with an alloy of low melting point. The suspension cap is screwed to the upper end of the orientation unit with commercial single-shot inclinometer. The electric motor from a submersible pump is filled with thin oil; the gear was taken from the starter of a Douglas DC-3 aircraft. Below the motor gear, there are three 150-mm-long antitorque skates. The core barrel has two flights made of rectangular steel bars (5 × 5 mm²) silver-soldered to the barrel. The pitch of the flights is 200 mm at the lower end and increases to 260 mm at the upper end. The increasing pitch is necessary to prevent clogging of the narrow space between the core barrel and the outer barrel (only 6 mm) by the ice chips. The chip chamber is emptied after each run through a hatch on the outer barrel.

The drill was used only once, in the summer of 1972, to bore a 415-m-deep hole into Bardarbunga on Vatnajokull glacier, Iceland—the deepest hole ever drilled by shallow drills. The drilling equipment was arranged in a pit (Figure 10.18). The speed of hoisting and lowering the drill was 0.3–0.5 m s⁻¹. The drilling and breaking off of the core took 5–7 min. The surface work with the drill after each run took about 15 min. The drill was designed to take a 2-m-long core on each run, but only once was a full length core obtained. The drill would usually cut about 0.3–0.4 m easily but after that the drilling speed would fall, soon coming to a halt. The mean length of the cores was 0.7–0.8 m.

At a depth of 34 m below the surface of the glacier, the water table was drilled through and after that the water level stayed the same despite increasing the depth of the hole. To avoid refreezing of ice chips on the surface of cutters and the core barrel, an antifreeze mixture was introduced to the bottom of the water-filled hole. A closed polyethylene bag containing about 180 mL of isopropyl alcohol was tied to the inside of the lower end of the core barrel. The bag was thus lowered to the bottom of the hole with the drill and without being damaged. When the drill started and the ice core penetrated into the core barrel, the bag burst and the vigorous stirring of the rotating core barrel mixed the alcohol with the water, lowering its freezing point. This technique brought quick and positive results: it was possible to drill 1.0–1.4 m in about 6 min. After drilling 1.2–1.6 m, the drilling speed would fall, probably because by then the alcohol mixture had been diluted so much that freezing on the bits would start. The drill had no difficulty drilling through numerous volcanic ash layers, the thickest of which was up to 9 cm. The quality of the ice cores was good. Usually the core was in one piece, and recovery was better than 99%.

FIGURE 10.18
Drilling pit at Vatnajokull glacier, Iceland, 1972. (From Árnason et al., *J. Glaciol.*, 13, 67, 1974. With permission of the International Glaciological Society.)

10.2.6.2 University of Copenhagen (UCPH) Drill

The University of Copenhagen (UCPH) drill was developed under the Greenland Ice Sheet Program (GISP) in the middle of the 1970s [Johnsen et al., 1980]. The total weight of the drilling equipment (Figure 10.19) is 300 kg, including a winch, cable, mast, electronic control system, generator, fuel, and packing materials. The largest dimension is 3.5 m. All equipment can be placed on two Nansen sledges to be towed with a Ski-Doo or it can be carried by a small plane to less accessible drill sites.

The UCPH drill has a classical design with a U-type double-core barrel. The jacket is 2.65 m long and reaches from the bottom of the antitorque section all the way to the drill head. On the inside, the outer core barrel has 20 closely spaced grooves (0.4 mm deep) parallel to the axis enhancing the upward transport of the chips. The core barrel is 2.35 m long. On the outside, it is provided with three lead auger flights. The cutters are mounted on a drill head fixed to the lower end of the inner core barrel. The width of the cutters is 13 mm, leaving a 1 mm clearance between the hole wall and the outer core barrel, and a 0.1 mm clearance between the core and the inner surface of the drill head (later this clearance was enlarged to 0.5 mm).

The drill is usually operated over a trench 0.3 m wide and 1.3 m deep (Figure 10.20). In the upright position, the device reaches from 2.3 m above to 1.2 m below the surface. The drill and mast can be turned into a horizontal position to facilitate the removal of chips and cores.

The motor-gear section in the drill is 290 mm long (ball-bearing support for the exit shaft included) and has a 97 mm diameter. The power consumption is 230 W no-load, and 450 W when cutting ice at a penetration rate of 0.5 m min^{-1}. The bottom of the motor-gear section is bolted to the outer core barrel, and the moment is transferred to the inner core barrel through a simple shaft, strong enough to withstand the maximum 20 kN pull from the cable.

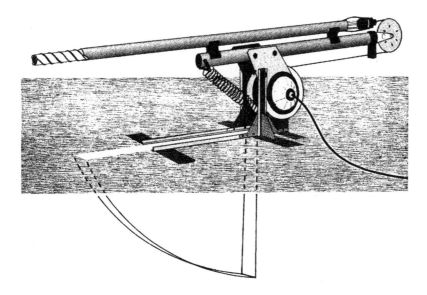

FIGURE 10.19
UCPH shallow drill in horizontal position for removal of core and chips. (From Johnsen et al., *J. Glaciol.*, 25, 91, 1980. With permission of the International Glaciological Society.)

FIGURE 10.20
UCPH drill at action, NEEM Camp, Greenland, 2008. (Photograph by H. Thing, NEEM ice core drilling project, http://www.photo.neem.dk/2008.)

The antitorque system has three leaf springs: 500 mm long, 20 mm wide, and 2 mm thick. A steel hammer block is mounted between the springs. It weighs 7 kg and glides along the three supporting rods over a distance of 100 mm. The breaking of a drilled core is done by ramming the block against the upper stopper. The same procedure serves to disengage the drill in case of sticking. At the upper part of the drill, the electromechanical cable is fixed by a Dynagrip termination to a hammer block through a system of ball bearings, which prevents cable twisting in case of antitorque failure. In such situations, three steel supporters on the rotating antitorque system pass a micro-switch on the cable, thus producing an alarm signal to the surface.

The winch is the heaviest item, weighing 140 kg, including a motor, a gear, 200 m of cable, and a wooden container (later the length of spooled cable was increased to ~350 m). It has a built-in 1-kW motor fixed to the cylinder. The motor-gear assembly is built principally the same way as the drill motor-gear section. The rotation of the winch can be adjusted continuously within the range ±60 rpm, giving a maximum hoisting or lowering speed of more than 1 m s^{-1}.

The preliminary version of the drill was tested in May 1976 at Dye 3, Greenland, where the BH was deepened to a depth of 100 m. Since this time, the UCPH drill has remained almost unchanged and has been known in Greenland as the ice drilling "workhorse." The deepest hole was drilled on the Renland Ice Cap in East Greenland during 9 days to a depth of 325 m in 1988 [Clausen and Stauffer, 1988].

10.2.6.3 Fast Electromechanical Lightweight Ice Coring System

The Fast Electromechanical Lightweight Ice Coring System (FELICS) was designed and manufactured by the Swiss small enterprise Icedrill.ch AG. The system is installed within 30 min by three people inside a commercial protective tent (Figure 10.21) and does not require trench excavation, since it is a nontilting drill [Ginot et al., 2002]. The separation of the barrels to extract ice cores and chips is performed in a vertical position. The system including power supply weighs 228 kg. The heaviest component is the cable drum with 200-m cable, which weighs 28 kg.

The drill's drive unit includes a 60 V–4.2 A DC-motor with a 28:1 reduction gear. Three spring-loaded skates are snugly installed into the slots of the drive-unit casing. The peculiarity differential of the drill is the absence of the jacket. The drive shaft is connected with the top of the chip chamber made from a thick-walled aluminum tube with two narrow machined-out parallel transporting spirals (their width is continuously increasing to the top). It has an outer diameter of 100 mm and an inner diameter of 78 mm. The core barrel was manufactured in the same manner; it is 0.95 m long and designed to get a maximum length of ice cores of 0.9 m. The chip chamber has two modifications differing by length: the longer one (1.2 m) allows for the recovery of the maximum length cores in high-density ice, and the short one (0.95 m) is used for firn with lower density. The high rotation velocity (220 rpm) of the core barrel ensures good chip transportation to the top of the chip chamber where they fall inside through two openings.

A very simple system was developed to couple the drive unit, the chip chamber, and the core barrel. Two spring-loaded pistons glide radially into two holes of the thick-walled barrel. To disengage the coupling, a U-shaped tool is pushed into tangential slots of the barrel.

FIGURE 10.21
FELICS inside the protective tent at Svalbard, March 2009. (Courtesy of F. Stampfli, icedrill.ch SA [http://www.icedrill.ch/].)

The cores and chips are extracted from the barrels by means of a simple core pusher and chip extractor. The integral cutting ring with two cutters is fixed at the lower end of the core barrel. One of the two teeth cuts 2 mm deeper but only a narrow ring around the core. The second tooth cuts the outer part of the annulus. There are no cutter shoes, as the rate of penetration is controlled with the winch speed.

The final version of the FELICS drill is equipped with a Schlumberger 4–18 PSS cable. It weighs only 0.097 kg m^{-1} and has four conduits of 0.22 mm^2. The winch is driven by a 0.42 kW winch motor (3000 rpm) coupled to a 25:1 gear reducer.

The FELICS drill was used mainly on high-altitude glaciers, in the European Alps and in the Andes (on Cerro Tapado, Chile at 5536 m a.s.l., on Illimani, Bolivia at 6300 m a.s.l., and on Chimborazo, Ecuador at 6250 m a.s.l.) with a typical average production of 20–25 m day^{-1}. In Bolivia at Illimani, bedrock was reached at a depth of 138.7 m over the course of only 6 days including installation.

10.2.6.4 Backpack Drill

The Backpack Drill is a smaller version of FELICS. It was developed for high-mountain or preparatory expeditions, and hence the total weight of the drilling equipment is only ~20 kg [Ginot et al., 2002]. This drilling system consists of a core barrel, a chip barrel and a drive unit. There is no winch (Figure 10.22). The core diameter is 57 mm, and the maximal core length is 0.9 m. The system is attached to an electric lifting cable 20 m in length, which is connected to a small control box and a battery pack (Figure 10.23). Drilling depth is limited to 15–18 m. The drill is possible to use in water-flooded ice, as the drive unit is sealed for a maximal pressure of 50 kPa. All parts in contact with ice and water are made from anodized aluminum stainless steel, polyethylene, and polyoxymethylene (POM).

Core dogs with horizontal mobility are used to cut the core by extracting knives through backward rotation of the drill. These core dogs are especially suitable for the first meters of drilling of snow and firn. Normal core catchers for denser ice are also available.

Power is provided by a small solar panel (SUNLINQ, 12 W) via a battery pack (36 V, 3.2 Ah). Drilling of 15-m firn cores requires about 2–4 working hours. Because of the low weight of the drill, sometimes it is difficult to drill through solid ice. In order to apply stronger pressure to the drill, an assembly of connectable stakes attached to the top of the motor is recommended.

FIGURE 10.22
Backpack Drill in stowed position. (Courtesy of Felix Stampfli, president of icedrill.ch SA [http://www.icedrill.ch/].)

FIGURE 10.23
Testing of Backpack Drill at the Jungfraujoch, Alps. (Courtesy of Felix Stampfli, president of icedrill.ch SA [http://www.icedrill.ch/].)

10.2.7 Cable-Suspended Electromechanical Drills with Near-Bottom Circulation

In 1947, in Bartlesville, OK, the first electromechanical cable-suspended drill with near-bottom circulation, the "Electrodrill," was developed by A. Arutunoff of Reda Pump Co. The drilling tests in sedimentary rocks resulted in a number of wells being drilled to as deep as ~400 m. Due to the insufficient power and the low WOB produced by the Electrodrill, penetration rates did not exceed 4.2 m h^{-1}. Moreover, the friction antitorque system caused numerous accidents with BH wall collapse and drill sticking resulting in the termination of these activities.

In 1964, the US Army Cold Regions Research and Engineering Laboratory (CRREL) modified Arutunoff's Electrodrill for glacial research (Ueda and Garfield, 1968). That was the turning point in ice deep drilling technology. Since the first CRREL drill was implemented, at least six different electromechanical drills with near-bottom fluid circulation have been designed in the USA, Denmark, Russia, France, Germany, Switzerland, and Japan for ice deep drilling. These included ISTUK (Gundestrup et al., 1984), KEMS [Kudryashov et al., 1994], PICO-5.2" [Wumkes, 1994], JARE [Fujii et al., 2002], Hans Tausen [Johnsen et al., 2007], and the DISC [Shturmakov et al., 2007].

In these types of drills, the only connection to the surface is through an electromechanical cable; all powered systems are contained within the downhole unit. The rotor of the downhole electric motor produces a rotation that is transmitted through the reducer to the core barrel with the drill head. Various types of electric motors have been used to drive electromechanical drills such as AC and DC motors operating in sealed chambers or directly in the drilling fluid [Talalay, 2003]. The output power of the motors is in the range of 0.3–2.2 kW usually, but in the first Electrodrill, it was 13 kW. In the latter case, the total power input reached 11–12 kW but less than 1 kW of power went into the actual cutting of ice.

Although some attempts to minimize counter torque were carried out, an antitorque system is still included in the downhole unit to prevent the spinning of the upper nonrotated part of the drill [Talalay et al., 2014c]. The antitorque system must allow the drill to move up and down during both drilling and tripping operations. Usually the antitorque system has blades of various designs that can engage with BH wall and hold the torque from the stator of the driving electric motor.

Unlike conventional rotary drilling where the fluid is normally pumped to the bottom of the hole from the surface, the electromechanical drills suspended on a cable use a bottom circulation system with a downhole pump and a chip chamber for filtering the fluid and collecting the cuttings [Talalay, 2006]. The circulation flow may be (1) in a normal direction where the flow is going down the inside of the core barrel and up between the core barrel and the hole wall or (2) in the reverse direction. The latter flow ensures better cleaning of the bottom cuttings and requires a lesser flow rate. Drill construction with the reverse circulation is also simpler, and that is why this type of circulation is normally used in deep ice electromechanical drills. The chips are temporarily stored in a chamber within the drill, which is periodically withdrawn to the surface and emptied. In the case of the CRREL drill, however, the ethylene glycol drilling fluid dissolves the ice chips; hence, there is no need to empty the chamber.

Some of the electromechanical drills use a single-core barrel and the flow goes from the pump into the space between the drill and the BH wall to the bottom of the drill. Passing through the openings in the drill head, the liquid picks up the cuttings and then flows upward between the core and the inner surface of the core barrel. Certain drills have a double-core barrel, and the cuttings are transported through the space between the outer and the inner core barrel that has three spiral flights. Then the fluid with cuttings goes through the pump into the chip chamber. The drill head is attached to the lower end of the core barrel. Typically, a drill head consists of the body, cutters mounted to the bottom side, shoes to control the rate of penetration, and core catchers [Talalay, 2014].

The surface equipment associated with electromechanical drills includes a tower with a suitable base to provide stability and a winch. The size of the winch depends primarily on the depth of penetration desired, and hence the length and diameter of the cable. Typically, tower and drill, when it is fully out of the hole, together can be rotated to a horizontal position at a comfortable working height for ease of removal of the ice core and the chips. An additional advantage of the tilting action is that it reduces the needed height of the tower above the surface, although it requires a slot cut into the glacier surface to accommodate the lower part of the tower when it is tilted to the vertical.

10.2.7.1 The CRREL Drill

The CRREL drill consists of a cutting head that has a 156 mm outside diameter and a 114 mm inside diameter; a core barrel; a gear reducer; a 2300 V 13 kW electric motor; a centrifugal pump rated at 303 L min^{-1} at a pressure of 360 kPa; a bailer; antitorque skates (two leaf springs were added to augment the skates); and a swivel section with slip rings at the top (Figure 10.24) [Ueda, 2007]. Drill rotation can be up to 225 rpm. The drill is 25 m long and weighs 1.18 t. The core barrel is a double-tube, swivel-head type capable of holding a 6-m core. Two types of cutting bits are used, one with plain steel teeth and one with diamond-encrusted teeth. The diamond bit is used in both the ice or silt and rock material. The drill is suspended from a double-armored electromechanical cable 25.4 mm in diameter, with 12 electrical conductors and weighing 2.1 kg m^{-1}.

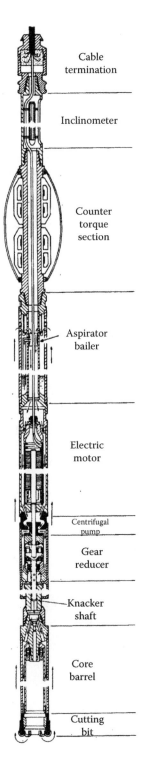

Cable
termination

Inclinometer

Counter
torque
section

Aspirator
bailer

Electric
motor

Centrifugal
pump

Gear
reducer

Knacker
shaft

Core
barrel

Cutting
bit

FIGURE 10.24
CRREL electro-mechanical drill. (From Ueda, H.T., *Ann. Glaciol.* 47, 2007. With permission of the International Glaciological Society.)

The system uses two different fluids in the BH: the lower part is filled with an aqueous ethylene glycol solution and the upper part with diesel fuel of arctic blend (DF-A) mixed with a densifier (trichlorethylene). The aqueous solution of ethylene glycol is sent down on each run in the bailer to dissolve the cuttings, and the spent solution is removed on each run via the bailer. An aspirating system was designed within the bailer to ensure that the more concentrated solution (sent down on each run) reached the cutting head. Any solution remaining would stay at the bottom of the BH, since its density would be higher than the second BH fluid.

In 1966, the hole at Camp Century, Greenland, was advanced by the CRREL electromechanical drill from 535 m, where thermal drilling had been terminated, to the bottom of the ice sheet, at 1387.5 m from the surface with average rate of 12.5–15 m min^{-1} [Ueda and Garfield, 1968]. The coring continued to 1391 m, until a worn bearing in the gear prevented further penetration. After reaching the bed at Camp Century, CRREL moved its drills immediately to Byrd Station in central West Antarctica [Ueda and Garfield, 1969b]. During the season of 1967–1968, drilling proceeded in two shifts, 24 h a day (Figure 10.25), and reached the wet bed at 2164 m on January 29, 1968. A layer of water estimated to be less than 0.3 m thick was found at the bed. The bed was penetrated to ~1 m, but no material was recovered. The drill was frozen in and lost following redrilling into the bed in 1969.

10.2.7.2 ISTUK Electromechanical Drill

The ISTUK electromechanical drill ("IS" means "ice" in Danish, while "TUK" means "drill" in Greenlander) was designed at the University of Copenhagen. It was powered by a rechargeable battery pack, and controlled by a microprocessor in the drill [Gundestrup et al., 1984]. The length and the weight of the drill were 11.5 m and 180 kg, respectively (Figure 10.26). The produced chips were mixed with hole liquid and sucked into three channels on the outside of the drill. Each channel ended in a piston pump in a storage chamber.

The drill cuts a hole with a diameter of 129.5 mm. The core diameter is 102.3 mm. The upper part of the drill, including the antitorque system, pressure chamber with motors, and a screw at the lower end of the pressure chamber, is prevented from rotating by three

FIGURE 10.25
Core from 1982 m depth, Byrd Station, 1968. (From Ueda, H.T., *Ann. Glaciol.*, 47, 2007. With permission of the International Glaciological Society.)

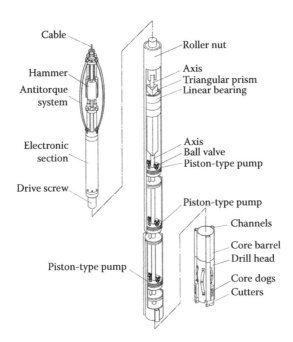

FIGURE 10.26
ISTUK Drill. (Modified from Gundestrup et al., *Proc. of the Second Int. Workshop/Symposium on Ice Drilling Technology*, 1984.)

leaf springs pressed against the hole wall. The cable is terminated with ball bearings in a weight inside the antitorque system to allow for rotation of the cable. The weight serves as a hammer, which can be used to break the core.

The motor rotations are transferred to the lower part of the drill and the drill barrel, through a hollow screw, a triangular shaft, and a linear bearing. The roller nut on top of the barrel engages the external thread on the screw and creates a linear motion of the barrel. This changes the distance between the discs and creates the pumping action. The diameter of the pump is 100 mm, which gives an effective volume of 7 L per chamber. With a core length of 2.3 m, the volume of the ice cut by each cutter is 3.8 L. The pitch of the screw is 4 mm, and the number of rotations is 234, giving a stroke length of 936 mm.

During 1979–1981, the ISTUK electromechanical drill was used to core to bedrock near Dye-3 in South Greenland. A mixture of Jet A-1 fuel with 10% perchloroethylene was used as a drilling fluid. Debris-containing ice first appeared at 2012.83 m depth and progressively increased in concentration with depth. At the final depth of 2037.63 m, the drill became stuck. The drill was left with tension in the cable and excess pressure in the hole during winter. By the summer of 1982, the drill had become loose, and after being raised to the surface, the last core was removed.

A new improved version of the ISTUK electromechanical drill was used at the Greenland Ice Core Project (GRIP) at the Summit of the Greenland Ice Sheet [Johnsen et al., 1994]. The drilling fluid was a mixture of Exxsol™D60 and Freon 113. After three summer seasons (1990–1992) the final depth of 3028.8 m was reached after penetrating 6.3 m of debris-containing ice. Drilling was stopped because the cutting knives were being destroyed by hitting gravel and stones close to the bottom. The last core sections were yellow with bedrock material (Figure 10.27).

FIGURE 10.27
The leader of the GRIP drilling, S. Johnsen, with the final piece of the debris-containing ice, 1992 (From Dansgaard, W and Gundestrup, N., *Endeavour*, 17, 1, 1993.)

10.2.7.3 KEMS Drill

The KEMS drill was designed in the Leningrad Mining Institute in the early 1980s (KEMS is the Russian abbreviation for "Core ElectroMechanical Drill"). The KEMS drill consists of a core chamber inside a single-walled barrel, a driving electric motor with gear reducer, a pump, an antitorque system, a hammer block, an electronics chamber, and a cable termination [Kudryashov et al., 1994, 2002]. The antitorque system consists of a lever system, a spring, and three skates. The core chamber comprises a drill head with three cutters and three core dogs, a nipple, a barrel, and a chip filter. The outer/inner drill head diameters in different modifications are 112–116/85–89 mm (KEMS-112) or 132–135/107 mm (KEMS-132 or KEMS-135); the corresponding outer/inner diameters of the core barrel are 108/99 mm or 127/117 mm; the core barrel length is 1.5–3.0 m.

The main difference between the KEMS drill and the other electromechanical drills is that it was equipped with two electric motors: one was a three-phase AC motor (2.2 kW) for rotation of the core barrel and the drill bit, and another was a DC motor (220 W) for driving of the rotary type pump. In later years, the same two-motor concept was implemented in the US DISC drill. The independent smoothly regulated electric drive of the pump provided continuous circulation of the hole liquid not only during drilling but also during other technological operations (BH cleaning, for example). The total drill length, depending on the length of the core barrel, is 7–12 m; the drill weight is 120–180 kg. The diameter of the six-conductor cable is 16.5 mm; it is wound up and down by a 7-kW electric winch.

The KEMS drill has been used extensively for deep drilling at Vostok Station, Antarctica, including coring of the deepest ice [Vasiliev, 2002; Vasiliev and Kudryashov, 2002]. On February 5, 2012, the drill reached subglacial Lake Vostok at a depth of 3769.3 m [Talalay, 2012; Lukin and Vasiliev, 2014]. The BH liquid level rapidly rose, and the drill was immediately recovered, but upon reaching the surface, the whole drill was filled and coated with refrozen water ice (Figure 10.28). In January 2013, the drill deployment found the first signs of frozen lake water (cork of bright white hard material) at the depth of ~3200 m (569 m from the bottom of the ice sheet). At the depth of 3385 m, the first crescent-shaped fragments of refrozen water ice were brought to the surface. From a depth of 3424 m, the drill began to recover the continuous full-diameter core, composed of refrozen subglacial water.

FIGURE 10.28
The refrozen Lake Vostok water recovered from the last run, February 5, 2012. (Courtesy of N.I. Vasiliev, National Mineral Resources University, St Petersburg, Russia.)

Hole 5G-1 is inclined from the vertical by several degrees, and during redrilling, the drill moved away from the axis of the main hole. The proportion of glacial ice core increased steadily with depth, and the subglacial refrozen water was completely gone from the core at the depth of 3458 m. By the end of the 2014–2015 field season, subglacial Lake Vostok was accessed for the second time at the depth of 3769.15 m (15 cm less than in 2012).

10.2.7.4 PICO-5.2" Electromechanical Drill

The PICO-5.2" electromechanical drill was designed and built in the Polar Ice Coring Office, University of Alaska–Fairbanks [Kelley et al., 1994; Wumkes, 1994; Stanford, 1994]. Typical drill head dimensions were 177.5 mm outer diameter and 137 mm inner diameter (Figure 10.29). The total length of the drill was 27.5 m, including a 6-m core barrel. The weight of the PICO-5.2" electromechanical drill was ~730 kg. Variable (0–1100 V) AC power for the drill was transmitted down an instrumented cable that used a Kevlar strength member. The motor section consisted of a 2.2 kW DC motor and gear reducer to reduce the speed of rotation of the inner core barrel and head assembly to 0–150 rpm. The head was driven at a typical rotation speed of 100 rpm. A progressive cavity type pump was used to move the drill fluid in a flow cycle from the drill head, upward between double-core barrel, through the pump, into and through the screens and back down to the head. The pumping rate was typically ~135 L min^{-1}.

A revolving carousel-type storage rack was developed to handle the drill string at the surface (Figure 10.30). This allowed the drill to be handled in a vertical mode with the advantage that core removal and screen cleaning could take place after the drill made the return trip down the hole. The carousel had eight storage positions, so two sets of screens and core barrels could be available. The core barrel containing the core was lowered to a horizontal position on a tilt table for core removal.

Between 1989 and 1993, the PICO-5.2 "electromechanical drill was used to acquire continuous ice core through the Greenland Ice Sheet at the Summit in support of the GISP-2 project, the Greenland Ice Sheet Program. For the first time, *n*-butyl acetate was

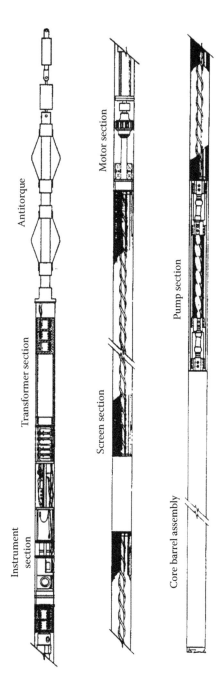

FIGURE 10.29
Schematic of PICO-5.2″ electromechanical drill. (Modified from Stanford, K.L., *Mem. Natl. Inst. Polar Res.*, 49, 1994.)

used as the drilling fluid. In July 1993, after five seasons of drilling, the drill penetrated several meters of debris-containing ice to reach the bedrock. The lower part of the PICO-5.2″ electromechanical drill was modified to accept a rock-drilling bit in order to penetrate the rock substrate under the ice. The well screen sections were removed along with the large-diameter core barrel and replaced with a weighted drive section

FIGURE 10.30
PICO drill inside GISP2 dome, Summit, Greenland, showing carousel. (From Bentley et al., *Drilling in Extreme Environments. Penetration and Sampling on Earth and Other Planets*, 2009. Copyright Wiley-VCH Verlag GmbH & Co., KgaA, Weinheim. Reproduced with permission.)

and small-diameter, diamond-tipped core barrel. A 1.55 m length of rock core with 33.4 mm diameter was recovered using a standard rock-coring diamond bit before the drilling was terminated.

10.2.7.5 JARE Electromechanical Drill

The JARE electromechanical drill was developed by the National Institute of Polar Research in Tokyo for the Japanese Antarctic Research Expedition (JARE). The JARE drill consisted of a core barrel, a chip chamber, a pressure-tight section, and an antitorque section (Figure 10.31) [Fujii et al., 2002]. Three cutters were attached to cut an ice core of 94 mm diameter, leaving a BH of 135 mm diameter. Ice chips were transported through the space between the core barrel and the outer jacket by three spiral flights attached to the core barrel. The chips were separated from liquid with a filter at the top of the chip chamber, and compacted by a booster (a spiral device for packing chips more tightly) to the density of 500 kg/m^3. Initially the JARE drill was designed to get 2.2 m of the core, but later it was redesigned to penetrate up to 3.84 m per core length. The circulation system was modified, and the jacket of the chip chamber was replaced with a special pipe perforated with small holes (45000 × Ø1.2 mm) for storing the cutting chips, while the liquid could easily pass through the perforations. The cutting chips, however, created a countercurrent in the chip chamber during the drill's ascent, leading to leakage of the chips from the chip chamber. A check valve and DC drill motor were adopted to prevent this from happening.

In the pressure-tight section, a drill computer, a DC brushless motor (270 V, 0.6 kW, 12000 rpm) and a planetary reduction gear (1:170) were installed. The antitorque section consists of three leaf springs. The winch is driven by the Toshiba VF-V3 inverter motor

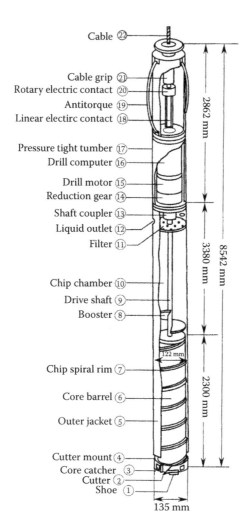

Cable ㉒

Cable grip ㉑
Rotary electric contact ⑳
Antitorque ⑲
Linear electirc contact ⑱

Pressure tight tumber ⑰
Drill computer ⑯

Drill motor ⑮
Reduction gear ⑭

Shaft coupler ⑬
Liquid outlet ⑫
Filter ⑪

Chip chamber ⑩

Drive shaft ⑨
Booster ⑧

Chip spiral rim ⑦

Core barrel ⑥

Outer jacket ⑤

Cutter mount ④
Core catcher ③
Cutter ②
Shoe ①

2862 mm

8542 mm

3380 mm

2300 mm

122 mm

135 mm

FIGURE 10.31
Schematic of JARE electromechanical ice core drill. (From Fujii et al., *Natl. Inst. Polar Res.*, 56, 2002.)

(11 kW, 200 V, 3-phase), which has constant torque of 70.1 N-m for motor pivot from 0 to 1500 rpm. The winch can be controlled from 0 to 1.5 m/s when hoisting and lowering the drill. A speed as low as 1 cm/min is achievable using a vernier dial when ice drilling. A steel armored cable of 7.72 mm diameter with seven conductors was used. The cable's weight in air is 0.246 kg/m.

Between 1995 and 1996, the JARE drilled a hole to the depth of 2503 m at Dome F, Antarctica (Figure 10.32). As the result of BH closure, the drill was stuck at the depth of 2250 m during hoisting operations. The new hole was started in 2002, and after 5 years, the depth of 3035.22 m was reached in 2007 (Figure 10.32). The drill was stopped because of problems with penetration in "warm ice" [Motoyama, 2007].

The JARE drill is currently being used by the Chinese National Antarctic Research Expedition (CHINARE) to drill a deep hole at Kunlun Station in the Dome A region, the highest plateau (~4100 m a.s.l.) in Antarctica [Fujita et al., 1994; Takahashi et al., 2002]. By the end of the 2015–2016 field season, the depth of the hole reached 654 m (Figure 10.33).

FIGURE 10.32
Dome F drilling trench, season 2005–2006. (Courtesy of H. Motoyama, National Institute of Polar Research, Tokyo, Japan.)

FIGURE 10.33
First ice cores obtained by JARE electromechanical drill at Kunlun Station, January 2013. (Courtesy of X. Fan, Jilin University, China.)

Drilling to the bedrock at the target depth of 3100 m is planned to be completed during a further three to four seasons.

10.2.7.6 The Hans Tausen Electromechanical Drill

The Hans Tausen electromechanical drill was designed for the European Project for Ice Coring in Antarctica (EPICA) and North Greenland Ice Core Project (NorthGRIP) at the University of Copenhagen. The drill was named after the Hans Tausen Ice Cap, Greenland, where it was first tested in 1995 [Johnsen et al., 2007]. The structure of the Hans Tausen drill is quite similar to the JARE drill (Figure 10.34). The electronic parts including the motor and gear section were imported from either the UCPH shallow drill or the ISTUK drill, and the antitorque section was used from the ISTUK drill. The drill incorporates an inner core barrel with spiral flights, a 100-mm core drill head scaled up from the UCPH shallow drill, and an outer barrel with inside grooves.

The 5.0-m long Hans Tausen drill was a prototype for the ~11 m long EPICA and NorthGRIP versions of the drill, which were mechanically identical to the Hans Tausen drill except for a much longer core barrel and chip chamber. Another difference between the Hans Tausen and longer versions of the drill is that the piston pump for fluid circulation is installed just above the inner auger core barrel instead of a booster. An axial piston

FIGURE 10.34
Major components of Hans Tausen drill. (From Johnsen et al., *Ann. Glaciol.* 47, 2007. Reprinted with permission of the International Glaciological Society.)

pump moves the drill fluid–chip slurry into a filtering-storage section. The cuttings are stored in a chip chamber right on top of the core barrel, equipped with a 30-mm-diameter hollow shaft (that doubles as a drive shaft) with several holes and a fine mesh screen clamped on the outside for filtering the chips. To improve separation of drilling fluid from the chips in entering the chip chamber, 18,000 holes of 1.4 mm diameter were added along the body of the chip chamber of the NEEM version of the drill (similarly to the JARE drill). This modification increased the filter area of the circulating liquid to include both the hallow shift and the chip chamber barrel itself.

A minimum clearance of 5.8 mm between the outer barrel and the BH wall was provided to ensure fast movement up and down the hole. Furthermore, valves at the ends of the chip chamber could be opened to allow fluid passage through the chip chamber and core barrel on the trip downhole.

Since 1995, almost 15 km of good-quality ice core have been drilled by the different versions of the Hans Tausen drill in Greenland, by NorthGRIP (3090 m, 1998–2004) and NEEM (2537 m, 2007–2010), and in Antarctica: Dome C (3270 m, 2000–2005); Berkner Island (998 m, 2002–2005); Kohnen Station (2774 m, 2001–2006); Talos Dome (1620 m, 2003–2008); James Ross Island (364 m, 2008); Fletcher Promontory (654 m, 2011–2012); Roosevelt Island (765 m, 2011–2013); and Aurora Basin North (303 m, 2013–2014). Twice the drill was stuck, once at NorthGRIP (1371 m, 1997) and once at Dome C (720 m, 1999). There was no way to recover drills from the hole. At Kohnen Station, the hole reached the subglacial water, and during the last run, the subglacial water was refrozen on the drill's surface. Normal drilling operations at NEEM in Northwest Greenland were declared terminated when further penetration was stopped by a stone embedded in the ice in the path of the drill head (Figure 10.35).

10.2.7.7 DISC Electromechanical Drill

The DISC electromechanical drill is the latest deep coring drill to be developed at the University of Wisconsin-Madison [Shturmakov et al., 2007]. The DISC drill has a single rotating core barrel with a fiberglass inner sleeve to support the core. Standoffs between the sleeve and the core barrel provide the space for the mixture of chips and drilling fluid to pass upward to the screen section. The core barrel and the screen barrel each comprise multiple short segments, 276 mm long, mechanically fastened together to make a barrel of the required length. This concept allows for adaptability and convertibility of core and screen barrel lengths. The modular design provides high flexibility for producing a 122-mm-diameter ice core with variable core lengths up to a design limit of 4 m. The maximum diameter of the drill is 157 mm, and the drill produces a hole 170 mm in diameter.

Separate and independent motors for the drill and the single-stage turbine pump allow cutter speeds from 0 to 150 rpm and pump rates from 0 to 400 L min^{-1}. The high pumping rate may alleviate problems when drilling in warm ice near the bed. The pumping system also helps to increase the tripping speed (the speed at which the drill travels up and down the hole), by permitting fluid to be pumped through the drill during tripping.

The instrument section consists of devices for power conditioning, communications, sensing, and control. Among other innovations are multiple sensors for monitoring the drilling process, including a navigation sensor, an accelerometer, and a barrel position sensor, and a high-speed data acquisition system, which allows real-time monitoring of more than 30 parameters for operational and scientific use.

The cable is ~15.2 mm in diameter and consists of galvanized plow steel wires on the outside to provide the mechanical strength needed to suspend the drill and

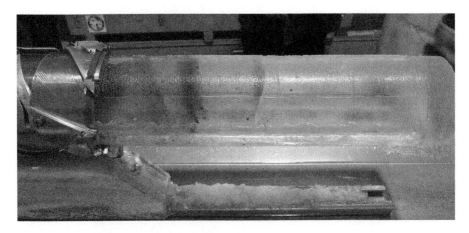

FIGURE 10.35
Ice core drill hit bedrock at NEEM, Greenland, July 2010. (Courtesy of K. Kawamura, NEEM ice core drilling project [http://www.photo.neem.dk/2010].)

copper-coated steel electrical conductors for the transmission of power. The center of the cable consists of optical fibers for high-speed data transmission between the surface and the drill.

The cable winch is electrically driven. It has two electric motors driving the drum: a 112 kW motor capable of raising the drill at speeds of up to 3 m/s and a 2.24 kW motor used for fine control while drilling and positioning the drill on the tower. The drum axis is parallel to the drill cable. The tilting 15.3-m tower utilizes modular truss construction for flexibility and portability.

All drilling and auxiliary equipment is installed and maintained inside a steel structure called the "Drill Arch" divided into two separate rooms. The larger of the two halves, measuring 30.5 m long × 9.1 m wide × 7.6 m tall, housed the entire DISC drill system as well as the main power distribution panels for the building (Figure 10.36). The smaller half of the structure, measuring 25.6 m long × 9.1 m wide × 4.6 m tall, housed the core processing equipment, including stations for measuring, cutting, and packing the ice cores as well as drying booths.

The U.S research community conducted a deep ice coring project in the West Antarctic Ice Sheet (WAIS) ice flow divide for six field seasons between 2007 and 2013. Continuous ice core samples were obtained between the snow surface and 3405 m depth. During the 2012–2013 austral summer, the DISC drill's newly designed Replicate Ice Coring System was utilized to collect nearly 285 m of additional high-quality core samples at depths of high scientific interest (Figure 10.37).

FIGURE 10.36
C. Bentley at WAIS Divide drill site, 2008. (Courtesy of T. Wendricks, University of Wisconsin-Madison.)

FIGURE 10.37
Replicate core recovered at WAIS Divide from a depth of 3001 m, December 2012. (Photograph by J. Johnson, *Ann. Glaciol*, 55, 68. Reprinted with permission of the International Glaciological Society.)

10.3 Thermal Drilling in Ice

10.3.1 Electro-Thermal Coring Drills

Ice coring electro-thermal (ET) drills (also referred to as thermal-electric or TE) were designed for continuous or discrete sampling of glaciers. Thermal ice coring drills were developed in the early 1960s and were practically obsolete in the 1980s (Table 10.1). The main limitations of these types of drills are relatively high power requirements, poor core quality, low production drilling rate, and relatively high weight. Ice coring electro-mechanical (EM) drills replaced ET drills in most applications. However, there are a few thermodynamic conditions in glaciers where the use of an antifreeze thermal electric drill (ATED) is more effective than EM drills.

Generally Earth glaciers are divided into three groups: polar or cold, temperate, and polythermal. Cold ice dominates Central Antarctica, Central Greenland, and high-altitude (>6500 m a.s.l.) glaciers. Temperate or pressure melting temperature (PMT) ice is common in mountains, and in Arctic and sub-Antarctic glaciers. Temperate ice in glaciers consists of 2–10% of water concentrated in crystal junction veins. Often this ice is water permeable. Polythermal glaciers present conditions when layers of cold ice overlay temperate, water-permeable layers of ice or vice versa. The thickness of these layers may be a few meters or a few kilometers.

Three types of ET drills were designed for coring in shallow, intermediate, and deep holes. A total of 20 ET drills were developed in the period 1964–2005. Most ET drills are not in use now, since the dry and fluid EM ice core drilling systems have become lighter and faster and produce cores of better quality. The general characteristics of some ET drills are presented in Table 10.1 and Figure 10.38.

The principal components of the cable-suspended ET or EM ice core drilling systems shown in Figure 10.39 are (1) downhole drill/sonde, (2) electro-mechanical cable, (3) winch, (4) hoisting mast, (5) power controller, and (6) power system.

The general design of the TE drills (downhole sonde) includes a hollow, electrically heated coring head, core barrel, fluid circulation system, and cable termination.

TABLE 10.1

Characteristics of ET Drills (Also See Figure 10.38)

Year	Drill Name [a]	Ice Temp., °C	Borehole Fluid	Reference
1964	CALTECH	PMT	Air–water	Shreve and Kamb, 1964
1974	UWS	PMT	Air–water	Taylor, 1976
1983	PICO	PMT	Air–water	Koci, 2002
1972–1989	ETB-3	PMT to –28°C	Air–water–EWS	Bogorodsky and Morev, 1984
1978	ETB-5	PMT to –60°C	EWS	Bogorodsky et al., 1984
1997–2009	ATED	PMT to –30°C	Air–water—EWS	Zagorodnov et al., 2005, 2014
1961–1964	CRREL (deep BH)	–30°C	DFA+TCE	Ueda and Garfield, 1968
1963–1966	CRREL Mk II	–30°C	Air	Ueda and Garfield, 1969a
1968–1974	Au. CRREL Mk II	–30°C	Air	Bird, 1976
1969–1972	TELGA	–6…–57°C	Air	Korotkevich and Kudryashov, 1976
1974–1981	TBZS	–6…–57°C	DFA+TCE	Vasiliev et al., 2007
1990–1993	TBS-112VCh	–6…–57°C	DFA+TCE	Vasiliev et al., 2007
1968	LGGE 1964	–30°C	Air–water–DFA	Gillet et al., 1976; Donnou et al., 1984
1984	LGGE 1984	–60°C	DFA+TCE	Augustin et al., 1988
1972	JARE 140 Mk II	–50°C	Air	Suzuki, 1976

PMT, Pressure melting temperature.

[a] ATED: antifreeze thermal-electric drill; ETB-3 and ETB-5 (double chamber) are the same type of ET drill. Au. CRREL Mk II: Australian Antarctic Expedition drill, version of the CRREL Mk II drill. BH: borehole. CRREL: Cold Regions Research and Engineering Laboratory, USA. CALTECH: California Institute of Technology, USA. DFA+TCE: mixture of diesel fuel arctic grade and trichloroethylene. ETB-3 and ETB-5: thermal electric drills, Arctic Antarctic Research Institute, Russia. EWS: ethanol–water solution. JARE 140 Mk II: Japan Antarctic Research Expedition ice coring drill, Japan. LGGE: Laboratory of Glaciology and Geophysics of Environment, France. PICO: Polar Ice Coring Office (now IDDO-IPDO at the University of Wisconsin), USA. TELGA, TBZS and TBS-112VCh: drills developed in Leningrad (St. Petersburg) Mining Institute (Mining University), USA. UWS: the University of Washington Seattle, USA.

In the course of the operation of ET drills, the coring head is heated; it melts ice and forms the ice core that is captured within the core barrel. After the length (typically 1–3 m) of the ice core is drilled, the winch lifts the drill to the surface. On the surface, the ice core is removed from the drill, the drill is serviced, and the next drilling run begins. Servicing of TE drills includes draining of the melt water that is removed from the kerf (Figure 10.40b). The servicing of ATED (Figure 10.40c) drills includes ice core removal and filling the drill with ethanol–water solution (EWS). EM drills can operate with the same hoisting system (Figure 10.39) as the TE drills. ET drills require an electro-mechanical cable of lower electrical resistance than the cable used in EM systems.

The following sections review three types of TE drills and present two design options that may increase the efficiency of ATED type drill performance at drilling sites that are difficult to reach (e.g., high-altitude glaciers), and may significantly reduce the logistic burden.

10.3.1.1 The Temperate Ice ET Drills

The first ET drills were designed for the study of temperate glaciers and intended to operate submerged in melt water that could completely or partially fill the hole bottom. These drills (Table 10.1: CALTECH; UWS; PICO; Figures 10.38 and 10.40a) are simple, lightweight, and require relatively low power. The drills can operate in temperate ice only and do not have bottom hole (BH) fluid circulation systems. They are suitable also for coring cold

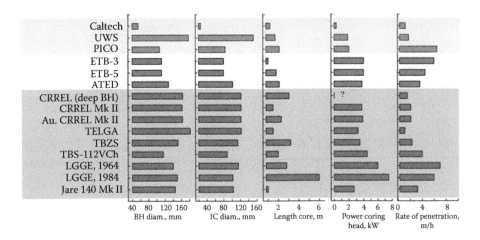

FIGURE 10.38
Major characteristics of thermal-electric ice core drills; for references and acronyms, see Table 10.1; the top three drills represent drills used in temperate glaciers only; the following three drills were used in cold, temperate, and polythermal glaciers; the lower nine drills operated only in cold ice; "?": value unknown.

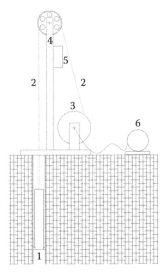

FIGURE 10.39
Electro-drill (cable-suspended) drilling system: 1, downhole drill/sonde; 2, electro-mechanical cable; 3, cable winch; 4, hoisting mast; 5, drill/winch controller; 6, source of electric power.

firn at any temperature. The drill is slightly longer than the ice core recovered during one drilling run. The drills of this type have an open top-core barrel and therefore have a low hydraulic drag, and while moving through a BH, they demonstrate high travel speed. The PICO ET drill equipped with a narrow cross-section coring head (Figure 10.40a) has a high penetration rate at relatively low applied power [Zagorodnov et al., 1994a]. The low logistic burden associated with this type of drill makes it possible to core high-altitude glaciers [Koci, 2002].

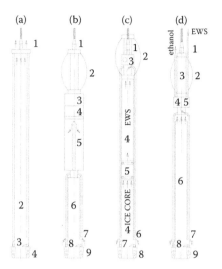

FIGURE 10.40

General schematic of the TE drills: (a) TE drill for temperate ice: 1, cable termination; 2, core barrel; 3, core catchers; 4, coring head; (b) conventional ET dry and fluid drill removes melt water from the kerf: 1, cable termination; 2, centralizer; 3, telemetry/control electronics; 4, vacuum (water) pump; 5, melt water container; 6, core barrel; 7, water sucking tubes; 8, core catchers; 9, coring head; (c) p-ATED: 1, cable termination; 2, centralizer; 3, EWS filling socket and valve; 4, core barrel/EWS container; 5, piston; 6, EWS injection tubes; 7, core catchers; 8, coring head; (d) new concept drill: 1, cable termination; 2, centralizer; 3, antifreeze container; 4, metering antifreeze pump (N1); 5, EWS circulation pump (N2); 6, top-open core barrel; 7, EWS injection tubes; 8, core catchers; 9, coring head.

10.3.1.2 The Cold Ice ET Drills

A few ET drills for cold ice coring were developed for operations in the Antarctic and Greenland ice sheets and other glaciers. These drills remove melt water from the kerf and store it in the container above the core barrel (Figure 10.40; Table 10.1: CRREL Mk II; Au. CRREL Mk II; TELGA; JARE 140 Mk II). The drills are slow (penetration rate <1.5 m/h), and require considerable power input (5–10 kW) and substantial time for servicing on the surface. All together this results in a low ice core production rate (~0.5–1 m/h). This type of drill includes water pumps and heating elements for sucking tubes and the water tank. Because the drill operates in air-filled BH, the travel speed is little affected by friction and determined mainly by winch motor power. Also these drills are long, operate with a heavy electro-mechanical cable, are not appropriate for regions with limited logistic support, and are not suitable for operation in temperate glaciers. The major disadvantage of these drills is poor ice core quality due to the thermoelastic stress caused by drill melting (Nagornov et al., 1994). Some of the well-known drilling sites where these drills were used include Dome C (905 m deep BH) [Lorius and Donnou, 1978]; Vostok Station (952 m and 2755 m deep BH) [Vasiliev et al., 2007]; and Mizuho Station (700 m deep BH) [Narita et al., 1994]. It should be noted that the 2755-m-deep BH at Vostok Station was filled with hydrocarbon-based antifreeze.

An air-filled BH is subject to the overburden ice pressure and rheological closure. The closure rate depends on the ice temperature, BH depth, and diameter [Talalay and Hooke, 2007; Talalay and Gandestrup, 1999]. The maximum operating depth of a dry BH reached with an ET drill in cold ice (below −50°C) is about 952 m; in warmer ice, the maximum depth is ~200 m at PMT. To increase the maximum depth of ice coring, it is necessary to compensate for the overburden ice pressure. This is achieved by filling BHs with a non-freezing fluid of greater density than that of ice. Filling BHs with hydrophobic or hydrophilic antifreeze compensates for the overburden ice pressure and permits ice coring virtually to any depth at any ice temperature.

A second generation of this type of ET drill was developed for fluid-filled BH operations. In general, these drills are similar in design to the TE drills described earlier (Figures 10.38 and 10.40b; Table 10.1: TBZS; TBS-112VCh; LGGE 1964; LGGE 1984). In these drills, the melt water is also pumped from the kerf, stored in the drill, and removed to the surface. The BH is filled with hydrophobic hydrocarbon-based nonfreezing fluid from the surface.

Apart from the benefits of having increased maximum depth capabilities, this type of ET drill shows limitations similar to those of dry hole TE drills, including the poor ice core quality [Nagornov et al., 1994]. The travel speed of these drills in fluid-filled BHs is lower than that in air-filled BHs; hence, the increased penetration rate does not increase the ice core production rate (TBS-112VCh; 0.5 m/h below 1500 m). Some of these types of ET drills are able to operate in both dry and fluid-filled BHs.

10.3.1.3 ET Drills for Cold, Temperate, and Polythermal Glaciers

The third type of ET drill is the antifreeze thermal-electric drill (ATED) (Table 10.1: ETB-3, ETB-5, and ATED; Figures 10.38 and 10.40c and d) [Bogorodsky and Morev, 1984; Bogorodsky et al., 1984; Zagorodnov et al., 1994a, 1994b, 2005]. It was developed as a light and portable drilling system to increase the ice coring capabilities in difficult-to-reach sites in high-altitude environments (Abramov Glacier, Pamir Mountains, 1972; [Ueda and Talalay, 2007]) including the coldest glaciers in Central Antarctica (at Komsomolskaya Station in Antarctica [Morev et al., 1988]).

In these drills, hydrophilic antifreeze (EWS) is injected into the kerf during penetration and mixes with the melt water (Figure 10.40c). When the ATED's coring head is powered, the ice core enters the core barrel, pushes the piston up, and displaces EWS to the kerf through the side channels. The mixture at ethanol concentration close to equilibrium remains in the BH [Bogorodsky and Morev, 1984; Bogorodsky et al., 1984]. The volume of injected EWS is approximately equal to the volume of the ice core recovered from the BH. On the surface, the EWS is pumped to the core barrel above the piston and delivered to the BH bottom with the drill.

The compliance between the freezing (equilibrium) temperature of EWS and ice temperature is achieved by appropriate ethanol concentration in the EWS delivered with the drill to the BH kerf. The lower the ice temperature, the higher the ethanol concentration in EWS that is required [Morev and Yakovlev, 1984; Zagorodnov et al., 1994a]. Since the ice core cross-section area is close to the melted kerf area, the concentration of ethanol in the drill is about twofold higher than the equilibrium concentration of ethanol in the BH at the given ice undisturbed temperature. When the injected EWS mixes with the melt water in a proportion of 1:1, the mixture concentration comes closer to the equilibrium concentration. Ethanol, as a base of the BH fluid, was found to be a low-environmental-impact antifreeze. EWS stays in the BH and provides long enough (≥ 11 months [Morev et al., 1988]) BH wall stability for deep drilling.

The drill has no powered pumps and only one moving part—a piston inside of the core barrel used for drilling fluid circulation. The fluid circulation in the BH is achieved through the weight of the drill only. The core barrel of the drill is also used as the antifreeze container and as the ice core barrel/container. Thus, the length of the ATED drill exceeds the length of one drilling run ice core by only 10%–20%. The ATED drill is equipped with an efficient coring head that provides 4–6 m/h penetration rate at a power input of about 4 kW. Relatively thin and light-weight steel armored coaxial cable (OD = 8.8 mm; 28 kg/100 m) allows for the reduction of weight and power of the hoist system. ATED type drills do not require heating of their structure and therefore need less power compared to the other ET drills for cold ice. Based on previous experience from the Austfonna ice core drilling operation, we estimate that the setup with a total weight of <1000 kg is capable of reaching ~1000 m depth in 500 drilling hours [Zagorodnov, 1988; Zagorodnov et al., 1998]. Lighter, smaller, and power-efficient ATED drills operate in smaller shelters, and require less fuel and smaller power systems compared to other drilling systems for cold ice.

A double-chamber ATED (ETB-5) was designed to drill a deep BH (>1000 m) in Central Antarctica [Bogorodsky et al., 1984]. In spite of successful drilling operation in cold ice at –50°C (Komsomolskaya Station, Antarctica), this drill is long, heavy, complex, and inferior to other fluid EM drills [Manevskii et al., 1983].

The last modification of the ATED drill was developed with an emphasis on improving the core quality, increasing the production drilling rate, and reducing the total weight of the drilling equipment and power system, which makes it suitable for high altitude (>6000 m a.s.l.) ice coring [Zagorodnov et al., 2005]. This drill, equipped with a 2-m-long core barrel, produces an ice core of 100–104 mm in diameter at a production drilling rate of 2 m/h and power input of ~4 kW. The ice core quality was improved because of the reduction of thermoelastic stresses and the bigger ice core diameter [Nagornov et al., 1994; Zagorodnov et al., 2005]. The weight of the 550-m drilling setup and power system was reduced to 300 kg. Up to 470-m-deep BHs were cored using a combination of dry hole EM and ATED drills at Bona-Churchill col (Alaska) and at Bruce Plateau, Antarctic Peninsula [Zagorodnov et al., 2002, 2005, 2012].

The ATED drills have demonstrated the cited capabilities at different polar and high-altitude glaciers at ice temperatures above –30°C [Koci and Zagorodnov, 1994; Morev et al., 1988; Ueda and Talalay, 2007; Zagorodnov et al., 1998, 2005]. The most useful capabilities of the ATED drills were realized in intermediate-depth BHs drilled through polythermal and shelf glaciers [Morev et al., 1988; Zagorodnov, 1988; Zagorodnov et al., 2005, 2014; Zotikov, 1979]. In some of these operations, ATED drills provided high ice core production rates of close to 420 m/week (assumed 20 h/day operation [Zagorodnov, 1988; Torsteinsson et al., 2002]).

10.3.2 Electro-Thermal Open BH Ice Drills: Hot Point Drills

The first ice electro-thermal drills, used for glaciers research, were designed for producing an open BH (no core or access BH) by melting ice with an electrically heated solid penetration tip. These types of ice drills are called "Hotpoint" (HP) after the company that mass-produced electric heaters and electric stoves [http://www.hotpoint.com]. The electric HP drill, developed in Arctic Antarctic Research Institute, was named the "thermo needle." The main advantages of HP drills compared to other types of drills used in glaciers research in the 1940s–1970s are high penetration rate ranging from 2 to 25 m/h, lightweight drilling setup, and a power source used for making small-diameter (20–120 mm) BHs. These thermal drills are cable suspended and do not require extension pipes. Electricity for the melting tip heater is provided via flexible cable.

HP drills have been used mainly for temperate glaciers research where BH filled with fresh water stays open for days, which allows interruption in the drilling process and reaming [Nizery, 1951; Gillet, 1975; Stacey, 1960; Sukhanov et al., 1974]. However, a few applications of special HP drills in cold ice have been also successful (Table 10.2). This type of drill is capable of making air-filled access BHs in cold snow and firn in Central

TABLE 10.2

Selected HP Type Ice Drills Specifications

Year	Melting Surface	Ice T, °C	Power, kW	BH Diameter, mm	Penetration Rate, m/h	Max Depth Reached, m	Reference
1945	S (Figure 10.45b)	?	?	?	2.07	?	Koechlin, France patent, in Nizery, 1951
1948–49	W (Figure 10.45a)	PMT	7.8	60–80	20–25	195	Nizery, 1951
1956–1959	S (Figure 10.45b)	PMT	2.5	130	1.8	157	Ward, 1961
1959	S (Figure 10.45b)	PMT	1.8; 2.3	41	9.2; 6.4	312	Stacey, 1960
1968	S (Figure 10.45b)	−29°C	<3.7	~110	~1.8	1005	Philberth, 1976
1970–2009	S (Figure 10.45b)	PMT	1.5–2	50	7–8	568[a]	Korotkevich and Kudryashov, 1976
1972	S (Figure 10.45b)	PMT	<0.37	20	8–13	51	Sukhanov et al., 1974
1973	S (Figure 10.45b)	PMT	<0.8	26	<18.6	<150	Gillet, 1975
1974	S (Figure 10.45b)	−12 to −5°C	2, 4.4	45, 89	7.6–8.2	300	Hooke, 1976; Hooke et al., 1980
1977–78	S (Figure 10.45b)	−54°C	1.5[b]	~56	5–8/	235	Gillet et al., 1976
1984–87	L (Figure 10.45c)	−10°C to PMT	1.6	30	14–16	140	Rado et al., 1988
1992–94	S (Figure 10.45b)	−24°C	2.5	80	2.2/0.9[c]	100	Zeibig and Delisle, 1994
2011	S (Figure 10.45b)	−8°C to PMT	1.8	56	<9	195	Zagorodnov et al., 2014
2014	L (Figure 10.45c)	-? To PMT	5	>250	?	30	Taylor, 2012
2014	S (Figure 10.45b)	−58°C	<5	~180	~1.8	n/a	Talalay et al., 2014d

[a] HP drilling started from 368 m depth, drilled with electro-thermal drill (ETB-3), VZ, not published.
[b] 1.5 kW melting tip power only; 0.35 kW water tank heater; 0.4 kW water suction pump.
[c] Penetration/production drilling rates
"?": Value unknown.

Antarctica or other polar glaciers to a depth of tens of meters. An unrecoverable hot point drill carrying and releasing a tethered power cable reached over 1000 m depth in cold ice in Greenland (Table 10.2). Here we present selected concepts of HP drills and research apparatus that use electrically heated melting tips to penetrate into glaciers.

10.3.2.1 Hot Point Drills: Design and Performance

Performance characteristics of selected HP drills are presented in Table 10.2. The fastest HP drill was built and used for glacier thickness and structure investigation in 1948 (Table 10.2 [Nizery, 1951]). The deepest continuous BH drilled with an HP drill reached a depth of 312 m in the early days of the scientific drilling of glaciers in 1959 (Table 10.2 [Stacey, 1960]). The most efficient, fast, and lightweight HP drill was built by Gillet [1975]. Over 3000 m of total cumulative depth was drilled with the HP drill designed by the Arctic and Antarctic Research Institute (AARI) [Morev, 1976; Korotkevich and Kudryashov, 1976; Sukhanov et. al., 1974].

A modern HP drilling setup is shown in Figure 10.41. The HP drill is a cable-suspended system where flexible electromechanical cable transmits electricity for the downhole drill-sonde. Most HP drills in the past were operated without a winch and hoist mast. New systems include an incremental encoder and load cell that permit continuous monitoring and control of drill parameters.

Axial asymmetry of heat flux from the HP melting tip and nonhomogeneous dust and air bubble concentration in ice tend to divert HP from the vertical pass. The control of melting tip pressure with fixed-adjusted brake permits pendulum verticality control of the HP drill (#4 in Figure 10.41d [Aamot, 1967a]). In addition, a bouncy chamber incorporated in the HP housing permits vertical stabilization of an HP drill [Aamot, 1968b]. To maintain drill vertical stability, some HP drills are equipped with centralizers (Figure 10.42) [Morev et al., 1984].

HP drills can be used as a tool for deployment of small sensors in combination with dry hole electromechanical (EM) drills. Dry hole EM drills are lightweight and fast (average production drilling rate in 200-m-deep BH is <8 m/h). They allow fast penetration through the coldest upper part of the glaciers and penetrate through relatively warm ice below [Zagorodnov et al., 2014].

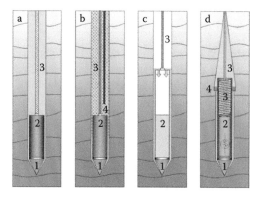

FIGURE 10.41
Prevention of refreezing of a BH drilled with hot point drills in cold ice: (a) heating cable, where 1 is melting tip, 2 is drill housing or extension pipe, and 3 is power cable; (b) antifreeze-assisted drilling, where 1 is melting tip, 2 is drill housing or extension pipe, 3 is power cable, and 4 is antifreeze deploying hose; (c) dry hole drilling-bailing, where 1 is melting tip, 2 is drill bailer, and 3 is power cable; (d) nonrecoverable on-board tethering cable probe (Philberth probe), where 1 is melting tip, 2 is instrumental section, 3 is winded power cable, and 4 is "pendulum pivot" heater.

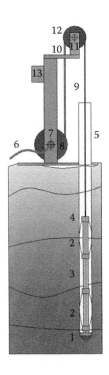

FIGURE 10.42

Hot point drill: 1, melting tip (hot point); 2, centralizer; 3, stem extension; 4, cable termination; 5, guidance tube; 6, power cable; 7, slip ring; 8, winch dram; 9, electro-mechanical cable; 10, load cell; 11, encoder; 12, top sheave; 13, controller.

A majority of HP drills were designed and used in temperate glaciers at PMT. However, a few successful drilling operations were conducted in polar glaciers at ice temperatures down to –29°C. There are four possible designs that allowed penetration with HP type ice drills into cold ice: (1) heating power cable; (2) antifreeze-assisted drilling; (3) bailing of melt water from lower portion of a BH; and (4) unrecoverable penetration with instrumented probe carrying tethering power cable (Figure 10.41; Table 10.2).

10.3.2.1.1 Antifreeze-Assisted Drilling

A total of seven BH were drilled by Hooke [1976] in Barns Ice Cap using an antifreeze-assisted drilling technique (Figure 10.41b). The deepest BH reached was 300 m. The BH in cold ice was kept open by injection of antifreeze (ethylene glycol) close to the HP drill. The lowest ice temperature drilled using this technique was –12°C [Hooke et al., 1980].

10.3.2.1.2 Dry Hole Drilling-Bailing

Dry hole drilling in cold ice is possible with the drill bailer (Figure 10.41c) [Zeibig and Delisle, 1994]. This type of HP drill makes it possible to produce dry BH in cold ice and recover a water sample. An HP is equipped with a hollow extension that functions as a bailer. A socking pump, low end valve, or top overflow is used to fill the bailer. At low ice temperatures, the sample water is heated with an electric heater. When the bailer is filled with water, it is removed from a BH to the surface and emptied, and then the drilling cycle is repeated.

The Zeibig and Delisle [1994] drill-bailer produced a BH of 80 mm in diameter and recovered 5.7 l of melt water per ~1.3 m deep drilling run. The drill bailer was equipped with a 0.4-kW top heater that allowed recovery of the research tool in case of ice formation on the BH wall. The typical penetration rate of the drill bailer in dense ice at temperature

–24°C was around 2.2 m/h. Due to bailing cycles, the production drilling rate achieved was closer to 0.9 m/h. A total of 48 BH were drilled in Antarctica using this technique, while the maximum depth reached was 100 m.

10.3.2.1.3 Philberth Probe

Another type of HP drill is an unrecoverable drill-sonde named the Philberth Probe (PP) after its inventor [Aamot, 1967b, 1968a, 1970a,b; Philberth, 1976]. During penetration, PP releases wires that freeze into ice. At the low end, the PP probe has a solid surface HP, sealed instrumentation canister carrying a commutator, and temperature, pressure, and inclination sensors (Figure 10.41d). Almost 80% of the probe length is occupied by the tether for power and data transmission. In 1968, two PP probes reached 218 m and 1005 m depth at Jarl-Joset Station in Greenland. At maximum depth, the *in situ* measured ice temperature was –30°C.

The latest PP type returnable probe is the Very-deep Autonomous Laser-powered Kilowatt-class Yo-yoing Robotic Ice Explorer (VALKYRIE) [Taylor, 2012; Xu et al., 2003]. This apparatus uses a ~5 kW infrared laser transmitted through coiled fiber optics. The laser light is used to heat HP and is also converted into electricity to power PP subsystems. In June 2014, VALKYRIE was field-tested in temperate glaciers to a depth of 30 m.

A new concept of returnable PP type probe for the study of subglacial lakes up to 3000 m depth was also proposed by Talalay et al. [2014]. RECoverable Autonomous Sonde (RECAS) is being considered for environmental exploration of Antarctic subglacial lakes. RECAS has lower and upper end electrically heated melting tips. The tethered electromechanical cable is spooled on a motor-driven winch that allowed pulling the RESAS to the surface when the research mission is complete. The winch allows controllable sinking of the RECAS and provides pendulum vertical stabilization. The apparatus can carry research equipment and fulfill water sampling requirements. The RECAS concept is considering autonomous operation of the probe for several months.

10.3.2.2 Keeping the BH Open

Freezing of melt water in a BH drilled with HP is relatively slow. The diameter of a BH just above the melting tip is higher than the melting tip diameter. Therefore, known closure (refreezing) rates (Figure 10.43) and HP penetration rates allow for estimating the time for safe removal of the HP drill before the hole becomes smaller than the probe.

Figure 10.44 shows the dependence of ice temperature on specific power required to prevent the freezing of melt water. The power was calculated based on experimental data presented in Figure 10.43. Comparison of experimental freezing rate with theoretical estimates shows a close match at low ice temperatures (<–30°C) and some discrepancies at higher temperatures (Humphrey and Echelmeyer, 1990; Zagorodnov et al., 2014). Prevention of a BH refreezing in cold ice with a heating cable (Figure 10.41a) can be realized with reasonable power (<10 kW) either for small-diameter shallow-depth holes or in greater depths (~100 m) at ice temperatures close to PMT.

10.3.2.3 Melting Tips for Hot Point Drills

The main element of the HP drill that makes possible penetration into ice is the electrically heated tip. Three types of HP melting tips were developed (Figure 10.45): (1) web melting surface (W in Table 10.2) [Nizery, 1951]; (2) solid melting surface (S in Table 10.2) [Ward, 1961; Korotkevich and Kurryashov, 1976; Sukhanov et al., 1974; Zagorodnov et al., 2014], and (3) melt (water) heat exchange (L in Table 10.2) [Rado et al., 1988; Taylor, 2012]. Normally, the faster the thermal melting drill penetrates, the less the ateral heat loss.

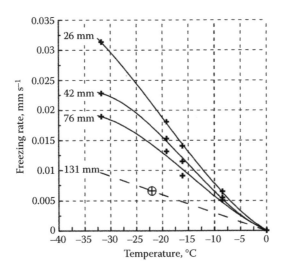

FIGURE 10.43
Borehole freezing rate as function of ice temperature and BH diameter at normal atmospheric pressure, fresh water ice. (Courtesy of V. Zagorodnov, experimental data.)

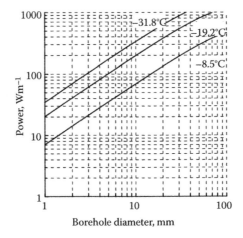

FIGURE 10.44
Specific power of the heating/power cable to prevent BH refreezing in cold ice at normal atmospheric pressure, fresh water ice. (Courtesy of V. Zagorodnov, experimental data.)

A web melting surface HT tip provides the highest penetration rate ever achieved with this type of ice drill at 20–25 m/h (Nizery, 1951). Although it has a simple design, it has not been used since its first deployment in 1948–1949. A prototype system to provide 7.8 kW of heat with bare nickel–chromium alloy (NiCr) wire (1.63-mm diameter wire has resistance of 0.54 ohms) needed a 120 A, 65 V power supply. The system therefore required very heavy lead cables (2 × 30 mm²; 2 kg/m) and a power generator (fuel consumption of 5 L/h).

Two more common approaches to the melting tip were developed: (1) a cartridge with single or multiple heaters in a heat conducting core and (2) tubular heaters molded into a heat-conducting core. All conventional big power (>20 W) cartridge heaters have a 6- to

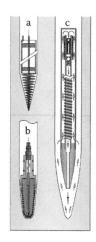

FIGURE 10.45
Hot point tips: (a) web (W) melting surface (open high-resistance wire; after Nazery, 1951). (b) Solid (S) melting surface. (From Zagorodnov et al., *J. Glaciol.*, 60, 223, 2014.) (c) Melt (water, L) heat exchange. (From Rado et al., *Ice Core Drilling. Proceedings of the Third International Workshop on Ice Drilling Technology, 1988.*)

9-mm-long cold section at the end; hence, they do not allow heat flux to concentrate at the melting tip. As a result, HP tips with cartridge heaters perform slower (<8 m/h).

In early designs, small-diameter (6–9 mm) tubular heaters were coiled and molded in either aluminum or copper core. Modern designs of the PH melting tip use microtubular heating elements. Gillet [1975] built a small-diameter (18 mm), low-power (0.8 kW), and fast (<18 m/h) HP tip, placing a microtubular heating element in a silver core.

Measured and estimated efficiency of HP melting tips is about 80–85% [Nizary, 1951; Ward, 1961, Hooke, 1976]. However, in ice-laden particles the efficiency of HP melting tips is reduced. At a certain concentration of particles, the HP drill stops penetrating [Sukhanov et al., 1974]. In addition, drill melting through low concentration of particles accumulates insolvable particles on the BH bottom and large particles could potentially deflect the HP drill.

The third type of HP melting tip was developed to overcome the problem of drilling ice laden with particles. This type of HP tip (Figure 10.45c; Table 10.2) is equipped with a high-resistance heating element coiled in a ceramic tube. An oscillating pump sucks melt water just above the nozzle tip and circulates it through the outside and inside of the ceramic tube with the heating element. The water is then injected into the kerf. Hot water is essentially used to melt ice. The water jet keeps particles suspended and partly accumulated in the filter inside of the HP. This type of HP drill was successfully used in Alpine glaciers where solid-tip HP drill operation was difficult. This type of HP drill was also tested in cold ice with antifreeze [Rado et al., 1988].

10.3.3 Hot-Water Drilling Systems

Hot-water drilling (HWD) is the fastest method to drill BH in glaciers. This type of ice drilling has been used for shallow, intermediate, and deep BH drilling in temperate and cold glaciers. In general, the HWD drilling systems include four major components: (1) source of heat (furnace and heat exchanger); (2) flexible hose; (3) nozzle tip; and (4) pump. Simple HWD systems were used for shallow BH drilling in temperate glaciers [Gilpin, 1974, 1975; Kasser, 1960; Taylor, 1984]. More powerful and better controlled setups were developed for drilling BHs in cold ice to depths of up to ~3000 m [Benson et al., 2014; Bindschadler et al., 1988; Browning et al., 1979; Craven et al., 2002; Engelhardt et al., 2000; Humphrey and Echelmeyer, 1990; Iken et al., 1977, 1988; Koci, 1994; Koci and Bindschadler, 1989; Makinson and Anker, 2014; Napoleoni and Clarke, 1978; Tulaczyk et al., 2014]. A schematic of the HWD system used for the construction of the Antarctic Muon and Neutrino Detector Array (AMANDA) is shown in Figure 10.46, and the downhole HWD-sonde is shown in Figure 10.47.

To calculate optimal drilling, a number of mathematical models were developed [Humphrey and Echelmeyer, 1990; Iken et al., 1977, 1988]. These models were based on assumptions that nozzle heat power is completely spent for cylindrical kerf formation and the heat loss from the hose is transmitted to the BH wall. The models were in agreement with measured parameters. New experimental data, obtained during the AMANDA project, allow for a more accurate calculation of the BH hole diameter [Koci et al., 1996].

The energy balance of the HWD is shown in Figure 10.48. The HWD melts ice by expelling hot water from the nozzle (Figures 10.46 and 10.48). The melt water in the BH is pumped by the downhole pump below firn ice transition (FIT) depth, to the buffer container. The high-pressure circulation pump, in turn, pumps water out of the buffer container to the heating system. The hot water is then moved down the hose to the drill's tip.

Following Iken et al. (1988) it is assumed that emitted water melts a cylindrical cavity of radius r_{BH} normal to the surface and the hose hanging vertically through the center of the BH. The BH axis z is vertically down. The data obtained during upward reaming demonstrate that the BH has a maximum diameter just above the drill. The power balance during HWD drilling is as follows:

$$Q = Q_1 + Q_{2\text{-}1} + Q_{2\text{-}2} \tag{10.1}$$

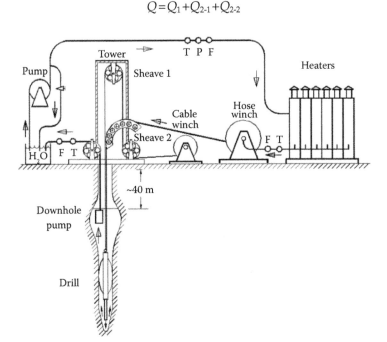

FIGURE 10.46
Schematics of the HWD drilling system: F, T, and P are flow, temperature, and pressure sensors, respectively.

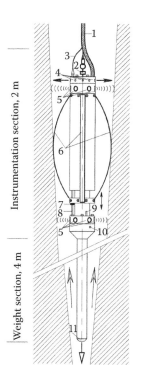

FIGURE 10.47
Schematic diagram of downhole HWD-sonde: 1, hose, 2, load cell; 3, data cable; 4, upper nozzles; 4, sonar transducers; 6, caliper springs; 7, magnet; 8, displacement transducer; 9, return water flow temperature sensor; 10, pressure transducer; 11, nozzle.

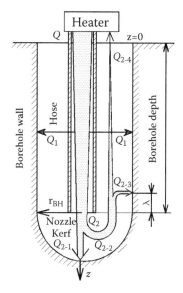

FIGURE 10.48
Energy distribution of the HWD: Q is effective heater output power, Q_1 is power loss due to heat conduction through the hose, Q_2 is power of water flow emitted through the nozzle; Q_{2-1} is power spent for kerf formation, and Q_{2-2} is power that comes up with return water flow (RWF); Q_{2-3} is power that enlarges the BH above the nozzle at depth interval λ; Q_{2-4} is the heat residue including fraction of the hose heat losses; r_{BH} is BH radius at nozzle depth.

where Q is effective heater output power, Q_1 is power loss due to heat conduction through the hose, Q_{2-1} is power spent for kerf formation, and Q_{2-2} is power that comes up with return water flow (RWF). The experimental data allows determination of the power distribution during HWD. The output heater power is $Q = \rho_w C_w M_w T_w(0)$, where ρ_w, C_w, M_w, and $T_w(0)$ are the density, heat capacity, mass flux, and temperature of water coming from the heater, respectively. This power partially dissipates through the lateral surface of the hose.

Temperature T_w decreases with depth as a depth function [Humphrey and Echelmeyer, 1990]:

$$T_w(Z) = T_{PMT} + T_w(0) \cdot e^{-z/\lambda} \tag{10.2}$$

where λ is an experimental parameter that determines the effective decay length of water temperature in the hose and T_{PMT} is the pressure melting temperature of ice at the kerf depth z. According to the experimental data, $\lambda \geq 3600$ m. Therefore, the relationship in Equation 10.2 can be assumed as linear.

Because the water temperature drops in the hose, the power at the nozzle and melting power depend on the BH depth and ice temperature distribution as follows:

$$Q_2 = \rho_w C_w M_w T_w(0) \cdot e^{-z/\lambda} \tag{10.3}$$

$$Q_{2-1} = \pi r_{BH}^2 V_{PR} \rho_i (L_i + C_i |T_i|) \tag{10.4}$$

where ρ_i, C_i, L_i, T_i are density, heat capacity, latent heat, and temperature of ice, and V_{PR} is penetration rate of hose feeding rate. The penetration rate of the HWD drill is as follows:

$$V_{PR} = \frac{K \rho_w C_w M_w T_w(0)}{\pi r_{BH}^2 \rho_i \left(L_i + C_i |(T_{PMT}(z) - T_i|(z)) \right)} \tag{10.5}$$

Three types of HWD systems are currently in use: (1) portable systems used for shallow drilling of temperate glaciers; (2) lightweight, fast, small BH diameter, intermediate-depth cold ice drilling systems; and (3) large-diameter, deep, cold ice, deep drilling systems (see Table 10.3).

10.4 Planetary Ice Drills

NASA's strategic goals and science objectives increasingly require future missions to focus on planetary bodies where life might have evolved. Since life, as we know it, requires water, the target bodies for such missions include those where water exists in large quantities. The target bodies, therefore, include the Mars polar caps, Europa (a moon of Jupiter), and Titan and Enceladus (moons of Saturn) [Rapp, 2007; Zacny and Cooper, 2006].

Although ice drilling and coring on Earth are relatively mature technologies, developing ice drilling systems for planetary applications would require either substantial modifications to the existing terrestrial systems or development of new systems [Bar-Cohen and Zacny, 2009; Zacny and Cooper, 2006; Zacny et al., 2008]. There is no doubt, though, that

TABLE 10.3

General Specification of HWD Systems

Type of HWD Drilling System	Portable	Lightweight	Large Diameter
Depth capabilities, m	30	1500	3500
Production drilling rate, m/h	10–30	50–100	30–80
Ice temperature, °C	PMT	–30	–55
BH diameter, m	0.05	0.1–0.3	<0.7
Weight, t (estimate)	0.020–0.1	0.5–5	50–100
References	Gillet, 1975; Heucke, 1999; Kasser, 1960; Verrall, and Baade, 1984	Craven et al., 2002; Engelhardt et al., 2000; Iken et al., 1977 and 1988; Humphrey and Echelmeyer. 1990; Napoleoni and Clarke, 1978; Makinson and Anker, 2014; Taylor, 1984	Browning et al., 1979; Koci and Bindschadler, 1989; Bindschadler et al., 1988; Benson et al., 2014; Tulaczyk et al., 2014

lessons learned from terrestrial ice drilling could be applied (with some limitations) to develop planetary drills.

As opposed to terrestrial systems, planetary drills need to be low mass and low volume to fit within the fairing and launch capabilities of a rocket. In addition, the drill needs to be low energy, since electrical (and thermal) energy is a function of distance from the Sun. All the planetary bodies under consideration are much further than Earth, and in turn, energy density is extremely low. Alternative power sources such as nuclear power–based systems require additional mass and volume within the spacecraft.

There are several approaches to accessing great depths in ice, and each one of them has advantages and disadvantages, as shown in Table 10.4. At a high level, there is a melting and mechanical approach. The major difference is that melting requires at least two orders of magnitude more power and energy than a mechanical probe. As such, melting probes require space-rated 20–30 kWe nuclear reactors whereas mechanical systems could use proven 100 We radioisotope thermal generators (RTGs). The availability of the power source is mission critical.

In a continuous drill string system, as the BH gets deeper, new drill sections need to be added. This approach has been successfully used in the oil and gas industry for over a century. The entire drill string is normally rotated by a motor above the surface or downhole motors, and the drilled cuttings are removed by circulating water, mud, or compressed air. Because of the requirement for circulating fluid and added mass for drill pipes, this approach is not suitable for planetary deep drilling. However, this approach is optimal if the target depth can be achieved by a single drill string. This is, for example, the case in the Icebreaker mission whose goal is to search for organics and specific biomolecules that would be conclusive evidence of life in the top one meter of Martian Northern Polar ice (Figure 10.49). The Icebreaker drill captures ice and ice-cemented ground samples in 10-cm-deep bites. The sample from each of the 10-cm-deep sections is then transferred to instruments for analysis. The drill is rotary-percussive to enable efficient and fast penetration even in extremely hard formations [Zacny et al., 2010., 2013a]. The system has been extensively tested in the Arctic, Antarctica, and Greenland (i.e., Mars analog locations) as well as in the Mars chamber and achieved a

TABLE 10.4

Approaches to Deep Drilling in Planetary Ice

Continuous Drill String		Coil Tubing		Wireline	Inchworm				Surface Laser, Microwave, etc.
Top drive	Downhole Mech. motor		Melt	Mech.	Power on surface		Power in a probe		
					Melt	Mech.	Melt	Mech.	
Honeybee and Swales drills	Blasthole rigs	Various coil tubing systems	Cryobot	Auto–Gopher	Valkyrie	Badger	Various	Inchworm deep drill system	ChemCam on MSL Curiosity
Zacny et al., 2008		Zimmerman et al., 2001		Zacny et al., 2013b	Stone et al., 2014	http://www.bxpl.com/		Elliott and Carsey, 2004	Myrick et al., 2004

FIGURE 10.49
The Icebreaker mission would search for organic biomarkers at the northern latitudes of Mars by drilling to at least 1 m depth below the surface and transferring samples to life detection instruments. (From McKay et al., *Astrobiology*, 13, 4, 2013. Courtesy of NASA Ames.)

1-1-100-100 performance metric; that is, it reached ~1 m in ~1 h with ~100 W of power and ~100 N WOB.

A coil tubing approach solves a problem related to assembling and disassembling of drill pipes. A flexible hose with a motor and a bit at its end is fed into the hole. Cuttings still need to be moved via gas or water (if a mechanical drill is used) or can be pumped out as water, if the drill bit is a melt probe. A melt probe prototype, called Cryobot, has been developed and tested by NASA JPL in Greenland [Zimmerman et al., 2002]. However, since it is a melt system, it requires more energy not only to melt the ice ahead of the probe but to keep all the pumping warm enough to prevent freezing. In addition, it is not known how the water pumping system would work over a kilometer of vertical distances. The melt probe would also have difficulty penetrating ice with sediments, permafrost, and rocks (if present).

In the wireline system, the drill is suspended on a tether and all the motors and mechanisms are built into a tube that ends with a drill bit. The tether provides the mechanical connection to a spacecraft on a surface as well as power and a data link. Upon reaching the target depth, the drill is retracted from a hole by a pulley system, which can be either on the surface or integrated into the top part of the drill itself. Generally, wireline systems involve the mechanical complexity of packaging motors and actuators into a slim tube. The weight on bit (WOB) is provided by anchoring the drill to the BH wall and use of an internal screw to push on the drilling mechanism and the drill bit itself. Hence, apart from initially starting the hole, the drill does not rely on the mass of the deployment platform. The main disadvantage of the wireline system is a possibility of borehole collapse. For this reason, the drilled environment is restricted to stable formations such as ice or ice-cemented grounds: the polar regions of Mars, Enceladus, Titan, and Europa. In addition, tripping times (retracting the drill to dump the cuttings and lowering it back in the hole) would add additional time to the mission.

There are also a number of inchworm approaches where the probe is either connected to the surface by a tether or fully contained, that is, with an integrated nuclear power source. In the inchworm approach, the probe either melts through the ice, which refreezes above the probe, or cuts through the ice and compacts the cuttings above the probe (Figure 10.50). The melt probe requires more energy not only to melt ice but also to keep water around the probe in liquid state (i.e., prevent freezing). Since water needs to refreeze above the probe to the same density as original ice, it needs to be pumped to the top and somehow cooled down in a controlled manner (warm water rises, and hence it would always end up on top, which is counter to what the melt probe requires to penetrate down). In mechanical systems, the ice needs to be transported along the length of the probe to the top and compacted. If the probe uses a tether (tether is paid out of the probe and freezes in the BH above the probe), ice needs to be compacted to greater densities than the original ice to account for the volume of the tether.

Other, nontraditional drilling technologies (laser, electron beam, microwave, etc.) could use a surface-mounted "drill" to advance the hole [Jerby et al., 2002; Ready, 1997]. For example, a laser could be progressively focused to sublimate water-ice out of the hole. However, this approach requires at least seven times more energy than melt probes (heat of vaporization is 2260 kJ/kg, and latent heat is 333 kJ/kg).

For planetary applications, the most promising approaches would include mechanical drills, since these are orders of magnitude more energy efficient than melting probes and in turn could be powered by existing power sources such as a radioisotope thermoelectric generator (RTG). RTGs have been used in planetary missions for decades. A 45-kg RTG on the MSL Curiosity mission, for example, provides approximately 100 W of electrical power and 2 kW of heat. Hence, a few RTGs could be sufficient to power a mechanical drill. On the other hand, to power a thermal drill, a nuclear reactor rated at ~30 kW would likely be required. In 2006, Project Prometheus, with a goal of developing a nuclear-powered system for space applications, was shut down after 3 years and costs of several hundred million dollars. It is difficult to determine the exact cost and schedule if the project was restarted, but it would be safe to assume this effort would be in the $B range and take over 10 years; that is, it might end up costing more and take more time than the mission itself.

Wireline drills allow testing of the drilling, sample capture approach, and downhole sensing with drill integrated instruments, but at the same time, allow drill recovery after the test. Depending on the mission requirements, deep drilling could be accomplished with the wireline system, or an inchworm system tethered or untethered, as shown in

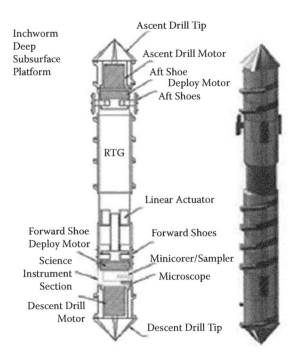

FIGURE 10.50
Schematic of the Inchworm Deep Drilling System (IDDS).

Figure 10.50. The following sections show an example of wireline systems developed for planetary exploration.

10.4.1 The Autonomous Tethered Corer Drill

The autonomous tethered corer (ATC) drill built by Raytheon-UTD weighs ~7 kg. A short auger conveys the cuttings to a reservoir in the top of the core barrel. Because of the short auger, the torque on the drill segment is somewhat reduced. The drill has been designed for low-power applications: 50–75 W. To start a drill hole, the initial WOB is provided by the weight of the surface deployment system. The drill system operates by gripping the sides of the BH with anchor shoes, drilling to the maximum stroke of 110 mm, and winching the probe to the surface to empty the cuttings and deposit the core. During initial testing, the ATC drilled 10 m in Texas limestone with an average power of 74 W and a rate of penetration of 20 cm/h [Zacny et al., 2008].

10.4.2 The JSC Wireline Drill

NASA's Johnson Space Center in collaboration with Baker Hughes Inc. and NASA Ames Research Center developed a coring wireline drill [Zacny and Cooper, 2006]. The drill has been tested numerous times on Elsmere Island in the Canadian High Arctic in frozen rock and ice (Figure 10.51). The drill is 45 mm in diameter and approximately 2 m long. It was designed to obtain a 25-mm-diameter and up to 15-cm-long core. However, the core breakoff, capture, and retrieval were done manually after the drill was pulled out of the hole. Drilled cuttings were moved up the hole using an auger and collected in a container

FIGURE 10.51
NASA JSC/Baker-Hughes wireline drill undergoing testing in the Arctic. (Courtesy of NASA JSC.)

on top of the core barrel. Once the drill penetrated to incremental target depth, the drill was winched to the surface and emptied of cuttings. The system was demonstrated to a depth of 2.2 m in Briar Hill Sandstone. Drilling rates approached 15 cm/h with mechanical power of ~20 W.

10.4.3 The Ultrasonic/Sonic Driller/Corer

Rotary drilling techniques are limited by the need for high force on the bit, and by an inability to efficiently duty cycle [Maurer, 1968, 1980]. To address these limitations, the NASA JPL developed the ultrasonic/sonic driller/corer (USDC) for potential planetary exploration missions [Bar-Cohen et al., 2001, 2008; Bao et al., 2003; Badescu et al., 2006a; http://ndeaa.jpl.nasa.gov/nasa-nde/usdc/usdc.htm]. The USDC is a penetration mechanism that is driven by low-frequency hammering action resulting from conversion of high-frequency vibration (Figure 10.52). The drill was developed to support the NASA search for existing or past life in the universe allowing probing and sampling of rocks, ice, and soil. This mechanism is driven by an ultrasonic piezoelectric actuator that impacts a bit at sonic frequencies through the use of an intermediate free mass. The fact that the USDC is driven by a piezoelectric actuator allows it to operate at extremely low temperatures that are expected to be encountered on Mars, Europa, Enceladus, and Titan.

The USDC was designed to produce both cores and powdered cuttings, operate as a sounder to emit elastic waves, and serve as a platform for sensors. It requires low axial force, thereby overcoming one of the major limitations of planetary sampling using conventional drills in low-gravity environments. The USDC is capable of drilling and coring in hard rocks, ice, and packed soil using relatively small force and relatively lightweight hardware.

The USDC consists of three key components: actuator, free mass, and bit (Bao et al., 2003). Unlike typical ultrasonic drills where the bit is acoustically coupled to the horn, in

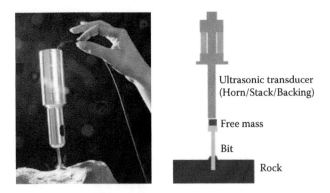

FIGURE 10.52
Photographic view of the USDC showing its ability to core with minimum axial force (left), and a schematic cross-section view (right).

the USDC the actuator drives the free mass that converts ultrasonic impacts to hammering at sonic frequencies. The actuator operates as an ultrasonic vibration mechanism that imparts energy to the free mass that in turn impacts the bit and produces stress impulses into the drilled medium. This impulse fractures rock when its ultimate strain is exceeded at the rock/bit interface. The actuator consists of a piezoelectric stack with backing structure for forward power delivery and a horn on its front for amplification of the generated displacements. The actuator is driven in resonance and held in compression by a stress bolt that prevents its fracture during operation. In the basic design, the piezoelectric stack has a resonance frequency of about 20 kHz.

The drive electronics is designed to maintain tuning of the actuator resonance in either software or hardware and thus ensure maximum electric current input. This tuning is required, since there are several factors that affect the resonance frequency including the decrease in the Q of the resonator and a slight shifting of the frequency. In addition to the need for tuning that is common to ultrasonic actuated mechanisms, the USDC requires attention to the impacts that cause time variations in the signal electric current. This effect is minimized using various control algorithms including hill climbing, extremum seeking, and others [Aldrich et al., 2006].

Following the development of the mechanism that drives the USDC, a series of novel designs were conceived and disclosed in NASA New Technology Reports and patents [e.g., Aldrich et al., 2006, 2008; Badescu et al., 2006b, 2009a, 2009b, 2009c; Bao et al., 2004, 2015; Bar-Cohen et al., 1999, 2001, 2003a, 2003b, 2005, 2010; Bar-Cohen and Sherrit, 2003; Dolgin et al., 2001; Sherrit et al., 2001, 2002, 2003, 2004b, 2005, 2006, 2008, 2009, 2010a, 2010b]. Some of the devices that were developed include the ultrasonic/sonic rock abrasion tool (URAT), ultrasonic/sonic gopher for deep ice drilling, the lab-on-a-drill, and many others. The USDC was demonstrated to drill ice and various rocks including granite, diorite, basalt [Schultz, 1993], and limestone. Under a NASA Mars Exploration Program Advanced Technologies (MEPAT) task (2004–2008), the drilling rate and the cuttings removal mechanism were enhanced by developing jointly with Honeybee Robotics a rotary/percussive mechanism (Figure 10.53).

The development of the USDC has been pursued on various fronts ranging from analytical modeling to field-testing as well as the implementation of improvements in support of a wide range of potential applications. While developing the analytical capability to predict and optimize its performance, efforts were made to enhance its ability to drill at a higher power and

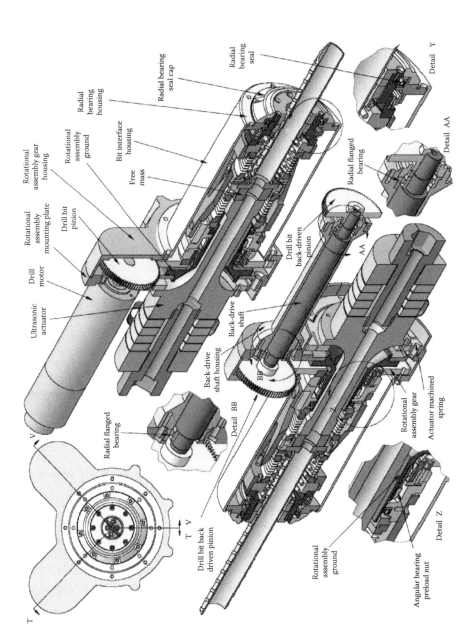

FIGURE 10.53
Under a NASA MEPAT task (2004–2008), the percussive mechanism of the USDC has been integrated with rotation capability to produce a rotary/percussive drill.

high speed. Operating the USDC as a percussive mechanism that is not rotated, sensors (e.g., thermocouple and fiber optics) were integrated into the bit to examine the BH during drilling.

10.4.4 Ultrasonic/Sonic-Gopher

Under a NASA Astrobiology for Science and Technology for Exploring Planets (ASTEP) task (2003–2006), a wireline percussive deep ice drill called the ultrasonic/sonic-gopher (U/S gopher) using piezoelectric actuation was developed at JPL [Doran et al., 2003; Bar-Cohen et al., 2004, 2012; Badescu et al., 2006a; Bar-Cohen and Zacny, 2009]. The operation of the U/S gopher was field-demonstrated at Mr. Hood, Oregon, as well as McMurdo Sound and Lake Vida in Antarctica. The goal of the task has been to establish the capability to investigate the presence of ice and fluids near the surface of Mars and the possibility of discharge of subsurface brines. Other benefits of the developed U/S gopher are its potential applicability to other icy bodies including Europa.

10.4.4.1 Analytical Modeling

The U/S gopher system was developed based on the ultrasonic/sonic driller/corer (USDC) mechanism, which is shown schematically in Figure 10.54 (for further details, see Section 10.4.3). The key components of this ice drill are a piezoelectric actuator [Tichý et al., 2010], a drill bit, a free mass, a pump with housing, and a set of preload weights. For modeling, the weights of the transducer housing, the pump and its housing, and the preload were combined and treated as the total weight of the transducer. The details of the analytical modeling of the USDC have been documented in previous publications [Sherrit et al., 2000; Bao et al., 2003; Sherrit et al., 2004a; Badescu et al., 2005].

10.4.4.2 Design and Fabrication of the U/S Gopher

The designed and fabricated U/S gopher system (Figure 10.55) consists of many components including a piezoelectric actuator, a drill bit, a free mass, a pump, preload weights, a housing, and a housing cap. Further, the transducer consists of a stack of piezoelectric rings maintained in compression using a stress bolt between a backing and the "dog-bone"

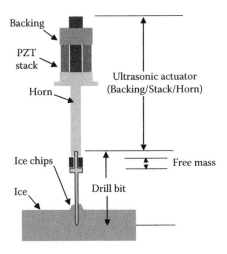

FIGURE 10.54
General schematic details of the USDC mechanism.

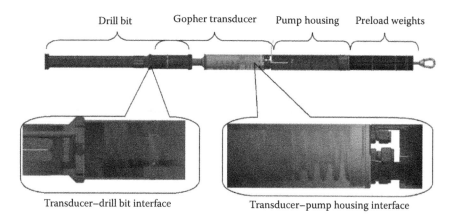

Drill bit   Gopher transducer   Pump housing   Preload weights

Transducer–drill bit interface   Transducer–pump housing interface

FIGURE 10.55
A general schematic view of the U/S gopher system design.

shaped horn. The smaller-diameter neck of the horn increases the amplitude of the vibrations generated by the piezoelectric stack. The horn head, having a larger diameter than the neck, acts as a hammer on impacting the free mass and provides a mounting fixture for attaching the drill bit. Also, the horn provides a mounting flange for the transducer housing. To minimize vibration "bleeding" from the actuator, the mounting flange was designed to coincide with the neutral plane of the transducer. A through-hole that was drilled through the actuator allowed for the transmission of compressed air down to the drill bit for ice chip removal.

A cavity in the drill bit was designed to produce cores while cutting the ice at the bottom by two concentric rings that were connected via radial teeth (Figure 10.56). Both rings and the teeth were made with sharp edges at their bottom ends. Using two side air tubes, the bit was designed to allow for compressed air to be delivered onto the cutting surface. While drilling, the bit creates an ice core that remains inside the bit cavity while ice chips are channeled away between the cutting rings of the bit cutting edge. The compressed air pushes the ice chips up the hole, along the outside of the bit, and they fall inside the coring bit cavity through an opening. When the cuttings cavity is full, the U/S gopher is pulled up from the holes and the cuttings are removed. The two side tubes along the bit were also designed to be used to sample liquid from the bottom of the hole when reaching wet ice or brine.

The U/S gopher was tested in the lab and at two field sites: on a glacier at Mt. Hood, Oregon, and in Antarctica. The lessons learned from the first field test led to design changes and additions that were made. These included a temperature sensor that was embedded into the housing cap for measuring the ambient temperature in the BH while drilling. Moreover, a sediment sampling bit was retrofitted with an adaptor to allow mounting it onto the actuator as a replaceable fixture. A drive electronics system was used consisting of a function generator and a power amplifier, and its frequency control was made to operate both manually and automatically. In the automatic mode, a digital oscilloscope, a computer, and an in-house developed software program were used. The oscilloscope reads the averages of the peak voltage and current across the actuator. The computer is used to run a control interactive software program that uses an in-house developed algorithm called "hill-climbing" [Aldridge et al., 2006]. This software program uses an extremum finding algorithm and maximizes the current through the actuator by controlling the frequency of the signal generated by the function generator.

FIGURE 10.56
The bit with the cutting rings at the bottom and a view of one of the side tubes.

10.4.4.3 Test Results and Lessons Learned

In preparation for the field test in Antarctica, a field test was conducted at Mt. Hood, Oregon, on a glacier with packed snow (Figure 10.57). The ambient temperature was above freezing; however, the glacier provided a good testbed for the first demonstration of U/S gopher capability. This test pointed out design flaws and inefficiencies that were corrected for the Antarctica field test. During this test session a total of 1.25 meters of drilling was accomplished in a total drilling of 5 h with an average drilling rate of 0.25 m/h. Heating of the transducer was observed and there were difficulties removing the ice chips due to limited pressure of the compressed air. There were also failures of actuator transducers mainly due to improper location of the electrodes soldering and heating due to inefficient frequency control. Prior to the departure to Antarctica, these problems were addressed in the final redesign of the U/S gopher hardware and software control.

In Antarctica, tests were performed at two locations: on sea ice and on a frozen lake. The sea ice tests were performed at McMurdo Sound, where the ambient temperature ranged from –15 to –5°C depending on the day; however, the sea ice temperature was not measured. Compressed air at the level of about 50–70 psi was used to remove the formed ice chips. The water vapors were filtered using a desiccant air filter. The tracking frequency control was initially done using the "hill-climbing" software program, and later it was done manually. Bits with and without core traps were tested, and a total depth of 35 cm was drilled. One of the drill bits was modified by removing four of the cutting teeth and sharpening the remaining ones. The drilling was halted at the depth of about 35 cm and could not be continued due to the hardness of the ice.

The lake ice tests were done at Lake Vida, which is located in the Dry Valleys, Antarctica. The ice layer thickness ranged from 15 to 20 m, and the temperature was from –5°C at the surface to –12°C at 16 m depth. An unheated Scott tent was dedicated to the field test of the U/S gopher. This allowed installation of electronic equipment and the U/S gopher

FIGURE 10.57
The field test at Mt. Hood.

with the supporting stand to be operated inside the tent while the power generator and air compressor were operated outside the tent.

Various duty cycle levels and drilling times were tested with both manual and automatic drive frequency control. Also, attempts were made to more accurately determine the drilling rate, and data provided by the driving frequency control program was saved and analyzed later in the lab at JPL. Air flow for removing the ice chips was used at all times during drilling where the air pressure in front of the filter was maintained at 70 psi. The initial tests at Lake Vida showed faster drilling than on the sea ice and then a reduced drilling rate rather than penetration halt was encountered. The deepest penetration that was reached at the end of this field test was 176 cm, which was deeper than the total length of the full U/S gopher assembly (Figure 10.58). As can be seen from Figure 10.58, the U/S gopher was completely buried in the drilled BH. This is an important milestone, since once the drill reaches this depth it can effectively drill as deep as logistically possible without other physical limitations (e.g., adding drilling tools). The in-house developed "hill-climbing" program control software was mainly used with fixed frequency to control the duty cycle. During the manual control of the frequency, it was found that the voltage–current phase matching approach provided better drilling performance than the current maximization algorithm. A duty cycle of 30% was used to limit the increase in temperature of the transducer.

Due to ice melting, it was observed that the chip removal mechanism worked up to approximately 2 or 3 cm ice penetration depth and needed the removal of the core pucks and chips in order to penetrate further. This melting caused the ice chips to get wet and stick to the bit surface above the cutting ring. Also, if the ice pucks were big enough they got packed inside the drill bit. This required frequent extraction of the U/S gopher from the BH for chips removal. After reaching the 1.76-m mark, an attempt was made to determine more accurate drilling using a duty cycle of 30% with 30-s running time for a total period of 30 min. At the end of this period, the U/S gopher froze inside the BH and could not be removed using the winch and "hammer out" mechanism. Eventually, the U/S gopher was removed by digging using chainsaws. A key lesson learned from the use of the percussive U/S gopher is that to improve cuttings removal it is essential to incorporate

FIGURE 10.58
The gopher in the drilled hole.

a rotation mechanism. This led to the concept of the auto-gopher [Bar-Cohen et al., 2008] that is described in the next section.

10.4.5 Auto-Gopher

The first wireline drill incorporating rotary percussion was called the auto-gopher [Bar-Cohen et al., 2008; Zacny et al., 2013b; Bar-Cohen et al., 2014]. This technology development project was funded by the NASA ASTEP program and executed by Honeybee Robotics and the NASA Jet Propulsion Laboratory. NASA JPL provided the ultrasonic-based percussive system for the drill. The auto-gopher drilled to 3 m depth in 10-cm intervals, that is, the drill had to be pulled out of the hole ~30 times to empty the cuttings catching bucket and to allow manual breakoff and retrieval of 10-cm-long and 5-cm-diameter cores. The entire process of drilling 10 cm depth, clearing the hole of cuttings and the core, and lowering the drill back into the hole took approximately 1 h. Of this, lifting the drill and lowering it back in a hole was the fastest, followed by removal of cuttings, drilling, and capture of a core. The auto-gopher was tested in 40-MPa gypsum rock in a nearby quarry to reduce field deployment costs (Figure 10.59). The strength of gypsum is similar to the strength of cryogenic ice.

10.4.6 Planetary Deep Drill

Building on this successful project, Honeybee Robotics has designed, built, and is currently laboratory testing a next-generation deep drill that will be field tested to a 100 feet (30 m) depth in the gypsum quarry. The planetary deep drill differs in many respects from the auto-gopher. To allow deep drilling, the tether diameter has to be reduced by reducing the number of wires. Hence, custom electronics have to be developed that fit small-diameter drill housing. Coincidently, processing data downhole addressed electrical losses along long wires. To reduce complexity related to core capture and delivery, the bit was changed from a coring system to a fully faced setup. Hence, the drill captures ice chips. To speed up the drilling process, reduce drilling energy, reduce sleeve friction, and reduce fluff factor of cuttings (and in turn pack more cuttings into the catch basket), a permanent hammering system was introduced. The same motor spins the auger and drives the percussive system. The major advancement was also integration of a custom-built microscope into the drill.

FIGURE 10.59
Honeybee Robotics/NASA JPL auto-gopher undergoing testing at a US gypsum quarry.

The microscope faces a side wall with a resolution of 0.25 micron per pixel. The microscope camera allows for a live video feed to the surface.

Table 10.5 summarizes specifications of the auto-gopher and the planetary deep drill. Pending successful tests, the planetary deep drill might be intergraded with the second set of anchors to allow inchworm capability, that is, it would be able to "walk" up and down inside the BH. The inchworming capabilities would allow precision positioning of a microscope anywhere along the BH length. In addition, this would be a technology demonstration for potential mission implementation of an inchworm deep drill system, as shown in Figure 10.50.

The planetary deep drill components, identified in Table 10.5, are as follows:

Instrument Bay: The instrument bay will house a microscope with a resolution of 0.25 micron per pixel and white and UV LED lights. The bay also includes temperature, humidity, and pressure measurements. The bay is notionally placed toward the top but could also be located lower down (underneath the forward anchor).

Rear and Front Anchors: The anchor uses a set of three compliant shoes to push against a BH and anchor itself to a hole with a force of 2 kN. This force is sufficient to provide a resistance to rotary torque from the cutting bit as well as vertical force from the WOB. The rear and front anchors could also be actuated together with the Z-stage to provide inchworming capabilities, as shown in Figure 10.50. The system uses a redundant actuator on the opposite side of each anchor system to prevent a "drill stuck in a hole" event. The actuator is on standby and, if required, is driven by the same driver/controller.

The Z-Stage and WOB System: The WOB drive is provided by an internally actuated ball screw and is designed for a WOB of 1000 N. An integrated load cell provides a force feedback for WOB control. The current stroke is 20 cm. However, to reduce the number of up and down trips, the Z-screw could be increased to 30 or 40 cm (and, in turn, the auger length would need to be increased).

The Rotary System: The rotary system uses a set of four actuators sharing a common shaft. The combined electrical power is 800 W for rotating a bit and an auger. The

TABLE 10.5

Comparison of the Two Wireline Drill Systems: Auto-Gopher and Planetary Deep Drill

	Auto-Gopher	Planetary Deep Drill
System mass, kg	22	30
Drill length, m	1.8	4.5
Hole diameter, mm	71	65
Core diameter, mm	60	N/A
Core length, mm	100	N/A
Max. power, W	500	800
Rot. vel., rpm	90	250
WOB, N	<200	<200
Core capture?	No	N/A
Internal electronics	No	Yes
Deployment system	Manual	Manual
Drilling automation	Yes	Yes
Core handling	None	N/A
Inchworm?	No	Yes

rotary system is overdesigned and allows for large torque margins of 25 N-m at 250 rpm.

The Hammer/Percussive System: The percussive system is based on a proven dog-clutch technology. It has been designed to provide 17 blows per revolution at 1 J per blow. The indexing has been designed to fit the number of teeth in such a way that it will take over 50 revolutions before the teeth impact the same location again.

The Bit and Auger System: The bit has been designed with a series of tungsten carbide cutters at zero rake angle. The cutters are staggered and allow for uninterrupted cuttings flow through strategically placed junk slots.

10.5 Conclusions

This chapter presented a number of ice drilling approaches used in terrestrial and planetary applications for scientific research. A large variety of drills developed to date are driven by different and very unique applications as well as different environmental conditions. For example, a drill developed for firn will not necessarily work well in cold ice.

Due to continued interest in obtaining high-quality cores from more challenging locations (e.g., greater depths or high-altitude glaciers), new drilling technologies are constantly being developed around the world. At the same time, the planetary ice drilling community is developing approaches that would allow penetration to great depths on extraterrestrial bodies covered with ice, such as Mars, Europa, Enceladus, and Titan. Although extraterrestrial drilling poses different challenges (e.g., limited power, mass, volume), the act of penetrating ice would be similar to approaches used on Earth. Hence, lessons learned and various technologies tested on Earth would be applicable (with various level of modifications) to other planetary bodies. There is no doubt that one day the ice drilling record will be broken not on Earth but on Mars.

Acknowledgments

Some of the research reported in this chapter was conducted at the Jet Propulsion Laboratory (JPL), California Institute of Technology, and Honeybee Robotics under a contract with the National Aeronautics and Space Administration (NASA). The authors would like to thank George Cooper, University of California, Berkeley; Chris P. McKay, NASA Ames Research Center, Division of Space Science, Moffett Field, CA; and Lori Shiraishi of Jet Propulsion Laboratory (JPL), California Institute of Technology, for reviewing this chapter and providing valuable technical comments and suggestions.

References

Aamot, H.W.C. 1967a. Pendulum steering for thermal probes in glaciers. *J Glaciol.* 6, 935–939.

Aamot, H.W.C. 1967b. The Philberth probe for investigating polar ice caps. CRREL Special Report 119. Cold Regions Research & Engineering. Laboratory, Hanover, New Hamphire.

Aamot, H.W.C. 1968a. Instrumented probe for deep glacial investigations. *J. Glaciol.* 7, 50, 321–328.

Aamot, H.W.C. 1968b. A buoyancy stabilized hot point drill for glaciers study. CRREL Technical Report 215.

Aamot, H.W.C. 1970a. Self-contained thermal probes for remote measurements within an ice sheet. In: *International Symposium on Antarctic Glaciological Exploration (ISAGE)*, Hanover, N.H., September 3–7, 1968; International Association of Scientific Hydrology Publication 86, pp. 63–68.

Aamot, H.W.C. 1970b. Development of a vertically stabilized thermal probe for studies in and below ice sheets. *J. Manuf. Sci. Eng.* 92(2), 263–268, doi:10.1115/1.3427727. Transactions of the ASME, paper no. 69-WA/UnT-3.

Aldrich, J., S. Sherrit, X. Bao, Y. Bar-Cohen, M. Badescu, Z. Chang. 2006. Extremum-seeking control and closed-loop quality factor monitoring for an ultrasonic/sonic driller/corer (USDC) driven at high-power, Proceedings of the SPIE Annual International Symposium on Smart Structures and Materials, San Diego, CA: SPIE, 6166-45.

Aldrich, J., Y. Bar-Cohen, S. Sherrit, M. Badescu, X. Bao, and J. Scott. 2008. Percussive Augmenter of Rotary Drills (PARoD) for Operating as a Rotary-Hammer Drill, NTR Docket No. 46550.

Árnason, B., H. Bjornson, P. Theodorsson. 1974. Mechanical drill for deep coring in temperate ice. *J. Glaciol.* 13 (67), 133–139.

Augustin, L., D. Donnou, C. Rado, A. Manouvrier, C. Girard, and G. Ricou. 1988. Thermal ice core drill 4000, ice core drilling, Proceedings of the Third International Workshop on Ice Drilling Technology, pp. 59–65.

Badescu, M., X. Bao, Y. Bar-Cohen, Z. Chang, S. Sherrit. 2005. Integrated modeling of the ultrasonic/sonic drill/corer - procedure and analysis results, Proceedings of the SPIE Annual International Symposium on Smart Structures and Materials, San Diego, CA: SPIE, 5764-37, March 6–10.

Badescu, M., S. Sherrit, A. Olorunsola, et al. 2006a. Ultrasonic/sonic gopher for subsurface ice and brine sampling: analysis and fabrication challenges, and testing results, Proceedings of the SPIE Smart Structures and Materials Symposium, Paper #6171-07, San Diego, CA.

Badescu, M., S. Sherrit, Y. Bar-Cohen, X. Bao, and S. Kassab. 2006b. Ultrasonic/Sonic Rotary-Hammer Drill (USRoHD), U.S. Patent No. 7,740,088, June 22, 2010. NASA New Technology Report (NTR) No. 44765.

Badescu, M., D.B. Bickler, S. Sherrit, Y. Bar-Cohen, X. Bao, and N.H. Hudson. 2009a. Rolling tooth core break-off and retention mechanism. NTR Docket No. 47354, Submitted on October 19, 2009. *NASA Tech Briefs.* 35(6), 2011, 50–51.

Badescu, M., S. Sherrit, Y. Bar-Cohen, X. Bao, and P.G. Backes. 2009b. Scoring dawg - core break-off and retention mechanism, NTR Docket No. 47355, Submitted on October 19, 2009. *NASA Tech Briefs*. 35(6), 2011, 48–49.

Badescu, M., S. Sherrit, Y. Bar-Cohen, X. Bao, and R.A. Lindemann. 2009c. Praying mantis - bending core break-off and retention mechanism, NTR Docket No. 47356, Submitted on October 19, 2009. NASA Tech Briefs, 35(6), 2011, 49–50.

Bao, X., Y. Bar-Cohen, Z. Chang, B.P. Dolgin, S. Sherrit, D.S. Pal, S. Du, and T. Peterson. 2003. Modeling and computer simulation of ultrasonic/sonic driller/corer (USDC). *IEEE Transaction on Ultrasonics, Ferroelectrics and Frequency Control (UFFC)*.50(9), 1147–1160.

Bao X., Y. Bar-Cohen, Z. Chang, S. Sherrit and R. Stark, 2004. Ultrasonic/Sonic Impacting Penetrator (USIP), NASA NTR No. 41666.

Bao X., S. Sherrit, M. Badescu, Y. Bar-Cohen, S. Askins, and P. Ostlund. 2015. Free-mass and interface configurations of hammering mechanisms, US Patent Number. 8,960,325 was filed on February 24, 2015.

Bar-Cohen, Y., S. Sherrit, B. Dolgin, X. Bao, Z. Chang, R. Krahe, J. Kroh, D. Pal, S. Du, T. Peterson. 2001. Ultrasonic/sonic driller/corer (USDC) for planetary application. *Proc. SPIE Smart Struct. Mater. 2001*. 529–551.

Bar-Cohen, Y., and K. Zacny (Eds.) 2009. *Drilling in Extreme Environments–Penetration and Sampling on Earth and Other Planets*. Hoboken, NJ: Wiley–VCH, ISBN-10: 3527408525, ISBN-13: 9783527408528, pp. 1–827.

Bar-Cohen, Y., S. Sherrit, B. Dolgin, T. Peterson, D. Pal and J. Kroh. 1999. Smart-ultrasonic/sonic driller/corer, U.S. Patent No. 6,863,136, March 8, 2005, NASA NTR No. 20856.

Bar-Cohen, Y., and S. Sherrit. 2003. Self-Mountable and Extractable Ultrasonic/Sonic Anchor (U/S-Anchor), NASA NTR No. 40827.

Bar-Cohen, Y., S. Sherrit and J.L. Herz. 2003a. Ultrasonic/sonic jackhammer (USJ), NASA New Technology Report (NTR), Docket No. 40771.

Bar-Cohen, Y., and S. Sherrit. 2003b. Thermocouple-on-the-bit a real time sensor of the hardness of drilled objects, NASA NTR No. 40132.

Bar-Cohen, Y., S. Sherrit, Z. Chang, L. Wessel, X. Bao, P.T. Doran, C.H. Fritsen, F. Kenig, C.P. McKay, A. Murray, and T. Peterson. 2004. Subsurface ice and brine sampling using an ultrasonic/sonic gopher for life detection and characterization in the McMurdo dry valleys, Industrial and Commercial Applications of Smart Structures Technologies Conference, SPIE Smart Structures and Materials Symposium, Paper #5388-32, San Diego, CA, March 15–18, 2004.

Bar-Cohen, Y., S. Sherrit, B. Dolgin, X. Bao and S. Askin. 2005. Ultrasonic/sonic mechanism of deep drilling (USMOD), U.S. Patent No. 6,968,910.

Bar-Cohen, Y., M. Badescu, and S. Sherrit. 2008. Rapid rotary-percussive auto-gopher for deep sub-surface penetration and sampling, NASA NTR No. 45949.

Bar-Cohen, Y., M. Badescu, and S. Sherrit. 2010. Acquisition and retaining granular samples via rotating coring bit, NASA NTR No. 47606.

Bar-Cohen, Y., M. Badescu, S. Sherrit, K. Zacny, G. Paulsen, L. Beegle, X. Bao. 2012. Deep drilling and sampling via the wireline auto-gopher driven by piezoelectric percussive actuator and EM rotary motor, Proceedings of the SPIE Smart Structures and Materials/NDE Symposium, San Diego, CA, March 12–15, 2012

Bar-Cohen, Y., M. Badescu, H.J. Lee, S. Sherrit, K. Zacny, G. L Paulsen, L. Beegle, and X. Bao. 2014. Auto-Gopher—A wireline deep sampler driven by piezoelectric percussive actuator and EM rotary motor, Proceedings of the ASCE Earth and Space 2014 Conference, Symposium 2: Exploration and Utilization of Extraterrestrial Bodies, held in St. Louis, MO, 27–29 October 2014.

Bazanov, L.D. 1961. Opyt kolonkovogo bureniya na lednikakh Zemli Frantsa-Iosifa [Core drilling experiment on glaciers of Franz Josef Land]. Issledovaniya lednokov i lednikovikh raionov. Akademiya nauk SSSR. Institut Geografii. Mezhduvedomstvennyi Komitet po Provedeniiu MGG [Investigations of Glaciers and Polar Regions. Academy of Sciences of USSR. Interdepartmental Committee on Realization of International Geophysical Year]. Vol. 1, pp. 109–114. (Text in Russian with English summary.)

Benson, T., J. Cherwinka, M. Duvernois, A. Elcheikh, F. Feyzi, L. Greenler, J. Haugen, A. Karle, M. Mulligan and R. Paulos. 2014. IceCube enhanced hot water drill functional description. *Ann. Glaciol.* 55(68), 105–114, doi:10.3189/2014AoG68A032.

Bentley, C.R., B.R. Koci, L.J.-M. Augustin, R.J. Bolsey, J.A. Green, J.D. Kyne, D.A. Lebar, W.P. Mason, A.J. Shturmakov, H.F. Engelhardt, W.D. Harrison, M.H. Hecht, V.S. Zagorodnov. 2009. Ice drilling and coring. In: *Drilling in Extreme Environments. Penetration and Sampling on Earth and Other Planets.* Y. Bar-Cohen, K. Zacny, eds. Weinheim: Wiley-VCH Verlag GmbH & Co., KGaA, pp. 221–308.

Bindschadler, R., B. Koci, A. Iken. 1988. Drilling on the Crary ice rise, Antarctica. *Antarctic J.U.S.* 5, 60–62.

Bird, I.G. 1976. *Thermal Ice Drilling: Australian Developments and Experience, Ice-Core Drilling,* J.F. Splettstoesser, ed. Lincoln, NE: University of Nebraska Press, pp. 1–18.

Bogorodsky, V.V. and V.A. Morev. 1984. Equipment and Technology for Core Drilling in Moderately Cold Ice. CRREL Special Report 84–34, 129–132.

Bogorodsky, V.V., V.A. Morev, V.A. Pukhov and V.M. Yakovlev. 1984. New Equipment and Technology for Deep Core Drilling in Cold Glaciers, CRREL Special Report 84–34, 139–140.

Browning, J.A., R.A. Bigl, and D.A. Somerville. 1979. Hot-water drilling and coring at Site J-9, Ross Ice Shelf. *Antarctic J.U.S.* 30(1) 1–6.

Casas, J.M., Sabat, F., Vilaplana, J.M., Pares, J.M., Pomeroy, D.M. 1998. New portable ice-core drilling machine: Application to tephra studies. *J. Glaciol.*, 44 (146), 179–181.

Clarke, G.K.C. 1987. A short history of scientific investigations on glaciers. *J. Glaciol. Spec.* 4–24.

Clausen H.B. and B. Stauffer. 1988. Analyses of two ice cores drilled at the ice-sheet margin in West Greenland. *Ann. Glaciol.* 10, 23–27.

Clow, G.D., and B. Koci. 2002. A fast mechanical-access drill for polar glaciology, paleoclimatology, geology, tectonics and biology. Ice drilling technology, Proceedings of the Fifth International Workshop on Ice Drilling Technology, Nagaoka, 30 October to 1 November 2000. *Mem. Natl. Inst. Polar Res.* 56, pp. 5–37.

Craven, M., A. Elcheikh, R. Brand, and N. Jones. 2002. Hot water drilling on the Amery Ice Shelf – the AMISOR project. *Mem. Natl. Inst. Polar Res.*, Special Issue No. 56, pp. 217–225.

Dansgaard, W., and N. Gundestrup. 1993. Greenland: a temptation and a challenge. *Endeavour.* New series. 17(1), 12–16.

Das, D.K., B.R. Koci, and J.J. Kelley (1992). Development of a thermal mechanical drill for sampling ice and rock from great depths. Polar Ice Coring Office, University of Alaska Fairbanks, USA. PICO Report TJC-104.

Dolgin B., S. Sherrit, Y. Bar-Cohen, R. Rainen, S. Askins, D. Sigel, D. Bickler, J. Carson, S. Dawson, X. Bao, Z. Chang, T. Peterson. 2001. Ultrasonic Rock Abrasion Tool (URAT), NASA NTR No. 30403.

Donnou, D., F. Gillet, A. Manouvrier, J. Perrin, C. Rado, and G. Ricou. 1984. Deep Core Drilling: Electro-Mechanical or Thermal Drill? CRREL Special Report 84–34, 81–84.

Doran, P.T., Y. Bar-Cohen, S. Sherrit, C. Fritsen, F. Kenig, C.P. McKay, and A. Murray. 2003. Life detection and characterization of subsurface ice and brine in the McMurdo dry valleys using an ultrasonic gopher: a NASA ASTEP project, Third International Conference on Mars Polar Science and Exploration, Alberta, Canada, (October 13–17, 2003).

Elliott, J., and F. Carsey. 2004. Deep subsurface exploration of planetary ice enabled by nuclear power, 2004 IEEE Aerospace Conference, Big Sky, MT, 10.1109/AERO.2004.1368103

Engelhardt, H, B. Kamb, and R. Bolsey. 2000. A hot-water ice-coring drill. *J. Glaciol.* 46(153), 341–345, doi:10.3189/172756500781832873.

Fujii, Y., N. Azuma.Y. Tanaka, et al. 2002. Deep ice core drilling to 2503 m depth at Dome Fuji, Antarctica. *Natl. Inst. Polar Res.* 56, 103–116.

Fujita, S., T. Yamada, R. Naruse, S. Mae, N. Azuma, and Y. Fujii. 1994. Drilling fluid for Dome F Project in Antarctica. *Natl. Inst. Polar Res.* 49, 347–357.

Geothermal History Reports (GTP)(2010).A history of geothermal energy research and development in the United States, 1976–2006, 2: Drilling (2010).

Gershman, R., and R.A Wallace. 1999. Technology needs of future planetary missions. *Acta. Astronautica*. 45(4-9), 329–335.

Gillet F. 1975. Steam, hot-water and electrical thermal drills for temperate glaciers. *J. Glaciol.* 14, 70, 171–179

Gillet, F., D. Donnou, and G. Ricou. 1976. A New Electrothermal Drill for Coring in Ice. In: *Ice-Core Drilling*, J.F. Splettstoesser, ed. Lincoln, NE: University of Nebraska Press, pp. 19–27.

Gilpin, R.R. 1974. The ablation of ice by a water jet. *Transactions of ASME* 2, 91–96.

Ginot, P., F. Stampfli, D. Stampfli, M. Schwikowski, H.W. Gaggeler. 2002. FELICS, a new ice core drilling system for high-altitude glaciers. Ice drilling technology, Proceedings of the Fifth International Workshop on Ice Drilling Technology, Nagaoka, 30 October to 1 November 2000. *Mem. Natl. Inst. Polar Res.* 56, pp. 38–48.

Goodge, J., J. Severinghaus. 2014. Construction of the rapid access ice drill (RAID) begins! Ice Bits Newsletter. *U.S. Ice Drilling Program*, 5–7.

Gundestrup, N.S., S.J. Johnsen, N. Reeh. 1984. ISTUK: a deep ice core drill system. Proceedings of the Second International Workshop/Symposium on Ice Drilling Technology. USA CRREL Spec. Rep. 84-34. Hanover, USA CRREL, pp. 7–19.

Heuberger, J.-C. 1954. *Glaciologie. Groenland. Vol. I: Forages Sur L'inlandsis. Jean-Charles.* Expéditions Polaires Françaises. V. Paris, Hermann & Cie. p. 68.

Heucke, E. 1999. A light portable steam-driven ice drill suitable for drilling holes in ice and firn. *Geografiska Annaler: Series A. Physi. Geography*, 81: 603–609. doi:10.1111/1468-0459.00088

Hodge, S.M. 1971. A new version of a steam-operated ice drill. *J. Glaciol.* 10(60), 387–393.

Hong, J., P. Talalay, M. Sysoev, and X. Fan. 2014. DEM modeling of ice cuttings transportation by electromechanical auger core drills. *Ann. Glaciol.* 55(68), 65–71(7).

Hooke R. 1976. University of Minnesota Ice Drill, In: *Ice-Core Drilling*, J.F. Splettstoesser, ed. Lincoln, NE: University of Nebraska Press, pp. 47–57.

Hooke, R. LeB., E.C. Alexander Jr., R.J. Gustafson. 1980. Temperature profiles in the Barnes Ice Cap, Baffin Island, Canada, and heat flux from the subglacial terrane. *Can. J Earth Sci.* 1980, 17(9), 1174–1188, 10.1139/e80–124.

Howorka, F. 1965. A steam-operated ice drill for the installation of ablation stakes on glaciers. *J. Glaciol.* 5(41), 749–750; Hughes H., "X D Drill", US Patent No. 930,758 (1909).

Humphrey, N., K. Echelmeyer. 1990. Hot-water drilling and bore-hole closure in cold ice. *J. Glaciol.* 36, 124, 287–289.

Ignatov, V.S. 1965. Experiment in the thermal drilling of holes in the ice at Vostok Station. In: *Information Bulletin Soviet Antarctic Expedition*, III, pp. 50–52. (Translation by Elsevier Publishing Company, Amsterdam, London, New York.)

Iken, A., H. Rothlisberger, and K. Hutter. 1977. Deep drilling with hot water jet. *Z. Gletschek, Glazialgeol.* 12(2), 143–156.

Iken, A., K. Echelmeyer, and W. Harrison. 1988. A light weight hot water drill for large depth: experiences with drilling on Jakobshavns Glacier, Greenland. In: *Ice Core Drilling*, C. Rado and D. Beaudoing, eds. Proceedings of the Third International Workshop on Ice Drilling Technology, Grenoble-France, 10–14 October 1988, pp. 123–136.

Jerby, E., V. Dikhtyar, O. Aktushev, and U. Grosglick. 2002. The Microwave Drill. *Science*. 298(5593), 587–589.

Johnsen, S.J., W. Dansgaard, N. Gundestrup, S.B. Hansen, J.O. Nielsen, N. Reeh. 1980. A fast lightweight core drill. *J. Glaciol.* 25(91), 169–174.

Johnsen, S.J., N.S Gundestrup, S.B Hansen, J. Schwander, and H. Rufli. 1994. The new improved version of the ISTUK ice core drill. Ice drilling technology, Proceedings of the Fourth International Workshop on Ice Drilling Technology, Tokyo, April 20–23, 1993. *Mem. Natl. Inst. Polar Res.* 49, 9–23.

Johnsen, S.J., S.B. Hansen, S.G. Sheldon, et al. 2007. The Hans Tausen drill: Design, performance, further developments and some lessons learned. *Ann. Glaciol.* 47, 89–98.

Just, G.D. 1963. The Jet piercing process, *Quarry Managers'. J. Inst. Quarrying Transactions*, 219–26.

Kapitsa, A.P. 1958. Opyt bureniya l'da v Antarktide s ochistkoi zaboya vozdukhom [Experiment in ice drilling with removal of cuttings by air]. Burenie geologorazvedochnih skvazhin kolonkovim sposobom s ochistkoi zaboya vosdukhom [Prospect Core Drilling with Removal of Cuttings by Air]. Moscow: Gosgeoltechizdat, pp. 78–81. (Text in Russian.)

Kasser, P. 1960. Ein leichter thermischer Eisbohrer als Hilfsgerät zur Installation von Ablationsstangen auf Gletschern. 45(1), 97–114.

Kelley, J.J., K. Stanford, B. Koci, M. Wumkes, V. Zagorodnov. 1994. Ice coring and drilling technologies developed by the Polar Ice Coring Office. Ice drilling technology, Proceedings of the Fourth International Workshop on Ice Drilling Technology, Tokyo, April 20–23, 1993. *Mem. Natl. Inst. Polar Res.* 49, 24–40.

Koci, B. 2002. A review of high-altitude drilling. *Mem. Natl. Inst. Polar Res.* (56), 1–4.

Koci, B., and R. Bindschadler. 1989. Hot-water drilling on crary ice rise, Antarctica. *Ann. Glaciol.* 12, 214.

Koci, B.R., and K.C. Kuivinen. 1984. The PICO lightweight coring auger. *J. Glaciol.* 30(105), 244–245.

Koci B., O. Nagornov, V. Zagorodnov, and J. Kelley. 1996. Hot water drilling of large diameter holes in cold ice. In: Proceedings of Fifth International Symposium on Thermal Engineering and Sciences for Cold Regions, Y. Lee and W. Hallett, eds., pp. 312–317, 501.

Koci, B.R., J.M. Sonderup. 1990. Evaluation of deep ice core drilling system. Polar Ice Coring Office, University of Alaska Fairbanks, PICO Tech. Rep. 90–1.

Koci, B. and V. Zagorodnov. 1994. The Guliya Ice Cap, China: Retrieval and return of a 308-m ice core from 6200 m altitude. *Mem. Nat. Insi. Polar Res.*(49), 371–376.

Koci, B.R. 1994. The AMANDA Project: Drilling precise, large-diameter holes using hot water. *Mem. Nat. Inst. Polar Res.*(49), 203–211.

Korotkevich, Ye.S., and B.B. Kudryashov. 1976. Ice sheet drilling by Soviet Antarctic Expeditions. Ice-core drilling, Proceedings of the Symposium, University of Nebraska, Lincoln, USA, 28–30 August 1974. Lincoln, University of Nebraska Press, pp. 63–70.

Koshima, S., T. Shiraiwa, M.A. Godoi, K Kubota, N. Takeuchi, and K. Shinbori. 2002. Ice core drilling Southern Patagonia ice field—Development of a new portable drill and field expedition in 1999. *Mem. Natl. Inst. Polar Res.*(56), 49–58.

Kudryashov, B.B., N.I. Vasiliev, P.G. Talalay. 1994. KEMS-112 electromechanical ice core drill. Ice drilling technology, Proceedings of the Fourth International Workshop on Ice Drilling Technology, Tokyo, April 20–23, 1993. *Mem. Natl. Inst. Polar Res.* 49, 138–152.

Kudryashov, B.B., N.I. Vasiliev, R.N. Vostretsov, et al. 2002. Deep ice coring at Vostok Station (East Antarctica) by an electromechanical drill. *Mem. Natl. Inst. of Polar Res.* 56, 91–102.

Lange, G.R. 1973. Deep rotary core drilling in ice. Cold Regions Research and Engineering Laboratory, Hanover, N.H. CRREL Tech. Rep. 94.

Lorius, C., and D. Donnou. 1978. A 905-meter deep core drilling at Dome C (East Antarctica) and related surface programs. *Antarct. J. U.S.* 13(4), 50–51.

Lukin, V.V., and N.I. Vasiliev. (2014). Technological aspects of the final phase of drilling borehole 5G and unsealing Vostok Subglacial Lake, East Antarctica. *Ann. Glaciol.* 55(65), 83–89.

Makinson K., and Anker P.G.D. 2014. The BAS ice-shelf hot-water drill: design, methods and tools. *Ann. Glaciol.* 55(68), 44–52, doi:10.3189/2014AoG68A030.

Manevskii, L.N., V.A. Morev, A.U. Nikiforov, V.A. Pukhov, and V.M. Yakovlev. 1983. Experimental drilling at Komsomilskaya Station [Experimental'noe burenie na stantcii Komsomolskaya]. *Bull. Sov. Ant. Exp.* 103, 71–73.

Maurer, W.C. 1968. *Novel Drilling Techniques,* ISBN-10: 0080036155, ISBN-13: 978-0080036151, New York, NY: Pergamon Press, pp. 1–114.

Maurer, W.C. 1980. *Advanced Drilling Techniques,* ISBN-10: 0878141170, ISBN-13: 978-0878141173, Tulsa, OK: Petroleum Pub. Co., pp. 1–698.

McAnerney, J.M. 1970. Tunneling in a subfreezing environment. In: *Rapid Excavation Problems and Progress*. Proceedings of the Tunnel and Shaft Conference, Minneapolis, Minn., 15–17 May 1968. Society of Mining Engineers of the American Institute of Mining, New York, NY: Metallurgy, and Petroleum Engineers, Inc., pp. 378–394.

McCallum, A. 2014. Cone penetration testing (CPT) in Antarctic firn: an introduction to interpretation. *J. Glaciol*. 60(219), 83–93.

McClung, D., and P.A. Schaerer. 2006. *The Avalanche Handbook*, 3rd Edition, Mountaineers Books, p. 342.

McKay et al. 2013. The Icebreaker Life Mission to Mars: A search for biomolecular evidence for life. *Astrobiology*. 13(4).

Mellor, M., and P.V. Sellman. 1976. General consideration for drill system design. Ice-core drilling: Proceedings of the Symposium, University of Nebraska, Lincoln, USA, 28–30 August 1974. Lincoln, University of Nebraska Press, pp. 77–111.

Miller, M.M. 1951. Englacial investigations related to core drilling on the Upper Taku Glacier. *Alaska. J. Glaciol*. 1(10), 579–581.

Miller, M.M. 1954. Mechanical core drilling in firn and ice (with a report on related investigations in the Taku Glacier, S.E. Alaska, 1950–53). Mimeographed report prepared for the E.J. Longyear Co., The Eastman Oil Well Survey Co., and the Geological Society of America, p. 72.

Morev, V.A. 1976. Elektrotermobury dlja skvazin v lednikovom pokrove. *Mat. Glacjol. Issl. Kronika Obsuzdenija*. 28,118–120.

Morev, V.A., V.A. Pukhov, V.M. Yakovlev, and V.S. Zagorodnov. 1984. Equipment and technology for drilling in temperate glaciers, CRREL Special Report 84-34, 125–127.

Morev, V.A., and V.A. Yakovlev. 1984. Liquid fillers for bore holes in glaciers. Proceedings of the Second International Workshop/Symposium on Ice Drilling Technology. CRREL Special Report 84-34, 133–135.

Morev, V.A., L.N. Manevskiy, V.M. Yakovlev and V.S. Zagorodnov. 1988. Drilling with ethanol-based antifreeze in Antarctica, ice core drilling. Proceedings of the Third International Workshop on Ice Drilling Technology, pp. 110–113.

Motoyama, H. 2007. The second deep ice coring project at Dome Fuji, Antarctica. *Scientific Drilling*. (5), 41–43.

Myrick, T., S. Frader-Thompson, J. Wilson, and S. Gorevan. 2004. *Development of an Inchworm Deep Subsurface Platform for in situ Investigation of Europa's Icy Shell*, The Workshop on Europa's Icy Shell: Past, Present and Future, February 6–8, 2004, Houston, TX: Lunar and Planetary Institute (LPI).

Nagornov, O.V., V.S. Zagorodnov, and J.J. Kelley. 1994. Effect of a heated drilling bit and borehole liquid on thermo-elastic stresses in an ice core. *Mem. Natl. Inst. Polar Res*. (49), 314–326.

Napoleoni, J.-G.P., and G.K.C. Clarke. 1978. Hot water drilling in a cold glacier. *Can. J. Earth Sci*. 15(2), 316–321.

Narita, H., Y. Fujii, Y. Nakayama, K. Kawada, and A. Takahashi. 1994. Thermal ice core drilling to 700 m depth at Mizuho Station, East Antarctica. *Mem. Natl. Inst. Polar Res*. 49, 172–183.

Nizery, A. 1951. Electrothermic rig for the boring of glaciers. *Trans. American Geophys. Union*. 32(1), 66–72.

Philberth, K. 1976. The thermal probe deep-drilling method by EGIG in 1968 at Station Jarl-Joset, Central Greenland. In: *Ice-Core Drilling*, J.F. Splettstoesser, ed. Lincoln, NE: University of Nebraska Press, pp. 117–131.

Ract-Madoux, M., and L. Reynaud. 1951. L'Exploration des Glaciers en Profondeur. Mémoires et Travaux de la Société Hydrotechnique de France. pp. 1, 89–98.

Rado, C, C. Girard, and J. Perrin. 1988. Electrochaude: Recent Development in Borehole Drilling, In: Ice Core Drilling. Proceedings of the Third International Workshop on Ice Drilling Technology, C. Rado, and D. Beaudoing, eds., International Glaciological Society, Grenoble, France, 10–14 April 1988. pp. 164–168.

Ragle, R.H., R.G. Blair, and L.E. Persson. 1964. Ice core studies of Ward Hunt Ice Shelf, 1960. *J. Glaciol*. 5(37), 39–59.

Rand, J. 1977. Ross ice shelf project drilling, October–December 1976. *Antarctic J.U.S.* 12(4), 150–152.

Rand J., and M. Mellor. 1985. Ice-coring augers for shallow depth sampling. US Army Cold Regions Research and Engineering Laboratory, CRREL Report 85–21.

Rapp, D. 2007. *Human Missions to Mars: Enabling Technologies for Exploring the Red Planet*, Springer/Praxis Publishing, ISBN: 3-540-72938-9, Appendix C, Water on Mars.

Rausch, D.O. 1958. Ice tunnel, TUTO area, Greenland, 1956. US Army Snow Ice and Permafrost Research Establishment, SIPRE Technical Report 44.

Ready, J.F. 1997. *Industrial Applications of Lasers*, 2nd Edition, New York, NY: Academic Press. ISBN-10: 0125839618, ISBN-13: 978-0125839617.

Rufli, H., B. Stauffer, and H. Oeschger. 1976. Lightweight 50-metercore drill for firn and ice. In: *Ice Core Drilling*, J.F Splettstoesser, eds. Lincoln, NE: University of Nebraska Press, pp. 139–153.

Schultz, R.A. 1993. Brittle strength of basaltic rock masses with applications to Venus. *J. Geophys. Res.* 98(E6), 10, 883, doi:10.1029/93JE00691.

Schytt, V. 1958 Glaciology II C. The inner structure of the ice shelf at Maudheim as shown by core drilling. Norwegian-British–Swedish Antarctic Expedition 1949–1952, Scientific Results, 4, 113–151.

Schwander, J., S. Marending, T.F. Stocker, and H. Fischer. 2014. RADIX: A minimal-resources rapid-access drilling system. *Ann. Glaciol.* 55(68), 34–38.

Sherrit, S., X. Bao, Z. Chang, B. Dolgin, Y. Bar-Cohen, D. Pal, J. Kroh, T. Peterson. 2000. Modeling of the ultrasonic/sonic driller/corer: USDC, 2000. *IEEE Int. Ultrason. Symp. Proc.* 1, 691–694,

Sherrit, S., S.A. Askins, M. Gradziel, B.P. Dolgin, Y. Bar-Cohen, X. Bao, and Z. Cheng. 2001. Novel ultrasonic horns for power ultrasonics. *NASA Tech Briefs.* 27(4), 54–55, NASA NTR No. 30489.

Sherrit, S., Y. Bar-Cohen, B. Dolgin, X. Bao, and Z. Chang. 2002. Ultrasonic crusher for crushing, milling, and powdering. NASA NTR No. 30682.

Sherrit, S., Y. Bar-Cohen, X. Bao, Z. Chang, D. Blake and C. Bryson. 2003. Ultrasonic/sonic rock powdering sampler and delivery tool. NASA NTR No. 40564.

Sherrit, S., M. Badescu, X. Bao, Y. Bar-Cohen, and Z. Chang. 2004a. Novel horn designs for power ultrasonics, Proceedings of the IEEE International Ultrasonics Symposium, Montreal, Canada, August 24–27.

Sherrit, S., X. Bao, Y. Bar-Cohen, and Z. Chang. 2004b. Resonance analysis of high temperature piezoelectric materials for actuation and sensing, SPIE Smart Structures and Materials Symposium, Paper #5388-34, San Diego, CA, March 15-18.

Sherrit, S., M. Badescu, Y. Bar-Cohen, Z. Chang, and X. Bao. 2005. Portable rapid and quiet drill (PRAQD), Patent disclosure submitted on Feb. 2006. U.S. Patent No. 7,824,247, November 4, 2010. NASA NTR No. 42131.

Sherrit, S., M. Badescu, and Y. Bar-Cohen. 2008. Miniature low-mass drill actuated by flextensional piezo-stack (DAFPiS), NASA NTR No. 45857.

Sherrit, S., Y. Bar-Cohen, M. Badescu, X. Bao, Z Chang, C. Jones, and J. Aldrich. 2006. Compact non-pneumatic powder sampler (NPPS), NASA NTR No. 43614.

Sherrit, S., X. Bao, M. Badescu, and Y. Bar-Cohen. 2009. Single piezo-actuator rotary-hammering (SPaRH) drill, NASA NTR No. 47216.

Sherrit, S., X Bao, M. Badescu, Y. Bar-Cohen, and P. Allen. 2010a. Monolithic flexure pre-stressed ultrasonic horns. A provisional patent application 61/362,164 was filed on July 8, 2010, NASA NTR No. 47610.

Sherrit, S., X. Bao, M. Badescu, Y. Bar-Cohen, P. Ostlund, P. Allen, and D. Geiyer. 2010b. Planar rotary piezoelectric motor using ultrasonic horns. A provisional patent was filed on July 7, 2011, NASA NTR No. 47813.

Shreve, R.L., and W.B. Kamb. 1964. Portable thermal core drill for temperate glaciers. *J. Glaciol.* 5(37), 113–117.

Shturmakov, A.J., D.A. Lebar, W.P. Mason, and C.R. Bentley. 2007. A new 122 mm electromechanical drill for deep ice-sheet coring (DISC): 1. Design concepts. *Ann. Glaciol.* 47, 28–34.

Smith, M., G. Cardell, R. Kowalczyk, and M.H. Hecht. 2006. The Cronos thermal drill and sample handling technology, the 4th International Conference on Mars Polar Science and Exploration, Davos, Switzerland, Abstract 8095, October 2–6.

Stacey, J.S. 1960. A prototype hotpoint for thermal boring on the Athabaska Glacier. *J. Glaciol.* 3, 783–786.

Stanford, K.L. 1994. Future technical developments for the Polar Ice Coring Office 13.2 cm ice core drill. Ice drilling technology, Proceedings of the Fourth International Workshop on Ice Drilling Technology, Tokyo, April 20–23, 1993. *Mem. Natl. Inst. Polar Res.* 49, pp. 57–68.

Stone, W., B. Hogan, V. Siegel, S. Lelievre, and C. Flesher. 2014. Progress towards an optically powered cryobot. *Ann. Glaciol.* 55(65), 1.

Sukhanov, L.A., V.A. Morev and I.A. Zotikov. 1974. Portable ice thermoelectric drills. [Portativnyye termoelektrobury]. *Data Glaciol. Stud.* 23, 234–238.

Suzuki, Y. 1976. Deep core drilling by Japanese Antarctic research expeditions. In: *Ice-Core Drilling*, J.F. Splettstoesser, ed. Lincoln, NE: University of Nebraska Press, pp. 155–166.

Suzuki Y. 1984. Light weight electromechanical drills. In: Proceedings of the Second International Workshop, G. Holdsworth, K.C. Kuivinen, and J.H. Randeds., Canada: Alberta, pp. 33–40.

Suzuki, Y. and K. Shimbori. 1985. Ice core drills usable for wet ice. *Natl. Inst. Polar Res. Mem.*(39), 214–218.

Suzuki, Y. 1994. *Development of Japanese mechanical drills. Personal reminiscences.* Proc. of the Fourth Int. Workshop on Ice Drilling Technology, Tokyo, 20–23 Apr., 1993. *Mem. Nat. Inst. Polar Res.*, 49, pp. 1–4.

Talalay, P.G., and N.S. Gundestrup. 2002. Hole fluids for deep ice core drilling. Ice drilling technology, Proceedings of the Fifth International Workshop on Ice Drilling Technology, Nagaoka, 30 October to 1 November 2000. *Mem. Natl. Inst. Polar Res.* 56, pp. 148–170.

Talalay, P.G. 2003. Power consumption of deep ice electromechanical drills. *Cold Reg. Sci. Technol.* 37, 69–79.

Talalay, P.G. 2006. Removal of cuttings in deep ice electromechanical drills. *Cold Reg. Sci. Technol.* 44, 87–98.

Talalay, P.G., R. LeB. Hooke. 2007. Closure of deep boreholes in ice sheets: A discussion. *Ann. Glaciol.* 47, 125–133.

Talalay, P. 2012. Russian researchers reach subglacial Lake Vostok in Antarctica. *Adv. Polar Sci.* 23(3). 176–180.

Talalay, P.G. 2014. Drill heads of the deep ice electromechanical drills. *Cold Reg. Sci. Technol.* 97, 41–56.

Talalay, P., Z. Hu, H. Xu, D. Yu, L. Han, J. Han, and L. Wang. 2014a. Environmental considerations of low-temperature drilling fluids. *Ann. Glaciol.* 55(65), 31–40.

Talalay, P.G., X. Fan, H. Xu, D. Yu, L. Han, J. Han, and Y. Sun. 2014b. Drilling fluid technology in ice sheets: Hydrostatic pressure and borehole closure considerations. *Cold Reg. Sci. Technol.* 98, 47–54.

Talalay P., X. Fan, Z. Zheng, J. Xue, P. Cao, N. Zhang, R. Wang, D. Yu, C. Yu, Y. Zhang, Q. Zhang, K. Su, D. Yang, and J. Zhan. 2014c. Anti-torque systems of electromechanical cable-suspended drills and test results. *Ann. Glaciol.* 5(68), 207-218. doi: 10.3189/2014AoG68A025.

Talalay, P.G., V.S. Zagorodnov, A.N. Markov, M.A. Sysoev, and J. Hong. 2014d. Recoverable autonomous sonde (RECAS) for environmental exploration of Antarctic subglacial lakes: General concept. *Ann. Glaciol.* 55, 65(8), 23–30.

Takahashi, A., Y. Fujii, N. Azuma, et al. 2002. Improvements to the JARE deep ice core drill. Ice drilling technology, Proceedings of the Fifth International Workshop on Ice Drilling Technology, Nagaoka, 30 October to 1 November, 2000. *Mem. Natl. Inst. Polar Res.* 56, pp. 117–125.

Taylor, M.R. 2012. Cryobots could drill into icy moons with remote fiber-optic laser power. Wired. 04.19.12.

Taylor, P.L. 1976. Solid-nose and coring thermal drills for temperate ice. In: *Ice-Core Drilling*, J.F. Splettstoesser, ed. Lincoln, NE: University of Nebraska Press, pp. 167–177.

Taylor, P.L. 1984. A hot water drill for temperate ice. U.S. Army CRREL Special Report 84 34, pp. 105–117.

Tichý, J., J. Erhart, E. Kittinger, and J. Prívratská. 2010. *Fundamentals of Piezoelectric Sensorics: Mechanical, Dielectric, and Thermodynamical Properties of Piezoelectric Materials*, Berlin, Germany: Springer, ISBN-10: 3540439668, ISBN-13: 978-3540439660, p. 160.

Tongiorgi, E., E. Picciotto, W. de Breuck, T. Norling, J. Giot, and F. Pantanetti. 1962. Deep drilling at Base Roi Baudouin, Dronning Maud Land, Antarctica. *J. Glaciol.* 4(31), 101–110.

Torsteinsson, T., O. Sigurðsson, T. Jóhannesson, G. Larsen. C Drücker, and F Wilhelms. 2002. Ice core drilling on the Hofsjökull ice cap. *Jökull.* 51, 25–41.

Treshnikov, A.F., (Ed.). 1960. Vtoraya kontinental'naya ekspeditsiya 1956–1958 gg. Obshchee opisaniye (Second Continental Expedition 1956–1958, General Description). *Trudy Sovetskoy antarkticheskoy ekspeditsii (Transactions of Soviet Antarctic Expedition)*, Vol. 8. (Text in Russian.)

Tsykin, E.N. 1962. Metodika izmereniya temperatury lednikov, primenyavshaysya Institutom Geografii AN SSSR v issledovaniyakh Mezhdunarodnogo geofizicheskogo goda [Methods of temperature measurements used by Institute of Geography at researches of International Geophysical Year]. Akademiya nauk SSSR. Institut geografii. Materialy gliatsiologicheskikh issledovanii [Academy of Sciences of the USSR. Institute of Geography. Data of Glaciological Studies]. Vol. 6, p. 113–127. USSR (Text in Russian.)

Tsykin, E.N. 1966. Udarno-kanatnoye bureniye lednikov [Cable-churn drilling in glaciers]. Akademiya nauk SSSR. Institut geografii. Materialy gliatsiologicheskikh issledovanii [Academy of Sciences of the USSR. Institute of Geography. Data of Glaciological Studies] Vol. 12, p. 239–248. (Text in Russian.)

Tulaczyk, S., J.A. Mikucki, and M.R. Siegfried, et al. 2014. WISSARD at Subglacial Lake Whillans, West Antarctica: Scientific operations and initial observations. *Ann. Glaciol.* 55(65), pp. 51–58, doi:10.3189/2014AoG65A009.

Ueda, H.T., and D.E. Garfield. 1968. Drilling through the Greenland ice sheet, CRREL Special Report 126, p. 7.

Ueda, H.T., and D.E. Garfield. 1969a. The USA CRREL drill for thermal coring in ice. *J. Glaciol.* 8(53), 311–314.

Ueda, H.T., and D.E. Garfield. 1969b. Core drilling through the Antarctic ice sheet. *CRREL Tech. Rep.* 231.

Ueda, H.T. 2007. Byrd Station drilling 1966–69. *Ann. Glaciol.* 47, 24–27.

Ueda, H.T., and P.G. Talalay. 2007. Fifty years of Soviet and Russian drilling activity in polar and non-polar ice: a chronological history. *CRREL Tech. Rep.* TR-07-20.

Varnado, S.G. (Ed.). 1980. Geothermal drilling and completion technology development program, annual progress report, October 1979 to September 1980, SNL Report SAND80-2179, SNL National Laboratories.

Vasiliev, N.I. 2002. Some features of ice drilling technology by a drill on a hoisting cable. *Mem. Natl. Inst. Polar Res.* 56, 136–141.

Vasiliev, N.I., and B.B. Kudryashov. 2002. Hydraulic resistance to movement of a drill suspended by a cable in a bore-hole in travels. *Mem. Natl. Inst. Polar Res.* 56, 142–147.

Vasiliev, N.I., P.G. Talalay, N.E. Bobin, V.K. Chistyakov, V.M. Zubkov, A.V. Krasilev, A.N. Dmitriev, S.V. Yankilevich, and V. Ya. Lipenkov. 2007. Deep drilling at Vostok station, Antarctica: History and recent events. *Ann. Glaciol.* 47(1), 10–23, doi:10.3189/172756407786857776.

Verrall, R., and D. Baade. 1984. A simple hot-water drill for penetrating ice shelves. CRREL Special Report 84–34, 87–94.

Ward, W.H. 1952. The glaciological studies of the Baffin Island Expedition, 1950. Part III: Equipment and techniques. *J. Glaciol.* 2(12), 115–121.

Ward, W.H. 1954. Studies in glacier physics on the Penny Ice Cap, Baffin Island, 1953. Part II: Portable ice-boring equipment. *J. Glaciol.* 2(16), 433–436/415.

Ward, W.H. 1961. Experiences with electro-thermal ice drills on Austerdalsbre. 1956–59. *J. Glaciol.* 4(32), 151–160.

World Meteorological Organization 2008. *Guide to hydrological practices*, Volume I: Hydrology—From Measurement to Hydrological Information, 6th edition, WMO-No. 168.

Wumkes, M.A. 1994. Development of the U.S. deep coring ice drill. Ice drilling technology, Proceedings of the Fourth International Workshop on Ice Drilling Technology, Tokyo, April 20–23, 1993. *Mem. Natl. Inst. Polar Res.* 49, 41–51.

Xu Z., C.B. Reed, G. Konercki, B.C. Gahan, R.A. Parker, S. Batarseh, R.M. Graves, H. Figueroa, and N. Skinner. 2003. Specific energy for pulsed laser rock drilling. *J. Laser Appl.* 15(1), 25–30, doi 10.2351/1.1536641.

Zacny, K., and G. Cooper. 2006. Considerations, constraints and strategies for drilling on Mars. *Planet. Space Sci.J.* 54(4),345–356, doi:10.1016/j.pss.2005.12.003.

Zacny, K., Y. Bar-Cohen, M. Brennan, et al. 2008. Drilling systems for extraterrestrial subsurface exploration. *Astrobiology J.* 8(3), doi:10.1089/ast.2007.0179.

Zacny, K., G. Paulsen, and B. Glass. 2010. *Field Testing of Planetary Drill in the Arctic*, Anaheim, CA: AIAA Space.

Zacny K., G. Paulsen, C.P. McKay, et al. 2013a. Reaching 1 m deep on Mars: The icebreaker drill. *Astrobiology*. 13(12),1166–1198. doi:10.1089/ast.2013.1038.

Zacny, K., G. Paulsen, B. Mellerowicz, et al. 2013b. *Wireline Deep Drill for Exploration of Mars, Europa, and Enceladus*, 2013 IEEE Aerospace Conference, Big Sky, Montana, March 2–9, 2013.

Zagorodnov, V.S. 1988. Antifreeze-thermodrilling of cores in Arctic sheet glaciers. Ice core drilling. Proceedings of the Third International Workshop on Ice Drilling Technology, pp. 97–109.

Zagorodnov, V.S., J.J. Kelley, and O.V. Nagornov. 1994a. Drilling of glacier boreholes with a hydrophilic liquid. *Mem. Natl. Inst. Polar Res.* (49), 153–164.

Zagorodnov V.S, V.A. Morev, O.V. Nagornov, J.J. Kelley, T.A. Gosink, and Koci B.R. 1994b. Hydrophilic liquid in glacier boreholes. *Cold Reg. Sci. Technol.* 22(3),243–251, doi:10.1016/0165-232X(94)90003-5.

Zagorodnov, V., L.G. Thompson, J.J. Kelley, B. Koci, and V. Mikhalenko. 1998. Antifreeze thermal ice core drilling: an effective approach to the acquisition of ice cores. *Cold Reg. Sci. Technol.* 28(3), 189–202.

Zagorodnov, V.S., L.G. Thompson, E. Mosley-Thompson, and J. Kelley. 2002. Performance of intermediate depth portable ice core drilling system on polar and temperate glaciers. *Mem. Natl Inst. Polar Res.* (56), 67–81.

Zagorodnov, V., L.G. Thompson, P. Ginot, and V. Mikhalenko. 2005. Intermediate depth ice coring of high altitude and polar glaciers with a light-weight drilling system. *J. Glaciol.* 51(174), 491–501.

Zagorodnov, V., O. Nagornov, T.A. Scambos, A. Muto, E. Mosley-Thompson, and S. Tyuflin. 2012. Borehole temperatures reveal details of 20th century warming at Bruce Plateau, Antarctic Peninsula. *Cryosphere*. 6, doi:10.5194/tc–6–675–2012.

Zagorodnov, V., S. Tyler, D. Holland, A. Stern, L.G. Thompson, C. Sladek, S. Kobs, and J.P. Nicolas. 2014. New technique for access-borehole drilling in shelf glaciers using lightweight drills. *J. Glaciol.* 60(223), 935–944, doi:10.3189/2014JoG13J211.

Zeibig, M., and G. Delisle. 1994. Drilling into Antarctic ice—the new BGR ice drill. *Polarforschung*. 62 (1994), 147–150.

Zimmerman, et al. 2002. The Mars '07 North Polar Cap Deep Penetration Cryoscout Mission, IEEE Aerospace Conf.

Zimmerman, W., R. Bonitz, J. Feldman. 2001. Cryobot: an ice penetrating robotic vehicle for Mars and Europa, IEEE Aerospace Conference, 2001, Big Sky, MT, 10.1109/AERO.2001.931722

Zotikov, I.A. 1979. Antifreeze-thermodrilling for core through the central part of the Ross Ice Shelf (J-9 Camp), Antarctic. CRREL Report 79–24.

11

Medicine and Biology: Technologies Operating at Extremely Low Temperatures

Alasdair G. Kay and Lilia L. Kuleshova

CONTENTS

11.1 Introduction

Medical and biological technologies operating at extremely low temperatures have been developed as part of the discipline of low temperature biology (cryobiology), and the wider field of bioengineering [Kuleshova and Hutmacher, 2008]. The direct aim of medical cryobiology is to utilize reduced temperatures to facilitate the long term preservation of cells, tissues, and organs with guaranteed phenotypic and genotypic stability. Cryobiology enables the induction of a state whereby "translational molecular motions are significantly arrested, marking the end of biological time" [Fahy et al., 1984].

Advances in medical technologies, such as organ transplantation, *in vitro* fertilization (IVF), or commercialization of cell therapies, are limited by the availability of consistent and safe viable materials for use. Reliable and reproducible protocols for the preservation of living materials are required to facilitate banking and subsequent supply. In addition to storage of biological material, low temperature therapies contribute effectively to the treatment of debilitating disorders and cancers.

Cryobiology covers a broad range of activities within a diversity of scientific aspects. Reference information specific to these individual areas can be accessed from multiple sources, including but not limited to, the journals of *Cryobiology*; *Cryoletters*; *Cell Preservation Technologies and Cryobiology*; and organizations including the Society for Cryobiology (SfC); the Society for Low Temperature Biology (SLTB); the Japan Society of Low Temperature Medicine (JSLTM); the Japanese Society for Cryobiology and Cryotechnology (JSCC); the European Alliance for Medical and Biomedical Engineering & Science (EAMBES); the International Society of Cryosurgery (ISC); and the American College of Cryosurgery (ACC).

"Low temperature" refers to temperatures below physiological mean (e.g., 34°C–37°C). "Extremely low temperatures" outlined in this chapter may be different from those described in materials science where scientists study and utilize significantly reduced temperatures, as low as –269°C or 4 K (the temperature of liquid helium). As opposed to scientific material presented in other chapters, cryobiology predominantly operates at temperatures of liquid nitrogen (–196°C or 77 K) or above. The reader may ask whether it would not be beneficial to further examine these lower temperatures. Temperatures below –196°C have been assessed in a cryobiological context; however, no significant benefits are found from applying such strategies for cryopreservation of cells or tissues. This chapter also describes the technology used for short-term preservation of organs at subzero or hypothermic temperatures (around 0°C).

11.2 Cryostorage and Mechanisms of Cryopreservation

The goal of cryopreservation technology is to achieve amorphous solidification of cells and preserve this state during long-term storage. Moreover, ice formation within the cells also has to be avoided on cooling and warming. Successful cryopreservation of cells or tissues allows for almost indefinite storage in effective stasis without loss of cell vitality [Mazur, 1984; Fahy et al., 1984; Hoffmann and Bischof, 2002]. At subzero temperatures, energy production reduces to around 10% of that at normothermia [Hoffmann and

Minor, 2014], and consequently cell degradation due to metabolic processes is reduced. Depending upon the cell type and location, the water content of a cell is 60%–85% of the total volume, some of which is bound to intracellular solutes or molecules [Mazur, 2004]. To achieve long periods of storage without damage to the cells (attributable to storage conditions), there must be no liquid state present. At temperatures approaching liquid nitrogen temperature, chemical reactions cannot occur due to insufficient thermal energy [Mazur, 1984].

Vitrification eliminates ice formation inside and outside cells on cooling, cryostorage, and warming and involves exposure of cells and tissue to aqueous solutions of special supporting chemicals, commonly at room temperature [Fahy et al., 1984; Kuleshova et al. 2004, 2007, 2009a]. Many of these chemicals, termed cryoprotectants (CPAs), form solid "glass" at –130°C [Boutron and Kaufman,1979; Boutron, 1990; Fahy et al., 1984; Luyet and Rasmussen, 1967; Kuleshova et al., 1999a; Kuleshova and Hutmacher, 2008; Macfarlane et al., 1987, 1991; Shaw, 1997]. The challenge in the cryopreservation of cells or tissues comes in careful preparation using the combined action of CPAs to be preserved without damage by vitrification rather than in the long-term survival of cells once cooled and stored [Kuleshova, 2009a]. Alternative ways exist to avoid lethal formation of ice inside the cells, namely, to cool the cells slowly enough in the presence of low-concentration CPA that freezable water is osmotically drawn out of the cell to freeze externally. The initial theory outlined in this section serves as the basis of this and describes the challenges in cryopreservation of cells and tissues faced by cryobiologists pursuing a slow cooling approach [Mazur and Leibo, 1972; Mazur, 1984].

Storage below –130°C of cells preserved in both ways ensures that only solid physical states exist, and can be achieved through transfer of cells or tissue for storage in liquid nitrogen, liquid nitrogen vapor phase (ranging from –145°C to_–156°C), or electrical freezers (–150°C) [Fahy et al., 1984; Pegg, 2009; Pegg et al., 2006a, 2006b, 2006c; Kuleshova and Hutmacher, 2008]. Vapor storage is of great benefit when large tissue samples need to be preserved. Heart valves for transplantation, for example, are widely stored in nitrogen vapor. Storage above the level of liquid nitrogen at the temperature of –145°C makes the samples less brittle and prevents them from cracking. Second, numerous viruses and microorganisms of infected materials previously stored in these liquid nitrogen tanks can survive and become a source of cross-contamination. Fluctuation of temperature during the opening of vapor storage tanks is not significant for large samples [Wood et al., 1999]. Small specimens warm significantly faster than large specimens. As storage temperature is higher in the vapor phase than in liquid nitrogen, the risk of accidental thawing or the effects of greater fluctuations in storage temperature may counteract the benefits of using this storage method for small samples. Some sealing strategies are more effective than others; therefore, extra protection, for example, wrapping straws used for cell vitrification in a plastic film or a double bag, may be needed to further reduce the risk of leakage/contamination of samples [Russell et al., 1997]. Developing protocols to prevent contamination of very rapidly cooled or vitrified biological material may prove more difficult. A strategy to minimize the likelihood of contamination while allowing specimens to be cooled by direct immersion in liquid nitrogen and warmed by direct immersion into a water bath was developed using a double straw arrangement as a simple strategy to prevent cross-contamination [Kuleshova and Shaw, 2000; Kuleshova et al., 2004, 2009a]. The major advantage of electrical freezers is in removing the necessity of refilling the storage container with liquid nitrogen, but the storage volume available is more limited.

11.3 General Principles of Vitrification

A deep understanding of physical chemistry was a foundation for the development of vitrification as a method of cryopreservation [Luyet and Rasmussen, 1967; MacFarlane et al., 1987, 1991]. A broad class of solutes are able to dissolve uncharged in water. The common factor in alcohols, amides, ketones, and other lower-molecular-weight solutes of interest is an ability to enter into hydrogen-bonding interactions with water. Generally, high-molecular-weight solutes tend to suppress ice nucleation, and therefore promote glass formation, more effectively than low-molecular-weight solutes. First, because of this high molecular weight, larger weight by weight (w/w) percentage solutions have small molar concentrations. Consequently, there is no colligative effect of these solutions on water. Second, the high molecular weight of the solute causes the viscosity of the solutions to rise rapidly, which is beneficial in achieving glass transition.

Chemicals such as propane diols, butane diols, sugars, and polymers are able to undergo vitrification and form a glass [Luyet and Rasmussen, 1967; MacFarlane et al., 1987, 1991; Fahy et al., 1984; Kuleshova and Hutmacher, 2008; Kuleshova, 2009a], yet research aims to find the optimal composition and conditions to achieve a vitreous state of moderately concentrated aqueous solutions [Shaw, 1997; Kuleshova et al., 1999a] and avoid the toxicity of CPAs. The glass-forming ability of the first group has been investigated for decades [Luyet and Rasmussen, 1967; Boutron and Kaufman, 1979; Boutron, 1990]. Low-molecular-weight agents (MW < 100 Da) were characterized for lowest total solute concentration required for vitrification calculated from the diagram of phase and physical transitions of CPAs based on differential scanning calorimetry thermograms and other physical methods [MacFarlane et al., 1991; Shaw et al., 1997; Kuleshova et al., 1999a]: 2,3-butandiol [Boutron, 1990], propylene glycol [Boutron and Kaufman, 1979], ethylene glycol [Rasmussen and Luyet, 1969], dimethyl sulfoxide [Rasmussen and Mackenzie, 1968], and glycerol [Rasmussen and Luyet, 1969]. For example, 1,2-propanediol can form a stable glass at a solute concentration ≥45% (w/w). Another popular CPA, ethylene glycol, is able to form a stable glass at 59% (w/w) solute concentration. If concentration is sufficient at a given cooling rate, crystallization may not occur and the solute remains wholly amorphous. It has been observed that ice crystals do not form during rapid cooling; however, as the cell returns to normothermic temperatures, it passes through the range $-100°C$ to $-40°C$ and devitrification occurs. Cubic ice may form from supercooled liquid, which can transform into hexagonal ice. Furthermore, recrystallization occurs with an increase of temperature, whereby small or isolated ice crystals combine to form large crystals [Luyet and Rasmussen, 1967; Boutron and Kaufman, 1979; Fahy et al., 1984; Boutron, 1990; Kuleshova and Hutmacher, 2008].

In a cryobiology context, vitrification is the cooling of a suspension or tissue exposed to a CPA solution in stepwise manner to below the glass transition point. Following glass transition, biological stasis is achieved such that the tissue does not appear to alter or degrade over time [Fahy et al., 1984]. The challenge with such a technique is to achieve exposure to high levels of CPA while minimizing CPA toxicity. Processes involve the design of a pre-equilibration protocol to allow the cells or tissue to adapt to the final vitrification solution as well as a subsequent dilution procedure. Toxicity can also be avoided through the addition of high-molecular-weight additives to solutions that have not only the advantage to penetrate the cells but additionally are able to dehydrate cells as described earlier. To participate in the vitrification process, two main classes of chemicals were described, namely, penetrating cryoprotectants with molecular weight (MW) less than 100 Da and nonpenetrating CPAs that are in turn divided into two classes, namely, sugars

(180 Da < MW < 600 Da) and polymers with MW up to hundreds of kDa [Luyet and Rasmussen, 1967; Kuleshova et al., 1999a, 2009a]. Although vitrification is a dynamic process, reduction of total solute concentration is required for vitrification; this is a challenge. It is essential to decrease cooling/warming rates by orders of magnitude to reduce the total solute concentration by 1%–2%. Ice crystals could still be formed on warming from supercooled liquid (devitrification) if the total solute concentration has reduced below the calculated level, as well as through recrystallization of small crystal nuclei, further damaging biological material. The phenomenon of devitrification is a major concern in cells. For this reason, warming has to be achieved faster than cooling.

Insufficient dilution of CPA during warming is an issue for larger tissues, as CPA toxicity increases with rising temperature. Retention of excess CPA in the tissue will have a toxic effect on the cellular component. A narrow range of CPA combinations and concentrations will allow a tissue to be cooled to super-low temperatures while inhibiting ice formation, will not be excessively toxic to cells, and will not be prone to excessive ice formation during devitrification. This range of concentrations, related to the temperature at which exposure occurs, will vary depending upon the tissue type, content, size, and cellularity.

Applications of vitrification to medical cryopreservation date back to the 1940s in the cryopreservation of spermatozoa and protoplasm [Hoagland and Pincus, 1942; Luyet and Gehenio, 1947a; Polge et al., 1949]. The technology has, however, seen an upsurge in use since Greg Fahy and William F. Rall reintroduced the method and applied it to embryo and organ preservation [Fahy and Hirsh, 1982; Rall and Fahy, 1985; Fahy, 1986].

Practical realization of vitrification technology broadly falls into two categories: those that utilize very rapid cooling and warming, usually in stepwise manner with a brief application of high-concentration CPA in the initial stages followed by very rapid reduction in temperature; and those that use controlled slow rate cooling with a gradual increase in CPA, best termed as "liquidus tracking" (LT). Vitrification of tissue-engineered constructs involving preservation of encapsulated clusters of cells, several layers of cells, and substrate or neo-tissue has several specific challenges described in Section 11.8 (Figure 11.1).

The challenge in vitrifying native tissue, rather than cell suspensions, by either method is in achieving both the correct CPA concentration throughout the tissue to be protective but not toxic, and controlling cooling temperature throughout a tissue. It is a complication

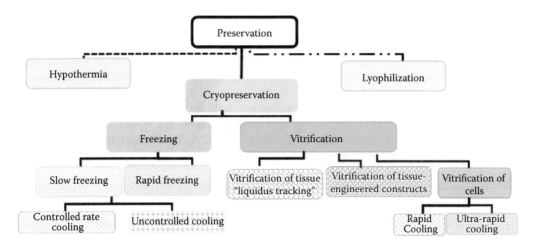

FIGURE 11.1
Schematic representation of medical and biological technologies operating at subzero and low temperatures.

that difficult-to-permeate deep regions of a tissue will be least responsive to environmental temperature variations. For large-volume solutions or tissues, this often means approximating control of both temperature and CPA concentration at deep levels within the tissue.

Vitrification by rapid and ultra-rapid cooling is a technique used predominantly for single cells like gametes, as well as for clusters of cells or multicellular organisms of small dimensions (embryos). Rapid cooling "freezing" is a method that was tested in the late 1980s for the cryopreservation of embryos using intermediate concentrations of CPA [Shaw et al., 1991]. This technique involves the direct plunging into liquid nitrogen of embryos held in straws. The supporting solutions used are commonly vitrified on cooling and devitrified on warming, which was identified as detrimental. Since this time, cryobiology progressed and the value of vitrification was established in the field of reproductive biology [Mukaida et al., 1998; Kuleshova et al., 1999b]. However, currently the majority of groups working in the field of assisted reproductive technology (ART) in a clinical setting are less focussed on historical ideation, borrowing the terminology "vitrification" due to the attractiveness of the concept while disputing evidence of devitrification. Several studies reported a substantial reduction of total CPA concentration by ~20% from that initially proposed for the vitrification of embryos [Rall and Fahy, 1985; Mukaida et al., 1998]. It is necessary to increase cooling/warming rates to about $10^{6}°C/min$ for vitrification, which is impractical and rather impossible taking into consideration the size of a sample holder and solution used to carry oocytes or embryos. Unfortunately the suggested current techniques reported by ART groups could not reach strict vitrification criteria, which can lead to nonlethal injury of valuable human gametes and embryos. Thus, these techniques are rapid freezing rather than vitrification.

Availability of transplant tissues and organs is a bottleneck in treating many potentially fatal disorders, such as cancers, renal failure, or liver or heart disease. The cryopreservation of organs or tissues would facilitate the building of biobanks to store healthy organs for transplantation. Organ preservation is discussed in Section 11.8; however, potentially the most appropriate technique for the cryopreservation of whole organs is vitrification (Figure 11.1).

11.4 Mechanisms of Freezing Injury

11.4.1 Mechanisms of Slow-Freezing Injury in Cells Intended for Cryopreservation

A majority of work investigating cryoinjury for therapeutic use has been performed *in vitro* owing to the need to focus on direct cell injury, without cell vitality being affected by immune reaction or vascular damage.

The "two-factor hypothesis" describes the potentially lethal effects of cooling rate of ice formation during temperature reduction and identified from the boundary parameters of lethality between which cell survival is possible. The hypothesis identified two factors of cell injury leading to loss of viability:

- Excessively rapid cooling rates lead to intracellular ice formation (IIF), which is lethal to the cell.
- Excessively slow cooling rates lead to a lethal increase in the concentration of solutes within cells as they dehydrate due to osmotic removal of water from the cell in response to the reduced availability of liquid water after extracellular ice formation, and associated "solution effects."

11.4.1.1 Formation of Intracellular Ice

When cells or tissues are cooled, supercooling occurs whereby the temperatures reached are below the melting point of the liquid component of the tissue, yet ice nucleation has not occurred. Typically, cells or tissues will supercool down to a temperature of –5°C. Below this temperature, ice will start to form spontaneously or due to crystallization typically in the range –10°C to –15°C [Mazur, 1984]. The cell will remain liquid, with an absence of IIF, to a temperature of around –15°C as ice formation is extracellular while cell contents are supercooled. At these temperatures, intracellular water may osmotically diffuse out of the cell due to increased extracellular solute concentration, which occurs as water "freezes out" of aqueous solutions. The risks of IIF and "solution effects" are the highest at temperatures ranging between –15°C to –40°C [Gao and Critser, 2000]. When cooling below –40°C there is reduced likelihood of ice formation and bound water will not freeze [Mazur, 1984; Walstra, 2002].

The tissue must pass through this potentially lethal temperature range twice, once during cooling and again during warming, and the rate at which the temperature of the tissue must traverse [Gao and Critser, 2000] is considered to be the greatest risk of lethal effects, rather than the ultra-low temperatures used for storage.

Rapid cooling rates lead to IIF as water is unable to diffuse out of the cell sufficiently fast to equilibrate with the surroundings, such that residual water in the cell crystallizes. This enables equilibrium as the osmotic balance is met as the intracellular solute (electrolyte) concentration increases to match the extracellular concentration. IIF causes cell death and increased pressures due to water flux may be sufficient, or analogous to those required to disrupt organelles and membranes [Muldrew and McGann, 1994]. It has been observed that crystals formed during rapid cooling tend to be small; however, as the cell returns to normothermic temperatures, it passes through the range –40°C to –15°C and recrystallization occurs, whereby isolated crystals combine to form large crystals. The effect of recrystallization is the rupture of membranes and damage to cell components [Mazur et al., 1972]. The primary methods of avoiding IIF are either through the use of CPAs of sufficient concentration to fully inhibit ice formation, or through a reduction in the cooling rate sufficient to allow water to osmotically migrate from the cell. However, if the cooling rate is too low, the latter situation may present problems through "solution effects" that are also lethal to the cell.

11.4.1.2 Solution Effects

There are many suggested solution effects that may be lethal or minimally detrimental to the cells within a tissue. Reduction in cooling rate allows supercooled water to diffuse out of the cell; however, excessive loss of water causes harmful cellular dehydration. This may cause binding of membrane proteins that were previously not in contact. On warming, the increased intracellular electrolyte concentration osmotically rehydrates the cell; however, the available volume is reduced and rupture may occur. This is referred to as the maximum cell-surface area hypothesis [Steponkus and Wiest, 1979]. Furthermore, the dehydration of cells can cause damage to membranes, internal architecture, and organelles. Denaturation of lipid protein complexes can also make cell membranes excessively permeable, causing swelling and rupture [Lovelock, 1957]. Similarly, the phase change of water to ice may cause the denaturation of membrane lipids or macromolecules crucial to cell function [Fishbein and Winkert, 1978]. Water loss from the cell may, in extreme cases, remove bound water that is otherwise inaccessible. Many tissues contain a component of nonsolvent, or inaccessible, water that may be either trapped in tissues or covalently bound within the cell and loss decreases cell vitality. [Jaffe et al., 1974; Maroudas and

Schneiderman, 1987; Muldrew et al., 1996; Elmoazzen et al., 2005; Sun, 1999]. A low cooling rate allows water to exit the cell without IIF; however, ice crystals that form external to the cell accumulate into larger blocks leaving high solute concentration liquid channels separating them. Cells accumulate in these liquid channels where they will be exposed to rising concentrations of solutes as the aqueous components freeze out and cell volume reduces, which, combined with the decreasing size of these channels adding physical pressures to the cell, could theoretically cause irreparable cell damage [Mazur, 1984]. These narrowing channels may also bring cells into direct close contact with other cells or the ice front, or allow interactions that are potentially harmful to the cell. Pegg and coworkers described a "packing effect" whereby the increased proportion of cell density in a system prior to freezing causes increased loss of cell vitality [Pegg and Diaper, 1988; Mazur and Cole, 1985]. Equally, damage caused by extracellular ice formation can also be both lethal to cells and damaging to the biomechanics of tissues through similar processes [Pegg et al., 2006c].

Osmotic stresses due to increases in solute concentration as water freezes out of aqueous solutions are a major cause of cell damage. Metabolic processes are vital in cells in a liquid state [Hoffmann and Minor, 2014]. During slow cooling or hypothermic temperature storage, metabolic damage may occur due to low oxygen (ischemic injury); cell starvation due to decreased availability of nutrients during storage; accumulation of toxic products through metabolism or from storage solutions; loss of necessary substrates during perfusion; and simple cold damage [Pegg et al., 1984]. Ischemia from tissue freezing will prompt tissue necrosis and is a mechanism of action in cryosurgery.

11.4.2 Mechanisms of Freezing Injury *In Vivo*

There exist many mechanisms leading to cryoinjury in tissues. The properties of ice formation may be used beneficially to selectively destroy tissues such as cancers. The mechanism by which these procedures cause cellular damage are now openly discussed and applied to cold-temperature therapies, particularly that of vascular damage within the tissue [Hoffman and Bischof, 2002; Fraser and Gill, 1967].

IIF inflicts extensive damage to cells. Minor freezing does not cause extensive cell damage but does prompt an immune response that aids recovery and reduces symptoms of disorders [Gage et al., 2009]. Slow freezing may cause lethal concentration of the ionic components within the cell or dissociation of the tissue [Han and Bischof, 2004; Edd and Rubinsky, 2006]. The severity of damage from application of a cryoprobe in cryotherapy depends largely on the minimum temperature reached and the rates of cooling and thawing. Repeated cycles of freeze-thawing and slower rates of thawing cause more extensive damage. Typically, malignant tissue may be treated with cycles of freeze thaw, while benign tissues may be treated with a single freeze thaw. The temperatures experienced over time for a tissue can be referred to as the "thermal history" of the tissue. This history can be described whether the tissue has undergone cryopreservation or cryosurgical therapy, and briefly comprises cooling rate, final temperature reached, storage/hold time, and warming rate [Hoffmann and Bischof, 2004; Gao and Critser, 2000; Mazur, 1963]. In contrast, with cryopreservation of a tissue the CPA content of the tissue/cell suspension affects cell vitality, in particular the addition of CPA, final concentration of CPA, length of exposure related to temperature, and the efficiency of clearing the warmed tissue of residual CPA [Hoffmann and Bischof, 2004]. Following treatment for around 24 h, inflammation and immune-mediated responses cause further cell death. When freezing is applied to a region, a cryogenic lesion forms comprising a region of coagulation necrosis that corresponds to the extent of tissue freezing that has occurred [Gage and Baust, 2007].

Benign tumors can be effectively destroyed using temperatures of –20°C to –30°C, while malignant tumors or neoplastic tissue requires lower temperatures of –40°C to –50°C or less. Prolonged freezing will involve an immune response but also prompt formation of a volume of coagulation necrosis and vascular stasis [Gage and Baust, 2007; Gage et al., 2009]. The destruction of vascular tissue is the primary site of injury induction and is vital for the treatment of cancerous tissues, particularly at the microvascular circulation level [Daum et al., 1987; Rabb et al., 1974]. The location of vascular damage can be extensive, but particular attention is on the vascular bed, or wall, where damage may occur due to mechanical injury to the vessel wall, injury to the cells lining the vessel walls, or injury through reperfusion or immune response post-thaw [Hoffmann and Bischof, 2002].

In cancers, continual rapid replication is necessary as cancerous cells require a steady supply of nutrients from blood. Often angiogenesis causes necrosis by drawing blood supply from surrounding tissues. The applied cryosurgical damage to the vascular system of tumors starves the cells and isolates the tumor, aiding the adjacent tissues and advancing the destruction of the tumor. The principle applied in general is that the freeze aspect of the cycle should be performed as rapidly as possible and the thaw stage performed as slowly as possible to maximize cell damage [Gage and Baust, 1998]. Vascular injury is the predominant surgical means to control difficult-to-access tumors [Fraser and Gill, 1967]. Subsequent to cryosurgery, however, an immune response initiated by the cryosurgical damage will destroy tissues, both healthy and compromised, due to the heightened sensitivity of the host immune system. This is referred to as "freezing-stimulated immunologic injury" [Ablin, 1995]. Equally, evidence suggests that sublethal cryosurgical injury may prompt the cell to initiate apoptotic mechanisms for programmed cell death [Baust et al., 2000].

In broad terms, these three types of damage describe the mechanisms of cryoinjury *in vivo* [Hoffmann and Bischof, 2004]. Knowledge of the mechanisms of cryoinjury is vital to the successful application of thermal therapies such as cryosurgery. The mechanisms described here fit broadly into all of the following chapters as these processes underlie both the successful cryostorage of cells and tissues and also the controlled ablation of diseased tissues.

11.5 Cell Cryopreservation in Mammals by Conventional Slow Freezing

Observation of natural survival through freeze thaw cycles dates back to 1670 in vinegar eels reported by Henry Power and subsequently in 1683 in frogs and fish reported by Robert Boyle [Thomson, 1964]. The targeted cryopreservation and successful revival of living cells developed from significant advances in 1949 with applications for the freezing of sperm using glycerol as a CPA [Polge et al., 1949] applied in cattle farming and livestock breeding, with human IVF adaptations following. The cryopreservation of gametes is discussed in more detail in Section 11.6. Improvements in cell survival came with the identification of dimethyl sulfoxide (Me$_2$SO) as an improved preservative with enhanced permeation in cells, in particular red blood cells [Lovelock and Bishop, 1959]. Me$_2$SO offers advantages and disadvantages as a CPA. Unlike polyvinylpyrrolidone (PVP), Me$_2$SO penetrates the cell, enabling reduced IIF, however making it more difficult to ensure complete removal of CPA post-thaw, although Me$_2$SO is not excessively toxic to cells at low concentrations [Pegg et al., 2006b, 2006c].

Many different cell types are cryopreserved for a range of applications [Fuller et al., 2004]. Cells may be stored for clinical use or reuse by research teams, banked for commercial sale, or archived as part of a cell library. Fertility treatments, discussed later, and commercial applications of freezing sperm have fueled early development of cell cryopreservation options.

The preservation of cells *in situ* is complicated by the need for CPA to permeate matrix and cell layers, potentially of varied types and sensitivities. In cell suspensions, there is rapid and even exposure to CPA, allowing accurate control of concentrations experienced by the cell, which may not be possible in all whole or partial tissues.

Cell preservation protocols aim to restrict the formation of intracellular ice crystals, using CPA to remove water from the cell and the surrounding extracellular matrix to inhibit ice crystallization [Fuller et al., 2004]. Mechanisms of damage to cells due to this process are described in Section 11.4. Clinical cell banking is used to preserve cells as characterized cell lines for response testing or as cell suspensions for pharmaceutical use in the production of biological or biotechnological products from cell substrates such as vaccines [Petricciani and Sheets, 2008; Knezevic et al., 2010]. Cell substrates are cell lines that produce target biologicals either naturally or through genetic modification, or have been inoculated with vaccine viral proteins. Viral replication utilizing cell architecture is used to produce large quantities of vaccine. Cell substrates are preferably rapidly replicating cells of animal origin or microbial cells such as fungi or yeasts [Knezevic et al., 2009]. Characterized diploid cells or cell lines capable of continuous culture without senescence or terminal differentiation can be effectively stored long term using cryopreservation. This facilitates consistently high-quality products derived from these cells that remain closely associated to a single original culture. Once a cell line has been isolated, characterized, and stored, the culture can be tested and confirmed negative for contaminations. This results in reduced post-thaw contaminations [Schiff, 2005].

The results from testing within cell lines can be variable; however, this variability can be counteracted by using confirmed cell lines from trusted sources. WHO retains stocks of useful cell lines, such as the Vero cell line (Vero 10-87). This cell line can be used as a seed to culture a master cell bank, particularly for use in the development of new vaccines [WHO website].

The first stem cell bank was set up in 2002 in the UK [MRC UK Stem Cell Bank], and is currently supplying 25 human embryonic stem cell (hESC) cell lines for research, fully characterized for differentiation capacity and surface markers representing cell "stemness" [NIBSC]. The cryopreservation of mammal cells in small scales is now a standardized process that is largely unchanged with the increased demand for cell therapies, and the loss of a significant proportion of cells is accepted as inevitable even though improvements to the process may be possible.

Cell lines are commonly preserved using Me_2SO as CPA. Cooling may be performed using an automated controlled rate freezer (Figure 11.2), giving monitored and accurately reproducible cryopreservation protocols that can be tailored to the individual sensitivities of the cell line being preserved, or a simple device to control heat transfer, such as the commonly used "Mr. Frosty." This technology uses a buffer of 100% isopropyl alcohol surrounding cryovials of cells, with the entire unit placed into a mechanical freezer at −80°C overnight prior to transfer to liquid nitrogen storage. This method gives approximate control and regulation of heat transfer to provide a cooling rate of ~1°C/min. Certain sensitive cells benefit greatly from more specifically defined cooling protocols, such as hESC and other stem cell types, which are preferentially cryopreserved using vitrification (Figure 11.1) [Hunt, 2011]. Freeze-drying technologies are only used for preservation of

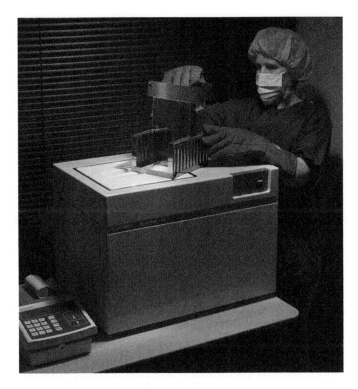

FIGURE 11.2
Automated cooling for cells and tissue using a controlled rate freezer (Kryo-560, Planer Plc). (Courtesy of Geoffrey Planer, CEO, Planer Plc. Middlesex, UK.)

nonsensitive cells. These are two-step processes, that is, vitrification, followed by freeze-drying; however, this is complex as many of the CPAs used to prepare cells are toxic, so lyophilized storage at temperatures above 0°C with these CPAs is challenging.

A freezing medium, often 10% Me2SO CPA in fetal bovine serum, is used to suspend 1.0×10^6 to 4.0×10^6 cells in specially designed cryovials. These are then cooled slowly using either a controlled rate freezer, or an approximation such as the Mr. Frosty. Following 24-h storage, cells cooled in this manner are transferred to liquid nitrogen at –196°C for long-term storage.

With the exception of advances in vitrification technologies, the basic protocols for cell cryopreservation remain largely unchanged. The establishment of cell banks has led to the creation of standardized protocols for the cryopreservation of mammalian cells such as defined by the European Collection of Cell Cultures.

Warming of cell suspensions is performed in a two-step process. Frozen cell suspensions are rapidly thawed by immersing in water from 37°C up to 42°C, and once liquid, the suspension is diluted using a dropwise process into the appropriate growth medium for cell expansion [Geraghty et al., 2014]. Diluted cell suspensions are seeded into adherent plastic tissue culture flasks and left for several hours prior to a complete medium change to remove residual CPA. The temperature in blood warmers is currently limited to a maximum of 42°C by the American Association of Blood Banks; however, no significant changes in blood cells have been found after exposure to 47°C for 1 h [Nienaber, 2003].

The demand for cell cryopreservation in research and clinical use has been driven by the need for consistent, characterized, early-stage cell lines to give results comparative for multiple studies or reliable clinical results. The clinical application of cell preservation facilitates cell screening, banking for storage, quality control, and maintenance of adequate stocks of cells for use and distribution of cells in a cryopreserved state [Karlsson and Toner, 2000]. In research laboratories, it is standard practice in regenerative medicine, when using cell lines from either commercial or primary sources, to routinely cryopreserve a stock (one to two ampoules) of early passage (early stage) cells for later use. The ability to store and supply cells has been integral to long-term or longitudinal studies of cell therapy. With the more recent classification and description of stem cells, particularly embryonic, cord blood, and iPS cells, the need to sensitively preserve cells at an early stage is evident.

11.6 Cryobiology as a Cornerstone of Assisted Reproductive Technology

Cryobiology is a cornerstone of assisted reproductive technology. There are several reasons for such a statement. A significant fall in female fertility occurs after 34 years in humans. Oocytes (female eggs) are the largest single cells in the human body. They age much faster than women plan to have children, which creates health concerns in modern society [Jolly et al., 2000]. A decline in sperm concentration and morphology attributed to socio-psycho-behavioral factors has also been documented over the past few decades. Reproductive cryobiology originated from the preservation of sperm more than 50 years ago and has gradually developed into a mature science over the past six decades [Polge et al., 1949; Holt et al., 1996, 1999; Leibo, 2002; Pickard and Holt, 2004]. Sperm cryobiology has been explored in three major directions—agriculture, laboratory animal medicine, and human clinical assisted reproduction. Cryopreserved human spermatozoa were first used in the 1950s [Sherman, 1964]. Complete description of the biophysical fundamentals of sperm cryobiology that may be of particular interest to readers with engineering physics backgrounds can be found in the review published by Critser's group [Benson et al., 2012]. There is a vast body of knowledge about the cryopreservation of sperm [Watson, 1995; Holt et al., 1996, 1999; Leibo, 2002; Benson et al., 2012; Morris et al., 2012]; thus, the main focus of this subchapter will be on the cryopreservation of oocytes, a subject of enormous interest and complexity.

Since 1999, when the birth of the first baby from an embryo derived from vitrified-warmed oocytes was reported by Kuleshova and coworkers in the journal of *Human Reproduction* [Kuleshova et al., 1999b], this concept has been recognized and implemented worldwide [Chian and Chao, 2009; Wood, 2012]. This first breakthrough in achieving ice-free low temperature preservation of human oocytes made an important advance in our knowledge and practice of the IVF process, as it removed the ethical problems associated with embryo cryopreservation. Over 500 births have been reported after the application of this strategy worldwide (Figure 11.3). This invention removed the problem of chromosome aberration associated with oocytes freezing and proved critical for oncology patients [Chian and Chao, 2009; Kuleshova, 2009a; Wood, 2012].

The first birth using vitrified oocytes is one of the main research achievements of the cryobiology community and is closely related to the ethical and social needs of society. Social implications of this invention involve 13% of the couples around the world. Couples with embryos frozen and stored in cryobanks face serious legal, ethical, and moral dilemmas. Just in the last two decades in the United Kingdom alone, 1.7 million embryos have been discarded as they are no longer needed by couples trying for children under IVF programs. In sharp contrast to embryo donation, vitrified surplus oocytes are not discarded, but usually donated by couples who have successfully had their babies from IVF and who

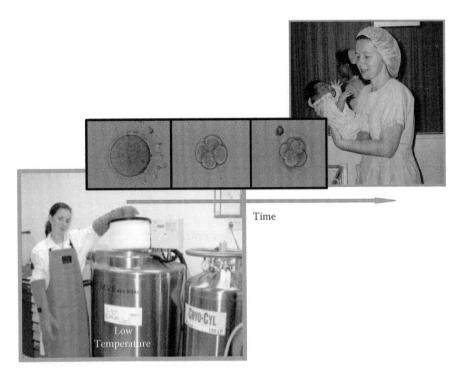

FIGURE 11.3
Embryo development and child birth following oocyte vitrification-warming and intracytoplasmic sperm injection.

want to help other couples. The main reason for donation lies in the fundamental difference between oocytes and embryos. Similar to red blood cells and bone marrow, oocytes are a person's cells, not an embryo developing into a new individual.

Since the concept was proven and quickly widespread, application of a vitrification strategy has led to conclusive, statistically confirmed results in recent years. The incidence of birth anomalies after vitrification of human oocytes does not differ from that observed in natural conception. The rates of ongoing pregnancy, embryo cleavage, and fertilization do not differ between vitrified and fresh oocytes. The difference in average survival rates after vitrification and after slow cooling shows that vitrification is superior to slow cooling [Kuleshova, 2009a]. Vitrification also resulted in significantly higher fertilization and embryo cleavage rates compared with slow cooling. The birth rate for slow cooling studies is approximately half of that observed in vitrification studies [Kuleshova, 2009a].

In clinical practice, the Practice Committee of the American Society for Reproductive Medicine estimates the live birth rate per vitrified-warmed oocytes at 4% in their early report. With time in embryo culture, embryo transferred efficacy is further improved [Patrizio and Sakkas, 2009]; equally the studies on human oocytes vitrification and comprehensive meta-analysis [Cobo and Meseguer, 2010; Cobo and Diaz, 2011] showed that complete outcomes are significantly higher. Finally, identical outcomes for vitrified and fresh oocytes have been demonstrated in an oocyte donation program [Nagy et al., 2009]. The Practice Committee is continuously updating their guidelines to assist the community and physicians with their decision-making process [Practice Committee of the American Society for Reproductive Medicine, 2013]. Recent clinical evaluation of the efficiency of donation programs demonstrated that pregnancy rates are similar with either fresh or vitrified oocytes and there is good evidence that the same is applicable in young patients. Different groups give the name of the method by the type of containers/carriers that has been used during exposure of oocytes to low temperature. The evaluation of holders employed for human oocyte vitrification revealed that they are not greatly different in their efficacy. More fundamental research regarding the vitrification concept is needed in the quest for a proper scientific approach to understanding the mechanism of success and injury to make any real progress in the field.

One of the first elegant methods that proved to be effective derived from the utilization of electron microscopy grids as miniature carriers. Current methods are numerous variations on the same strategy, mainly employing immersion of biological material directly into LN_2 with a minimum amount of vitrification solution (VS), while VSs currently in use suffer from some limitations. A major concern is the total solute concentration of proposed VSs. Ice crystal formation may take place inside and outside of cells and not be visible as solutions have insufficiently low concentration to support a stable amorphous state. Overall, it can be seen from the analysis of all reports on vitrification of human oocytes that lead to child birth that there is no distinct link between the success rate and the type of container used. The basis of success is a correspondence between equilibration and dilution steps that allow sufficient oocyte dehydration in combination with penetration and subsequent removal of CPAs post-warming. The composition of VS is also important. Ethylene glycol (EG)-based vitrification solutions are shown to be effective for human oocyte and embryo vitrification [Kuleshova et al., 1999b, 2009a; Yoon et al., 2003; Cha et al., 2011; Mukaida et al., 1998, 2001]. This idea was found to be valuable by others during application to porcine and bovine oocytes. Overall, despite some breaches of the protocols designed, many oocytes survived cryopreservation and produced good-quality embryos after fertilization and resulted in many healthy live births.

In a cryobiology context, human oocytes are unique cells. In contrast to other mammalian oocytes, some parameters of human oocytes at the same stage of maturation vary for different infertility patients. This may be a vital factor due to which vitrification of mammalian oocytes, except possibly for swine species, has been relatively successful for a number of years. It should be also clarified that success rates for cryopreserved embryos continued to be higher than for cryopreserved oocytes in most mammals [Fuller, 2009]. The challenges of human oocytes cryopreservation, development of protocols, and equipment for control rate cooling protocols have been comprehensively described by Fuller [2009].

Oncology patients are benefitting significantly from the invention of an effective strategy for oocyte vitrification [Kuleshova et al., 1999b; Chian and Chao, 2009]. The gonadotoxicity of chemotherapy is a well-established fact. The main strategy for preservation of fertility in young female oncology patients or those without partners is vitrification of human oocytes rather than embryos [Chian and Chao, 2009]. It is known that hormone stimulation is required for generation and collection of several mature oocytes simultaneously. The procedure covers a substantial period while the concern for cancer patients is the time constraints. Shorter protocols take only 3–4 days and result in a few immature oocytes ready for collection and *in vitro* maturation. Even with the rise in popularity of the vitrification concept, child birth following cryopreservation of immature oocytes is found to be difficult to achieve for reasons related to the substructure of immature oocytes, and it is an inappropriate practice for patients with a poor state of health or those undergoing severe operations. Therefore, human oocyte maturation followed by a preservation approach was adopted as a working hypothesis, giving the first fruitful outcomes in recent years. In general, there are two strategies of preservation for fertility in female oncology patients, both involving low temperature preservation by vitrification of either ovarian tissue or oocytes. Although ovarian tissue can also be preserved effectively, the overall procedure can have complications for patient health. Collection of tissue requires major surgery, which is suboptimal and may also cause inflammation in some cases. Thus, while vitrification of *in vivo* matured oocytes still remains more effective than vitrification of *in vitro* human matured oocytes, it is believed and it is our view that it is the most promising approach in the preservation of fertility of female cancer patients. In our time, embryos at all stages of development have been preserved by vitrification [Mukaida et al., 1998, 2001; Kuleshova and Lopata, 2002]. However, freezing of embryos is still common practice. Preimplantation genetic diagnosis and/or screening allow the assessment of the genetic health of an embryo before transferring it into the uterus. These techniques require the removal of cellular material in order to perform genetic analysis. Freezing of human embryos after biopsy is less effective than vitrification, which is in turn under development. An important issue to mention here is cryobiology research in the context of the preservation of fertility in male oncology patients. All methods of interest are based on cryopreservation and cryostorage at deep cryogenic temperatures (–196°C), irrespective of approaches explored for fertility preservation in prepubertal male oncology patients or spermatozoa preservation for adult male oncology patients.

Reproductive cryobiology related to mammals demonstrates that humans and wildlife share common environmental and genetic challenges. Comprehensive overviews published by Bill Holt and Paul Watson over the years provide understanding about the status of international conservation efforts for wildlife and the preservation of genetic resources of mammals in captivity as well as other breeding programs [Watson and Holt, 2001; Holt et al., 1996, 1999, 2008]. Frozen-thawed spermatozoa, similarly to human sperm banking, have become an integral component of animal agriculture and laboratory animal genome banking. The protocols, particularly for cryopreservation of oocytes, and testicular and ovarian tissue, are still in a stage of optimization and cannot readily be extrapolated to other species' reproductive tissue and cells [Leibo and Songsasen, 2002]. Human reproductive cryobiology is only part of this fascinating science.

From a background in physics and chemistry, an issue unsolved for decades has been resolved, that is, effective preservation of human oocytes has been achieved [Kuleshova et al., 1999b]. Subsequent years of successful vitrification of human oocytes have confirmed the validity of this idea. This research resulted in immediate improvements in the quality of life revealing the inevitability of bioethics in modern society, taking into account the ethical aspects of cryobiology research along with discussions on AIDS and the genome project, discarding embryos and surrogacy and dilemmas faced by scientists in today's highly challenging global environment.

11.7 Vitrification of Tissue: The LT Method

It has been shown that controlled rate cooling of large tissue volumes results in a temperature gradient within the tissue or solution, whereby the temperature experienced on the outer surface is lower than that experienced within inner tissue during cooling and higher during warming, with compressive tensions due to freezing causing extensive damage [Rubinsky et al., 1980]. This temperature gradient encourages ice formation in predominantly liquid regions of tissue where there is less tissue organization, such as blood vessels [Hunt, 1984] and articular cartilage [Pegg et al., 2006a]. There may also be a concentration gradient in CPA between the outer surface and the deep tissue due to diffusion characteristics and kinetics of CPA uptake. Permeation of CPA is particularly an issue in larger tissues where surface tissue absorbance of CPA may equilibrate with environmental CPA, while deep tissue equilibration is incomplete at the stage where freezing temperatures are reached. This leads to increased propensity for ice formation in deep tissue. Attempts to increase deep tissue permeation of CPA by increasing surface concentrations risk CPA toxicity causing fatal damage to surface layers of tissue and are ineffective in raising final equilibration coefficients [Carsi et al., 1985]. Attaining a balance between surface CPA toxicity and deep tissue freeze damage is a complex issue.

LT is a vitrification method that uses balanced control of CPA concentration against temperature during cooling and warming [Pegg et al., 2006c]. The goal of LT is to promote cell survival at low temperatures experienced by tissue, using exposure to the minimal amount of CPA necessary to inhibit ice formation at the temperature being experienced only as that temperature is reached. This is facilitated due to the diminution in toxicity of high-concentration CPA at low temperatures [Matheny et al., 1969]. For vitrification of bulky tissue, as previously described, a very high concentration of CPA is required, levels toxic to most tissues at even a brief-duration exposure in a previtrified state. If high levels of CPA are reached only at very low temperatures, toxicity is reduced and vitrification can occur with minimal cellular damage, potentially as little as 5% [Wang et al., 2007]. By gradually increasing CPA content during the cooling process, rather than prior to cooling, the solution composition matches tracks defined by the "liquidus line," which marks the transition between complete solution and ice crystallization/melting point, to beyond the glass transition temperature, achieving vitrification [Pegg, 1986]. Ice crystallization is fully inhibited if the degree of tracking is sufficiently precise. This enables ultra low temperature cooling with the level of CPA exposure experienced by the cell not exceeding that experienced during conventional preservation methods [Pegg et al., 2006c].

Vehicle solutions for CPA during cooling contain a complex electrolyte component to provide nutrients to cells and tissues, protect from damage during cooling, vitrify, and balance osmolarity and pH across a wide range of temperatures. The LT method also controls for changes in electrolyte concentration as temperature decreases [Farrant, 1965].

The concept of LT was first proposed by Farrant modeling on a NaCl-only electrolyte system. The first successful application of the LT method was in 1972 with smooth muscle [Elford and Walter, 1972]. The LT method was not substantially revisited until the research group of Professor David Pegg applied the technique to the vitrification of cartilage [Pegg et al., 2006c]. This group demonstrated the possibility of a technique that would allow clinical-grade living articular cartilage grafts to be stored for off-the-shelf use by orthopedic surgeons, potentially as an alternative to prosthetic total knee replacement surgeries. This is discussed further in Section 11.10. Cartilage is a particularly appropriate tissue to apply the LT technique to because it is uniform in structure with broadly speaking only one tissue composition involved. Standard cryopreservation techniques are particularly ineffective due to the high water content in cartilage and the increased susceptibility of chondrocytes to ice nucleation, preferentially in the lacunae, causing irreparable damage to the cells [Pegg et al., 2006b].

In LT, the variables to be optimized are as follows:

- Kinetics of permeation for the chosen CPA
- The duration of exposure to CPA, determined by cooling rate
- The final concentration of CPA, determined by the eutectic temperature of the system applied.

LT will typically use CPA that exhibits good uptake by cells and tissue, such as Me_2SO. Penetration rates are partially determined by the extraction of water as Me_2SO experiences less rapid osmotic diffusion. It is vital, therefore, to maintain osmotic balance to avoid a large change in cell volume or rapid dehydration of the cell.

LT may be achieved through the transfer of tissue between CPA solutions of increasing concentrations, or the increase in CPA concentration for a single solution bath in which the tissue remains static. Both methods require preparation of accurate solution concentrations precooled to the correct temperatures, and both require good mixing techniques. The timing (and duration) of tissue transfer or CPA increase must be handled precisely and at the correct temperature, so the risk of operator error is large. For these reasons, an effective automated process is preferred and one strand of current investigations is looking at an automated system for LT [Wang et al., 2007]. At present, the only commercially available automated LT system is produced by Planer Plc (Sudbury-on-Thames, London). This system provides a programmable approach to LT where the operator is able to control the rate of cooling/warming aligned with the exposure concentration of CPA. This is achieved through a calculated application of high concentration CPA to a constant-volume solution in the sample holder, such that target CPA is achieved accurately at the temperature instructed. A Planer 560 controlled rate freezer (Planer Products, London, UK) is used with standard Delta-T software to control temperature from built-in resistance thermometers, while a purpose-built control unit (CfgPid) designed by Planer Plc controls pumps that adjust the inflow of high-concentration CPA to achieve a temperature-matched target in the medium surrounding the tissue.

The issue of surface temperature to inner temperature gradient is important as it is likely that the deep tissue will remain vitrified, and therefore not available for CPA reduction, to higher temperatures than surface tissue. For this reason, most LT protocols for warming will contain a "devitrification" step that holds the tissue at a low temperature but above the melting point of the CPA solution, to allow free movement of CPA during warming.

LT presents a promising technique for the vitrification of living organs. CPA can be delivered through perfusion to complex organ tissues using intact vascular systems that run throughout the body of the tissue. While the toxicity and rate of diffusion remain significant issues, particularly in complex tissues, the promise of LT as a vitrification method for long-term storage of organs is an enticing one. The current state of organ preservation is discussed fully in Section 11.9.

11.8 Cryobiology as an Integral Aspect of Tissue Engineering

Developing effective cryopreservation strategies to enable off-the-shelf availability of complex cell-containing constructs and neo-tissues is necessary to realize their clinical potential. It is vital to provide an effective link between tissue engineering and biomaterial science and its biomedical application. The promises of tissue engineering rely upon the ability to physically distribute ready-to-use products of regenerative medicine without compromising their quality. The current research and practical interest is to establish how complex tissue-engineering principles can be sustained during cryopreservation to facilitate biomedical application of cell-containing constructs for applications in tissue repair and regeneration, drug development, and diagnostics. To date this emerging subject has received little consideration outside the work of a few groups, one of them being Kuleshova's laboratory [Kuleshova et al., 2004, 2007, 2009b, 2013; Kuleshova and Hutmacher,

2008; Gouk et al., 2011; Magalhaes et al., 2008, 2009; Magalhães et al., 2012; Wu et al., 2007, 2015; Wen et al., 2009; Bhakta et al., 2009].

There are several challenges in the cryopreservation of tissue-engineered constructs (TECs):

- Preservation of the integrity of constructs
- Maintaining high cell viability after cryopreservation
- Maintaining cell function
- Preserving cell proliferation and differentiation potential (particularly if stem cells are involved)
- Preserving cell–cell interactions

This leads to several specific challenges:

- The hydrogel matrix that allowed migration of cells should retain its original properties after cryopreservation.
- Maintaining the cell attachment ability to substrate is essential for cell-scaffold systems.

There are important differences between the two main methods of cryopreservation in a tissue-engineered context. Vitrification eliminates ice crystal formation during the course of a cooling and warming process not only inside the cells but also inside biomaterials [Kuleshova et al., 2004, 2007, 2009b; Kuleshova and Hutmacher, 2008; Magalhães et al., 2008, 2009, 2012; Wu et al., 2007, 2009]. The freezing process permits ice outside cells in the medium and in the biomaterial, because the concentration of supporting CPA is usually from 1 M to 3 M. This is an advantage over freezing, as vitrification provides amorphous solidification, resulting in less mechanical stress to the cells and TECs.

Several areas of regenerative medicine can benefit from the development of new integrated vitrification strategies through the efforts of groups such as Kuleshova's: (i) hepatic tissue engineering, (ii) tissue engineering of bone and cartilage, and (iii) neural tissue engineering. Cryopreservation of tissue-engineered blood vessels has been developed by of Song and coworkers [Dahl et al., 2006].

It is a challenge to design biomaterials with properties that will not be transformed during interaction with chemicals (e.g., CPA), cooling to cryogenic temperatures, and subsequent warming. Research on development of cryopreservation principles in tissue engineering allied with the exploration of biodegradable biomaterials, serving as permissive substrates for cell growth, differentiation, and biological function, aims to complete this vital step [Kuleshova et al., 2004, 2007; Kuleshova and Hutmacher, 2008; Wu et al., 2007; Wen et al., 2009]. It is believed that an integrated approach such as this would facilitate the medical applications of tissue engineering, stem cells, and biomaterials.

The development of biologically inspired nanoscale materials, including hydrogels that mimic the *in vivo* environment, is a promising strategy and an integral part of future developments. In the biomaterial context, the structure–property relationships of polymeric materials must be explored to ultimately allow the design of a new process for preservation by vitrification [Wen et al., 2009; Kuleshova et al., 2004; Wu et al., 2007]. Research into the modification of collagens for enhancing their properties, cell assembly encapsulation, and cryopreservation has been undertaken for the benefit of hepatic tissue engineering [Wu et al., 2007]. Methylated and, in turn, galactosylated collagen-based matrices with superior properties have been synthesized. Although galactosylated collagen results in a less fibrous structure in the shell of capsules with conjugation of a synthetic terpolymer, the work of Kuleshova's group was able to develop a vitrification process that did not affect the integrity of fragile constructs or the structure and chemical properties of collagen. This allowed interactions with receptors of hepatocytes post-cryopreservation. Taking into consideration the acidity of Me_2SO at the high concentrations required for vitrification, it is hypothesized that high acidity may impair the chemical property of collagens and subsequently affect cell migration of hepatocytes and/or viability and proliferation rate in stem cells [Heng et al., 2006]. The results of our studies proved that strategies based on this hypothesis are valid. Freezing allows ice formation, and thus results in rupture of the structure of construct itself, which leads to nearly half of tissue-engineered collagen-based capsules becoming broken [Heng et al., 2004].

Through the exploration of high-molecular-weight polymers as CPA, a strategy for the vitrification of large hepatocyte-containing constructs has been effectively established. A tissue-like structure with high levels of liver-specific functions was formed post-cryopreservation on collagen-coated polyethylene terephthalate fabricated into a thin film. It has been illustrated that the loss in cell viability of post-frozen thawed hepatocytes is due to apoptosis, which follows cytoskeletal disruption. Therefore, the question of why the cryopreservation by vitrification succeeded where freezing protocols generally fail has been answered. Through this comparative approach, effective preservation by vitrification of a bioartificial liver device was developed [Magalhaes et al., 2009, 2012].

It was demonstrated that a combination of mesenchymal stem cells (MSCs) and a hydrogel scaffold can be used successfully for the purpose of formation of new cartilaginous tissue or for regeneration of bone. A stable amorphous state of MSCs growing in alginate–fibrin beads used for bone regeneration has also been maintained during the vitrification-warming cycle [Bhakta et al., 2009]. The alginate–fibrin is a hydrogel, and so has a very high water content and brittleness at low temperatures (–196°C). This result is encouraging in a biomaterial context.

Nanofibrous TECs have been developed for biomedical applications such as cartilage, bone, arterial blood vessels, heart, and nervous system. Fabrication of polymeric nanofibrous poly(caprolactone) (PCL)-gelatin nanofibrous materials involving electrospinning is an area of growing interest. In earlier research, the success of tissue-engineered strategies combined with preservation of MSC-seeded nanofibrous scaffolds fabricated from PCL-gelatin has been demonstrated by Wen and coworkers [Wen et al., 2009]. The excellent results of all assessments after vitrification showed that a vitrification approach is effective in cryopreserving these PCL-gelatin fibrous TECs, retaining high cell viability and the capability to proliferate and differentiate while maintaining structural integrity. It has been demonstrated that cartilage defects can be restored through transplantation of TECs containing MSCs. A reliable strategy for the preservation of hydrogel substrates, serving as a base for MSC growth, is an important current task [Kuleshova et al., 2013]. Cryopreservation of stem cell cultures with their 3D culture support system adds to the flexibility of clinical scheduling and facilitates continuous cell expansion, permitting their effective utilization in the field of regenerative medicine.

Consequently, structural integrity is a high priority in the preservation of TECs at low temperatures. Cell populations can only function effectively as a tissue when the constructs they reside in remain intact; cracks in the constructs can severely affect the functionality of TECs. Numerous reports have demonstrated that the vitrification protocol, which includes a brief three-step exposure to a VS composed of EG and sucrose (40% v/v EG, 0.6 M sucrose), immersion into LN_2, warming, and gradual removal of the VS at room temperature, effectively preserves the integrity of dissimilar scaffolds and TECs as a whole. The promising results may be attributed to the regular shape of the constructs. For example, the more uniform heat distribution in thin disc-shaped TECs prevents crack formation and propagation [Wen et al., 2009]. It has been demonstrated that synthetic bio-degradable polymers support the growth of neo-tissue and mitigate any thermal expansion constraints of the CPA in the nanofibrous microstructure. It is also believed that the texture of nanofibrous scaffold, where the pore size of the scaffold is larger than the size of CPA molecules employed, facilitates the penetration of the VS [Wen et al., 2009]. A physical transition in the VS enclosing the construct, opposed to a phase transition that occurs during "freezing," apparently plays an important role in preserving the integrity of the TECs.

The duration of exposure to CPA plays a significant role in the success of vitreous cryopreservation. This, together with the appropriate conditions employed, effectively induces vitrification in both the supporting scaffold as well as formed neo-tissue. A "layer-by-layer" approach that constructs a tissue by sandwiching cell layers supported by scaffold layers may be adopted to generate bulky TECs. Vitrification of bulky TECs can be advocated while the length of exposure has to be extended correspondingly. This recommendation is based on (i) the results obtained on 2D, 3D, thin, and thick TECs during decades of our research, in addition to (ii) understanding gained in the area of bulky native tissue cryopreservation through the consistent efforts of Pegg's group, including an original "LT" method [Pegg et al., 2006c] discussed earlier.

Prior to incorporating neuronal stem cells into biomaterials, a series of comparative studies on known approaches to cryopreservation have been undertaken using neurospheres as model systems [Tan et al., 2007; Kuleshova et al., 2009b]. Vitrification is the opposite of conventional freezing and completely maintains the neuronal specific function. The process allowed the maintenance of the structural integrity of 3D neurospheres, suggesting that it may be applicable to other structural cultures such as nerve bridges and matrixes for filling lesions. It was confirmed that cryopreservation of NSCs as 3D clusters is beneficial, drawing a parallel with other 3D tissue-engineered cell cultures.

Cryopreservation of tissue-engineered blood vessels has been successfully attempted by Song et al. [2000a, 2000b; Dahl et al., 2006]. Tissue-engineered blood vessels have been developed as an alternative source of vascular tissue for use in bypass surgery. Cells grown onto a thick polyglycolic acid (PGA) mesh were preserved by freezing and vitrification methods. The contractility results for vitrified TECs were more than 80% of fresh controls and, in contrast, the results for frozen samples were only 11% of fresh controls. Vitrification has been explored with great success in the application to native vascular grafts and other tissues for years as described in previous subchapters [Pegg et al., 2006c; Song et al., 2000a, 2000b; Pichugin et al., 2006]. Yet significantly, Song and coworkers have revealed that it is possible to design a stepwise vitrification procedure for the cryopreservation of a tissue-engineered implant.

In the last decade we have provided evidence that a vitrification strategy is indispensable for the successful preservation of cell-containing TECs [Kuleshova et al., 2004, 2007, 2009b; Kuleshova and Hutmacher, 2008; Magalhaes et al., 2008, 2009, 2012; Tan et al., 2007; Wu et al., 2007, 2015; Wen et al., 2009; Bhakta et al., 2009], since it is able to mantain the attachment ability of cells to the substrate [Wen et al., 2009; Magalhaes et al., 2009, 2012], intactness of the cell membrane [Magalhaes et al., 2009], and interactions of cells in clusters [Tan et al., 2007]. Also demonstrated is the superiority of vitrification in the retention of viability and metabolic function in preserved cells, and proliferation and differentiation potential of 3D cultures of adult stem cells derived from a variety of sources [Kuleshova et al., 2004, 2007, 2009b, 2013; Kuleshova and Hutmacher, 2008; Tan et al., 2007; Bhakta et al., 2009; Wen et al., 2009; Wu et al., 2007, 2015]. Properly evaluated composition and concentration of

VSs and the cooling/warming rates do not impair the integrity or quality of the hydrogels involved, permitting free migration, aggregation, or proliferation of cells post-vitrification [Kuleshova et al., 2004; Wu et al., 2007; Bhakta et al., 2009]. The subsequent goal is to develop methods of cryopreservation for TECs made of different biodegradable biomaterials with varied geometry.

In summary, an encouraging start strongly supports the view that vitrification concepts for cryopreserving TECs have great potential for tissue engineering and will have a high impact on the application of tissue-engineered implants and novel stem cell–containing devices.

11.9 Organ Preservation

Organ cryopreservation, and tissue banking (Section 11.11), has developed in response to increased demands for tissue. The cryopreservation of organs is a major challenge due to the sensitivity of living organs and the complexity of these tissues. A study of naturally occurring freeze coping strategies has informed researchers about the techniques that may be applied to tissues and organs to achieve vitality after freezing. Kidneys were the first organ to survive cooling and rewarming with cooling to room temperature (1955), and then in 1958, cooling to refrigerated temperatures of 2°C–8°C (1958) [Owens et al., 1955; Stueber et al., 1958]. From here the field has progressed through rigorous research and trial from simple hypothermic storage in specialized solutions through to continuous machine perfusion techniques (Gallinat et al., 2013).

11.9.1 Solutions Used for Short-Term Preservation

Metabolic processes, particularly mitochondrial metabolism and energy production, are vital in maintaining homeostasis in cells [Hoffmann and Minor, 2014]. Metabolic damage can occur through ischemic injury due to low oxygen concentration, cell starvation due to decreased availability of nutrients during storage, accumulation of toxic products through metabolism or from storage solutions, loss of necessary substrates during perfusion, and simple cold damage [Pegg et al., 1981, 1984].

Preservation solutions are used for cold static storage or pulsatile perfusion. Early perfusates were used to flush kidneys for static cold storage. These included Sacks II [Sacks et al., 1973; Beck, 1979], Collins C3 [Beck, 1979], TP-II [Toledo-Pereyra, 1983], and hypertonic citrate [Marshall et al., 1977]. For successful cryopreservation of living organs, preservation solutions include sugars, osmotic agents, and/or buffers to minimize the damage caused by swelling or contraction of the cells. Solubility of oxygen (O_2) during cooling reduces O_2 availability. Ischemia or low O_2 leads to the production of free radicals, which prompt an increased immune response on transplantation and re-establishment of the blood supply in the host, termed reperfusion. Cell membranes may fail due to lipid peroxidation leading to failure of the graft [Suong-Hyu Hyon, 2011]. The inclusion of antioxidants will soak up free radicals [Pegg et al., 1984; Rolles et al., 1984, 1989]. Colloids must also be added to support the vascular bed [Pegg et al., 1984]. Immune reactions are controlled through the use of immunosuppressant treatments. The first solution to be effectively applied to organ (kidney) storage was Eurocollins [Collins et al., 1969; Eurotransplant Foundation Annual Report, 1976]. Eurocollins enables matching of extracellular and intracellular electrolytes, both having high potassium and phosphate, to ensure isotonicity and iso-osmolality [Mühlbacher et al., 1999]. Marshall's solution improved on this with the inclusion of citrate as a buffer to reduce swelling and enhance energy availability [Marshall, 1997].

In 1986, Jim Southard et al. developed University of Wisconsin (UW) solution, commercially marketed now as "Viaspan." Included were raffinose and lactobionate to maintain osmotic balance and reduce hypothermic cell swelling; a colloid carrier (hydroxyethyl-starch [HES]) to reduce vascular damage; and antioxidants (glutathione, allopurinol, and adenosine). UW solution was developed for pancreatic cryopreservation but has since been widely applied, most prominently for the liver and kidney [Southard and Belzer, 1993, 1995].

More recently, "Unisol-CV," a proprietary CPA medium similar in composition to CPTES, has been shown to be superior to Eurocollins solution as a CPA vehicle solution [Taylor et al., 2001]. "Custodiol-N" builds on increased knowledge of human physiology to include a wider range of amino acids and iron chelators to combat cold-induced injury to cells and hypoxic ischemia. Early results suggest that Custodiol-N improves graft quality through injury reduction [Rauen et al., 2008; Bahde et al., 2008; Stegemann et al., 2010].

11.9.2 Organ Perfusion

In a cold-tolerant organism, a pulse rate is maintained even at very low temperatures. Observation of continued circulation at low temperatures in the animal world led to the exploration of organ perfusion techniques. Perfusion may be used to flush the existing blood or fluids from an organ, to remove wash solutions and introduce preservation solutions, or for continuous perfusion such as used during brief transport [Pegg, 1981]. Organ perfusion must replace blood flow with storage solutions to deliver solutes and oxygen while maintaining osmotic balance. Perfusion solutions must be pure and uncontaminated prior to use [Pegg, 1981]. UW solution is currently popular for continuous organ perfusion [Kosieradzki et al., 1999]. In early research, the primary choice for perfusate was cryoprecipitated plasma [Sterling et al., 1971; Beck, 1979; Veller et al., 1994; Matsuno et al., 1994; Marshall, 1997]. Later studies used plasma protein fraction [Toledo-Pereyra, 1983; Alijani et al., 1985], solutions containing 5% albumin [Halloran and Aprile, 1987], silica-gel plasma perfusate [Mozes et al., 1985; Merion et al., 1990], and plasmanate [Halloran and Aprile, 1987; Jaffers and Banowsky, 1989; van der Vliet et al., 2001]. There are various issues that must be addressed in choice of perfusate, such as cation and anion content; osmolarity; pH; pressure; temperature; filtration; and substrate inclusions, such as oxygen, to support any continual metabolism in process during perfusion [Pegg, 1981].

Perfusion may be single-pass or continuous/pulsatile using a variety of pumps that may be peristaltic, reciprocating, or continuous [Pegg, 1981]. Pulsatile perfusion allows increased storage times and an estimated 20% reduction in the incidence of delayed graft function, giving an opportunity to perform viability testing for the organ and ensure medical clearance, thus expanding the pool of potential donors to include clinically marginal donors [Wight et al., 2003]. Delayed graft function is costly due to increased hospital stays and the need for continued dialysis and is linked with poorer long-term outcomes [Cecka and Terasaki, 1995]. Belzer, a leading pioneer in organ preservation science, was the first to develop a machine for the perfusion of kidneys at low temperatures during transport. However, in the late 1970s to mid-1980s several studies showed no improvement in machine perfusion over static storage, and consequently, largely for financial reasons, static storage dominated [Clark et al., 1974; Opelz and Terasaki, 1976, 1982; van der Vliet et al., 1983]. The use of pulsatile perfusion machines is reported at only 27% in the United States and 7% in the UK [Wight et al., 2003]. The most frequently used machine for kidney perfusion is the Waters MOX 100 [Wight et al., 2003], although also in clinical use are the Belzer LI 400 [Sterling et al., 1971; Beck, 1979], the Gambro [Halloran and Aprile, 1987; Marshall et al., 1977; van der Vliet et al., 2001], and the Nikiso APS-02 [Matsuno et al., 1994; Wight et al., 2003].

Hoffmann and Minor estimates that 33%–50% of organs received for clinical use are from marginal donors. Consequently, many organs retrieved do not function fully, with estimates of only 20%–40% of kidneys functioning adequately without requiring dialysis support [Hoffmann and Minor, 2014]. While clinical use remains focused on static storage, most current research is on continuous pulsatile perfusion, so strong is the evidence that this will ultimately provide the best solution for longer-term storage and transport of organs for transplantation. The overall progress of hypothermic machine perfusion preservation technologies and multiple organ perfusion *in situ* for clinical use have been comprehensively described by Taylor and colleagues [Taylor and Baicu, 2010].

Currently, an increased interest in matching physiological conditions for storage of tissue grafts has led to heightened interest in normothermic storage of tissues. As discussed in Section 11.10.4, corneas are currently banked at normothermic temperatures and articular cartilage may also benefit from natural temperatures for storage. No hypothermic damage is incurred in the tissue and liver storage has shown 83% success after 20-h storage where cold storage gave no positive outcomes [Brockmann et al., 2009; Jamieson et al., 2011; op den Dries et al., 2013]. There is a good deal of encouraging optimism in the field of organ transplantation, and increased funding is helping support the progress toward an effective solution.

11.10 Contribution of Cryobiology to *In Vitro* Toxicology and Drug Pharmaceutical Development

Exploitation of *in vitro* cell culture systems has proven to be essential in a number of studies of cell biological and physiological processes for over a century. Cryobiology is a vital part in *in vitro* toxicology and pharmaceutical drug development, since the use of cryopreserved primary cells with later reintroduction into culture is a valuable approach to study the problems of clinical relevance, especially those related to studies of cell toxicity and disease screening. However, as with any tool, it is subject to some limitations. For instance, the isolated hepatocyte is particularly convenient for studying the kinetics of hepatic drug uptake and excretion because the hepatocytes can be rapidly taken from the culture in an incubator. Since they show various features of the intact liver, isolated liver cells have also proved valuable for investigating drug metabolism. Yet, they also show important differences such as loss of membrane specialization and some degree of cell polarity. Surplus hepatocytes in small quantities are relatively easy to obtain. Therefore, tissue-engineered 2D or 3D constructs involving primary hepatocytes are widely appreciated as they sustain liver properties for an extended time, up to 2 weeks. Furthermore, a vitrification strategy has been developed by us that ensures long-term performance of cryopreserved units [Kuleshova et al., 2004; Wu et al., 2007; Magalhaes et al., 2008, 2012], paving the way for the use of hepatocyte- and other primary cell-based tissue-engineered constructs for *in vitro* drugs testing or as an integral part of *in vitro* models containing interconnected cultures for metabolic studies.

Cerebral organoids or mini brains are more advanced tissue-engineered systems that have the potential to model development and neurodegenerative conditions [Lancaster et al., 2013]. Alternatively, cerebral organoids can be used to grow specialized regions of brain tissue intended for their transplantation into areas of neurodegeneration as therapeutic treatment. The goal of cryobiology is to develop strategies of preservation for both applications.

In context of native tissue intended for physiological testing, metabolism studies, and drug screening, cryopreservation of cerebral and liver tissues has been of long-term interest. A vitrification strategy involving an elegant cassette design for liver slices in a number of species was suggested in the mid-1990s [Ekins et al., 1996]. Microscopic examination of rodent brain slices showed generally good to excellent ultrastructural and histological preservation after vitrification. Severe damage in frozen-thawed central villous explants has been reported as an approach to explore drugs in pregnancy disorders [Huppertz et al., 2011]. The study focused on traditional CPA, specifically Me_2SO, and found that a concentration of 3 M is best for maintaining explant viability, morphological integrity, and protein release during cryopreservation. The tested parameters were similar between controls and samples that were cryopreserved, placed on cryostorage, and transported to other locations, demonstrating the possibility of cryostoring explants and the logistics for functional studies.

Higher species require much more consideration for ethical approval. Even though it is possible to assess drugs in rodents *in vivo*, it is considered to be less applicable information. Consequently, cryopreserved rabbit, dog, swine, and canine vascular tissues are often used for drug testing. Further advances in the vitrification strategy involving vascular tissue have been undertaken. Improved tissue functions in vascular tissues cryopreserved using a vitrification approach were observed compared to a standard freezing method. Functional recovery of veins and arteries are usually evaluated by contractile responses and endothelium-independent relaxant responses post-thawing or post-vitrification. The maximum contractions achieved in vitrified vessels were >80% of fresh matched controls with similar drug sensitivities, whereas frozen blood vessels exhibited maximal contractions below 30% of controls and decreases in drug sensitivity [Song et al., 2000a, 2000b]. To date, there is difficulty finding an appropriate *in vitro* model to study human adult cardiac cell biology. A simple model to study the changes at the cellular level is in the culture of cardiomyocytes. Isolation and expansion of human cardiomyocyte progenitor cells from cardiac surgical waste or, alternatively, from fetal heart tissue is one option. Preservation is carried out based on outdated cryopreservation protocols utilized in hospital practice. Therefore, to overcome various issues related to progression of their cryopreservation is challenging. A couple of decades are a typical gap between clinical setting and developing the science of cryobiology. This is one reason why effective cryopreservation of cardiomyocytes has not been developed yet. Another option is emerging, that is, the vitrification of embryonic stem cells and induced pluripotent stem cells that are able to differentiate into cardiomyocytes *in vitro*. As yet, the whole process is still experimental. To conclude, the vitrification strategy is playing an increasingly important role in drug screening and *in vitro* toxicology studies.

11.11 Tissue Banking

In order to supply human tissue for medical treatments, it is necessary to establish tissue stocks. Biobanks exist as discrete units operating from acquisition, processing, and storage to on-demand supply of tissue grafts for surgical treatment and/or research purposes. Initial tissue banks were orthopedic banks supplying bone for surgical grafting [Strong, 2000]. Bone and skin grafts represent the most commonly used cryopreserved allografts in clinical use [Tomford and Mankin, 1999]. The science of transplantation has seen a rapid increase in application since the 1950s leading to an increased demand for tissues for transplantation. Although early reports exist of xenograft (1968) [Filipović-Zore, 2000] and autograft (1820)

[Chase and Herndon, 1955; Walther, 1821], the first peer-reviewed report of allografting was published by Sir William MacEwen on the reconstruction of the infected humerus in a 4-year-old child using the tibia of a child with rickets [MacEwen, 1881]. During World War II, a book by Inclan and articles by Wilson and Bush & Garber highlighted the demand for bone banks to supply stored bone to treat victims of war injuries [Inclan, 1942; Wilson, 1947; Bush and Garber, 1948]. The availability of tissue through the establishment of tissue banks has facilitated increased tissue transplantation, leading to a demand for rapid tissue supply.

Tissue and cell banking in its current form is tailored more closely to surgical requirements. Tissues routinely banked include bone, articular cartilage and osteochondral allografts, cornea, skin, tendon, heart valves, and vascular grafts utilizing hypothermic or ultra-low temperatures to maintain tissue stability.

11.11.1 Bone

Bone preserves as a natural scaffold that is biocompatible with the human body. It can be frozen in the absence of CPA without controlled rate cooling as the need to maintain cell viability is not a priority. A bone graft is selected for mechanical, osteoconductive, osteoinductive, or osteogenic properties. Osteoconductive materials function as a scaffold for the development of bone, while osteoinductive materials promote the recruitment of immature or progenitor cells and develop these down the osteoblast lineage to initiate new bone growth, termed osteogenesis [Albrektsson and Johansson, 2001].

Femoral head and shaft are commonly cryopreserved. Femoral heads are retrieved from living donors during hip replacement surgery or from deceased donors. Living donors are screened before surgery and again 4–6 months post-retrieval to ensure medical clearance, during which time the graft is stored fresh frozen in quarantine [Delloye et al., 2007]. Processing is undertaken for depletion of soft tissue, cellular content, and bone marrow, to remove donor-specific tissue that may prompt increased immune response or transmit infection. Irradiation of grafts from deceased donors is therefore routinely performed [Lavernia et al., 2004]. Removal of marrow and soft tissue components increases the osteoconductive capacity of the bone [Aspenberg, 1993]. Bone that has been freeze-dried and gamma-irradiated can be stored at room temperature. The primary use of cryopreserved femoral head allografts is cut or ground to aid osteogenic healing in knee or hip revision surgery. Grafts that have been stored fresh frozen, regardless of irradiation status, and grafts freeze-dried without gamma irradiation, are seen to be a reliable replacement of the fresh bone. However, freeze-dried, irradiated bone has reduced mechanical stability due to breakdown of collagen chains in response to ionization [Cornu et al., 2001; Dziedzic-Goclawska et al., 2005]. Freeze-drying has the effect of making bone more brittle unless rehydrated before use while fresh-frozen bone retains the mechanical properties of fresh bone [Delloye et al., 1991]. As result, freeze-dried bone tends to be applied to small defects, while frozen bone is used for larger repairs. Frozen bone allografts may typically be stored for up to 5 years at <–79°C and still be viable for use.

11.11.2 Articular Cartilage and Osteochondral Allografts

Articular cartilage varies from most other transplant tissues in that it is avascular, naturally hypoxic, has no lymphatic system, and relies on diffusion of nutrients and oxygen to nourish the cellular component. Autografts or allografts can be used with postoperative success at 5–10 years [Shasha et al., 2002]. Cartilage has a low cellular component, around 5%, and high water content, around 80%, making ice crystallization a major risk during cold preservation [Pegg et al., 2006c]. Cartilage is immunoprivileged with negligible immune response likely post-transplant; the main source of immune rejection of osteochondral grafts is directed toward subchondral bone.

The major limiting factors in the use of osteochondral grafts are the prohibitive costs of allografts and the availability of suitable tissue [Pegg, 2009]. Osteochondral grafts must be size-matched to the recipient, and retrospective studies show that graft contouring must match surrounding tissue for complete integration [Koh et al., 2006; Patil et al., 2008]. These factors and the need for undamaged replacement tissue increase the requirement for a long-term storage methodology for cartilage allografts.

Clinical cartilage storage protocols currently specify refrigeration at temperatures ~4°C. After 28 days of hypothermic storage, cell survival is detrimentally affected [Allen et al., 2005]. These grafts show sustained clinical success [Shasha et al., 2002]; however, cell viability and functionality are variable and it still remains preferential to implant fresh stored grafts within 14 days of removal from the donor [Allen et al., 2005]. Normothermic storage (~34°C) promotes increased functionality and cell viability in comparison to hypothermic storage [Pallante et al., 2009; Stoker et al., 2012]; however, the risk of bacterial infection increases. At hypothermic temperatures, enzyme function may be inhibited and the cellular sodium pump inactivated. While beneficial for short periods, long-term effects may include swelling due to water osmotically entering the cell through the membrane. An attempt to mimic the graft physiological environment allows the tissue to continue metabolic activity and cell function; however, metabolic activity results in the production of waste products such that it becomes necessary to refresh medium under normothermic conditions. Storage at normothermic conditions maintains chondrocyte vitality preferentially to bone cells, which may be beneficial in reducing immune responses [Bastian et al., 2011].

Standard preservation techniques at <–2°C cause the formation of both extracellular and intracellular ice crystals in cartilage chondrons [Pegg et al., 2006b]. Crystallization of ice is both directly and indirectly causal in the fatal damage sustained by cells during cryopreservation and consequently must be inhibited to ensure cell survival post-cryopreservation [Mazur, 1984; Pegg et al., 2006a]. Most studies in this field are examining the use of vitrification when applied to cartilage to facilitate a consistent supply of appropriate allografts.

The lack of vascularity and density of the extracellular matrix (ECM) network of collagen fibrils in cartilage makes CPA penetration and inhibition of ice formation a particular challenge [Pegg et al., 2006c]. CPAs inhibit ice crystallization; however, the toxicity of CPAs may severely limit successful cryopreservation [Pegg et al., 2006c]. As ice crystals accumulate, fissures, cracks, and holes form in the tissue, reducing biomechanical function. The vitrification method for storage of tissues is discussed in Sections 11.3 and 11.4. Chondrocytes serve to maintain matrix formation by synthesizing components of matrix structure, such as collagen and proteoglycans. It is therefore important to measure the functionality of cartilage, commonly through uptake of radiolabeled sulfate, in addition to cell viability through membrane integrity and measurement of metabolic activity [Pegg et al., 2006c].

Osteochondral plugs and cartilage slices have been successfully vitrified with post-vitrification cell survival quoted above 80% [Brockbank et al., 2010]. Cartilage is exposed to high-concentration CPA in a preconditioning step to allow diffusion of CPA into the tissue, which is then rapidly cooled (e.g., –43°C/min) to below the glass transition point to avoid prolonged exposure to toxic CPA. Post-vitrification, following a slow warming step to –100°C, tissue is rapidly warmed by placing specimens in a water bath at room temperature until tissue and solution begin to soften, to avoid devitrification and recrystallization of ice [Jomha et al., 2012]. The standard 5-mL vial used for tissue provides an estimated average warming rate of ~–100°C/min. High cooling rates as well as warming rates (typically up to 250°C/min) are achievable in small samples of cartilage; however, accurate control of temperature in a large allograft is difficult to achieve. Cartilage without subchondral bone has little clinical application as grafting is not possible. In the repair of large lesions, the orthopedic surgeon measures the contouring of the graft and cuts an osteochondral graft that approximates the size and curvature of the lesion site. Whole or hemi-condyles are necessary to achieve this, while osteochondral dowels have limited clinical application. LT targets the challenge of producing a GMP-compliant, clinically applicable method for preservation of cartilage utilizing a procedure shown to maintain cell function, based on a radioactive sulfate uptake assay demonstrating incorporation of sulfate in gag synthesis, at levels of 75%–95% in ovine articular cartilage and >70% in human tissue slices [Wang et al., 2007].

11.11.3 Tendon and Meniscus

Tendon or ligament grafts are generally from an autologous source using tissue harvested at the time of repair. A revision surgery where damaged tissue occurs in commonly harvested regions, or to avoid donor site morbidity, promotes the use of frozen or cryopreserved allograft tendons [Suhodolčan et al., 2013]. Around 20% of tendon repairs in the United States utilize stored allografts [Buchmann et al., 2008; Cohen and Sekiya, 2007; Mascarenhas et al., 2010] with the most commonly used allografts being Achilles, patellar, and semitendinosus tendons. Commercial sources of fascia lata, rotator cuff, tibialis posterior, tibialis anterior, and gracilis grafts are available [Robertson et al., 2006].

Tendon and meniscus allografts are typically deep frozen in gauze or in saline with antibiotics [Ochi et al., 1995] without prior chemical treatment or CPA. Storage is initially at −80°C, and tendons may undergo freeze-thaw cycles for transit, inventory, and short- and long-term storage [Sterling et al., 1995]. For long-term storage, grafts can be adequately stored long term at −80°C [Robertson et al., 2006; Wascher et al., 1999]. Cryopreservation and freeze-drying are alternate storage methods for tendons and meniscus [Arnoczky et al., 1998; Jackson and Simon, 1992; Robertson et al., 2006]. Cryopreservation of meniscus usually uses glycerol or Me_2SO as a cryoprotective agent and ultimately stores in liquid nitrogen at −196°C [Milton et al., 1990; Mickiewicz et al., 2013]. Cryopreserved meniscus retains between 4% and 54% viability in cells [Gelber et al., 2009; Milton et al., 1990].

11.11.4 Cornea

Corneal transplant is common for repair of many disorders associated with corneal blindness such as keratoconus and corneal scarring. Corneal disease is the most common cause of blindness [Wilson, 1980] but can be treated with restoration of sight using corneal graft transplantation. As with cartilage allografts, corneal transplantation provokes little immune response and therefore makes an ideal tissue for allografting. In the United States, there were 46,196 corneal transplants in 2011 with over 95% successful at restoring vision, and in the United Kingdom in 2013–2014, 5440 cornea were issued and 3313 graft procedures completed, with 91%–94% survival of the graft 1 year post-transplantation. At 4°C, corneas are stored in a hydroxyethyl piperazineethanesulfonic acid (HEPES)–buffered tissue culture medium supplemented with chondroitin sulfate, dextran, and standard antibiotics [Lass et al., 1992], allowing cornea to be stored for 14 days to facilitate testing and organization of the surgical procedure [Chu, 2000; Naor et al., 2002]. Storage at 4°C has been shown to disrupt the F-actin cytoskeleton and tight junctions, which could have the knock-on effect of increasing corneal swelling [Hsu et al., 1999], while storage at organ culture temperatures does not present this issue [Crewe and Armitage, 2001]. Corneal grafts in the UK are stored at 34°C for a period of up to 28 days [Armitage and Easty, 1997; Ehlers et al., 1999]. The availability of corneal grafts and increased storage has allowed grafting surgery to become elective and scheduled instead of emergency transplantation [Chu, 2000]. Initial tests using cryopreserved cornea to extend the storage period has prompted loss of viable endothelial cells and reduced success in grafting [Van Horn et al., 1970; Bourne, 1978].

11.11.5 Skin and Amniotic Membrane

Human dermis forms the underlying matrix of skin comprising collagen, elastin, and glycos-aminoglycans (GAGs) synthesized by fibroblasts. The upper stratified epidermis has a rapidly regenerating avascular epithelium barrier surface populated predominantly by keratinocytes. Skin can be retrieved from deceased donors and cryopreserved for allografting. The primary uses of banked skin are for treating burns, injuries, and skin conditions such as ulcers. Open or full-thickness wounds, where both the dermis and the epithelium are lost, benefit from treatment with skin allografts. Without the barrier protection of the epithelium, wounds are prone to infection and dehydration leading to necrosis. Equally, without the dermis full-thickness skin cannot easily regenerate. Burn victims experience a great deal of pain and potentially hypothermia, symptoms ameliorated by the application of skin allografts. Skin grafts may come from an autologous or allogeneic source. A full-thickness skin graft will include epidermal and dermal layers with subcutaneous base tissue present, while partial-thickness, or "split-thickness," grafts contain epidermis and a limited thickness of dermal base. Modern grafts may also comprise bioengineered cell scaffolds populated with keratinocytes retrieved from the host or an allogeneic donor, bioengineered artificial skin, or decellularized dermis [Badiavas et al., 2002]. In severe burn victims, a combination approach will be taken with surviving autologous donation sites harvested for autologous grafts that will not be rejected, while allografts are used to protect wound regions for short-term recovery of damaged and undamaged host skin for further graft harvesting [Hermans, 2011].

Skin grafts are commonly stored for 10 days at hypothermic temperatures in saline solutions, prompting reductions in metabolic activity and oxygen consumption while maintaining graft integrity [Knapik et al., 2013]. Skin cryopreservation is performed using controlled rate freezing, with variable pattern cooling in Me$_2$SO. Storage options using glycerol or lyophilization do not maintain cell viability. Reviews of clinical studies suggest that there is no requirement for living cells for a skin graft to function and allow epithelial repair [Hermans, 2011; Kagan, 1998; Bravo et al., 2000] and decellularized skin is in clinical use, although it may be that growth factors and signalling proteins secreted by living cells in the graft aid healing [Mansbridge, 2008].

Studies remain divided on the effects of cryopreservation on skin. Biomechanical properties may be altered by cryopreservation with Me$_2$SO, while Young's modulus may remain unchanged [Wood et al., 2014], suggesting that cryopreservation may produce a suboptimal graft. However, clinical results with cryopreserved grafts have been positive, and the benefit of maintaining stocks of allografts without resort to fresh-only grafts may be sufficient to offset the damage caused by cryopreservation.

11.11.6 Heart Valves

Valves for use in replacement surgery are broadly divided into mechanical heart valves (MHV) and bioprosthetic heart valves (BHV). The trend in the United States and Europe has been toward greater use of tissue rather than mechanical valves [Singhal et al., 2013]. Today, the most commonly used BHVs are those from human cadavers (homograft), porcine aortic valves, and calf pericardium [Siddiqui et al., 2009]. Human heart valves have been stored and utilized as allografts since 1962 [O'Brien et al., 2001]. The success of heart valve allografts has previously involved the presence of a population of viable cells within the grafts at the time of transplantation, requiring sensitive cryopreservation protocols to address this need. Consequently, relatively elaborate preservation protocols were developed to deliver final post-freeze temperatures of around –150°C or storage in LN₂ while retaining cell viability. More recently, however, acellular heart valves have been successfully applied in heart valve replacements, demonstrating that allograft scaffolds created from decellularized valves with no living cell component or DNA content function extremely well as transplanted allografts with host cells repopulating the acellular graft and returning to normal function and maintenance [da Costa et al., 2010; Konuma et al., 2009; Iop et al., 2014]. Decellularization is generally undertaken using detergent washes (e.g., SDS/SDC), enzyme digestion with detergent wash (e.g., Triton with EDTA and enzymes RNase and DNase), or purely enzymatic washing (e.g., trypsin, RNase, DNase) [Khorramirouz et al., 2014].

The introduction of acellular valve allografts greatly simplifies the cryopreservation procedures. The traditional method of cryopreservation involved dissection of the heart followed by immersion in nutrient medium with CPA. This was then cooled in a stepwise procedure at fixed rates with the protocol in use dependent upon the tissue bank [Birtsas and Armitage, 2005; Armitage et al., 2005]. Specific protocols are developed by individual banks and patented for commercial protection. For example, Cryolife, Inc., patented protocol from warming using –0.01°C/min to 4°C; –1.5°C/min to –3°C; –95°C/min to –140°C; holding at –140°C for 1 min; 20°C/min to 100°C; holding at 100°C for 6 min; 10°C/min to –70°C; 20°C/min to –26°C; holding at –26°C for 2 min; –1°C/min to –80°C, then transfer to LN₂ for long-term storage, with subsequent warming protocol comprising a substantial thaw step in a sterile saline water bath between 37 and 42°C, then removal of CPA through dilutions of CPA at 7.5% and 5% for 1 min each prior to transfer to nutrient medium. This example protocol estimates post-thaw retained viability at >70% of fresh control tissue [McNally et al. 1987, US Patent 4,890,457, 1990].

A consistent study of freeze-drying bovine pericardium suggests that treatment of tissue with chemical substances appears to prevent harmful calcification of the matrix [Aimoli et al., 2007]. Recent work of this group reviewed the reduction of the inflammatory effect post-implantation while no anticipated reduction of calcification after lyophilization was found [Maizato et al., 2013]. Without the requirement for a retained living cell component, decellularized heart valves can be refrigerated in glycerol or sucrose, cryopreserved using vitrification techniques as discussed earlier, or freeze-dried [Aimoli et al., 2007; Maizato et al., 2013]. Yet, cryopreserved and stored homografts at vapor phase of liquid nitrogen have been and still remain the gold standard.

11.11.7 Tissue Banking: Summary

The notable feature of most banked tissues is the ability to graft the tissue without the inclusion of viable cells, or with simple tissues where cell viability is easy to maintain through cryopreservation protocols. Each tissue type requires individual, distinct cryopreservation protocols with various options for cooling rates, CPA choice and concentration, manual or automated cryopreservation, and storage temperatures and durations [Wusteman and Hunt, 2004]. Equally, if not more importantly, protocols have to be defined for the warming of the tissues that will not allow further damage to be inflicted, and again these protocols will be specific to the tissue undergoing cryopreservation. The cryopreservation of whole, complex organs remains an elusive challenge discussed separately.

11.12 Cryosurgery

Low temperatures may be applied medically to achieve the destruction of cells, notably tumors. This is termed cryoablation, cryotherapy, or cryosurgery. Cryosurgery is an effective method for the removal of damaging or diseased tissue without incurring the injury that invasive resectional surgery would cause as cryosurgery is a noninvasive procedure [Zhao and Chua, 2013]. Cryotherapy procedures use freezing to destroy tissue, incidentally providing anesthetic effect and avoiding bleeding, but tissue remains *in situ* to be resorbed by the body rather than being excised [Onik, 1996]. Cryotherapy may target partial or complete ablation of tissue, depending upon the nature and likelihood of spread, particularly in cancers. An ablative procedure that removes an entire tissue, organ, or gland is defined by the critical temperature (critical isotherm protocol, CIP) commonly addressed as reliable at approximately −40°C or 233 K [Mazur, 1984; Bischof et al., 1997; Hoffmann and Bischof, 2002; Desai and Gill, 2002; Zhao and Chua, 2013], however also cited as higher for some tissues (e.g., renal tissue at −19.4°C [Desai and Gill, 2002]); and the duration or cycles of application, and monitoring of the "freeze front" as this CIP temperature progresses through tissue is key to successful surgery [Rewcastle et al., 1998; Rubinsky, 2000; Otten and Rubinsky, 2000; Edd and Rubinsky 2006].

11.12.1 Development and Applications

"Low temperature" cold compresses were used in Egypt (circa 2500 BC) to reduce infection, inflammation, and bleeding and to treat acute injuries such as skull fractures. Ice as a treatment is also recorded by Hippocrates (circa 460–370 BC) [Bleakley, 2013] and later by Dominique-Jean Larrey in local anesthetic prior to limb amputation [Larrey, 1832; Skandalakis et al., 2006]. Dr. James Arnott applied low temperature treatment to cancers to reduce inflammation and potentially damage cancer cells applying ice, crushed with saline solutions at a temperature of −18°C or lower, as treatment for breast, cervical, and skin tumors [Arnott, 1850, 1851]. Liquid air, temperature −190°C, was used clinically for the treatment of skin cancers in 1889 by Dr. Campbell White in New York [White, 1899] and later in 1907 by Whitehouse, who published the positive clinical outcomes from 15 skin cancers treated with spray or swabbing with liquid air [Whitehouse, 1907]. Pusey used solid carbon dioxide in the form of "carbon dioxide snow" to treat epithelial cancers [Pusey, 1907]. In the same year, Bowen and Towle reported the use of liquid air for the treatment of vascular lesions [Bowen and Towle, 1907]. Major developments in cryotherapy followed on from technological improvements, notably in the ability to produce liquid gases at extremely low temperatures (air, oxygen, and nitrogen) and the development of the dewar flask for transport of cryogens.

In 1910, solid carbon dioxide (~−78.5°C) became popular for cryotherapy on multiple conditions mainly at the skin surface, but also tumors of the bladder [Bracco, 1990], overtaking the use of liquid air and outlasting the use of liquid oxygen (−182.9°C). Solid carbon dioxide was in clinical use for the treatment of multiple conditions mainly on the skin surface, but also for tumors of the bladder [Bracco, 1990]. Liquid nitrogen (LN_2) was first applied in clinical treatments in 1950 by Dr. Ray Allington for the treatment of noncancerous skin disorders [Allington, 1950] and is currently the predominant means for delivering low temperatures for cryopreservation, storage, and cryotherapy. The ability to accurately apply LN_2 to a localized region enabled the treatment of small surface defects. In 1938, hollow surgical instruments to deliver ice and saline for the treatment of tumors were designed and built by Temple Fay and George Henny [Fay and Henny, 1938]. Temple Fay reported the observation that cancers preferentially developed in regions of higher body temperature and less at lower temperatures such as in limbs, reinforcing the efficacy of cryosurgery [Alzaga, 2009; Henderson and Fay, 1963; Smith and Fay, 1939].

Early methods for the application of liquid nitrogen to the skin focused on the use of cotton buds as a cryogen carrier. In 1961, Irving Cooper and Arnold Lee designed a cryosurgical probe circulating LN_2 under pressure through the center of the cannula to the tip for cooling. The central conduit is surrounded by a space that returns nitrogen vapor for removal while the inner layers are encased in a vacuum insulation layer for cold retention [Cooper and Lee, 1961]. This system was converted into a sealed circuit, or closed system, by Douglas Torre in 1965, enabling safer, aseptic use of the system on skin disorders, including carcinomas [Torre, 1968]. These developments allowed the eventual introduction of a commercial, handheld cryosurgery probe developed by Setrag Zacarian and Michael Byrne [Zacarian, 1973]. Current cryoprobes use either liquid nitrogen or argon gas to cool the tip. Cryosurgery probes apply a single freeze or repeated cycles of freeze-thaw to tissue to promote the intracellular crystallization of ice to inflict extensive damage to cells, discussed earlier.

Cryosprays apply aerosol liquid nitrogen to a treatment site such as brain tumors or the eye condition trichiasis [Hamlin, 1969, 1971; Fraunfelder and Petursson, 1979]. The application of cryosprays has developed, most recently with the potential for the treatment of esophageal cancers, discussed later [Cash et al., 2007; Dumot et al., 2009]. Cryosprays used for skin disorders were shown to be as effective as earlier cotton bud methods [Ahmed et al., 2001].

The major limitations in the use of cryosurgery had been the inability to fully monitor the extent or severity of freezing and the lack of precise control of the freeze region [Rubinsky, 2000]. A solution to the problem of monitoring treatment and damage was described in the 1980s with the development of an ultrasound system that can be operated during surgery to positively locate the precise location of the tumor or lesion, monitor the insertion of the probe, and then track the progress of the ice front as the tissue freezes. This enables the surgeon to accurately freeze tissue that is being targeted for removal. This system was initially demonstrated in liver and prostate cancers [Onik et al., 1988]. Ultrasound was a cheap and convenient method for tracking cryosurgery [Onik et al., 1991]. The moving front of ice formed during cryosurgery can be monitored due to the differing velocities of sound waves as they travel through either ice or water in tissues [Rubinsky, 2000]. Ultrasound is limited in penetration depth and the clarity of the image produced, particularly through frozen regions where resolution is impossible. For cryosurgical treatment of potentially recurrent or metastatic disorders such as in cancer, it is important to fully remove the malignant tissue without residual remnants [Edd and Rubinsky, 2006]. With variable size and shape lesions, accurate monitoring and control are vital [Rubinsky, 2000]. The development of a vacuum-insulated probe of small diameter that could be accurately placed into a small region increased the versatility of cryosurgical techniques for treating unusual shapes [Baust and Chang, 1995]. Various imaging systems can be applied to monitoring the movement of a freeze front within tissue. These systems include magnetic resonance imaging (MRI) and nuclear magnetic resonance (NMR), computerized tomography scanning (CT), and electrical impedance tomography (EIT) that all examine the topography of an imaged region to give accurate representations of a treatment area. MRI or NMR produces a 3D image of the region by applying alternating magnetic force that exerts upon protons in the tissue causing deflection and relaxation. The deflection occurs under the influence of magnetic forces, while the relaxation follows administration and varies greatly between frozen and unfrozen regions. There are multiple forms of MRI related to the rate of imaging and the orientation at which images are taken, but all forms are applicable to monitoring the movement of an ice front during cryosurgery [Rubinsky, 2000]. The CT scan is a form of x-ray scan with the benefit of developing a 3D image that includes details of soft tissues within the body. EIT measures the impedance of electrical signals through tissue, with impedance significantly increasing in frozen tissue [Otten and Rubinsky, 2000; Edd and Rubinsky, 2006]. EIT allows the surgeon to evaluate pre- and post-surgery the volume of tissue affected by the cryotherapy, and therefore assess whether sufficient tissue has been treated to ensure complete eradication of malignant tissue without excess damage to surrounding tissues [Edd and Rubinsky, 2006]. EIT has the benefit that, as cell death occurs and membrane permeability increases, impedance drops, therefore enabling this method to measure directly the loss of cell viability in a region undergoing cryotherapy [Edd and Rubinsky, 2006]. To further control the loss of cell viability due to over-extension of the freeze front, heat may be applied to healthy tissues to counteract freezing by impeding the cooling function of the cryoprobe in the region of targeted healthy tissue. One example of this would be insulation or proactive rectal warming or urethral warming during cryoablation of the prostate, where cell vitality is adversely affected in the anterior wall of the rectum or sloughing experienced in the urethra owing to cryoablation of the prostate. Warming of the rectal tissue protects against the progression of the ice front and prevents damage to healthy tissue during ablation of the cancerous gland, while a urethral warming catheter prevents lethal effects on cells by maintaining temperatures above the freezing point [Chen and Pu, 2014; Favazza et al., 2014; Bischof et al., 1997]. This type of tissue warming technique may be applied elsewhere to elicit a similar defense.

Examples of external conditions that could be treated with cryotherapy include UV-related disorders such as skin cancers and solar keratosis; interdigital, or Morton's, neuroma; and warts, moles, freckles, and skin tags. Internal conditions treated using cryosurgery are extensive, but include multiple cancers such as prostate, liver, cervical, lung, retinal, and oral cancers; fibroma; and plantar fasciitis, although only in severe cases and less commonly now as alternative treatments predominate. In both external and internal treatments, the biggest impact in terms of patient benefit is on the treatment of tumors.

11.12.2 Cryosurgery in Cancer Treatment

Treatment of cancerous tumors using cryosurgery was minimally used prior to the 1990s and introduction of ultrasound monitoring systems. Tumors of the skin may be treated distinctly from tumors in internal locations, as skin cancers may be treated through topical application of liquid nitrogen or probes, whereas accessing and freezing internal lesions must be more precise to avoid detrimental damage to surrounding tissues. Cryosprays may be used for skin cancers and have recently been investigated for esophageal cancer [Cash et al., 2007; Dumot et al., 2009]. Associated with cancer therapies, cryosprays have also been useful in treating chronic radiation proctitis, which occurs as a side effect following radiation therapy for colorectal, prostate, or gynecological cancers. Standard treatments to remove the affected mucosa typically cause tissue damage, whereas cryoablation is able to remove the mucosa without deeper tissue damage [Hou et al., 2011]. Mechanisms for targeting cancer tissue with cryotherapy are described in Section 11.4.

Cryosurgical probes are less useful for metastatic tumors. Cryosurgery can be applied as a treatment for large solid tumors or localized disorders. Initially it was believed that

–20°C was sufficiently low to prompt cryolysis in any cell [Cooper, 1964] based mainly on frostbite investigations and animal experimentation, not human *in vivo* testing [Gage and Baust, 1998].

11.13 Conclusion

In this chapter, we have attempted a brief overview of the medical applications of low-temperature biology. Medical cryobiology has been practiced across four millennia and across the span of scientific applications; an entire book could be dedicated to the subject without covering every aspect. What we have presented here reflects upon the current and previous state of the art for low temperature applied to medicine. What remains, and perhaps presents the most exciting and optimistic viewpoint, is what is yet to come in the field.

Cryobiology is a developing science, like so many disciplines in a broader field of biomedical engineering. The potential for low temperature biology to solve some of the key issues in medicine and effective therapies for hard-to-treat conditions, most notably, the ability to store living organs for transplantation, could revolutionize medicine, and cryobiology, in particular vitrification, offers a solution to this problem. Similarly, advances in complex tissue-engineered solutions for medical treatments will ultimately rely upon cryobiology to enable a consistent and reliable supply. Developments in biomaterials science will complement this process. Advances in cell therapies also require stocks of clinically suitable cells to be maintained, and biobanking is becoming an increasingly vital commodity to national health services. Pharmaceutical options require testing prior to clinical introduction, and at these trial phases, the application of cryobiology to sample storage is again instrumental in facilitating clinical introduction of novel therapeutic options.

In summary, the advancement of a discipline in science requires three developments. These are (i) the development of innovative technologies to facilitate existing concepts; (ii) the development of new concepts; and (iii) application of the fundamental sciences to existing problems. Many historical developments in cryobiology have been achieved by observation of the natural world and the lateral application of those observations to alternative problems. In the last seven decades, significant advances have been made in knowledge and technology. A deeper understanding has been achieved to give greater insight into physiological function, interactions, and mechanisms through research in fields related to, or dependent upon, cryobiology. This presents us with a wealth of new knowledge regarding how to treat complex medical issues. Many challenges remain to be overcome in low temperature biology, but with continuing dedication and interest from a new generation of researchers, and the wealth of knowledge already achieved in the field, these challenges will be met and overcome so that the discipline will fully achieve its potential.

Acknowledgments

The authors would like to thank Kenneth Diller, The University of Texas at Austin, TX; Brian Grout, University of Copenhagen, Denmark; and David Pegg, University of York, UK, for reviewing this chapter and for providing valuable technical comments and suggestions.

References

Ablin R.J. An appreciation and realization of the concept of cryoimmunology. In: Onik G., Rubinsky B., Watson G., Ablin R.J. (eds), *Percutaneous Prostate Cryoablation*. St. Louis, MO: Quality Medical Publishing, Inc. (1995) p. 136.

Ahmed I., Agarwal S., Ilchyshyn A., Charles-Holmes S., Berth-Jones J. Liquid nitrogen cryotherapy of common warts: Cryo-spray vs. cotton wool bud. *Br. J. Dermatol.* 144(5) (2001) pp. 1006–1009.

Aimoli C.G., Nogueira G.M., Nascimento L.S., Baceti A., Leirner A.A., Maizato M.J., Higa O.Z., Polakiewicz B., Pitombo R.N.M., Beppu M.M. Lyophilized bovine pericardium treated with a phenethylamine-diepoxide as an alternative to preventing calcification of cardiovascular bioprosthesis: Preliminary calcification results. *Artif. Organs.* 31 (2007) pp. 278–283.

Albrektsson T., Johansson C. Osteoinduction, osteoconduction and osseointegration. *Eur. Spine J.* 10(Suppl. 2) (2001) pp. 96–101 [Review].

Alijani M.R., Cutler J.A., DelValle C.J., Morres D.N., Fawzy A., Pechan B.W., Helfrich G.B. Single-donor cold storage versus machine perfusion in cadaver kidney preservation. *Transplantation.* 40(6) (1985) pp. 659–661.

Allen R.T., Robertson C.M., Pennock A.T., Bugbee W.D., Harwood F.L., Wong V.W., Chen A.C., Sah R.L., Amiel D. Analysis of stored osteochondral allografts at the time of surgical implantation. *Am. J. Sports Med.* 33(10) (2005) pp. 1479–1484.

Allington H.V. Liquid nitrogen in the treatment of skin diseases. *Calif. Med.* 72 (1950) pp. 153–155.

Alzaga A.G., Salazar G.A., Varon J. Resuscitation great. Breaking the thermal barrier: Dr. Temple Fay. *Resuscitation.* 69(3) (2006) pp. 359–364.

Armitage W.J., Dale W., Alexander E.A. Protocols for thawing and cryoprotectant dilution of heart valves. *Cryobiology.* 50(1) (2005) pp. 17–20.

Armitage W.J., Easty D.L. Factors influencing the suitability of organ-cultured corneas for transplantation. *Invest. Ophthalmol. Vis. Sci.* 38(1) (1997) pp. 16–24.

Arnoczky S.P., McDevitt C.A., Schmidt M.B., Mow V.C., Warren R.F. The effect of cryopreservation in canine menisci: A biochemical morphologic and biochemical evaluation. *J. Orthop. Res.* 6 (1988) pp. 1–12.

Arnott J. *On the Treatment of Cancer by the Regulated Application of an Anaesthetic Temperature.* London: Churchill (1851).

Arnott J. Practical illustrations of the remedial efficacy of a very low or anaesthetic temperature. I. In cancer. *Lancet* 2 (1850) pp. 257–259.

Aspenberg P. A new bone chamber used for measuring osteoconduction in rats. *Eur. J. Exp. Musculoskeletal Res.* 2 (1993) pp. 69–74.

Badiavas E.V., Paquette D., Carson P., Falanga V. Human chronic wounds treated with bioengineered skin: Histologic evidence of host-graft interactions. *J. Am. Acad. Dermatol.* 46(4) (2002) pp. 524–530.

Bahde R., Palmes D., Gemsa O., Minin E., Stratmann U., de Groot H., Rauen U., Spiegel U. Attenuated cold storage injury of rat livers using a modified HTK solution. *J. Surg. Res.* 146 (2008) pp. 49–56.

Bastian O., Pillay J., Alblas J., Leenen L., Koenderman L., Blokhuis T. Systemic inflammation and fracture healing. *J Leukoc Biol.* 89(5) (2011) pp. 669–673. doi: 10.1189/jlb.0810446. Epub 2011 Jan 4 [Review].

Baust J., Chang Z. Underlying mechanisms of damage and new concepts in cryosurgical instrumentation. In: Baust J., Chang Z. (eds), *Cryosurgery: Mechanism and Applications.* Paris: Institut International de Froid (1995) pp. 21–36.

Baust J.M., Van Buskirk, Baust J.G. Cell viability improves following inhibition of cryopreservation-induced apoptosis. *In Vitro Cell Dev Biol Anim.* 36(4) (2000) pp. 262–270.

Beck T.A. Machine versus cold storage preservation and TAN versus the energy charge as a predictor of graft function posttransplantation. *Transplant. Proc.* 11 (1979) pp. 459–464.

Benson J.D., Woods E.J., Walters E.M., Critser J.K. The cryobiology of spermatozoa. *Theriogenology.* 78(8) (2012) pp. 1682–1699.

Bhakta G., Lee K.H., Magalhaes R., Wen F., Gouk S.S., Hutmacher D.W., Kuleshova L.L. Cryoreservation of alginate-fibrin beads involving bone marrow derived mesenchymal stromal cells by vitrification. *Biomaterials.* 30 (2009) pp. 336–343.

Birtsas V., Armitage W.J. Heart valve cryopreservation: Protocol for addition of dimethyl sulphoxide and amelioration of putative amphotericin B toxicity. *Cryobiology.* 50(2) (2005) pp. 139–143.

Bischof J.C., Merry N., Hulbert J. Rectal protection during prostate cryosurgery: Design and characterization of an insulating probe. *Cryobiology.* 34 (1997) pp. 80–92.

Bleakley C.M. Acute soft tissue injury management: Past, present and future. *Phys. Ther. Sport.* 14(2) (2013) pp. 73–74. doi: 10.1016/j.ptsp.2013.01.002.

Bourne W.M. Penetrating keratoplasty with fresh and cryopreserved corneas. Donor endothelial cells survival in primates. *Arch. Ophthalmol.* 96 (1978) pp. 1073–1074.

Boutron P., Kaufman A. Stability of the amorphous state in the system water-1,2-propanediol. *Cryobiology.* 16 (1979) pp. 557–568.

Boutron P. Levo- and dextro-2,3-butanediol and their racemic mixtures: Very efficient solutes for vitrification. *Cryobiology.* 27 (1990) pp. 55–69.

Bowen J.T., Towle H.P. Liquid air in dermatology. *Med. Surg. J.* 157 (1907) p. 561.

Bracco D. The historic development of cryosurgery. *Clin. Dermatol.* 8(1) (1990) pp. 1–4.

Bravo D., Ridgley T.H., Gibran N., Strong D.M., Newman-Gage H. Effect of storage and preservation methods on viability in transplantable human skin allografts. *Burns.* 26 (2000) pp. 367–378.

Brockbank K.G., Chen Z.Z., Song Y.C. Vitrification of porcine articular cartilage. *Cryobiology.* 60(2) (2010) pp. 217–221.

Brockmann J.M., Reddy S.F., Coussios C.P., Pigott D.F., Guirriero D.M., Hughes D.P., Morovat A.P., Roy D.F., Winter L.M., Friend P.J.M. Normothermic perfusion: A new paradigm for organ preservation. *Ann. Surg.* 250 (2009) pp. 1–6.

Buchmann S., Musahl V., Imhoff A.B., Brucker P.U. Allografts for cruciate ligament reconstruction. *Orthopade.* 37 (2008) pp. 772–778.

Bush L.F., Garber C.Z. The bone bank. *JAMA.* 137(7) (1948) pp. 588–594.

Carsi B., Lopez-Lacomba J.L., Sanz J., Marco F., Lopez-Duran L. Cryoprotectant permeation through human articular cartilage. *Osteoarthritis Cartilage.* 12(10) (2004) pp. 787–792.

Cash B.D., Johnston L.R., Johnston M.H. Cryospray ablation (CSA) in the palliative treatment of squamous cell carcinoma of the esophagus. *World J. Surg. Oncol.* 5 (2007) p. 34.

Cecka J.M., Terasaki P.I. The UNOS scientific renal transplant registry. *Clin. Transplant.* 1 (1995) p. 1–18.

Cha S.K., Kim B.Y., Kim M.K., Kim Y.S., Lee W.S., Yoon T.K., Lee D.R. Effects of various combinations of cryoprotectants and cooling speed on the survival and further development of mouse oocytes after vitrification. *Clin. Exp. Reprod. Med.* 38(1) (2011) pp. 24–30.

Chase S.W., Herndon C.H. The fate of autogenous and homogenous bone grafts: A historical review. *J. Bone Joint Surg. Am.* 37(4) (1955) pp. 809–841.

Chen C.-H., Pu Y.-S. Proactive rectal warming during total-gland prostate cryoablation. *Cryobiology.* 68 (2014) pp. 431–435.

Chian R.C., Chao Y.X. Clinical evidence of oocyte vitrification. In: Borini A., Coticchio G. (eds), *Preservation of Human Oocytes.* London, UK: Informa Healthcare (2009) pp. 245–255.

Chu W. The past twenty-five years in eye banking. *Cornea.* 19(5) (2000) pp. 754–765 [Review].

Clark E.A., Terasaki P.I., Opelz G., Mickey M.R. Cadaver-kidney transplant failures at one month. *N. Engl. J. Med.* 291 (1974) pp. 1099–1102.

Cobo A., Diaz C. Clinical application of oocyte vitrification: A systematic review and meta-analysis of randomized controlled trials. *Fertil. Steril.* 96(2) (2011) pp. 277–285.

Cobo A., Meseguer M., Remohi J., Pellicer A. Use of cryo-banked oocytes in an ovum donation programme: A prospective, randomized, controlled, clinical trial. *Hum. Reprod.* 25 (2010) pp. 2239–2246.

Cohen S.B., Sekiya J.K. Allograft safety in anterior cruciate ligament reconstruction. *Clin. Sports Med.* 26 (2007) pp. 597–605.

Collins G.M., Bravo-Shugarman M., Terasaki P. Kidney preservation for transplantation. Initial perfusion and 30 hours' ice storage. *Lancet* 2 (1969) pp. 1219–1222.

Cooper I.S. Cryobiology as viewed by the surgeon. *Cryobiology.* 1 (1964) pp. 44–54.

Cooper I., Lee A. Cryostatic congelation: A system for producing a limited controlled region of cooling or freezing of biological tissue. *J. Nerv. Ment. Dis.* 133 (1961) pp. 259–263.

Cornu O., Banse X., Docquier P.L., Luyckx S., Delloye C. Effect of freeze-drying and gamma irradiation on the mechanical properties of human cancellous bone. *J. Orthop. Res.* 18 (2001) pp. 426–431.

Crewe J.M., Armitage W.J. Integrity of epithelium and endothelium in organ-cultured human corneas. *Invest. Ophthalmol. Vis. Sci.* 42(8) (2001) pp. 1757–1761.

da Costa F.D., Costa A.C., Prestes R., Domanski A.C., Balbi E.M., Ferreira A.D., Lopes S.V. The early and midterm function of decellularized aortic valve allografts. *Ann. Thoracic Surg.* 90(6) (2010) pp. 1854–1860.

Dahl S.L., Chen Z., Solan A.K., Brockbank K.G., Niklason L.E., Song Y.C. Feasibility of vitrification as a storage method for tissue-engineered blood vessels. *Tissue Eng.* 12(2) (2006) 291–300.

Delloye C., Cornu O., Druez V., Barbier O. Bone allografts: What they can offer and what they cannot. *J. Bone Joint Surg. Br.* 89(5) (2007) pp. 574–579 [Review].

Delloye C., De Halleux J., Cornu O., Wegmann E., Buccafusca G.C., Gigi J. Organizational and investigational aspects of bone banking in Belgium. *Acta Orthop. Belg.* 57(Suppl. 2) (1991) pp. 27–34.

Desai M.M., Gill I.S. Current status of cryoablation and radiofrequency ablation in the management of renal tumors. *Curr. Opin. Urol.* 12 (2002) pp. 387–393.

Dumot J.A., Vargo J.J. 2nd, Falk G.W., Frey L., Lopez R., Rice T.W. An open-label, prospective trial of cryospray ablation for Barrett's esophagus high-grade dysplasia and early esophageal cancer in high-risk patients. *Gastrointest. Endosc.* 70(4) (2009) pp. 635–644.

Dziedzic-Goclawska A., Kaminski A., Uhrynowska-Tyszkiewicz I., Stachowicz W. Irradiation as a safety procedure in tissue banking. *Cell Tissue Bank.* 6 (2005) pp. 201–219.

Edd J.F., Rubinsky B. Detecting cryoablation with EIT and the benefit of including ice front imaging data. *Physiol. Meas.* 27 (2006) pp. S175–S185.

Ehlers H., Ehlers N., Hjortdal J.O. Corneal transplantation with donor tissue kept in organ culture for 7 weeks. *Acta Ophthalmol. Scand.* 77(3) (1999) 277–278.

Elford B.C., Walter C.A. Effects of electrolyte composition and pH on the structure and function of smooth muscle cooled to $-79°C$ in unfrozen media. *Cryobiology.* 9 (1972) pp. 82–100.

Elmoazzen H.Y., Elliott J.A., McGann L.E. Cryoprotectant equilibration in tissues. *Cryobiology.* 51(1) (2005) pp. 85–91.

Eurotransplant International Foundation, Eurotransplant Foundation Annual Report, Leiden, Netherlands 1976.**

Fahy G.M. Vitrification: A new approach to organ cryopreservation. *Prog. Clin. Biol. Res.* 224 (1986) pp. 305–335 [Review].

Fahy G.M., Hirsh A. Prospects for organ preservation by vitrification. In: Pegg D.E., Jacobsen I.A., Halasz N.A. (eds), *Organ Preservation, Basic and Applied Aspects.* Lancaster: MTP Press (1982) pp. 399–404.

Fahy G.M., MacFarlane D.R., Angell C.A., Meryman H.T. Vitrification as an approach to cryopreservation. *Cryobiology.* 21(4) (1984) pp. 407–426.

Farrant J. Mechanism of cell damage during freezing and thawing and its prevention. *Nature.* 205 (1965) pp. 1284–1287.

Favazza C.P., Gorny K.R., King D.M., Rossman P.J., Felmlee J.P., Woodrum D.A., Mynderse L.A. An investigation of the effects from a urethral warming system on temperature distributions during cryoablation treatment of the prostate: A phantom study. *Cryobiology.* 69(1) (2014) pp. 128–133.

Fay T., Henny G.C. Correlation of body segmental temperature and its relation to the location of carcinomatous metastasis: Clinical observations and response to methods of refrigeration. *Surg. Gynecol. Obstet.* 66 (1938) pp. 512–524.

Filipović-Zore I., Katanec D., Sušić M., Dodig D., Mravak-Stipetić M., Knezović-Zlatarić, D. Bone morphogenetic proteins—New hope in the reconstruction of bone defects in the stomatognathic area. *Acta Stomatol. Croat.* 34(3) (2000) pp. 319–324.

Fishbein W.N., Winkert J.W. Parameters of biological freezing damage in simple solution: Catalase. II. Demonstration of an optimum-recovery cooling-rate curve in a membraneless system. *Cryobiology.* 15(2) (1978) pp. 168–177.

Fraunfelder F.T., Petursson G.J. The use of liquid nitrogen cryospray for treatment of trichiasis. *Ophthalmic Surg.* 10(8) (1979) pp. 42–46.

Fuller B.J. The rational basis for controlled rate slow cooling. In: Borini A., Coticchio G. (eds), *Preservation of Human Oocytes.* London, UK: Informa Healthcare (2009) pp. 25–35.

Fuller B.J., Lane N., Benson E.E. *Life in the Frozen State*, 1st edition. Boca Raton, FL: CRC Press LLC (2004) p. 672.

Fuller B.J., Paynter S.J., Watson P. Cryopreservation of human gametes and embryos. In: Fuller B.J., Lane N., Benson E.E. (eds), *Life in the Frozen State*, 1st edition. Boca Raton, FL: CRC Press LLC (2004) pp. 505–539.

Gage A.A., Baust J. Mechanisms of tissue injury in cryosurgery. *Cryobiology.* 37(3) (1998) pp. 171–186 [Review].

Gage A.A., Baust J.G. Cryosurgery for tumors. *J. Am. Coll. Surg.* 205(2) (2007) pp. 342–356 [Review].

Gage A.A., Baust J.M., Baust J.G. Experimental cryosurgery investigations *in vivo. Cryobiology.* 59(3) (2009) pp. 229–243. doi: 10.1016/j.cryobiol.2009.10.001 [Review].

Gallinat A., Moers C., Smits J.M., Strelniece A., Pirene J., Ploeg R.J., Paul A., Treckmann J. Machine perfusion versus static cold storage in expanded criteria donor kidney transplantation: 3 year follow-up data. *Transplant. Int.* 26 (2013) e52–e53.

Gao D., Critser J.K. Mechanisms of cryoinjury in living cells. *ILAR J.* 41(4) (2000) pp. 187–196 [Review].

Gelber P.E., Gonzalez G., Torres R., Giralt N.G., Caceres E., Monllau J.C. Cryopreservation does not alter the ultrastructure of the meniscus [Internet]. *Knee Surg. Sports Traumatol. Arthrosc.* 17(6) (2009) pp. 639–644.

Geraghty R.J., Capes-Davis A., Davis J.M., Downward J., Freshney R.I., Knezevic I., Lovell-Badge R., Masters J.R., Meredith J., Stacey G.N., Thraves P., Vias M., Cancer Research UK. Guidelines for the use of cell lines in biomedical research. *Br. J. Cancer.* 111(6) (2014) pp. 1021–1046. doi: 10.1038/bjc.2014.166.

Gouk S.S., Loh Y.F., Kumar S.D., Watson P.F., Kuleshova L.L. Cryopreservation of mouse testicular tissue: Prospect for harvesting spermatogonial stem cells for fertility preservation. *Fertil. Steril.* 95 (2011) pp. 2399–2403.

Halloran P., Aprile M. A randomized prospective trial of cold storage versus pulsatile perfusion for cadaver kidney preservation. *Transplantation.* 43 (1987) pp. 827–832.

Hamlin H. Abeyance of haemorrhage by (LN$_2$) "cryospray" during meningioma removal. *Neurochirurgia (Stuttg.).* 14(3) (1971) pp. 115–117.

Hamlin H. Aid of LN$_2$-cryospray in extirpation of brain tumors. *Int. Surg.* 52(4) (1969) pp. 285–286.

Han B., Bischof J.C. Direct cell injury associated with eutectic crystallization during freezing. *Cryobiology.* 48 (2004) pp. 8–21.

Henderson A.R. Temple Fay M.D. Unconformable crusader and harbinger of human refrigeration. *J. Neurosurg.* 20 (1963) pp. 627–634.

Heng B.C., Ye C.P., Liu H., Toh W.S., Rufaihah A.J., Yang Z., Bay B.H., Ge Z., Ouyang H.W., Lee E.H., Cao T. Loss of viability during freeze-thaw of intact and adherent human embryonic stem cells with conventional slow-cooling protocols is predominantly due to apoptosis rather than cellular necrosis. *J. Biomed. Sci.* 13 (2006) pp. 433–445.

Heng B.C., Yu H., Ng C.S. Strategies for the cryopreservation of microencapsulated cells. *Biotechnol. Bioeng.* 85(2) (2004) pp. 202–213.

Hermans M.H. Preservation methods of allografts and their (lack of) influence on clinical results in partial thickness burns. *Burns.* 37(5) (2011) pp. 873–881.

Hoagland H., Pincus G. Revival of mammalian sperm after immersion in liquid nitrogen. *J. Gen. Physiol.* 25(3) (1942) pp. 337–344.

Hoffmann N.E., Bischof J.C. The cryobiology of cryosurgical injury. *Urology* 60(2 Suppl. 1) (2002) pp. 40–49.

Hoffmann N.E., Bischof J.C. Mechanisms of injury caused by in vivo freezing. In Fuller B.J., Lane N., Benson E. (eds). *Life in the Frozen State.* Boca Raton, FL: CRC Press (2004) pp. 455–481.

Hoffmann T., Minor T. New strategies and concepts in organ preservation. *Eur. Surg. Res.* 54(3–4) (2014) pp. 114–126.

Holt W.V. Cryobiology, wildlife conservation and reality. *Cryo Letters.* 29 (2008) pp. 43–52.

Holt W.V. Fundamental aspects of sperm cryobiology: The importance of species and individual differences. *Theriogenology.* 53(1) (2002) 47–58.

Holt W.V., Bennett P.M., Volobouev V., Watson P.F. Genetic resource banks in wildlife conservation. *J. Zool. Lond.* 238 (1996) pp. 531–544.

Holt W.V., Pickard A.R. Role of reproductive technologies and genetic resource banks in animal conservation. *Rev. Reprod.* 4(3) (1999) pp. 143–150.

Hou J.K., Abudayyeh S., Shaib Y. Treatment of chronic radiation proctitis with cryoablation. *Gastrointest. Endosc.* 73(2) (2011) pp. 383–389. doi: 10.1016/j.gie.2010.10.044. Erratum in: *Gastrointest. Endosc.* 73(5) (2011) p. 1073.

Hsu J.K., Cavanagh H.D., Jester J.V., Ma L., Petroll W.M. Changes in corneal endothelial apical junctional protein organization after corneal cold storage. *Cornea.* 18(6) (1999) pp. 712–720.

Hunt C.J. Cryopreservation of human stem cells for clinical application: A review. *Transfus. Med. Hemother.* 38(2) (2011) pp. 107–123.

Hunt C.J. Studies on cellular structure and ice location in frozen organs and tissues: The use of freeze-substitution and related techniques. *Cryobiology.* 21(4) (1984) pp. 385–402. ISSN 0011-2240, doi: 10.1016/0011-2240(84)90077-4.

Huppertz B., Kivity V., Sammar M., Grimpel Y., Leepaz N., Orendi K., Pekarski I., Meiri H., Gonen R., Lubzens E. Cryogenic and low temperature preservation of human placental villous explants— A new way to explore drugs in pregnancy disorders. *Placenta.* 32(8) (2011) pp. 611–615.

Inclan A. The use of preserved bone graft in orthopaedic surgery. *J. Bone Joint Surg.* 24 (1942) pp. 81–96.

Iop L., Bonetti A., Naso F., Rizzo S., Cagnin S., Bianco R., Dal Lin C., Martini P., Poser H., Franci P., Lanfranchi G., Busetto R., Spina M., Basso C., Marchini M., Gandaglia A., Ortolani F., Gerosa G. Decellularized allogeneic heart valves demonstrate self-regeneration potential after a long-term preclinical evaluation. *PLoS One.* 9(6) (2014) p. e99593.

Jackson D.W., Simon T. Biology of meniscal allograft. In: Mow V.C., Arnoczky S.P., Jackson D.W. (eds), *Knee Meniscus: Basic and Clinical Foundations.* New York, NY: Raven Press (1992) pp. 141–152.

Jaffe F.F., Mankin H.J., Weiss C., Zarins A. Water binding in articular cartilage of rabbits. *J. Bone. Joint. Surg. [Am.]* 56 (1974) pp. 1031–1039.

Jaffers G.J., Banowsky L.H. The absence of a deleterious effect of mechanical kidney preservation in the era of cyclosporine. *Transplantation.* 47 (1989) pp. 734–736.

Jamieson R.W., Zilvetti M., Roy D., Hughes D., Morovat A., Coussios C.C., Friend P.J. Hepatic steatosis and normothermic perfusion-preliminary experiments in a porcine model. *Transplantation.* 92 (2011) pp. 289–295.

Jolly M., Sebire N., Harris J., Robinson S., Regan I. The risks associated with pregnancy in women aged 35 years or older. *Hum. Reprod.* 15(11) (2000) pp. 2433–2437.

Jomha N.M., Elliott J.A., Law G.K., Maghdoori B., Forbes J.F., Abazari A., Adesida A.B., Laouar L., Zhou X., McGann L.E. Vitrification of intact human articular cartilage. *Biomaterials.* 33(26) (2012) pp. 6061–6068. doi: 10.1016/j.biomaterials.2012.05.007.

Kagan R.J. Human skin banking: Past present and future. In: Phillips G.O., Strong D.M., von Versen R., Nather A. (eds), *Advances in Tissue Banking,* Vol. 3. Singapore: World Scientific (1998) pp. 297–321.

Karlsson J.O.M., Toner M. Cryopreservation: Foundations and applications in tissue engineering. In: Lanza R., Langer R., Vacanti J. (eds), *Principles of Tissue Engineering*, 2nd edition. New York, NY: Academic Press (2000) pp. 293–307.

Karow A.M., Webb W.R. Tissue freezing: A theory for injury and survival. *Cryobiology.* 2 (1965) pp. 99–108.

Khorramirouz R., Sabetkish S., Akbarzadeh A., Muhammadnejad A., Heidari R., Kajbafzadeh A.M. Effect of three decellularisation protocols on the mechanical behaviour and structural properties of sheep aortic valve conduits. *Adv. Med. Sci.* 59(2) (2014) pp. 299–307.

Knapik A., Kornmann K., Kerl K., Calcagni M., Contaldo C., Vollmar B., Giovanoli P., Lindenblatt N. Practice of split-thickness skin graft storage and histological assessment of tissue quality. *J. Plast. Reconstr. Aesthet. Surg.* 66(6) (2013) pp. 827–834.

Knezevic I., Stacey G., Petricciani J., Sheets R. On behalf of the WHO study group on cell substrates. Report of the WHO Study Group on Cell Substrates for the Production of Biologicals, 22–23 April 2009, Bethesda, USA. *Biologicals.* 38 (2010) pp. 162–169.

Koh J.L., Kowalski A., Lautenschlager E. The effect of angled osteochondral grafting on contact pressure: A biomechanical study. *Am. J. Sports Med.* 34 (2006) pp. 116–119.

Koh J.L., Wirsing K., Lautenschlager E., Zhang L.O. The effect of graft height mismatch on contact pressure following osteochondral grafting: A biomechanical study. *Am. J. Sports Med.* 32 (2004) pp. 317–320.

Konuma T., Devaney E.J., Bove E.L., Gelehrter S., Hirsch J.C., Tavokkol Z., Ohye R.G. Performance of CryoValve SG decellularized pulmonary allografts compared with standard cryopreserved allografts. *Ann. Thorac. Surg.* 88(3) (2009) pp. 849–854.

Kosieradzki M., Danielewicz R., Kwiatkowski A., Polak W., Wegrowicz-Rebandel I., Wałaszewski J., Gaciong Z., Lao M., Rowiński W. Rejection rate and incidence of acute tubular necrosis after pulsatile perfusion preservation. *Transplant. Proc.* 31(1–2) (1999) pp. 278–279.

Kuleshova, L. L., MacFarlane, D. R., Trounson, A. O., Shaw, J. M. Sugars exert a major influence on the vitrification properties of ethylene glycol-based solutions and have low toxicity to embryos and oocytes. *Cryobiology* 38 (1999a) pp. 119–130.

Kuleshova L., Gianoroli L., Magli C., Ferraretti A., Trounson A. Birth following vitrification of small number of human oocytes. *Hum. Rep.* 14(12) (1999b) 3077–3079.

Kuleshova L.L., Gouk S.S., Hutmacher D.W. Vitrification as a prospect for cryopreservation of tissue-engineered constructs. *Biomaterials.* 28 (2007) pp. 1585–1596.

Kuleshova L.L., Hutmacher D.W. Cryobiology. In: van Blitterswijk C. (ed), *Tissue Engineering*, 1st edition. London: Elsevier (2008) pp. 363–401.

Kuleshova L.L., Lopata A. Vitrification can be more favorable than slow cooling. *Fertil. Steril.*78(5) (2002) pp. 449–454.

Kuleshova L.L., MacFarlane D.R., Trounson A.O., Shaw J.M. Sugars exert a major influence on the vitrification properties of ethylene glycol-based solutions and have low toxicity to embryos and oocytes. *Cryobiology.* 38 (1999a) pp. 119–130.

Kuleshova L.L., Shaw J.M. A strategy for rapid cooling of embryos within a double straw to eliminate the risk of contamination during cryopreservation and storage. *Hum. Rep.* 15 (2000) pp. 2604–2609.

Kuleshova L.L., Wang X.W., Wu Y.N., Zhou Y., Yu H. Vitrification of encapsulated hepatocytes with reduced cooling and warming rates. *Cryo Letters.* 25(4) (2004) pp. 241–254.

Kuleshova L.L. Fundamentals and current practice of vitrification. In: Borini A., Coticchio G. (eds), *Preservation of Human Oocytes*. London, UK: Informa Healthcare (2009a) pp. 36–41.

Kuleshova L.L., Tan C.K.F., Magalhães R., Gouk S.S., Lee K. H., Dawe G.S. Effective cryopreservation of neuronal stem cells or progenitor cells without serum or proteins by vitrification. *Cell Transplant.* 18 (2009b) pp. 135–144.

Kuleshova L.L., Gouk S.S., Wen F., Wang X.W., Wu Y.N., Magalhães R., Lee X.W., Dawe G.S., Lee E.H., Watson P.F., Yu H. Cryopreservation in the future. *Cryo Letters.* 34(2) (2013) p. 175.

Lancaster M.A., Renner M., Martin C.A., Wenzel D., Bicknell L.S., Hurles M.E., Homfray T., Penninger J.M., Jackson A.P., Knoblich J.A. Cerebral organoids model human brain development and microcephaly. *Nature.* 501 (2013) pp. 373–379.

Larrey D.J. *Memoires de chirugie militaire et compagnies*. Philadelphia, PA: Carey & Lea (1832) pp. 1812–1817.

Lass J.H., Bourne W.M., Musch D.C., Sugar A., Gordon J.F., Reinhart W.J., Meyer R.F., Patel D.I., Bruner W.E., Cano D.B. Soon H.K., Maguire L.J., and Laing R.J. A randomized, prospective, double-masked clinical trial of optisol vs. dexsol corneal storage media. *Arch. Ophthalmol.* 110 (1992) pp. 1404–1408.

Lavernia C.J., Malinin T.I., Temple H.T., Moreyra C.E. Bone and tissue allograft use by orthopaedic surgeons. *J. Arthroplas.* 19(4) (2004) pp. 430–435.

Leibo S.P., Songsasen N. Cryopreservation of gametes and embryos of non-domestic species. *Theriogenology.* 57 (2002) pp. 303–326.

Lovelock J.E. The denaturation of lipid-protein complexes as a cause of damage by freezing. *Proc. R. Soc. Lond. B.* 147 (1957) pp. 427–434.

Lovelock J.E., Bishop M.W.H. Prevention of freezing damage to living cells by dimethyl sulphoxide. *Nature.* 183 (1959) pp. 1394–1395.

Luyet B., Rasmussen D. Study by differential thermal analysis of temperature instability of rapidly cooled solutions of glycerol, ethylene glycol, sucrose and glucose. *Biodynamica.* 10 (1967) 167–191.

Luyet B.J., Gehenio P.M. Thermoelectric recording of ice formation and of vitrification during ultra-rapid cooling of protoplasm. *Fed. Proc.* 6(1 Pt 2) (1947a) p. 157.

Luyet B.J., Gehenio P.M. Thermo-electric recording of vitrification and crystallization in ultra-rapid cooling of protoplasm. *Biodynamica.* 6(109–112) (1947b) pp. 93–100.

Macewen W. Observations concerning transplantation of bone illustrated by a case of inter-human osseous transplantation, whereby over two-thirds of the shaft of a humerus was restored. *Proc. R. Soc. Lond.* 32 (1854–1905) pp. 232–247 (1881-01-01).

MacFarlane D.R. Physical aspects of vitrification in agues solutions. *Cryobiology.* 24 (1987) pp. 181–185.

MacFarlane D.R., Forsyth M., Barton C.A. Vitrification and devitrification in cryopreservation. In: Steponkus P.L. (ed), *Advances in Low-Temperature Biology*, Vol. 1. London: JAI Press (1991) pp. 221–277.

Magalhaes R., Anil Kumar P.R., Wen F., Zhao X., Yu H., Kuleshova L.L. The use of vitrification to preserve primary rat hepatocyte monolayer on collagen-coated poly(ethylene-terephthalate) surfaces for a hybrid liver support system. *Biomaterials.* 30 (2009) pp. 4136–4142.

Magalhães R., Nugraha B., Pervaiz S., Yu H., Kuleshova L.L. Influence of cell culture configuration on the post-cryopreservation viability of primary rat hepatocytes. *Biomaterials* 33 (2012) pp. 829–836.

Magalhaes R., Wang X.W., Gouk S.S., Lee K.H., Ten C.M., Yu H., Kuleshova L.L. Vitrification successfully preserves hepatocyte spheroids. *Cell Transplant.* 17 (2008) pp. 813–828.

Maizato M.J., Taniguchi F.P., Ambar R.F., Pitombo R.N., Leirner A.A., Cestari I.A., Stolf N.A. Behavior of lyophilized biological valves in a chronic animal model. *Artif. Organs.* 37(11) (2013) pp. 958–964.

Mansbridge J. Skin tissue engineering. *J. Biomater. Sci. Polym. Ed.* 19(8) (2008) pp. 955–968.

Maroudas A., Schneiderman R. Free and exchangeable or trapped and non-exchangeable water in cartilage. *J. Orthopaed. Res.* 5 (1987) pp. 133–138.

Marshall V. Preservation by simple hypothermia. In: Collins G.M., Dubernard J.M., Land W., Persijn G.G. (eds), *Procurement, Preservation and Allocation of Vascularized Organs*. The Netherlands: Springer (1997) pp. 115–129. doi: 10.1007/978-94-011-5422-2_13.

Marshall V.C., Ross H., Scott D.F., McInnes S., Thomson N., Atkins R.C. Preservation of cadaveric renal allografts-comparison of flushing and pumping techniques. *Proc. Eur. Dial. Transplant. Assoc.* 14 (1977) pp. 302–309.

Mascarenhas R., Tranovich M., Karpie J.C., Irrgang J.J., Fu F.H., Harner C.D. Patellar tendon anterior cruciate ligament reconstruction in the high-demand patient: Evaluation of autograft versus allograft reconstruction. *Arthroscopy* 26(Suppl. 9) (2010) pp. S58–S66.

Matheny J., Karow A.M. Jr., Carrier O. Jr. Toxicity of dimethyl sulfoxide and magnesium as a function of temperature. *Eur. J. Pharmacol.* 5(2) (1969) pp. 209–212.

Matsuno N., Sakurai E., Tamaki I., Uchiyama M., Kozaki K., Kozaki M. The effect of machine perfusion preservation versus cold storage on the function of kidneys from non-heart-beating donors. *Transplantation.* 57 (1994) pp. 293–294.

Mazur P. Freezing of living cells: Mechanisms and implications. *Am. J. Physiol.* 247(3 Pt 1) (1984) pp. C125–C142 [Review].

Mazur P. Principles of cryobiology. In: Fuller B.J., Lane N., Benson E. (eds), *Life in the Frozen State.* Boca Raton, FL: CRC Press (2004) pp. 3–65.

Mazur P., Cole K.W. Influence of cell concentration on the contribution of unfrozen fraction and salt concentration to the survival of slowly frozen human erythrocytes. *Cryobiology.* 22 (1985) pp. 509–536.

Mazur P., Leibo S.P., Chu E.H.Y. A two-factor hypothesis of freezing injury. *Exp. Cell. Res.* 71 (1972) pp. 345–355.

McNally R.T., Heacox A., Brockbank K.G., Bank H.L. (Inventors). *Method for Cryopreserving Heart Valves.* Cryolife, Inc., Assignee. Jan 2, 1987, US Patent 4,890,457.

Merion R.M., Oh H.K., Port F.K., Toledo-Pereyra L.H., Turcotte J.G. A prospective controlled trial of cold storage versus machine-perfusion preservation in cadaveric renal transplantation. *Transplantation.* 50 (1990) pp. 230–233.

Mickiewicz P., Binkowski M., Bursig H., Wróbel Z. Preservation and sterilization methods of the meniscal allografts: Literature review. *Cell Tissue Bank.* 15 (2013) pp. 1–11.

Milton J., Flandry F., Terry G. Transplantation of viable, cryopreserved menisci. *Trans. Orthop. Res. Soc.* 15 (1990) p. 220.

Morris G.J., Acton E., Murray B.J., Fonseca F. Freezing injury: The special case of the sperm cell. *Cryobiology.* 64 (2012) pp. 71–80.

Mozes M.F., Finch W., Reckard C.R., Merkel F.K., Cohen C. Comparison of cold storage and machine perfusion in the preservation of cadaver kidneys: A prospective, randomized study. *Transplant. Proc.* 17 (1985) pp. 1474–1477.

Mühlbacher F., Langer F., Mittermayer C. Preservation solutions for transplantation. *Transplant. Proc.* 31(5) (1999) pp. 2069–2070 [Review].

Mukaida T., Nakamura S., Tomiyama T., Wada S., Kasai M., Takahashi K. Successful birth after transfer of vitrified human blastocysts with use of a cryoloop containerless technique. *Fertil. Steril.* (76) (2001) pp. 618–620.

Mukaida T., Wada M., Takahashi K., Pedro P.B., An T.Z., Kasai M. Vitrification of human embryos based on the assessment of suitable conditions for 8-cell mouse embryos. *Hum. Rep.* (13) (1998) pp. 2874–2879.

Muldrew K., McGann L.E. The osmotic rupture hypothesis of intracellular freezing injury. *Biophys. J.* 66(2 Pt 1) (1994) pp. 532–541.

Muldrew K., Sykes B., Schachar N.S., McGann L.E. Permeation kinetics of dimethyl sulfoxide in articular cartilage. *Cryo Letters.* 17 (1996) pp. 331–340.

Nagy Z.P., Chang C.C., Shapiro D.B., Bernal D.P., Elsner C.W., Mitchell-Leef D., Toledo A.A., Kort H.I. Clinical evaluation of the efficiency of an oocyte donation program using egg cryobanking. *Fertil. Steril.* (92) (2009) pp. 520–526.

Naor J., Slomovic A.R., Chipman M., Rootman D.S. A randomized, double-masked clinical trial of Optisol-GS vs Chen Medium for human corneal storage. *Arch. Ophthalmol.* 120(10) (2002) pp. 1280–1285.

Nienaber L.N. Evaluation of red blood cell stability during immersion blood warming. *S. Afr. J. Anaesth. Analg.* 9(4) (2003) pp. 11–15.

O'Brien M.F., Harrocks S., Stafford E.G., Gardner M.A., Pohlner P.G., Tesar P.J., Stephens F. The homograft aortic valve: A 29-year, 99.3% follow up of 1022 valve replacements. *J. Heart Valve Dis.* 10 (2001) pp. 334–344.

Ochi M., Sumen Y., Jitsuiki J., Ikuta Y. Allogeneic deep frozen meniscal graft for repair of osteochondral defects in the knee joint [Internet]. *Arch. Orthopaed. Trauma Surg.* (1995) pp. 260–266.

Onik G. Cryosurgery. *Crit. Rev. Oncol. Hematol.* 23 (1996) pp. 1–24.

colon carcinoma. Preliminary results. *Cancer.* 67(4) (1991) pp. 901–907.

Onik G., Cobb C., Cohen J., Zabkar J., Porterfield B. US characteristics of frozen prostate. *Radiology.* 168(3) (1988) pp. 629–631.

Onik G., Cooper C., Goldenberg H.I., Moss A.A., Rubinsky B., Christianson M. Ultrasonic characteristics of frozen liver. *Cryobiology.* 21 (1984) pp. 321–328.

Onik G., Gilbert J., Hoddick W., Filly R., Callen P., Rubinsky B., Farrel L. Sonographic monitoring of hepatic cryosurgery in an experimental animal model. *Am. J. Roentgenol.* 144 (1985) pp. 1043–1047.

Onik G., Rubinsky B., Zemel R., Weaver L., Diamond D., Cobb C., Porterfield B. Ultrasound-guided hepatic cryosurgery in the treatment of metastatic colon carcinoma. Preliminary results. *Cancer.* 67(4) (1991) pp. 901–907.

Op den Dries S., Karimian N., Sutton M.E., Westerkamp A.C., Nijsten M.W.N., Gouw A.S.H., Wiersema-Buist J., Lisman T., Leuvenink H.G.D., Porte R.J. Ex vivo normothermic machine perfusion and viability testing of discarded human donor livers. *Am. J. Transplant.* 13 (2013) pp. 1327–1335.

Opelz G., Terasaki P.I. Advantage of cold storage over machine perfusion for preservation of cadaver kidneys. *Transplantation.* 33 (1982) pp. 64–68.

Opelz G., Terasaki P.I. Kidney preservation: Perfusion versus cold storage—1975. *Transplant. Proc.* 8 (1976) pp. 121–125.

Otten D.M., Rubinsky B. Cryosurgical monitoring using bioimpedance measurements—A feasibility study for electrical impedance tomography. *IEEE Trans. Biomed. Eng.* 47(10) (2000) pp. 1376–1381.

Owens J.C., Prevedel A.E., Swan H. Prolonged experimental occlusion of thoracic aorta during hypothermia. *Arch. Surg.* 70 (1955) pp. 95–97.

Pallante A.L., Bae W.C., Chen A.C., Görtz S., Bugbee W.D., Sah R.L. Chondrocyte viability is higher after prolonged storage at 37 degrees C than at 4 degrees C for osteochondral grafts. *Am. J. Sports Med.* 37(Suppl. 1) (2009) pp. 24S–32S.

Patil S., Butcher W., D'Lima D.D., Steklov N., Bugbee W.D., Hoenecke H.R. Effect of osteochondral graft insertion forces on chondrocyte viability. *Am. J. Sports Med.* 36 (2008) pp. 1726–1732.

Patrizio P., Sakkas D. From oocyte to baby: A clinical evaluation of the biological efficiency of *in vitro* fertilization. *Fertil. Steril.* 91 (2009) pp. 1061–1066.

Pegg D.E. *Perfusion Technology.* In: *Organ Preservation for Transplantation*, 2nd edition, Karow Jr., A.M., Pegg D.E. (eds). New York and Basel: Marcel Dekker Inc. (1981).

Pegg D.E. Equations for obtaining melting points and eutectic temperatures for the ternary system dimethyl sulphoxide/sodium chloride/water. *Cryo Letters.* 7 (1986) pp. 387–394.

Pegg D.E. Chapter 8: Cryopreservation. In: *Essentials of Tissue Banking, Galea G. (Ed.).* UK: Springer (2009) p. 109–121.

Pegg D.E., Diaper M.P. The mechanism of injury to slowly-frozen erythrocytes. *Biophys. J.* 54 (1988) pp. 471–488.

Pegg D.E., Foreman J., Rolles K. Metabolism during preservation and viability of ischemically injured canine kidneys. *Transplantation.* 38(1) (1984) pp. 78–81.

Pegg D.E., Wang L., Vaughan D. Cryopreservation of articular cartilage. Part 3: The liquidus tracking method. *Cryobiology.* 52 (2006c) pp. 360–368.

Pegg D.E., Wang L., Vaughan D., Hunt C.J. Cryopreservation of articular cartilage. Part 2: Mechanisms of cryoinjury. *Cryobiology.* 52 (2006b) pp. 347–359.

Pegg D.E., Wusteman M.C., Foreman J. Metabolism of normal and ischaemically injured rabbit kidneys during perfusion for 48 hrs at 10°C. *Transplantation.* 32 (1981) pp. 437–443.

Pegg D.E., Wusteman M.C., Wang L. Cryopreservation of articular cartilage. Part 1: Conventional cryopreservation methods. *Cryobiology.* 52 (2006a) pp. 335–346.

Petricciani J., Sheets R. An overview of animal cell substrates for biological products. *Biologicals.* 36 (2008) pp. 359–362.

Pichugin Y., Fahy G.M., Morin R. Cryopreservation of rat hippocampal slices by vitrification. *Cryobiology.* 52(2) (2006) pp. 228–240.

Pickard A.R., Holt W.V. Cryopreservation as a Supporting Measure in Species conservation; "Not the Frozen Zoo!" In: Fuller B.J., Lane N., Benson E.E. (eds), *Life in the Frozen State*, 1st edition. Boca Raton, FL: CRC Press LLC (2004) pp. 393–413.

Polge C., Smith A.U., Parkes A.S. Revival of spermatozoa after vitrification and dehydration at low temperatures. *Nature*. 164(4172) (1949) p. 666.

Practice Committee of the American Society for Reproductive Medicine, the Society for Reproductive Medicine and Society for ART. Mature oocyte cryopreservation: A guideline. *Fertil. Steril*. 99(1) (2013) pp. 37–43.

Pusey W.A. The use of carbon dioxide snow in the treatment of nevi and other lesions of the skin. *J. Am. Med. Assoc*. 49 (1907) pp. 1354–1356.

Rabb J.M., Renaud M.L., Brandt P.A., Witt C.W. Effect of freezing and thawing on the microcirculation and capillary endothelium of the hamster cheek pouch. *Cryobiology*. 11(6) (1974) pp. 508–518.

Rall W.F., Fahy G.M. Ice-free cryopreservation of mouse embryos at –196°C by vitrification. *Nature*. 313 (1985) pp. 573–575.

Rasmussen D., Luyet B.J. Complementary study of some nonequlibrium phase transmissions in frozen solution of glycerol, etyleneglycol, glucose and sucrose. *Biodynamica*. 10 (1969) pp. 319–313.

Rasmussen D.H., Mackenzie A.P. Phase diagram for the system water-dimethylsulphoxide. *Nature*. 220 (1968) pp. 1315–1317.

Rauen U., Wu K., Witzke O., de Groot H. Custodiol-N—A new, mechanism-based organ preservation solution. *Cryobiology*. 57 (2008) p. 331.

Robertson A., Nutton R.W., Keating J.F. Current trends in the use of tendon allografts in orthopaedic surgery. *J. Bone Joint Surg. Br*. 88(8) (2006) pp. 988–992.

Rolles K., Foreman J., Pegg D.E. Preservation of ischemically injured canine kidneys by retrograde oxygen persufflation. *Transplantation*. 38(2) (1984) pp. 102–106.

Rolles K., Foreman J., Pegg D.E. A pilot clinical study of retrograde oxygen persufflation in renal preservation. *Transplantation*. 48(2) (1989) pp. 339–342.

Rubinsky B. Cryosurgery. *Annu. Rev. Biomed. Eng*. 2 (2000) pp. 157–187.

Rubinsky B., Cravalho E.G., Mikic B. Thermal stresses in frozen organs. *Cryobiology*. 17(1) (1980) pp. 66–73.

Russell P.H., Lyaruu V.H., Millar J.D., Curry M.R., Watson P.F. The potential transmission of infectious agents by semen packaging during storage for artificial insemination. *Anim. Reprod. Sci*. 47 (1997) pp. 337–342.

Sacks S., Petritsch P., Kaufman J. Canine kidney preservation using a new perfusate. *Lancet*. 301(7811) (1973) pp. 1024–1028.

Schiff L.J. Review: Production, characterization, and testing of banked mammalian cell substrates used to produce biological products. *In Vitro Cell Dev. Biol. Animal*. 41 (2005) pp. 65–70.

Shasha N., Aubin P.P., Cheah H.K., Davis A.M., Agnidis Z., Gross A.E. Long-term clinical experience with fresh osteochondral allografts for articular knee defects in high demand patients. *Cell Tissue Bank*. 3(3) (2002) pp. 175–182.

Shaw J.M., Diotallevi L., Trounson A.O. A simple rapid 4.5 M dimethyl-sulfoxide freezing technique for the cryopreservation of one-cell to blastocyst stage preimplantation mouse embryos. *Reprod. Fert. Develop*. 3(5) (1991) pp. 621–626.

Shaw J., Kuleshova L., MacFarlane D., Trounson A. Vitrification properties of solutions of ethylene glycol in saline containing PVP, Ficoll or dextran. *Cryobiology*. 35 (1997) pp. 219–229.

Sherman J.K. Research on frozen human semen: Past, present and future. *Fertil. Steril*. 15 (1964) pp. 485–499.

Siddiqui R.F., Abraham J.R., Butany J. Bioprosthetic heart valves: Modes of failure. *Histopathology*. 55(2) (2009) pp. 135–144.

Singhal P., Luk A., Butany J. Bioprosthetic heart valves: Impact of Implantation on Biomaterials. *ISRN Biomaterials* 2013 (2013) p. 14. Article ID 728791, http://dx.doi.org/10.5402/2013/728791.

Skandalakis P.N., Lainas P., Zoras O., Skandalakis J.E., Mirilas P. To afford the wounded speedy assistance: Dominique Jean Larrey and Napoleon. *World J. Surg*. 30(8) (2006) pp. 1392–1399.

Smith L.W., Fay T. Temperature factors in cancer and embryonal cell growth. *JAMA*. 113 (1939) pp. 653–660.

Song Y.C., Hagen P.O., Lightfoot F.G., Smith A.C., Taylor M.J., Brockbank K.G.M. *In vivo* evaluation of the effects of a new ice-free cryopreservation process on autologous vascular grafts. *J. Invest. Surg*. 13(5) (2000b) pp. 279–288.

Song Y.C., Khirabadi B.S., Lightfoot F., Brockbank K.G., Taylor M.J. Vitreous cryopreservation maintains the function of vascular grafts. *Nat. Biotechnol*. 18(3) (2000a) pp. 296–299.

Southard J.H., Belzer F.O. The University of Wisconsin organ preservation solution: Components, comparisons, and modifications. *Transplant. Rev*. 7 (1993) pp. 176–190.

Southard J.H., Belzer F.O. Organ preservation. *Annu. Rev. Med*. 46 (1995) pp. 235–247 [Review].

Stegemann J., Hirner A., Rauen U., Minor T. Use of a new modified HTK solution for machine preservation of marginal liver grafts. *J. Surg. Res*. 160 (2010) pp. 155–162.

Steponkus P.L., Wiest S.C. Freeze-thaw induced lesions in the plasma membrane. In: Lyons J.M., Graham D.G., Raison J.K. (eds), *Temperature Stress in Crop Plants: The Role of the Membrane*. New York, NY: Academic Press (1979) pp. 231–253.

Sterling J.C., Meyers M.C., Calvo R.D. Allograft failure in cruciate ligament reconstruction. Follow-up evaluation of eighteen patients. *Am. J. Sports Med*. 23 (1995) pp. 173–178.

Sterling W.A., Pierce J.C., Hutcher N.E., Lee H.M., Hume D.M. A comparison of hypothermic preservation with hypothermic pulsatile perfusion in paired human kidneys. *Surg. Forum*. 22 (1971) pp. 229–230.

Stoker A., Garrity J.T., Hung C.T., Stannard J.P., Cook J. Improved preservation of fresh osteochondral allografts for clinical use. *J. Knee Surg*. 25(2) (2012) pp. 117–125.

Strong D.M. The US navy tissue bank: 50 years on the cutting edge. *Cell Tissue Bank*. 1(1) (2000) pp. 9–16.

Stueber P., Kovacs S., Koletsky S., Persky L. Regional renal hypothermia. *Surgery*. 44 (1958) pp. 77–83.

Suhodolčan L., Brojan M., Kosel F., Drobnič M., Alibegović A., Brecelj J. Cryopreservation with glycerol improves the *in vitro* biomechanical characteristics of human patellar tendon allografts. *Knee Surg. Sports Traumatol. Arthrosc*. 21(5) (2013) pp. 1218–1225.

Sun W.Q. State and phase transition behaviors of quercus rubra seed axes and cotyledonary tissues: Relevance to the desiccation sensitivity and cryopreservation of recalcitrant seeds. *Cryobiology*. 38(4) (1999) pp. 372–385.

Suong-Hyu Hyon. In: Dr. Marcia Spear (ed), *Cryopreservation of Skin Tissues for Skin Grafts*, In: *Skin Grafts—Indications, Applications and Current Research*. ISBN: 978-953-307-509-9, doi: 10.5772/23268. InTech (2011).

Tan F.C.K., Lee K.H., Gouk S.S., Magalhaes R., Poonepalli A., Hande M.P., Dawe G.S., Kuleshova L.L. Optimization of cryopreservation of stem cells cultured as neurospheres: Comparison between vitrification, slow-cooling and rapid cooling freezing protocols. *Cryo Letters*. 28 (2007) pp. 445–460.

Taylor M.J., Baicu S.C. Current state of hypothermic machine perfusion preservation of organs: The clinical perspective. *Cryobiology*. 60(Suppl. 3) (2010) pp. S20–S35. doi: 10.1016/j.cryobiol. 2009. 10.006.

Taylor M.J., Campbell L.H., Rutledge R.N., Brockbank K.G.M. Comparison of Unisol (TM) with EuroCollins solution as a vehicle solution for cryoprotectants. *Transplant. Proc*. 33 (2001) pp. 677–679.

Thomson J.R. Current applications of biological science: Part V0. *Cryobiology. Bios*. 35(4) (1964) pp. 202–207.

Toledo-Pereyra L.H. Renal hypothermic storage with a new hyperosmolar colloid solution. *Bol. Assoc. Med*. 75 (1983) pp. 347–350.

Tomford W.W., Mankin H.J. Bone banking. Update on methods and materials. *Orthop. Clin. North Am*. 30(4) (1999) pp. 565–570 [Review].

Torre D. Cutaneous cryosurgery. *J. Cryosurg*. 1 (1968) pp. 202–209.

Ullmann E. Tissue and organ transplantation. *Ann. Surg*. 60 (1914) pp. 195–219.

van der Vliet J.A., Kievit J.K., Hene R.J., Hilbrands L.B., Kootstra G. Preservation of non-heart-beating donor kidneys: A clinical prospective randomised case-control study of machine perfusion versus cold storage. *Transplant. Proc.* 33 (2001) p. 847.

van der Vliet J.A., Vroemen J.P., Cohen B., Lansbergen Q., Kootstra G. Preservation of cadaveric kidneys. Cold storage or machine perfusion? *Arch. Surg.* 118 (1983) pp. 1166–1168.

Van Horn D.L., Hanna C., Schultz R.O. Corneal cryopreservation. II. Ultrastructural and viability changes. *Arch. Ophthalmol.* 84 (1970) pp. 655–667.

Veller M.G., Botha J.R., Britz R.S., Gecelter G.R., Beale P.G., Margolius L.P., Meyers K.E., Thompson P.D., Meyers A.M., Myburgh J.A. Renal allograft preservation: A comparison of University of Wisconsin solution and of hypothermic continuous pulsatile perfusion. *Clin. Transplant.* 8 (1994) pp. 97–100.

Walstra P. Chapter 15: Crystallization. In: *Physical Chemistry of Food.* ISBN: 978-0-8247-9355-5. CRC Press (2002).

Walther P.F.v. Wiederanheilung der bei der Trepanation ausgebrochenen Knochenscheibe. *J. Chir. Augen Heilkunde.* 2(8) (1821). pp. 571–583.

Wang L., Pegg D.E., Lorrison J., Vaughan D., Rooney P. Further work on the cryopreservation of articular cartilage with particular reference to the liquidus tracking (LT) method. *Cryobiology.* 55 (2007) pp. 138–147.

Wascher D.C., Becker J.R., Dexter J.G., Blevins F.T. Reconstruction of the anterior and posterior cruciate ligaments after knee dislocation. Results using fresh-frozen nonirradiated allografts. *Am. J. Sports Med.* 27(2) (1999) pp. 189–196.

Watson P.F. Recent developments and concepts in the cryopreservation of spermatozoa and the assessment of their post-thawing function. *Reprod. Fertil. Dev.* 7(4) 1995 pp. 871–891.

Watson P.F., Holt W.V. (eds). *Cryobanking the Genetic Resource.* London: Taylor and Francis Inc. (2001) pp. 317–348.

Wen F., Magalhães R., Gouk S., Bhakta G., Lee K., Hutmacher D., Kuleshova L. Vitreous cryopreservation of nanofibrous tissue engineered constructs generated using mesenchymal stromal cells. *Tissue Eng. Part C Methods.* 15 (2009) pp. 105–114.

White A.C. Liquid air: Its application in medicine and surgery. *Med. Rec.* 56 (1899) pp. 109–112.

Whitehouse H. Liquid air in dermatology: Its indications and limitations. *JAMA.* 49 (1907) pp. 371–377.

Wight J., Chilcott J., Holmes M., Brewer N. The clinical and cost-effectiveness of pulsatile machine perfusion versus cold storage of kidneys for transplantation retrieved from heart-beating and non-heart-beating donors. *Health Technol. Assess.* 7(25) (2003) pp. 1–94.

Wilson J. (ed) *World Blindness and its Prevention.* Oxford: Oxford University Press (1980).

Wilson P.D. Experiences with a bone bank. *Ann. Surg.* 126 (1947) pp. 932–946.

Wood J.M., Soldin M., Shaw T.J., Szarko M. The biomechanical and histological sequelae of common skin banking methods. *J. Biomech.* 47(5) (2014) pp. 1215–1219.

Wood M.J. The problems of storing gametes and embryos. *Cryo Letters.* 20 (1999) pp. 155–158.

Wood M.J. Vitrification of oocytes. *Obstet. Gynaecol.* 14 (2012) pp. 45–49.

Wu Y., Yu H., Chang S., Magalhães R., Kuleshova L.L. Vitreous cryopreservation of cell-biomaterial constructs involving encapsulated hepatocytes. *Tissue Eng.* 1 (2007) pp. 649–658.

Wu Y, Wen F, Gouk SS, Lee EH, Kuleshova L. Cryopreservation Strategy for Tissue Engineering Constructs Consisting of Human Mesenhymal Stem Cells and Hydrogel Biomaterials. *CryoLetters.* 36(5) (2015) pp. 325–335.

Wusteman M., Hunt C.J. Chapter 19. The scientific basis for tissue banking. In: Fuller B.J., Lane N., Benson E.E (eds), *Life in the Frozen State.* Boca Raton, FL: CRC Press (2004).

Yoon T.K., Kim T.J., Park S.E., Hong S.W., Ko J.J., Chung H.M., Cha K.Y. Live births after vitrification of oocytes in a stimulated *in vitro* fertilization-embryo transfer program. *Fertil. Steril.* 79(6) 2003 pp. 1323–1326.

Zacarian S.A. *Cryosurgery of Tumors of the Skin and Oral Cavity.* Springfield, IL: Charles C. Thomas (1973).

Internet References

MRC: UK Stem Cell Bank: http://www.mrc.ac.uk/research/facilities/stem-cell-bank/.
NIBSC: http://www.nibsc.org/science_and_research/advanced_therapies/uk_stem_cell_bank/
     cell_lines/available_human_cell_lines.aspx.
Planer Plc: http://planer.com/.
WHO: http://www.who.int/biologicals/vaccines/cell_substrates/en/.

12

Low Temperature Electronics

Nathan Valentine and Patrick McCluskey

CONTENTS

12.1 Introduction to Low Temperature Environments and Applications

Low temperature technology has allowed scientists and engineers to delve into a new realm of material science and physics, advancing material capabilities and system functionality. It first received prominence during World War II when it was discovered that metals that had been subjected to extremely low temperatures were more resistant to wear.

Since that time the field of low temperature physics has expanded into many fields such as electronics where low temperatures are used to produce high-efficiency electronic systems. The present chapter is a discussion of the physics and applications of electronics subjected to low temperatures, beginning with a background discussion, followed by information on the performance of electronics at low temperatures, and finally a discussion of both device- and circuit-level packaging.

12.1.1 Definition of Low Temperature and Motivation to Study Low Temperature Electronics

There are several reasons to study the effects of low temperature on electronics. Operation at low temperatures can improve the performance of devices. At lower temperatures, there is less thermal noise and thus semiconductors with lower energy band gap can be used, thereby increasing efficiency. Another reason to operate at low temperatures is because it is necessary to integrate electronics into systems that experience these low temperatures. Finally, low temperature electronics are studied to observe and use phenomena such as the Josephson effect and superconductivity and to better understand negative effects such as charge trapping.

Military standards limit the operating temperature for conventional electronics ranges from –55°C to 125°C. The scope of this chapter is to discuss electronics and their operation at temperatures colder than this range. There are three classes of low temperature generally defined by the mechanism by which they are achieved. The first are those systems that are cooled by liquid nitrogen; thus, they are operated at temperatures of approximately 77 K (or –196°C). The second are those that are achieved by mechanical cooling and can be operated at temperatures around 20 K (or –253°C). The final are those that operate around the temperature of liquid helium, 4.2 K (or –269°C) [Kirschman, 1986]. Applications in this range are generally limited to research purposes; thus, the scope of this chapter will be focused on practical applications in extreme exploration such as polar and space, as well as cryogenic computing. These applications generally experience temperatures in the range of the first two classes.

12.1.2 Polar Exploration

Polar exploration has long been of human interest. The extreme weather conditions in the area have made it difficult and, in many instances, impossible to explore via direct-on-surface methods. The average daily temperature in August for the American Amundsen-Scott station at the South Pole is –60°C, making it difficult for exploration by humans.

Due to the increasing interest in mitigating and monitoring the effects of climate change on Earth, several organizations have been studying the polar regions. One example of a polar exploration robot is NASA's Greenland Rover or Goddard Remotely Operated Vehicle for Exploration and Research (GROVER) (Figure 12.1), used to gather snow accumulation data using a ground-penetrating radar during its 5-week, 18-mile mission [Trisca et al., 2013]. The rover was exposed to temperatures of –30°C during its mission. However, temperatures in Greenland can reach as low as –60°C; therefore, the rover had to be designed with the proper circuitry and materials to withstand these worst-case operating conditions.

Another robotic application in a polar system is the Yeti (Figure 12.2). Originally designed by engineers at Dartmouth College for operation in Greenland, Yeti is a lightweight autonomous robot outfitted with a ground-penetrating radar sensor designed to detect crevasses masked by snow [Trautmann et al., 2009]. Yeti drives ahead of convoys

FIGURE 12.1
NASA GROVER during its mission in Greenland. (Courtesy of NASA.)

FIGURE 12.2
Yeti crevasse sensing autonomous. (Courtesy of Laura E. Ray, professor of engineering at Dartmouth College and Yeti project leader.)

that deliver essential supplies to outposts in remote regions of the polar ice caps, and identifies these gorges allowing the convoys to reroute avoiding costly and potentially life-threatening situations. The original design was to withstand sustained temperatures of −40°C in up to 60 mph winds. After a successful launch in Greenland, the Yeti is now being used in Antarctica, where the conditions are harsher still [Oskin, 2013].

12.1.3 Space Exploration

Often referred to as the final frontier, space exploration and the associated technology provide a challenge to engineers, as they must design systems to operate under extreme conditions on both ends of the temperature scale. Systems studying permanently shadowed lunar regions, other moons in the solar system, and comets can be expected to experience temperatures as low as 3–90 K [Cressler and Mantooth, 2012]. In addition to ensuring functionality at extreme low temperatures, systems such as the space station and other orbiting satellites must be able to withstand a large temperature cycling range. For example, the International Space Station experiences temperatures oscillating between –157°C and 121°C [Price et al., 2001].

Satellites with infrared imagers, which will operate at low temperatures, such as the James Webb telescope (Figure 12.3), are becoming essential tools as scientists look for further knowledge of the formation of the universe following the Big Bang. James Webb is designed to have infrared sensors that operate at temperatures as low as 7 K to provide low-noise, high-resolution images [NASA, 2014a]. To achieve such temperatures, new technologies such as advanced radiation heat shielding and cryocoolers have been implemented. Other infrared telescopes have been developed such as the European Space Agency's Planck Space Observatory and Herschel Space Observatory, which are designed to operate at temperatures as low as 0.1 and 1.4 K, respectively [ESA, 2014].

Other space exploration units such as NASA's Voyager spacecraft, must also endure low temperatures [NASA, 2012]. Systems still operate on both of these craft, although, due to their distance from the sun, their power source can no longer operate subsystems such as heaters. Consequently Voyager 1, which was originally designed to operate at a temperature of –35°C, is now operating at –79°C.

12.1.4 Cryogenic Computing

While polar and space exploration technologies experience low temperatures by their inherent environmental conditions, fields such as cryogenic supercomputing exploit the more advantageous phenomena of low temperature environments to improve system performance. Current technologies use enhancements to traditional semiconductor technologies to achieve higher speeds in high-performance computing applications [Top500, 2014].

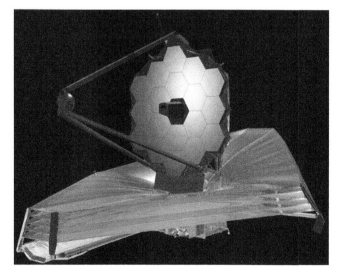

FIGURE 12.3
Artist's depiction of James Webb telescope [NASA, 2014b]. (Courtesy of NASA.)

As cryogenic technology becomes more affordable, it is expected that future supercomputers will be based on superconducting phenomena.

To achieve higher computing speeds in complementary metal-oxide semiconductor (CMOS)–based computers, a decrease in distance that the signals must travel and increased trace density are required; thus, such systems attempt to minimize the length and size of signal conductors. However, this has the adverse effect of increasing the RC delay for the system, as resistance is inversely proportional to area [Keyes et al., 1970]. At low temperatures, crystalline structures experience less lattice vibrations from thermal excitation; consequently, metals experience a decrease in resistivity when they are used at lower temperatures. Properties such as these can mitigate the increase in RC delay, thus allowing for a reduction in system size and an increase in the system speed [Keyes et al., 1970]. Decreasing the temperature also reduces thermal noise, allowing for the use of narrow band gap semiconductors, increasing system efficiency and speed. Due to their abundance and inert nature, liquid nitrogen and helium remain the most commonly used coolants in cryogenic supercomputers. One example of a cryogenic supercomputer is the ETA Systems Inc. ETA-10. This computer was first available in 1987 and was designed to operate at 90 K using liquid nitrogen in its internal cooling system [Carlson et al., 1989].

New cooling technologies are being developed to replace supercooled baths for cryogenic computers, such as quantum well on-chip cooling. Quantum well on-chip cooling has been shown to reduce chip temperature to 45 K [Anthony, 2014]. Due to the extreme costs and limited efficiency gains of cooling traditional CMOS systems, the Intelligence Advanced Research Projects Agency started a 5-year program called Cryogenic Computing Complexity designed to research superconducting technologies for use in supercomputers. Such a technology could provide a lower cost alternative to CMOS technologies in terms of cooling requirements [Manheimer, 2015].

12.2 Low Temperature Electronic Devices

To facilitate the operation of systems at extreme temperatures, extensive research on the operation of electronic devices in these environments has been documented by many groups. In many instances, traditional systems such as silicon CMOS experience performance gains when decreasing the temperature to approximately 77 K. Here, for silicon CMOS, benefits of decreased temperature tend to level off and phenomena such as freeze-out begin to deteriorate the system performance. The following section provides a brief overview of the many differing types of electronics that have been found or designed to operate in low temperature environments.

12.2.1 SiGe Transistors and Diodes

Traditional silicon-based CMOS devices experience increased transconductance, increased charge carrier mobility, decreased turn-on times, and reduced thermal noise when operating at decreased temperatures as low as 77 K; however, at lower temperatures, impurity freeze-out begins to decrease carrier concentrations and increase threshold voltages appreciably [Kirschman, 1986]. As such it is necessary to find a switching device that can operate stably at temperatures lower than 77 K. Silicon-based bipolar transistors were found to have very poor electrical characteristics in cryogenic regions, primarily due to low emitter-base injection efficiency [Lengeler, 1974]. Early research focused on finding a heterojunction bipolar transistor (HBT) with promising results. Heterojunction bipolar transistors are bipolar transistors that have different semiconductor materials in the emitter-base region than the rest of the transistor.

Silicon germanium HBTs have developed as the leading candidate for transistors in low temperature operation. The devices show increased performance down to sub-Kelvin temperatures [Cressler and Mantooth, 2012]. Current gain, transconductance, cutoff frequency, and switching frequency all show improved characteristics when cooled to cryogenic temperatures. Figure 12.4 shows peak cutoff frequency and peak current gain performance with varying temperature. In each of the formulations for these properties, the kT term favorably affects the overall parameter at decreasing temperature [Cressler and Mantooth, 2012]. The use of a silicon–germanium alloy, which has a smaller energy band gap, in the base region reduces the potential barrier between the emitter-base regions, thus amplifying the collector current. The exact band gap of the base region depends on the relative concentration of germanium and silicon in the alloy and is often graded within the device; the energy band gap of the SiGe layer will be between 0.67 and 1.1 eV, the intrinsic band gaps of the respective materials [Streetman and Banerjee, 1995].

In addition to their performance at cryogenic temperatures, SiGe HBTs have been shown to operate with minimal performance degradation at high temperatures and in radiation-intense environments. SiGe has been called a temperature-invariant material. These characteristics make SiGe ideal for applications such as space exploration in which systems can experience extreme temperature swings. As such SiGe is the leading candidate for extreme environment systems [Cressler and Mantooth, 2012].

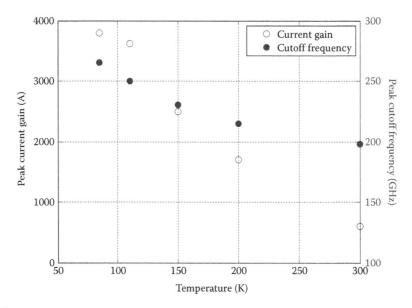

FIGURE 12.4
Peak current gain and peak frequency versus temperature for third-generation SiGe HBT. (Data from Cressler, J.D. and Mantooth, H.A. (Eds.), *Extreme Environment Electronics*, CRC Press, London, NY, 2012.)

12.2.2 Other Low-Bandgap Transistors and Diodes

Many other materials have been investigated for use in semiconductor junctions in low-temperature environments [Kelm, 1968; Forrest and Sanders, 1978; Lo and Leskovar, 1984; Prance et al., 1982; Keyes, 1977a, 1977b]. III–V semiconductors, namely InAs and InSb, are the most useful in low temperature applications due to their relatively small band gap; at room temperature, their operation can be dominated by intrinsic carriers. An important quality of high-speed bipolar transistors is that they have both high electron mobilities and high hole mobilities. While InAs and InSb have very high electron mobilities, they have very poor hole mobilities (Table 12.1), limiting their performance in high-speed applications. The reduction in noise and increase in mobility due to decreased lattice scattering have made GaAs an attractive material for preamplifiers that operate at cryogenic temperatures [Lo and Leskovar, 1984; Prance et al., 1982].

Early research also investigated the use of germanium in low temperature applications [Kelm, 1968; Keyes, 1977a]. Germanium has an intrinsic mobility that is five times that of silicon due to its small band gap (0.67 eV), which gives germanium a lower thermal voltage (Figure 12.5). Because of this property, germanium was of high interest during the search for a high-speed transistor in the 1960s. Germanium also showed excellent electrical characteristics in cryogenic applications, avoiding carrier freeze-out, which had been affecting silicon devices, due to its relatively low thermal voltage. Additionally, a lower thermal voltage allows for decreased voltage requirements when operating a germanium device, reducing the power dissipation. However, low-band-gap materials need to be properly cooled to ensure that the intrinsic behavior of the semiconductor does not dominate its operating characteristics. High-speed devices also needed to be contained in small packages to minimize signal propagation time. Small package sizes also make it difficult to remove heat. Ultimately, silicon remained the dominant material of choice for high-speed transistors [Keyes, 1977a].

12.2.3 Optical Devices

Optical devices such as light-emitting diodes (LEDs), optocouplers, and solid-state lasers are essential for modern signal transmission. While GaAs semiconductors did not become dominant for low temperature applications, variants of GaAs have become prevalent in infrared LEDs. The spectral power density (SPD) of an AlGaAs infrared LED at 300 and 77 K is shown in Figure 12.6 [Camin, 2006]. Several key conclusions can be made by analyzing this graph. First and most prevalent is that the output power of the LED is increased by a factor of 10. This observation can be explained by an increase in the quantum efficiency of the semiconductor at lower temperatures. Another observation is

TABLE 12.1

Energy Gap, Dielectric Constant, and Mobilities of Semiconductors at 77 K

Quantity	Si	Ge	GaAs	InAs	InSb	GaN	4H-SiC
E_g (eV)	1.15	0.73	1.5	0.4	0.23	3.4	3.3
μ_n	1.2	4	25	12	100	1	0.95
μ_n (doped 10^{17})	0.1	0.4	1	2	5	0.4	0.11
μ_p	0.6	4	0.4	0.4	1	0.2	0.9
μ_p (doped 10^{17})	0.05	0.4	0.1	0.05	0.2	0.11	0.6

Sources: From Keyes, R., *Comments on Solid State Physics*, 8(2), 37–46, 1977b; Lutz, J. et al., *Semiconductor Power Devices*, 17–75, 2011; Gaskill, D.K. et al., *Properties of Group III Nitrides*, Edgar, J. (Ed.), *EMIS Data Reviews Series, N11*, Wiley, pp. 101–116, 1995; Shatter, W.J. et al., *Institute of Physics Conference Series*, 137, 155, 1994.

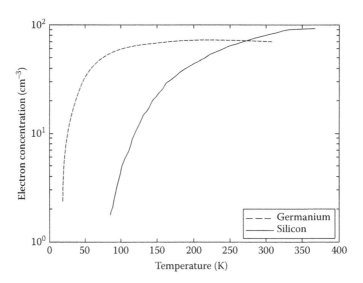

FIGURE 12.5
Difference between the trapping of electrons on donors in germanium and silicon. Specimens contain ~10^{17} donor impurities. (Based on Keyes, R., *Comments on Solid State Physics*, 8(3), 47–53, 1977a.)

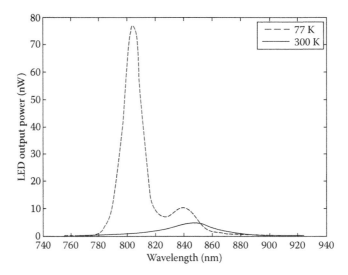

FIGURE 12.6
Spectral power density of AlGaAs LED under 1 mA bias. (Based on Camin, D., *IEEE Transactions on Nuclear Science*, 53(6), 3929–3933, 2006.)

that the peak wavelength has shifted to a shorter more energetic wavelength. As temperature decreases, the band gap of the semiconductor increases as described by the Varshni equation [O'Donnell and Chen, 1991]. A second peak appears when the LED is cooled; this is expected due to the presence of impurities. Finally, the full-width half-maximum for the cooled peaks is less than the 300 K peak, which can also be predicted by the Varshni equation [Camin, 2006]. The same AlGaAs study showed an increase in the turn on forward voltage drop, which is to be expected as temperature and forward voltage drop have a negative correlation [Hambley et al., 2008]. The observations found in Camin [2006] have also been observed in other studies [Cao and LeBoeuf, 2007; Carr, 1965; Wauters et al., 2009].

Another type of optical device is solid-state lasers, which have several advantages at cryogenic temperatures. The first is that at low temperatures, the laser threshold is decreased, meaning that the device can operate at more efficient voltages [Fan et al., 2007]. Another is that early diode-pumped solid-state lasers simply needed to be cooled to cryogenic temperatures to operate [Keyes and Quist, 1964]. In addition to the first two advantages, thermo-optic effects are reduced at cryogenic temperatures [Fan et al., 2007].

LEDs are one critical part of optocouplers, the second being a sensing mechanism, often a photodiode. Photodiodes showed a wider variety of behaviors depending on the device type. Standard Si p-n photodiodes experienced a decrease in responsivity at lower temperatures, which can be explained by a decrease in carrier concentration (Figure 12.7) [Camin, 2006]. Contrary to the standard Si p-n photodiode, Si p-i-n photodiodes exhibited increased or consistent performance at low temperatures. A Si p-i-n diode was tested at 300 and 77 K, and it was determined that the responsivity of the photodiode to an LED remained consistent (Figure 12.7). The consistent behavior can be explained by the large intrinsic region that remained unaffected by the decrease in temperatures. Furthermore, for both devices the noise of the photodiodes was reduced due to a decrease in dark current at cryogenic temperatures [Camin, 2006]. Again, the findings in Camin [2006] were consistent with other studies [Zhang et al., 1997; Wauters et al., 2009].

12.2.4 Timing Devices

As with the previous types of electronics, oscillators show improvements in performance when operated at cryogenic temperatures [Luiten et al., 1995; Giles et al., 1989]. For crystal oscillators, sapphire appears to be the dominant crystal. Research in the field has shown that cryogenic oscillators are more stable than atomic clocks in short-term applications. Specifically liquid helium cooled superconducting cavity masers have been shown to yield high quality factors, ~10^9 while outputting high power, 10^{-9} W relative to other oscillators that have shown a quality factor of 10^8 and output power of 10^{-12} W [Wang, 1988].

Superconducting cavity masers inherently have the requirement that they be cooled to liquid helium temperatures for operation. While these devices may be superior to others

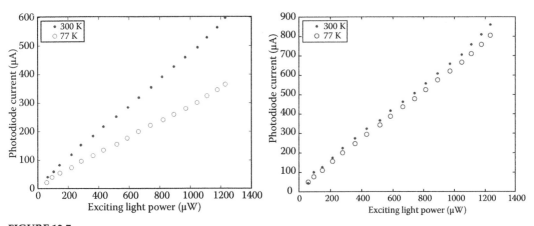

FIGURE 12.7
Responsivity of Si p-n diode (left) and responsivity of Si p-i-n photodiode (right) at 300 and 77 K. (Data from Camin, D., *IEEE Transactions on Nuclear Science*, 53(6), 3929–3933, 2006.)

at this temperature, systems used at low temperature but not necessarily below 4 K need to utilize other devices. GaAs FETs are widely used for clocks in 1–10 GHz applications due to their relatively low noise in a range of temperatures below 300 K [Kirschman 1986; Lo and Leskovar, 1984].

12.2.5 MEMS and Sensor Devices

Micro-electromechanical systems (MEMS) have rapidly achieved market integration by the application of fabrication techniques developed for processing silicon for electrical circuitry to mechanical designs [Kovacs, 1998]. These devices are superior to their conventional counterparts in terms of their size and per unit cost [Kovacs, 1998]. Nonetheless, operation of MEMS devices at cryogenic temperatures is largely unstudied [Attar et al., 2011]. Early research in the area of cryogenic RF switching MEMS found an increase in the required actuation voltage with decreasing temperatures, following the conventional wisdom of mechanical stiffening at low temperatures [Noel et al., 2008; Goldsmith and Forehand, 2005]. However, a later study [Attar et al., 2011] showed that the actuation voltage was more dependent on the mechanical structure of the device, and in fact, it is possible to get a positive or negative correlation with a decrease in temperature if the proper structure is selected.

Other applications of MEMS are microcantilevers, which are used in atomic force microscopy (AFM) and chemical sensing devices. Operating the AFM microcantilevers at cryogenic temperatures increases the resolution, as it reduces thermal noise and overcomes issues related with tissue softness at room temperature [Park et al., 2007]. To increase the range of measurable parameters for cryogenic temperatures, heated microcantilevers can be used in AFM. Using these cantilevers would allow for measurements of local heating on freezing processes, studying low temperature calorimetry, and other similar metrics [Park et al., 2007]. It is important to note when designing such a system that electrical resistance of the heated microcantilevers increases with decreasing temperature (Figure 12.8).

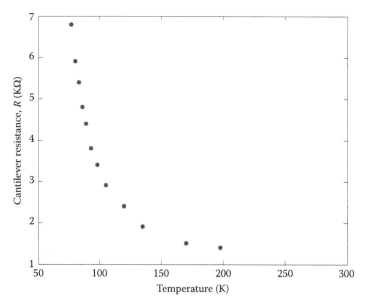

FIGURE 12.8
Heated cantilever electrical resistance versus temperature. (Data from Park, K. et al., *Journal of Applied Physics*, 101(9), 094504.)

In terms of sensing, infrared detectors experience similar characteristics to the photodiodes discussed in Section 12.2.3. That is, p-i-n diode infrared detectors experience a reduction in dark currents from thermal noise [Camin, 2006]. This is essential on telescopes such as the James Webb telescope, which use infrared detectors to observe faint infrared signatures to gather data on the beginning of the universe [NASA, 2014a].

12.3 Low Temperature Device Packaging

The packaging of the electronics within a system is essential to ensure the integrity and operation of the circuitry. For low temperature device packaging, perhaps the most prominent difficulty, due to the large difference in processing and operating temperatures, is matching the coefficient of thermal expansions (CTEs) of the materials making up the system [Cressler and Mantooth, 2012]. As materials of discrete devices are in intimate contact, systems that are not made up of materials with similar thermal expansion coefficients are prone to warping fracture and fatigue (Figure 12.9). The following sections discuss the various issues and potential solutions related to the selection of components of device packaging by examining several case studies [Shapiro et al., 2010; Sivaswamy et al., 2008; Tudryn et al., 2006].

12.3.1 Wire Bonding

For wire bonding in cryogenic applications, the material most often chosen is gold wire, bonded using a thermosonic bonding method [Sivaswamy et al., 2008; Shapiro et al., 2010; Tudryn et al, 2006]. The industry standard for military and space applications are 1-mil-diameter wires. Several commercial industries have switched to 0.7-mil wires to reduce cost; however, this size produces reliability concerns [Shapiro et al., 2010]. Additionally, wire bonds greater than 2 mil show fatigue failures at the ball bonds [Shapiro et al., 2010]. There are a wide range of results for wire bonds suggesting that

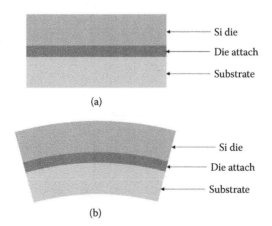

(a)

(b)

FIGURE 12.9
Exaggeration of warpage due to differing CTE. (a) The temperature is the bonding temperature, (b) after cooldown. The CTE of the Si die is less than that of the substrate. (Based on Cressler, J.D. and Mantooth, H.A. (Eds.), *Extreme Environment Electronics*, London, NY: CRC Press, 2012.)

their reliability in cryogenic operation may be application specific. Sivaswamy et al. have reported a case study of a system-in-package device to be thermal cycled from –180°C to 120°C that uses Au wire bonds. Early finite element analysis (FEA) showed that the wire bonds were relatively low stress and their reliability was of low concern [Sivaswamy et al., 2008]. Shapiro et al. also created a device with Au wire bonds that was thermal cycled from –180°C to 120°C; however, this study showed that wire bonds were the primary failure mechanism and, because of this, recommended flip chip for low temperature applications [Shapiro et al., 2010]. Tudryn et al. created a device very similar to that of Shapiro et al.; however, several of their bonds were aluminum. They also found that the wire bonds caused all observed failures, and in cases where the wire bonds had not failed, they showed strong evidence of stress [Tudryn et al, 2006]. Interestingly, although all bonds appeared stressed, the Au wire bonds failed before the Al ones. Therefore, the reliability of wire bonds at low temperatures exhibit varying results, although Al bonds may somewhat enhance reliability. Flip chip packaging appears to be a more reliable alternative.

12.3.2 Flip Chip

Due to the potential reliability issues associated with wire bond interconnections, flip chip is a leading candidate for the transmission of signals from the package to the die in reliable high thermal cycle cryogenic devices [Sivaswamy et al., 2008; Shapiro et al., 2010; Tudryn et al., 2006]. Of primary interest when investigating flip chip assemblies are the solder bumps used to connect the die to the substrate. Sivaswamy et al. employed a flip chip assembly using a In50Pb50 high-compliance solder [Sivaswamy et al., 2008]. A SiGe die was used with a Si substrate, materials that have highly similar CTE. The high-compliance solder in combination with the thermal expansion match of the die and substrate leads the authors to use no underfill (Figure 12.10). Previous studies had shown that the thermal expansion properties of the underfill are not desirable at low temperatures and can even increase stress in the system [Rahim et al., 2005].

Ulrich et al. and Yamamoto et al. separately designed microprocessors in multichip module packages using a flip chip assembly that can reliably be immersed in liquid nitrogen for performance enhancement at cryogenic temperatures [Ulrich and Rajan, 1996; Yamamoto, 1991]. Ulrich et al. used a silicon die joined with conductive epoxy-based solder bumps to the device's substrate. Yamamoto et al. used a GaAs die attached via an indium-based solder to a multilayer ceramic substrate to be described in Section 12.3.4.

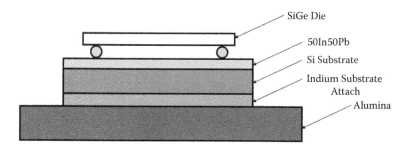

FIGURE 12.10
Schematic of flip chip assembly. (Based on Sivaswamy, S. et al., *Electronic Components and Technology Conference, 2008 (ECTC 2008)*, 58, 2044–2050, 2008.)

12.3.3 Die Attach

The use of several different types of die attach materials has been studied for cryogenic applications including solders, silicones, and epoxies [Kirschman et al., 1999; Rahim et al., 2005; Shapiro et al., 2004, 2010; Sivaswamy et al., 2008; Tudryn et al., 2006; Yamamoto, 1991]. Several considerations should be made when selecting a solder for die attach. The solder should have no phase transitions in the expected operating temperature range. The solder should be highly compliant to help reduce thermo-mechanical stresses. Finally, the solder should have minimal intermetallic growth with the metals used in traces or wire bonds of the device [Shapiro et al., 2004].

Indium solder exhibits many of these traits and has been used both as a die attach in wire bond assemblies and a substrate attach. As mentioned previously, an indium–lead alloy has been used for flip chip solder bumps. Indium remains malleable even at low temperatures, which can reduce residual stresses [Sivaswamy et al., 2008]. Additionally, indium does not experience a phase shift in the expected operating region for these devices, −180°C to 120°C, and indium is somewhat resistant to intermetallic growth with gold, which can cause failure [Shapiro et al., 2010]. Although Sivaswamy et al. used a pure indium solder, Tudryn et al. recommend the use of In80Pb15Ag5 [Tudryn et al., 2006]. Shapiro et al. [2010] provided a comprehensive list of so-called "soft" solders, which are primarily indium and lead based, and "hard" solders, which are primarily gold based [Kirschman et al., 1999]. Although the hard solders are not as compliant, they have the lowest melting temperature, which gives them the lowest "zero-stress" temperature, reducing the overall stress as the temperature is decreased. One concern with using these solders is that due to their relatively low melting points, efforts must be made to ensure that additional processing to the assembly does not reflow the solder.

In addition to studying solders, Shapiro et al. investigated the use of a silver-filled epoxy and a silicone-based adhesive [Shapiro et al., 2004]. Here it was observed that all samples that used a combination of silver-filled epoxy die attach with a low temperature co-fired ceramic (LTCC) were able to survive their stress test of 1500 thermal cycles between −180°C to 120°C. The silicone-based die attach also survived the stress test, although it should be noted that all samples prepared with a parylene survived the stress test, suggesting that the encapsulant material is of greater importance in determining reliability. Furthermore, only wire bonds were used here and they were the primary failure mode in this study, further suggesting that there is a strong dependence between reliability and the encapsulant [Shapiro et al., 2004].

12.3.4 Substrates and Leadframes

The selection of substrates and leadframes is highly dependent on the thermal expansion properties of the materials within the device. AlN is an ideal substrate material, as it has a CTE between silicon and many ceramics used for encapsulation AlN acts as a buffer layer between the Si die and an Al_2O_3 ceramic encapsulant. Sivaswamy et al. conducted an FEA analysis on both combinations mentioned and found that the interface between the substrate and the package was lifetime limiting (Figure 12.10) [Sivaswamy et al., 2008]. This is likely due to the fact that the substrate and package have a larger contact area than the substrate and die; therefore, this interface will also have the portion of attach that is furthest from the neutral point, giving it the largest strain. Another example of a good match is between Si and Si_3N_4, which has a CTE that also matches very well with a SiGe die and thus would make a good candidate for cryogenic applications [Sivaswamy et al., 2008].

Shapiro et al. tested a variety of other substrate materials such as polyimide, alumina, and LTCC. Polyimide was tested, as it is currently the most common space flight circuit board material. Alumina has been widely used in noncryogenic applications, so it was selected to determine its properties at low temperatures. The results of the –180°C to 120°C test showed failures in polyimide and alumina substrates. All samples prepared with an LTCC substrate and silver-filled epoxy attach survived the thermal cycle testing, which was expected after an FEA analysis showed that this combination of materials had the lowest stress [Shapiro et al., 2004]. The results from this study were confirmed in Tudryn et al. [2006].

Yamamoto et al. chose a GaAs die that has a CTE twice that of silicon and consequently had to fabricate a multilayer ceramic substrate to match this CTE [Yamamoto, 1991]. The authors sintered a zirconia powder with a borosilicate glass using an alumina film in between to prevent the zirconia from reacting with the glass at 1000°C. This produced a substrate with a CTE within 10% of GaAs. Thermal simulations of the die and substrate were performed to compare the maximum stresses using both alumina and the multilayer ceramic. It was found that the multilayer ceramic had a maximum stress of only 6.7 N/mm² compared to the alumina substrate, which had a maximum stress of 50 N/mm².

12.3.5 Encapsulants

Encapsulants have a large effect on the reliability of cryogenic devices. Due to the fact that at lower temperatures, materials become more brittle, plastic encapsulants are generally replaced with ceramics, which are more resistant to property changes due to thermal variations, thus improving device reliability [Cressler and Mantooth, 2012]. The most common ceramic used for packaging is alumina [Sivaswamy et al., 2008].

Another key consideration when designing discrete cryogenic ceramic electronics is the fill or encapsulation inside the ceramic housing. Shapiro et al. studied the use of an epoxy, a silicone, and parylene for such a function [Shapiro et al., 2004]. The epoxy and silicone provide a glob-top-like protection. Parylene is a thin-film covering that is deposited via a vapor deposition processing step. Interestingly, the parylene showed the best performance under thermal cycling, as all samples prepared with parylene survived the thermal stress test. One possible explanation for this is that the two glob-top-style encapsulants could have embrittled at cryogenic temperatures affecting the wire bonds in the sample [Shapiro et al., 2004]. While it would seem that parylene should then be used, it should also be noted that due to its thin-film nature, parylene does not provide mechanical support to the wire bonds. Thus, a tradeoff must be made and the application of the device should be considered when selecting an encapsulant.

12.4 Low Temperature Circuit Packaging

Once the design of the individual components has been completed, the next logical step is to design a board that can facilitate the interactions between multiple discrete devices as well as communicate with systems external to those on the board. The following sections are discussions of the various issues and potential solutions related to the selection of materials and components for board or so-called second-level packaging.

12.4.1 Boards and Flex Circuits

The printed circuit board (PCB) provides a mechanical structure for mounting of discrete packages, and facilitates the electrical connection between the devices and any external systems. Fink et al. completed a comprehensive mechanical strength study on the various materials used to make PCBs: testing PCB 2301 Multilayer Board (MLB) with epoxy glass, which is an FR-4-type material, PCB thermount, and PCB 2302 MLB polyimide glass [Fink et al., 2008]. In the study, samples were tested at room temperature, liquid nitrogen, and liquid helium for mechanical properties such as Young's modulus, yield strength, elongation at rupture, and ultimate tensile strength. Select results of the study are shown in Figure 12.11 for comparison. The 2301 MLB epoxy glass was determined to be the superior material for cryogenic PCBs, clearly showing the strongest ultimate and yield strengths. The Thermount PCB showed the weakest strengths for all categories.

In addition to testing PCB materials, Fink et al. studied various conformal coatings to determine their mechanical properties at low temperatures [Fink et al., 2008]. It was determined that there was a general trend among all of the seven different coatings tested that there is an increase in strength and modulus. Failure analysis of the coatings showed a

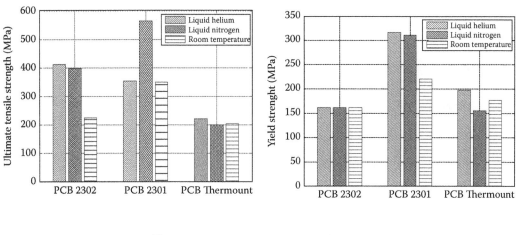

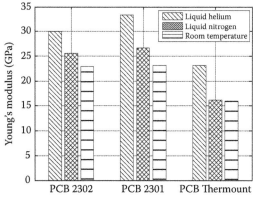

FIGURE 12.11
(Top left) Ultimate tensile strength, (top right) yield strength, and (bottom) Young's modulus of tested PCB materials. (Based on data from Fink, M. et al., *Cryogenics*, 48(11), 497–510, 2008.)

strong decrease in elongation. It is expected that due to the wide range of coatings tested, these characteristics can be extrapolated to most coatings. Fink et al. found that while the differences were minimal, Solithane 113 was superior in terms of its mechanical characteristics [Fink et al., 2008].

12.4.2 Solders

To facilitate a connection between the board and the component at cryogenic temperatures, a solder with the desired characteristics needs to be identified. The desired characteristics are likely application dependent. A general rule to follow would be to use a solder that remains soft at low temperatures to allow for compliance. In their study, Fink et al. tested a large variety of solders: 63Sn37Pb, 62Sn36Pb2Ag, 96Sn4Ag, 50In50Pb, 70Pb30In, 96.8Pb1.5Ag1.7Sn, and 96.5Sn3Ag0.5Cu [Fink et al., 2008]. The results of their testing are shown in Figure 12.12.

Lead-free tin-based solders did not perform well at low temperatures, showing a decrease in all strength parameters. Additionally, these components showed a drastic increase in

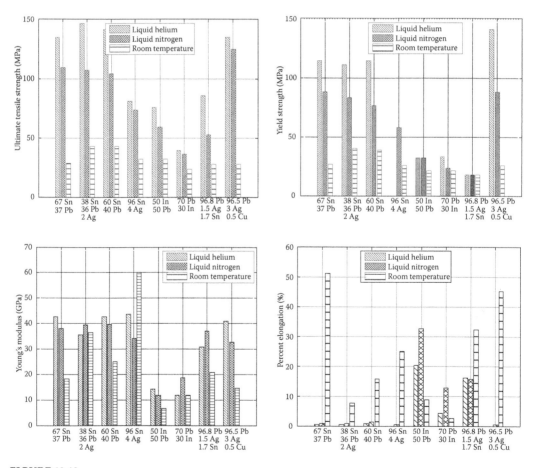

FIGURE 12.12
(Top left) Ultimate tensile strength, (top right) yield strength, (bottom left) Young's modulus, and (bottom right) percent elongation at failure of tested solder alloys. (Based on data obtained from Fink, M. et al., *Cryogenics*, 48(11), 497–510, 2008.)

brittle behavior, failing with very little elongation. This presents a difficulty when dealing with restrictions such as the restriction on the use of hazardous substances (RoHS), which attempts to remove substances such as lead from electronics. Many industries such as aerospace and military have temporary exemptions to this policy, and a significant development of lead-free solders will need to take place before such a restriction can be placed on these industries.

Lead- and indium-based solders perform very well in cryogenic testing. While these solders tend to show low yield strengths, they allow for elongation at cryogenic temperatures unlike the lead-free solders, which were highly brittle. Generally speaking, the ability to deform is desirable at low temperatures, allowing for compliance between materials of differing CTEs. One difficultly with using indium in solder is that its low melting temperature makes it susceptible to forming intermetallic compounds [Shapiro et al., 2010].

12.4.3 Passives Components: Capacitors, Inductors, and Resistors

Passive components are used in circuitry to condition signals, removing noise and helping to extract relevant information. A significant amount of research exists for capacitors at cryogenic temperatures [Teyssandier and Prêle, 2010; Patterson et al., 1998; Teverovsky, 2006]. Commercially available capacitors are often categorized into three classes: class 1, class 2, and class 3, which describe how the capacitance of the components changes with temperature or the tolerance of the temperature within a specified temperature range. Class 1 capacitors are typically the most robust in their performance and show linear trends with temperature. More information can be found on the classes of capacitors in Pan and Clive [2010]. A chart of the performance of capacitors at cryogenic temperatures is shown in Figure 12.13. Not surprisingly NP0-type capacitors showed little to no change in capacitance even at 4 K and thus should be strongly considered for systems that must show temperature invariance.

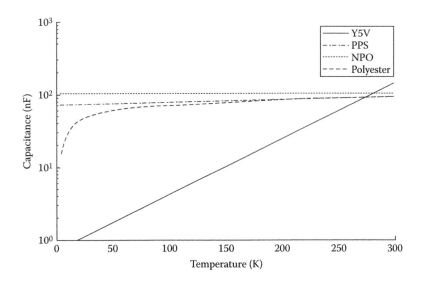

FIGURE 12.13
Performance of various classes of ceramic capacitors at cryogenic temperatures. (Data from Teyssandier, F. and Prêle, D., *Ninth International Workshop on Low Temperature Electronics (WOLTE9)*, 2010.)

For tantalum capacitors, a reduction in capacitance and an increase in equivalent series resistance were observed at 77 K [Teverovsky, 2006]. Additionally, the breakdown voltage was increased at low temperatures; however, for step surge tests, the breakdown voltage decreases, suggesting an increase in reliability for high-impedance cryogenic applications but a decrease in reliability for low-impedance cryogenic applications [Teverovsky, 2006]. Work completed by Patterson et al. suggests that polypropylene, polycarbonate, and mica-based capacitors show excellent stability at 77 K; however, tantalum capacitors showed an increase in dielectric loss, confirming the results in Teverovsky [2006] and Patterson et al. [1998].

Gerber et al. studied an array of cores used for inductors in switch mode power supplies (SMPS) [Gerber, 2004]. They tested four types of cores: three powered cores, molypermalloy cores (MPCs), high-flux cores (HFCs), and Kool Mu cores (KMCs), as well as a solid ferrite cores. The results of the study are shown in Figure 12.14. MPC showed the greatest stability in both frequency and temperature variation. HFC also showed consistent behavior across frequency and temperature; however, there were slight variations with temperature. Both the ferrite and KMC showed decreases in inductance with increasing temperature but showed stable inductance with frequency variation.

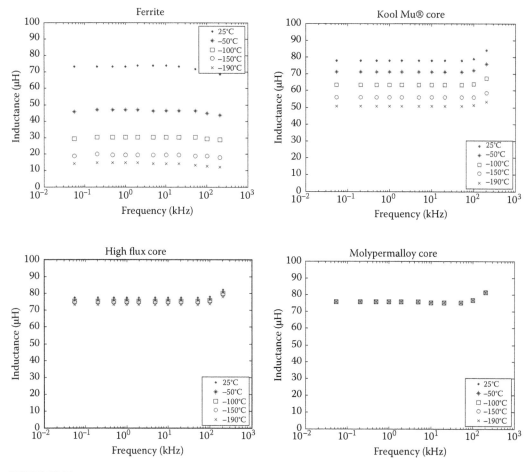

FIGURE 12.14
Core inductances at varying temperatures and frequencies. (Data from Gerber, S.S., *IECEC'02. 2002 37th Intersociety IEEE Energy Conversion Engineering Conference*, 249–254, 2004.)

TABLE 12.2

Performance of Resistors at Cryogenic Temperature [Patterson et al., 2001]

Type	Value (Ω)	Resistance (Ω) at 25°C	Resistance (Ω) at −190°C	Change in Resistance (%) at −190°C
Metal film	10	10.00	9.99	0.0
	1 K	999.15	1001.86	0.3
Wirewound	10	9.70	9.62	−0.9
	1 K	984.80	979.31	−0.6
Thin film	33	33.07	34.32	3.8
	1 K	995.41	1007.88	1.3
Thick film	100	99.99	105.42	5.4
	1 K	998.70	1003.22	0.5
Carbon film	10	9.96	10.46	5.1
	1 K	980.30	1035.83	5.7
Carbon composition	15	14.65	16.34	11.6
	1 K	1013.29	1296.54	28.0
Ceramic composition	10	9.49	10.99	15.8
	1 K	993.09	1167.51	17.6
Polymer film	10	10.00	10.48	4.9
	1 K	996.20	1037.06	4.1

Resistors are the final major passive component to be considered for second-level packaging. Patterson et al. studied the behavior of a wide variety of resistor types at 25°C and −190°C for electrical characterization in planetary exploration [Patterson et al., 2001]. The results of their study are shown in Table 12.2. Key findings from the research suggest that metal film, wirewound, thin-film, and thick-film resistors all showed temperature-independent results. Carbon film, carbon composition, ceramic composition, and polymer film all showed significant increases in resistance as temperature was decreased.

12.5 Low Temperature Housing

In this final section, third-level packaging will be considered. Again, CTE is of great importance; however, considering many cryogenic operations involve space flight, weight will be of paramount importance. Maximizing the strength to weight ratio of materials is one of NASA's key research areas [NASA, 2002]. The following sections very generally discuss issues and potential solutions related to the selection of materials for electronics housings or so-called third-level packaging.

12.5.1 Housings

Materials used for housing cryogenic electronics must be able to withstand low temperatures without producing large stresses. For space applications, these materials must also be as light and strong as possible to reduce costs. Using modern technology, it costs approximately $10,000 to put a pound of material into Earth orbit [NASA, 2014c]. Missions going further distances, such as to Mars and beyond can be expected to cost significantly more.

One category of low-density material being researched is metal matrix composites, which are materials composed of two distinct materials, one of which must be a metal [Rawal, 2001; Zweben, 1992; Ibrahim et al., 1991]. Metal–matrix composites can be used to tailor a material to attain properties that are otherwise not attainable by a pure substance. For example, an Al/B metal matrix composite has a higher ultimate tensile strength at approximately 1100 MPa, than titanium, which has a strength of 950 MPa, while having approximately half the density: 2.7 g/cc for Al/B, 4.4 g/cc for titanium.

For structures that must be kept thermally isolated from the environment, such as the interior of the International Space Station, which must maintain a livable temperature in both –160°C and 120°C conditions, thermal blankets called multilayer insulation (MLI) are employed (Figure 12.15). These are commonly composed of layers of either Kapton® or Mylar® separated by thin polyester netting. The Kapton and Mylar prevent radiation heat transfer, and the thin netting prevents conduction between the layers of the blanket [Savage, 2003].

While density and thermal insulation are priorities in housing space electronics, another key consideration in design is communications. Some internal electronics systems may need to communicate wirelessly to external systems; the housing should be designed to facilitate these communications while at the same time blocking noise from external sources.

12.5.2 Wires and Cables

Metal conductors used for wiring exhibit several phenomena at low temperatures. The general trends for a variety of metals are shown in Figure 12.16 [NIST, 2015]. Several distinct regions arise when observing this graph. First, it is seen that above 150 K all metals

FIGURE 12.15
Multilayer insulation "opened" to see various internal layering [Rossie, 2015]. (Courtesy of John Paul Rossie, Director, Aerospace Educational Development Program.)

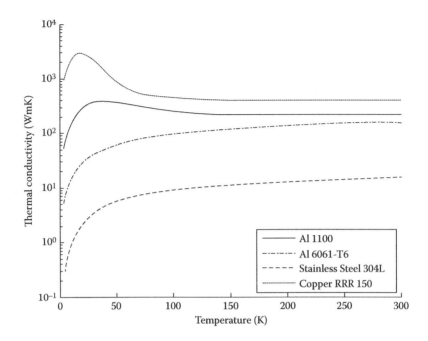

FIGURE 12.16
Thermal conductivity versus temperature for selected metals. (Data obtained from NIST, Material Properties, NIST Material Measurement Laboratory, Cryogenic Technologies Group, 2015.)

shown exhibit fairly constant thermal conductivity. The second key observation is the enhanced thermal conductivity that many metals exhibit around 20 K. As the thermal conductivity of a metal is mostly due to the transport of electrons through the metal, the reduced scattering at these temperatures allows the electrons to move most freely, increasing the thermal conductivity of the material. Finally, the conductivities of all metals approaches 0 at 0 K; this is due to electrons no longer being able to move at these low temperatures.

Metals also exhibit increased hardness, yield strength, tensile strength, modulus of elasticity, and fatigue resistance when temperature is lowered [Hurlich, 1968]. Where metals differ at low temperature is their failure signatures. Some metals lose their ductility properties at low temperatures and experience brittle failure. Other metals can remain soft at low temperatures and exhibit ductility in failure. Therefore, the choice of a metal is application dependent and may ultimately come down to which type of failure is more acceptable.

12.5.3 Connectors

Minimal research exists on the topic of electrical connectors at cryogenic temperatures. Several patents exist on the topic but these do not have much detail regarding the physical nature of connectors at low temperatures [Tighe, 2000; Otte and Fischer, 1973]. Connectors are likely to be made up of similar materials as those found in Section 12.5.2, and readers are referred to that section for metallic material considerations. In cryogenic applications, there appear to be two key considerations for connector design: electrical connection and thermal connection. Electrical connection is the more obvious and trivial of the two. Thermal connection is important for systems that must be cooled to extremely low temperatures. A designer must note the ability of the connector to participate in heat transfer. A design should strive to minimize the transfer of heat to these systems using adequate materials.

12.6 Summary

Low temperature applications for electronics systems are growing with continued space and polar exploration in addition to supercomputing technologies. Many types of electronic devices were shown to have improved operating characteristics such as transistors, LEDs, and oscillators. However, electronics engineers need to design electronics systems with packages that can reliably operate in cryogenic environments. Systems in low temperature environments should utilize small band gap semiconductors such as SiGe or germanium. Efforts should be made to match the CTEs of the materials used in the device packaging to reduce thermo-mechanical stresses. Flip chip provides the most reliable die to substrate bonding. Gold is the most reliable metal for applications that require wire bonding. Compliant "soft" indium- or lead-based solders and die attaches should be utilized to minimize residual stresses at cryogenic temperatures. Finally, due to their brittle nature at low temperatures, plastic encapsulants should be avoided where possible, in favor of ceramic packages.

Acknowledgments

The authors would like to thank Wayne Johnson, Tennessee Tech University, Cookeville, TN; Colin Johnston, Oxford University, UK; and Mike Hamilton, Auburn University, AL, for reviewing this chapter and for providing valuable technical comments and suggestions.

References

Anthony S. Cryogenic on-chip quantum electron cooling leads towards computers that consume 10x less power. Extreme Tech, 2014. Available at: http://www.extremetech.com/extreme/189999-on-chip-quantum-wells-create-cryogenic-electrons-computers-that-consume-10x-less-power.

Attar S. S., S. Setoodeh, R. Al-Dahleh, and R. R. Mansour. Cryogenic performance of gold-based and niobium-based RF MEMS devices. In *European Microwave Integrated Circuits Conference (EuMIC)*, 2011, Manchester, United Kingdom, October 2011, pp. 672–675. IEEE, 2011.

Camin D. Cryogenic behavior of optoelectrionic devices for the transmission of analog signals via fiber optics. *IEEE Transactions on Nuclear Science* 53, no. 6 (2006): 3929–3933.

Cao X. A. and S. F. LeBoeuf. Current and temperature dependent characteristics of deep-ultraviolet light-emitting diodes. *IEEE Transactions on Electron Devices* 54, no. 12 (2007): 3414–3417.

Carlson D. M., D. C. Sullivan, R. E. Bach, and D. R. Resnick. The ETA 10 liquid-nitrogen-cooled supercomputer system. *IEEE Transactions on Electron Devices* 36, no. 8 (1989): 1404–1413.

Carr W. N. Characteristics of a GaAs spontaneous infrared source with 40 percent efficiency. *IEEE Transactions on Electron Devices* 12, no. 10 (1965): 531–535.

Cressler, J. D. and H. A. Mantooth (Eds.). *Extreme Environment Electronics*. London, NY: CRC Press, 2012.

ESA. Space Science/Our Activities/ESA. European Space Agency (ESA), 2014.

Fan T. Y., D. J. Ripin, R. L. Aggarwal, J. R. Ochoa, B. Chann, M. Tilleman, and J. Spitzberg, Cryogenic Yb 3+-doped solid-state lasers. *IEEE Journal of Selected Topics in Quantum Electronics* 13, no. 3 (2007): 448–459.

Fink M., T. Fabing, M. Scheerer, E. Semerad, and B. Dunn. Measurement of mechanical properties of electronic materials at temperatures down to 4.2 K. *Cryogenics* 48, no. 11 (2008): 497–510.

Forrest, S. R. and T. M. Sanders Jr. GaAs junction field effect transistors for low-temperature environments. *Review of Scientific Instruments* 49, no. 11 (1978): 1603–1604.

Gaskill, D. K., L. B. Rowland, and K. Doverspike. Electrical transport properties of AlN, GaN and AlGaN. In *Properties of Group III Nitrides*, Edgar J. (Ed.). *EMIS Data Reviews Series, N11*. INSPEC, The Institution of Electrical Engineers, London, 1995, pp. 101–116.

Gerber S. S. Performance of high-frequency high-flux magnetic cores at cryogenic temperatures. In *IECEC'02. 2002 37th Intersociety IEEE Energy Conversion Engineering Conference*. 2004, Washington, WA, USA, July 2004, pp. 249–254.

Giles A. J., S. K. Jones, D. G. Blair, and M. J. Buckingham. A high stability microwave oscillator based on a sapphire loaded superconducting cavity. In *Proceedings of the 43rd Annual Symposium on Frequency Control, IEEE* (1989): 89–93.

Goldsmith C. L. and D. I. Forehand. Temperature variation of actuation voltage in capacitive MEMS switches. *IEEE Microwave and Wireless Components Letters* 15, no. 10 (2005): 718–720.

Hambley A. R., N. Kumar, and A. R. Kulkarni. *Electrical Engineering: Principles and Applications*. Upper Saddle River, NJ: Pearson Prentice Hall, 2008.

Hurlich A. Low temperature metals. In *1968 Summer Study on Superconducting Devices and Accelerators*. Upton, NY: Brookhaven National Laboratory, June–July, 1968.

Ibrahim I. A., F. A. Mohamed, and E. J. Lavernia. Particulate reinforced metal matrix composites—A review. *Journal of Materials Science* 26, no. 5 (1991): 1137–1156.

Kelm E. C. Operation of a germanium FET at low temperatures, *Review of Scientific Instruments* 39 (1968): 775–776.

Keyes R. Semiconductor devices at low temperatures. *Comments on Solid State Physics* 8, no. 3 (1977a): 47–53.

Keyes R. Low temperature high mobility transistor materials. *Comments on Solid State Physics* 8, no. 2 (1977b): 37–46.

Keyes R. J. and T. M. Quist, Injection luminescent pumping of CaF2: U3+ with GaAs diode lasers. *Applied Physics Letters* 4, no. 3 (1964): 50–52.

Keyes, R. W., E. P. Harris, and K. L. Konnerth. The role of low temperatures in the operation of logic circuitry. *Proceedings of the IEEE* 58, no. 12 (1970): 1914–1932.

Kirschman R. K. *Low-Temperature Electronics*. New York, NY: IEEE Press, 1986, p. 500. No individual items are abstracted in this volume.

Kirschman R. K., W. M. Sokolowski, and E. A. Kolawa, Die attachment for –120 C to +20 C thermal cycling of microelectronics for future Mars Rovers: An overview. In *ASME International Technical Conference and Exhibition on Packaging and Integration of Electronic and Photonic Microsystems* (InterPACK), Maui, Hawaii, June 1999.

Kovacs G. T. A. *Micromachined Transducers Sourcebook*. New York, NY: WCB/McGraw-Hill, 1998.

Lengeler B. Semiconductor devices suitable for use in cryogenic environments. *Cryogenics* 14, no. 8 (1974): 439–447.

Lo C. C. and B. Leskovar. Cryogenically cooled broad-band GaAs field-effect transistor preamplifier. *IEEE Transactions on Nuclear Science* 31, no. 1 (1984): 474–479.

Luiten A. N., A. G. Mann, M. E. Costa, and D. G. Blair. Power stabilized cryogenic sapphire oscillator. *IEEE Transactions on Instrumentation and Measurement* 44, no. 2 (1995): 132–135.

Lutz J., H. Schlangenotto, U. Scheuermann, and R. Doncker. Semiconductor properties. *Semiconductor Power Devices* (2011): 17–75.

Manheimer M. Cryogenic Computing Complexity. Intelligence Advanced Research Projects Activity (IARPA). http://www.iarpa.gov/index.php/research-programs/c3, visited July 1, 2015.

NASA. The Right Stuff for Super Spaceships, NASA, www.NASA.gov, 2002.

NASA. Voyager Instrument Cooling After Heater Turned Off. NASA, http://www.jpl.nasa.gov/news/news.php?release=2012-017, 2012.

NASA. The James Webb Space Telescope—About the Webb., http://jwst.nasa.gov/about.html, 2014a.

NASA. The James Webb Space Telescope—Images of the Spacecraft, http://jwst.nasa.gov/images.html, 2014b.

NASA. Advanced Space Transportation Program Fact Sheet, NASA, www.NASA.gov, 2014c.

NIST. Material Properties, NIST Material Measurement Laboratory, Cryogenic Technologies Group, http://www.nist.gov/mml/acmd/structural_materials/cryogenicmatprop.cfm, 2015.

Noel J. G., A. Bogozi, Y. A. Vlasov, and G. L. Larkins. Cryogenic pull-down voltage of microelectromechanical switches. *Journal of Microelectromechanical Systems* 17, no. 2 (2008): 351–355.

O'Donnell K. P. and X. Chen. Temperature dependence of semiconductor band gaps. *Applied Physics Letters* 58, no. 25 (1991): 2924–2926.

Oskin B. Yeti Robot Finds Cracks in Antarctic Ice. Discovery News. Discovery Communications, LLC (2013).

Otte R. F. and C. L. Fischer. Cryogenic Connection Method and Means. U.S. Patent 3,740,839, June 26, 1973.

Pan Ming-Jen and Clive A. Randall. A brief introduction to ceramic capacitors. *IEEE Electrical Insulation Magazine* 26, no. 3 (2010): 44–50.

Park K., A. Marchenkov, Z. M. Zhang, and W. P. King. Low temperature characterization of heated microcantilevers. *Journal of Applied Physics* 101, no. 9 (2007): 094504.

Patterson R., A. Hammoud, and S. Gerber, Performance of various types of resistors at low temperatures. NASA Glenn Res. Center, Cleveland, OH: GESS Rep. NAS3-00142 (2001).

Patterson R. L., A. Hammond, and S. S. Gerber. Evaluation of capacitors at cryogenic temperatures for space applications. Conference Record of the 1998 IEEE International Symposium on Electrical Insulation, vol. 2, (1998) pp. 468–471. *IEEE*, 1998.

Prance, R. J., A. P. Long, T. D. Clark, and F. Goodall. UHF ultra-low noise cryogenic FET preamplifier. *Journal of Physics E: Scientific Instruments* 15, no. 1 (1982): 101.

Price S., T. Phillips, and G. Knier. Staying Cool on the ISS. NASA Science, March 21, 2001.

Rahim M. K., J. C. Suhling, D. S. Copeland, M. S. Islam, R. C. Jaeger, P. Lall, and R. W. Johnson. Die stress characterization in flip chip on laminate assemblies. *IEEE Transactions on Components and Packaging Technologies* 28, no. 3 (2005): 415–429.

Rawal S. P. Metal-matrix composites for space applications. *Journal of Materials* 53, no. 4 (2001): 14–17.

Rossie. Multi-Layer Insulation for Satellites and Other Spacecraft Image. http://www.rossie.com/mli.htm, visited July 1, 2015.

Savage C. J. Thermal control of spacecraft. In *Spacecraft Systems Engineering*, 4th edition (2003), Hoboken, N.J.: Wiley,: 357–394.

Shapiro A. A., S. X. Ling, S. Ganesan, R. S. Cozy, D. J. Hunter, D. V. Schatzel, M. M. Mojarradi, and E. A. Kolawa. Electronic packaging for extended mars surface missions. *Proceedings of the 2004 IEEE Aerospace Conference* vol. 4, pp. 2515–2527. IEEE, 2004.

Shapiro A. A., C. Tudryn, D. Schatzel, and S. Tseng. Electronic packaging materials for extreme, low temperature, fatigue environments. *IEEE Transactions on Advanced Packaging* 33, no. 2 (2010): 408–420.

Shatter, W. J., H. S. Kong, G. H. Negley, J. W. Palmour, *Institute of Physics Conference Series* 137 (1994): 155.

Sivaswamy S., R. Wu, C. Ellis, M. Palmer, R. W. Johnson, P. McCluskey, and K. Petrarca. System-in-package for extreme environments. In *Electronic Components and Technology Conference, 2008* (ECTC 2008). 58, pp. 2044–2050. *IEEE*, 2008.

Streetman B. G. and S. Banerjee. Solid state electronic devices. Vol. 2. Englewood Cliffs, NJ: Prentice-Hall, 1995.

Teverovsky A. Performance and reliability of solid tantalum capacitors at cryogenic conditions, Greenbelt, MD: NASA Goddard Space Flight Center, 2006.

Teyssandier F. and D. Prêle. Commercially available capacitors at cryogenic temperatures. In Ninth International Workshop on Low Temperature Electronics (WOLTE9), Guaruja, Brazil, June 2010.

Tighe T. S. Flex cable connector for cryogenic application. U.S. Patent 6,045,396, issued April 4, 2000.

TOP500 Lists November 2014. Toplist 500—November 2014. Top 500, Nov. 2014.

Trautmann E., L. Ray, and J. Lever. Development of an autonomous robot for ground penetrating radar surveys of polar ice. In IEEE/RSJ International Conference on Intelligent Robots and Systems (IROS), pp. 1685–1690. *IEEE*, 2009.

Trisca G. O., M. E. Robertson, H. Marshall, L. Koenig, and M. A. Comberiate. GROVER: An autonomous vehicle for ice sheet research. In *AGU Fall Meeting Abstracts*, vol. 1, p. 0691, San Francisco, California, December 2013.

Tudryn C. D., B. Blalock, G. Burke, Y. Chen, S. Cozy, R. Ghaffarian, D. Hunter, M. Johnson, E. Kolawa, M. Mojarradi, D. Schatzel, and A. Shapiro. Low temperature thermal cycle survivability and reliability study for brushless motor drive electronics. *IEEE Aerospace Conference*, 2006 p. 37. *IEEE*, 2006.

Ulrich R. K. and S. Rajan. Thermal performance of an MCM flip-chip assembly in liquid nitrogen. IEEE Transactions on Components, Packaging, and Manufacturing Technology, Part A 19, no. 4 (1996): 451–457.

Wang R. T. Operational Parameters for the Superconducting Cavity Maser. Pasadena, CA: California Institute of Technology, 1988.

Wauters F., I. S. Kraev, M. Tandecki, E. Traykov, S. Van Gorp, D. Zákoucký, and N. Severijns. Performance of silicon PIN photodiodes at low temperatures and in high magnetic fields. *Nuclear Instruments and Methods in Physics Research Section A: Accelerators, Spectrometers, Detectors and Associated Equipment* 604, no. 3 (2009): 563–567.

Yamamoto H. Multichip module packaging for cryogenic computers. IEEE International Sympoisum on Circuits and Systems, pp. 2296–2299. *IEEE*, 1991.

Zhang, Y. M., V. Borzenets, N. Dubash, T. Reynolds, Y. G. Wey, and J. Bowers. Cryogenic performance of a high-speed GaInAs/InP pin photodiode. *Journal of Lightwave Technology* 15, no. 3 (1997): 529–533.

Zweben C. Metal-matrix composites for electronic packaging. *Journal the Minerals, Metals & Materials Society* 44, no. 7 (1992): 15–23.

13

A Cryogenic Axion Dark Matter Experiment (ADMX/ADMX-HF)

Gianpaolo Carosi and Karl van Bibber

CONTENTS

13.1 Introduction: Axions as Dark Matter

Modern cosmological research has revealed a fascinating if puzzling picture of our universe. Only one third of the energy density of the universe is in the form of matter, most of which is of unknown composition that we term *dark matter*. The remaining two thirds of the energy density is accounted for by an even more mysterious *dark energy*, responsible for driving the acceleration of the universe's expansion. What comprises the dark matter is one of the premier questions in all of science today. One candidate that has received much theoretical and experimental attention is the weakly interacting massive particle (WIMP), a heavy particle (10–1000 times the proton mass) arising from the theory of supersymmetry. To date however, no evidence for such particles has been seen either in large underground detectors, or at the Large Hadron Collider.

Another compelling candidate is the axion, arising from the best-motivated theory to explain the absence of charge-parity symmetry (CP)-violating effects in the strong interaction [Peccei & Quinn, 1977; Weinberg, 1978; Wilczek, 1978]. Unlike the WIMP, the axion is extremely light, perhaps a trillionth of an electron mass, and has extraordinarily weak interactions with radiation and matter.

Very light axions would have been produced prolifically at the time of the Big Bang, and with intrinsic mass of only a few microelectronvolts (μeV), could provide the entire dark matter of the universe. Numerical simulations bear out that cold (i.e., nonrelativistic) dark matter aggregates into dense structures, providing the deep gravitational potential wells that ordinary baryonic matter falls into, and condenses into gas clouds, stars, galaxies, and clusters. One refers to the dark matter around galaxies as their "halo"; that associated with our own Milky Way has a local density of 300–450 MeV/cm^3, or roughly a half of a proton mass per sugar-cube volume of space. Such axions would be highly nonrelativistic, having only the virial velocity of the galaxy, that is, $\beta \approx 10^{-3}$. Further details on axions and their role as dark matter can be found in the following review and lecture series [Graham et al., 2015; Kuster et al., 2008].

The theory behind the production of axions leads to a linear relation between the coupling and the mass of the axion in which the lighter the axion, the more weakly it couples to ordinary particles such as photons. The mass is unknown and was originally thought to be hundreds of keV, which would have been easily detected in beam-dump experiments and nuclear reactors [Bradley et al., 2003]. Lighter axions were originally thought to couple so weakly as to be undetectable and were dubbed "invisible axions." In 1983, Pierre Sikivie proposed an elegant experiment that would rely on the resonant conversion of dark matter axions into detectable microwaves in the presence of a large magnetic field [Sikivie, 1983]. Initial prototype experiments based on this dark matter "haloscope" concept were built and operated at the University of Florida [Hagmann et al., 1990] and at Brookhaven [Wuensch, 1989] in the late 1980s. The success of these early relatively simple experiments led to the construction of the Axion Dark Matter eXperiment (ADMX) in the mid-1990s. ADMX became the first experiment sensitive enough to detect plausible dark matter axions. It ran at Lawrence Livermore National Laboratory (LLNL) from 1996 to 2010, at which point it was moved to the University of Washington (UW) where it is undergoing a major upgrade. In addition, in 2010 a second experimental system was constructed at Yale University. Dubbed ADMX-High Frequency (ADMX-HF) it had a smaller but more powerful magnet and was designed to operate in parallel to ADMX at a higher axion mass sensitivity. In the following sections, we will cover the design parameters of the microwave cavity experiments as well as details of the ADMX and ADMX-HF experiments.

13.2 The Microwave Cavity Experiment

Axions in the 1–1000 μeV mass range possess couplings to photons that are so weak as to defy conventional strategies for production and detection, such as accelerator-based experiments. However, like its heavier cousin the neutral pion, π^0, it can decay into two photons: $a \rightarrow \gamma\gamma$. While its spontaneous decay lifetime vastly exceeds the age of the universe, as shown by Pierre Sikivie, an axion can resonantly convert into a single monochromatic photon of the same energy (rest mass + kinetic) in a microwave cavity permeated by a strong magnetic field, and with current technology, the transition rate can be made large enough for a feasible dark matter search [Sikivie, 1983]. A schematic representation of the experiment is shown in Figure 13.1. The conversion power is given by:

$$P_{SIG} = g_{a\gamma\gamma}^2 B^2 V Q C \left(\frac{\rho_a}{m_a} \right) \tag{13.1}$$

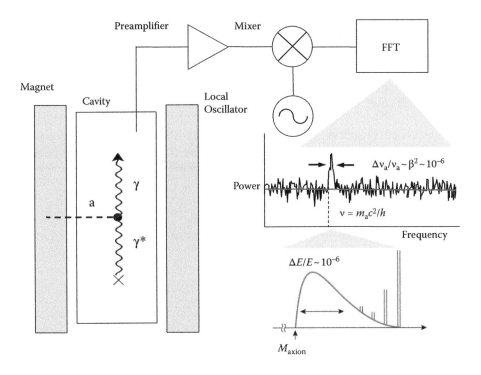

FIGURE 13.1
Schematic of the microwave cavity experiment. The axion resonantly converts into a weak microwave signal within a strong magnetic field; the signal is mixed down and a power spectrum derived by a fast Fourier transform. The signal is Doppler broadened by O (10^{-6}) due the motion of axions in the halo. The expected signal may possess fine structure due to late-infall axions into the halo.

where $g_{a\gamma\gamma}$ is the axion–photon coupling constant, B the magnetic field, V the cavity volume, Q the cavity quality factor, C the cavity form factor, and m_a and ρ_a the axion mass and its local density, respectively.

The expected signal power for current experiments is exceedingly tiny, of the order of 10^{-23} watts for ADMX, and to make the search more challenging, the cavity must be tuned in very small overlapping steps to cover the three decades of the axion's allowed mass range, corresponding to f = GHz to f = THz frequencies. The conversion condition being $hf = m_a c^2 (1 + O(\beta^2))$, where h is Planck's constant and c the speed of light.

As a signal processing problem, the sensitivity of the experiment to detect or exclude axions of particular mass and coupling constant (m_a, $g_{a\gamma\gamma}$) depends not only on maximizing the signal power, but on reducing the noise background as well. The experimental sensitivity is governed by the Dicke radiometer equation, which expresses the signal-to-noise ratio as:

$$\frac{s}{n} = \frac{P_{\text{SIG}}}{P_{\text{NOISE}}} = \frac{P_{\text{SIG}}}{k_B T_{\text{SYS}}} \sqrt{\frac{t}{B}} \qquad (13.2)$$

where P_{SIG} is the axion signal given by Equation (13.1), k_B is Boltzmann's constant, T_{SYS} is the system temperature, t is the integration time, and B is the bandwidth. The integration time is typically of the order of 80 seconds, and the bandwidth is typically ~1 kHz. The background can be expressed in terms of an equivalent system noise temperature, $T_{\text{SYS}} = T + T_N$, where T is the physical temperature of the microwave cavity and represents the blackbody photon contribution, and T_N is the noise equivalent temperature of the amplifier and receiver chain. All amplifiers contribute noise, and furthermore, linear amplifiers are subject to an irreducible minimum, the standard quantum limit (SQL), given by $k_B T_{SQL} = hf$. Thus the search for axionic dark matter naturally becomes a low temperature experiment; not only must one lower the physical temperature to the absolute minimum to reduce the blackbody contribution, but for most all amplifiers, their intrinsic noise also goes down as they are cooled.

Since the mass of the axion is unknown one would like to maximize the experiment's frequency scan rate. At a given signal-to-noise ratio, the logarithmic scan rate (which would be constant, as the axion linewidth increases as a function of frequency) is given as:

$$\frac{1}{f}\frac{df}{dt} \propto \left(B_0^2 V\right)^2 \cdot \frac{1}{T_{SYS}^2} \tag{13.3}$$

For a fixed scan rate, however, the sensitivity of the microwave experiment goes as:

$$g_{a\gamma\gamma}^2 \propto \left(B_0^2 V\right)^{-1} T_{SYS} \tag{13.4}$$

From these equations, we can see the premium one achieves by lowering the system noise temperature. Ideally one would like to search all of plausible axion model space indicated in the blue band of Figure 13.2b as quickly as possible.

The contributions to the noise temperature from the amplifier chain is usually dominated by the first-stage amplifier (T_1) giving a total system noise temperature of $T_{SYS} \approx T + T_1$. A subsequent amplifier stage (for example, with intrinsic noise temp T_2) is scaled by the gain of the amplifier preceding it (G_1), adding an additional (but subdominant) contribution of T_2/G_1. For the case of a first-stage amplifier with $T_1 = 0.1$ K and a gain $G_1 = 100$ (or 20 dB) followed by an amplifier with a $T_2 = 2$ K, the additional contribution from the second amplifier stage to the overall system noise is only $T_2/G_1 = 0.02$ K. As a result, it is crucial that the first-stage amplifier adds the minimum amount of noise to the system.

Figure 13.2 represents the setup from teh Rochester-BNL-FNAL experiment and the present exclusion region for axions constituting our galactic halo. The remainder of this chapter will describe the current and future microwave cavity experiments and technologies to improve their sensitivity by both boosting signal power and reducing noise.

13.3 ADMX

13.3.1 ADMX Experimental Layout

The ADMX was the first apparatus to have enough sensitivity to detect realistic axion models (indicated in the blue band of Figure 13.2b). ADMX, which began operations at LLNL in the mid-1990s, consists of a large-volume high-field magnet with a re-entrant bore. The main magnet consists of a superconducting solenoid with an 8-tesla maximum field and an approximately 500-henry inductance. The magnet bore is approximately 0.5 m in diameter with a 1 m long high-field region. The cryostat that holds the magnet has an open re-entrant bore that allows for insertion of an experimental insert. This cryostat holds up to 1000 L of LHe, which covers the magnet, keeping the superconducting coil cold

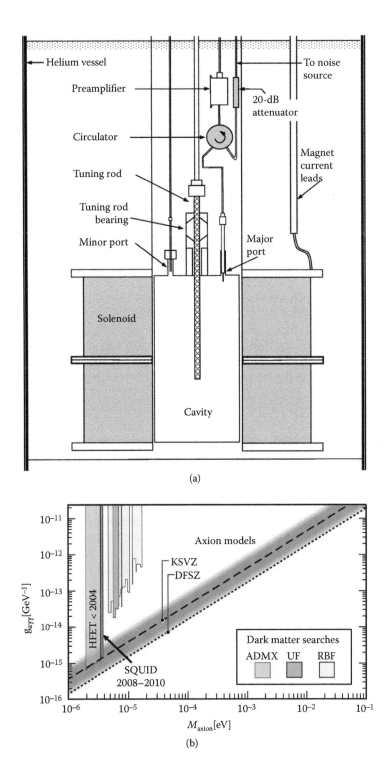

FIGURE 13.2

(a) The Rochester-BNL-FNAL experiment [Wuensch et al., 1989]. (b) Exclusion region (M_{axion}, $g_{a\gamma\gamma}$) for axions constituting our local halo. UF represents another early experiment at the University of Florida [Hagmann et al., 1990].

as well as providing a thermal sink in the event of a magnet quench. Due to the inductance of the coil, it takes approximately 24 hours to ramp the magnet to full field (or back down to zero field).

The re-entrant bore of the cryostat allows for insertion of the experimental insert without the need to warm the main magnet. This insert contains a large-volume tunable resonant microwave cavity that is placed in the center of the magnetic field. This microwave cavity consists of a 1 m long, 0.4 m diameter copper-coated stainless steel cylinder. Although the cavity supports an infinite number of resonant modes, the axion only couples to the ones whose E-field aligns with the solenoid's axial B-field (transverse magnetic modes). The form-factor C is a dimensionless number best described qualitatively as the overlap of the cavity's electric field at a particular mode with the external magnetic field. Only the TM_{0n0} modes have a nonzero form factor. The TM_{010} mode has the largest ($C \sim 0.68$ for an empty cavity) and is typically the one that the experiment tracks to maximize potential axion-to-photon conversion power. The resonant frequency of the TM_{010} mode is adjusted by translating two tuning rods from the edge of the cylinder to the center. This raises the resonant frequency in the case of metal tuning rods or lowers it in the case of dielectric rods. The current ADMX cavity has a tuning range of ~0.5–1 GHz by employing two 5 cm diameter copper tuning rods.

The quality factor (Q) of the microwave cavity is also directly related to the power generated from axion–photon conversions in the cavity (see Equation 13.1). The quality factor of a particular resonant mode is defined as the ratio of the stored energy (U) over the power loss (P_l) during one cycle at the central frequency (ω) of the mode ($Q = \omega U/P_l$). It is proportional to the cavity volume (V) over the volume occupied by the E&M fields in the walls ($S\delta$), where S is the surface area and δ is the skin depth into the wall. For an empty right-circular cylinder cavity of radius R and length d, the quality factor $Q = d/(1+d/R)\delta$. During operations Q is measured as the ratio of center frequency over the full width half maximum (FWHM) of a mode and is thus an indication of the sharpness of the resonant peak.

Cavities used in ADMX have been constructed with low RF surface resistivity metals such as pure oxygen-free copper in order to maximize their Q at cryogenic temperatures. The cavities are made out of 300-series stainless steel, which is largely nonmagnetic and then plated with a thin layer of pure oxygen-free high-conductivity copper. The copper is then annealed at moderate temperatures (typically 400°C) in vacuum for 8–10 hours in order to increase the copper grain sizes and lower resistivity. The stainless steel gives the cavity high mechanical strength relative to the soft annealed copper on the surface. The skin depth of ordinary metal is given by $\delta = \sqrt{\dfrac{\rho}{\pi \mu f}}$ where ρ is the resistivity, μ is the permeability, and f is the frequency. However at cryogenic temperatures the classic skin depth for pure copper is replaced by the anomalous skin depth δ_a, which occurs due to the long ballistic mean free path of electrons at cryogenic temperatures. The anomalous skin depth is given by:

$$\delta_a = \left[\frac{\sqrt{3}c^2 m_e v_f}{8\pi^2 \omega n e^2} \right]^{\frac{1}{3}}$$

(13.5)

where m_e and e are the electron mass and charge, n the electron density, ω the angular frequency, and v_f the Fermi velocity of the metal. For pure copper, $n_{cu} = 8.5 \times 10^{22}\,\text{cm}^{-3}$ and $v_f = 1.57 \times 10^8\,\text{cm/s}$, which gives $\delta_a = 2.8 \times 10^{-5}\,\text{cm}\left(f[GHZ] \right)^{-1/3}$.

For the ADMX cavity operating at cryogenic temperatures in the anomalous skin-depth regime, the quality factor is typically $Q \sim 10^5$. The signal from dark matter axions is also expected to have a finite width due to the added kinetic energy term of axions moving in the halo relative to the detector. It is assumed to be a Maxwellian distribution with typical galactic velocities of $\beta = v/c \sim 10^{-3}$, which results in an energy spread of $\Delta E \sim 10^{-6}$, or an equivalent $Q_{axion} \sim 10^6$. As a result, one would like to make a cavity with a Q as close to 10^6 as one can but not higher as one would only be sampling a portion of the axion linewidth and would thus lose signal. Although the Q of current copper-coated cavities are an order of magnitude lower than the expected axion line shape, there are currently R&D efforts underway to improve this using superconducting thin films, which we will briefly describe later in this chapter.

A side benefit of using the stainless steel substrate comes from the occurrence of eddy currents that can be generated in the cavity due to the changing B-field, either from the magnetic field being raised or lowered or from vibrations of the cavity during operations. By making the bulk of the cavity out of stainless steel, which has a resistivity several orders of magnitude higher than copper at cryogenic temperatures, the potential heating from these eddy currents can be minimized.

13.3.2 ADMX Data Collection

The ADMX experiment operates by first sweeping the cavity with RF power through a weakly coupled antenna port. This swept signal is picked up by a second receiving antenna that transmits the RF generated in the cavity through multiple cryogenic and room-temperature amplifiers before the signal is mixed down to audio frequencies via a double-heterodyne receiver chain. The receiving antenna position is adjusted mechanically until its impedance matches that of the cavities. This is known as "critical coupling" and maximizes the transmission of signals generated in the cavity to the amplifiers. After the cavity's resonant frequency and quality factor are verified, the swept power is switched off and the intrinsic signal of the cavity (containing both blackbody and any axion generated photons) is integrated for approximately 80 seconds. Then, the cavity's resonant mode is tuned to a new center frequency (usually in ~1 kHz steps) and the measurement repeated. Each 80 sec long voltage time series undergoes a fast Fourier transform, which turns the time-series data into a power spectrum. An axion candidate would show up as an excess power signal above the noise floor. Generally the threshold is set high enough above the noise floor so that statistical fluctuations do not generate too many candidates. Candidates that appear above the threshold are rescanned. If they are statistical in nature, they will disappear with subsequent rescans. If they are persistent then they can be correlated with potential external RF sources that might have evaded the careful electrical and RF isolation of the insert. If no external source can be found, a simple test is to vary the magnetic field. Power from an axion-generated signal should vary as B^2 and this relation can be considered a "smoking gun" as other backgrounds would be unlikely to follow such behavior. To date no such persistent signal has been detected.

13.3.3 ADMX Cryogenic Amplifiers

The initial phase of ADMX, which ran from 1996 to 2003, used first-stage amplifiers based on GaAs heterostructure field-effect transistors (HFETs). These transistor amplifiers were built by the National Radio Astronomy Observatory (NRAO) and are generally used in radio telescopes. They have cryogenic noise temperatures as low as 1.5–2 K and typical gain of order 20 dB. Details of the HFET amplifier design for ADMX can be found in Bradley [1999]. Each amplifier consists of two HFET transistors configured in a balanced design, which allows for 50-ohm input and output impedances over a broad frequency range without requiring lossy ferrite circulators (Figure 13.3).

During initial ADMX operations, cooling was provided by continuously pumping on liquid helium (^{4}He). This lowered the system's physical temperature to $T \sim 2$ K which, when combined with the first-stage amplifier noise temperature of 2 K, led to a $T_{SYS} \sim 4$ K. This data run achieved sensitivity to axions with couplings at KSVZ sensitivity over the frequency range of 400–800 MHz. The results were published in the study by Hagmann et al. [1998]. Although this was the first time plausible dark matter axion model space had been explored, it quickly became clear that lower noise amplifiers were required in order to search the full parameter space in a reasonable amount of time.

As a result, the HFET amplifiers were replaced with microstrip superconducing quantum interference device (SQUID) amplifiers (MSA) that could approach the quantum limit ($T_{SQL} \sim hf/k_B$ or ~48 mK at 1 GHz). These amplifiers consist of two Josephson junctions (JJ) in parallel with a superconducting loop. The introduction of a constant current bias (I_B) produces a voltage across the loop of ~10 µV. When the magnetic flux that threads the loop is varied, the voltage oscillates and the MSA acts like a flux-to-voltage transducer. Microwave signals can be coupled into the SQUID through a microstrip consisting of a square superconducting loop isolated from the superconducting washer underneath by an insulating layer (Figure 13.4). The MSA noise temperature is typically half its physical temperature until it begins to approach the quantum limit. Further details on the design and performance of MSA amplifiers can be found in the studies by Clarke et al. [2006] and Mück et al. [1998, 2001]. Their initial integration and performance in ADMX can be found in the study by Asztalos et al. [2011].

As the MSA is based on superconducting technology, it requires a magnetic field-free region in which to operate. As a result, a second B-field cancellation or "bucking" coil was introduced into ADMX, which negates the B-field approximately 2 m above the center of the main magnet. This bucking magnet is housed in a liquid helium reservoir located directly above the cavity, which cools the superconducting coil and provides quench protection. The bucking magnet itself consists of two counter-wound sections on a common

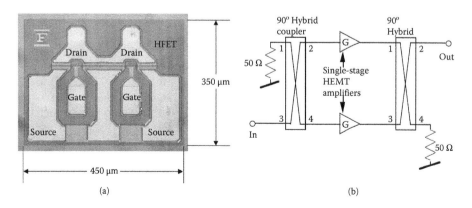

(a)   (b)

FIGURE 13.3

(a) Microphotograph of an HFET amplifier. (b) Balanced amplifier design, which allows for a broad 50-ohm input/output impedance [Bradley et al., 2003].

mandrel to minimize the mutual inductance with the main magnet and limit the net axial force to less than 400 N. The bore of the reservoir is open, allowing for placement of the MSA amplifier in the field minimum region, which is canceled to <1 mT on-axis. Additional layers of B-field canceling μ-metal and superconducting shields allow for an essentially field-free region where the MSA can operate.

In order to exploit the full potential of these amplifiers, they, along with the ADMX cavity, would need to be cooled to ~100 mK, which would require replacing the pumped ^{4}He system with a dilution refrigerator. Due to the expense and technical challenge of implementing both the MSA and dilution refrigerator, a phased approach was undertaken in which the MSA amplifiers were first demonstrated to operate in ADMX but with the 1.5 K physical temperature provided by the pumped ^{4}He pot that was attached directly to the cavity and cooled the MSA amplifier via a cold finger inserted in the bore of the bucking magnet. This "1 K" pot sourced its ^{4}He from the bucking magnet reservoir (see Figure 13.5). This was the ADMX Phase I experimental configuration that took data between 810 and 890 MHz (indicated by the gray band in Figure 13.2), from 2008 to 2010, and proved that MSA amplifiers could be operated successfully in the proximity of an 8-tesla magnet. The results of ADMX Phase I can be found in the studies by Asztalos et al. [2010] and Hoskins et al. [2011].

13.3.4 ADMX Near-Term Operating Plan

Directly after the ADMX Phase I data run the main magnet was moved from LLNL to the Center for Experimental Nuclear Physics and Astrophysics (CENPA) at the University of Washington. This was to prepare it for the ADMX "Generation 2" project, which would involve the installation of a dilution refrigerator to achieve the ultimate sensitivity. The dilution refrigerator operates with a mixture of ^{3}He/^{4}He and exploits the isotopes fermionic (^{3}He) and bosonic (^{4}He) nature to achieve temperatures in the milli-Kelvin range. The dilution refrigerator for ADMX was commissioned by Janis Corporation and is specified to have 800 μW of cooling power at 100 mK. The experimental insert was completely

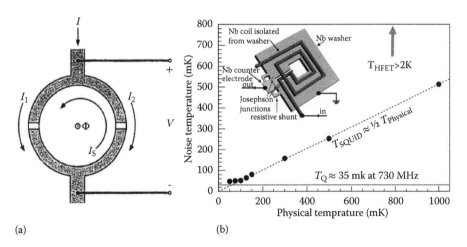

(a)   (b)

FIGURE 13.4

(a) Schematic of a SQUID loop with two Josephson junctions. (b) Noise contribution from an MSA amplifier as a function of physical temperature with a more realistic drawing of the microstrip SQUID geometry inserted [Bradley et al., 2003; Aszalos et al., 2011].

Field cancellation coil

SQUID amplifier package

Refrigeration

Antennas

8 T magnet

Microwave cavity

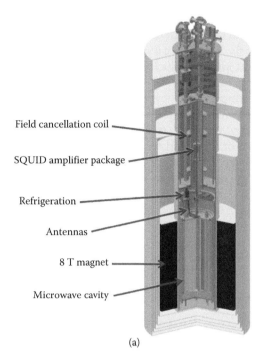

(a)

(b)

FIGURE 13.5

(a) Schematic of the ADMX detector system. The insert shows the microwave cavity with the cutout of the LHe reservoir that contains the field cancelling magnet. The entire insert (including cavity, reservoir, and shielding) can be removed from the cryostat that contains the main magnet (blue). (b) The ADMX cryogenic insert being removed from the bore. The copper thermal shields and the reservoir for the field cancellation magnet are visible.

redesigned with an eye toward the stringent thermal requirements for the dilution refrigerator to reach 100 mK. The bucking magnet's liquid helium reservoir will provide a 4.2 K thermal shield followed by a second "1 K" shield that would surround the cavity and dilution refrigerator system and be cooled by a separately pumped ^{4}He pot with a base temperature of 1.2 K. This nested set of thermal shields will reduce the expected radiation thermal load on the dilution refrigerator to <1 μW. Heat loads from necessary stainless steel supports for the cavity along with connectors for the RF lines, motion control, and sensors will be limited to <300 μW.

Currently the motion control for both the tuning rods and the antennas use external stepper motors that are connected through gearboxes thermally sunk to the 1 K shield. The rotary gearboxes consist of two sets of anti-backlash worm gears in series to achieve a large 19,600:1 gear reduction. The frictional heat generated from the gearbox motion will primarily be intercepted by the 1 K plate, minimizing the added heat load on the dilution refrigerator. Additionally rotary-drive-based commercial piezoelectric motors are being investigated as potential replacements to the gearboxes in order to further reduce the heat load on the dilution refrigerator.

In addition to the dilution refrigerator, the ADMX CENPA site was upgraded with the installation of a Linde L1410 helium reliquifier. This allowed the main magnet and the bucking magnet liquid helium reservoirs to operate in a closed-loop system that captures the boil-off gas. Without this system ADMX would boil off ~100 L a day in liquid helium. Given this consumption rate and the current price of liquid helium, this reliquefaction system has already passed the break-even point to justify its purchase.

The redesigned ADMX system performed an initial data run in the summer of 2014 at pumped ^{4}He temperatures ($T \sim 1.5$ K) as the dilution refrigerator was being constructed. The dilution refrigerator has just completed a lengthy commissioning process and has passed all tests satisfactorily; the first data run should begin in summer 2016. Its initial scan region, for the first year of data taking, will be between 0.5 and 1 GHz with sensitivity to pessimistic DFSZ axion couplings. Subsequent data runs are planned and will explore axion masses in the region from 1–40 μeV (0.25–10 GHz) (see Figures 13.2b and 13.6).

13.4 ADMX-HF (High Frequency)

A second platform of the ADMX collaboration was constructed and commissioned by groups from Yale (where it is sited), U of Colorado, UC Berkeley, and LLNL. The experiment fulfills a dual purpose. First, it is an *innovation test bed* to explore new concepts and develop beyond-state-of-the-art technologies that can improve the sensitivity of the microwave cavity experiment, and extend its mass reach into the 5–25 GHz range (approximately 20–100 μeV). Second, it serves as a *pathfinder* to take first data in that range, with projected sensitivity to the middle of the band of axion models.

While of smaller volume than ADMX, it is a highly capable experiment as it has incorporated a dilution refrigerator (VeriCold Technologies, Ismaning, Germany) from the outset. The magnet is a 17.5 cm diameter × 40 cm long NbTi superconducting solenoid with a 9.4-T central field, from Cryomagnetics Inc, Oak Ridge, Tennessee. The magnet was designed to have an extremely uniform field in the central region with strict limits on the radial component at the position of the walls of the microwave cavity, $B_r < 50$ G. This specification envisions the possibility of a cavity with a Type II superconducting thin film on all cylindrical surfaces of the cavity (i.e., those parallel to the magnetic field), to boost the Q of the cavity and thus conversion power by an order of magnitude, as will be described in the following section.

In its initial configuration, the microwave cavity is stainless steel, electroplated with high-purity copper and annealed in the same manner as the ADMX cavity. The cavity is a right circular cylinder of 10.2 cm inner diameter × 25.4 cm length, with a 5.1 cm diameter copper-plated tuning rod that pivots inside the cavity between the wall and the center, thereby spanning the frequency range 3.6–5.9 GHz in the TM_{010} mode used in the experiment.

The experiment utilizes Josephson parametric amplifiers (JPAs), which readily achieve quantum-limited performance, and are tunable over a broad frequency range, the current unit spanning 4.7–6.5 GHz. Figure 13.7 displays the setup in various stages of assembly.

The experiment completed its commissioning and performed a series of short engineering data runs in 2015. All systems performed as expected; the experiment was operated at a base temperature of ~ 100 mK, and the equivalent noise temperature of the JPA receiver was consistent with approximately four times the quantum limit, that

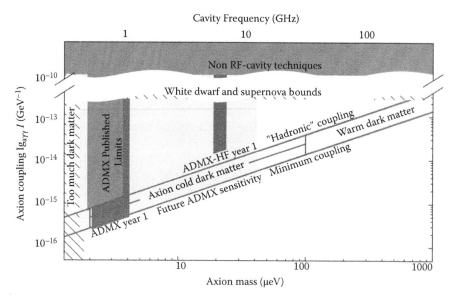

FIGURE 13.6
Projections for the first year of data-taking with both ADMX and ADMX-HF along with planned future ADMX "Generation 2" project sensitivity in the years following.

FIGURE 13.7
The ADMX-HF experiment. (a) The microwave cavity with the tuning rod pivoted to the outermost position (below; 3.6 GHz), and innermost position (top; 5.9 GHz). (b) The cavity suspended from the gantry showing both the rotary tuning motion, as well as the linear motions for the variable antenna coupling, and the tuning vernier. (c) The gantry within its thermal shields, suspended from the dilution refrigerator, ready to be lowered into the magnet. (d) The completed experiment from below deck. The data acquisition and control systems occupy the room above the grating.

is, $T_N \sim 4T_{SQL} = 1.1$ K at a frequency of 5.8 GHz. Insufficient cooling of the tuning rod is suspected in the elevated noise temperature, for which an improved thermal link has been designed for the next run. A long data-taking run began in late January 2016, which will continue until June. Identical data images are being archived at both Yale and Berkeley, where the analysis will be performed. Further details of ADMX-HF can be found in the study by Shokair [2014].

13.5 Future Developments

Two potentially high-impact technology developments will be explored by ADMX-HF in the near future.

13.5.1 Thin-Film Superconducting Cavities

There is a premium on improving the quality factor of the cavity, as the conversion power and the scan rate depend linearly on Q. The loaded quality factor, that is, critically coupled with the antenna and receiver, of the copper cavities used in ADMX and ADMX-HF is typically $Q_L \sim 10^{(4-5)}$, whereas the intrinsic width of the axion signal is $Q_a \sim 10^6$. Thus, an order of magnitude improvement would represent a significant enhancement of capability (it does not behoove us to make the bandwidth of the cavity narrower than the axion signal). Recently, Xi et al. have demonstrated that thin films of Type-II superconductors, for example, $Nb_xTi_{1-x}N$, are superconducting up to >100 GHz, even in a 10-T magnetic field parallel to the surface [Xi et al., 2010]. An R&D program has begun to produce thin-film superconducting (TFS) cavities by means of RF plasma deposition, and successful films have been made and tested in planar coupon tests, which exhibit suitably high critical temperatures, $T_C \sim 12.3$ K (Figure 13.8). The next steps will be to make uniform films on cylindrical substrates, that is, prototype cavity bodies and tuning rods, test RF performance in a magnetic field, and finally fabricate a full-size 5 GHz cavity for actual operation.

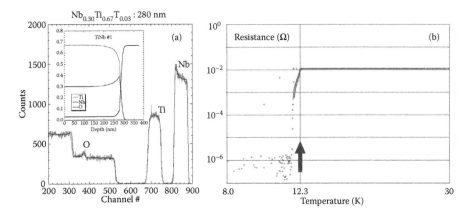

FIGURE 13.8
(a) Rutherford back-scattering (RBS) profile of a NbTiN film, showing that bulk oxidation is under control. (b) Resistivity versus temperature for this film, for which $T_C \sim 12$ K.

13.5.2 Squeezed Vacuum States

Reducing the system noise temperature is another frontier for the microwave cavity experiment. The Colorado/JILA group is providing JPAs for the current configuration of ADMX-HF, which are demonstrably operating at or near the quantum limit. However, sophisticated and elegant techniques exist to circumvent the SQL, which goes under the rubric of squeezed vacuum states. A complete discussion exceeds the scope of this short review, but in summary, noise can be decomposed into two independent quadratures (i.e., sine- and cosine-like), and a transformation performed that rotates fluctuations into a quadrature that is not measured. This group has demonstrated a system on the bench that has achieved an effective noise temperature of $k_B T_N \sim hf\Phi/4$ [Mallet et al., 2011], and a system of this design will be incorporated into the experiment and its *in situ* performance evaluated.

13.5.3 Other Axion Searches

In addition to the ADMX and ADMX-HF efforts in the United States, other experimental axion dark matter searches are coming online internationally. This includes the recently formed Center for Axion and Precision Physics (CAPP) in Daejeon, South Korea, which is in the early stages of constructing a microwave cavity haloscope search similar to ADMX. In addition, there are experimental efforts to look for axions or axion-like particles produced from the sun (CAST, IAXO) as well as from intense laser sources (ALPS II). These experiments are generally far less sensitive to axions as they require the axions to be first produced by an intense photon source (the sun or lasers) and then reconverted back to detectable photons. Although the details of these experiments are beyond the scope of this chapter, further details can be found in the study by Kuster [2008].

13.6 Summary and Conclusions

The microwave cavity experiment is now well poised to discover or exclude axions as a major component of the dark matter of the universe. The sensitivity of this search and its possible success build on a century of discovery and innovation in low temperature science and technology to both reduce backgrounds and maximize signal power: superconducting magnets and microwave cavities, dilution refrigerators, and quantum-limited or even sub-quantum limited amplifiers.

Acknowledgments

This work was supported under the auspices of the National Science Foundation, under grants PHY-1067242, and PHY-1306729, the Heising-Simons Foundation under grant 2014-182, and the auspices of the U.S. Department of Energy by Lawrence Livermore National Security, LLC, Lawrence Livermore National Laboratory under Contract DE-AC52-07NA27344. LLNL-BOOK-666403.

The authors would like to thank Georg Raffelt, Max Planck Institute for Physics, Munich, Germany; and William Wester, Fermilab, Batavia, IL, for reviewing this chapter and for providing valuable technical comments and suggestions.

References

Asztalos, S.J., G. Carosi, C. Hagmann, D. Kinion, K. van Bibber, M. Hotz, L. J. Rosenberg, G. Rybka, J. Hoskins, J. Hwang, P. Sikivie, D.B. Tanner, R. Bradley, and J. Clarke. SQUID-based microwave cavity search for dark-matter axions. *Phys. Rev. Lett.* 104(2010): 041301.

Asztalos, S.J., G.Carosi,, C. Hagmann, D. Kinion, K. van Bibber, M. Hotz, L. J. Rosenberg, G. Rybka, A. Wagner, J. Hoskins, C. Martin, N.S. Sullivan, D.B. Tanner, R. Bradley, J. Clarke. Design and performance of the ADMX SQUID-based microwave receiver. *Nucl. Instr. Meth. A* 656(2011): 39.

Bradley, R. Cryogenic, low-noise, balanced amplifiers for the 300–1200 MHz band using heterostructure field-effect transistors. *Nuc. Phys. B* 72(1999): 137.

Bradley, R., J. Clarke, D. Kinion, L. J. Rosenberg, K. van Bibber, S. Matsuki, M. Mück, P. Sikivie. Microwave cavity searches for dark-matter axions. *Rev. Mod. Phys.* 75(2003): 777.

Clarke, J., A. Lee, Mück, and P. Richards. SQUID voltmeters and amplifiers. In *The SQUID Handbook: Applications of SQUIDs and SQUID Systems*, Vol. II, John Clarke and Alex I. Braginski (Eds.), (2006), p. 1.

Graham, P.W., Irastorza, I.I., Lamoreaux, S.K., Lindner, A., van Bibber, K.A. Experimental searches for axions and axion-like particles. *Annu. Rev. Nucl. Part. S.* 65 (2015): 485.

Hagmann, C., P. Sikivie, N. S. Sullivan, and D. B. Tanner. Results from a search for cosmic axions. *Phys. Rev. D* 42(1990): 1297.

Hoskins, J., J. Hwang, C. Martin, P. Sikivie, N. S. Sullivan, D. B. Tanner, M. Hotz, L. J. Rosenberg, G. Rybka, A. Wagner, S. J. Asztalos, G. Carosi, C. Hagmann, D. Kinion, K. van Bibber, R. Bradley, and J. Clarke. Search for nonvirialized axionic dark matter. *Phys. Rev. D.* 84(2011), 121302.

Kuster, M., Raffelt, G., Beltran, B. (eds). *Axions: Theory, Cosmology, and Experimental Searches*. Berlin, Germany: Springer-Verlag, (2008).

Mallet, F., M. A. Castellanos-Beltran, H. S. Ku, S. Glancy, E. Knill, K.D. Irwin, G. C. Hilton, L. R. Vale, and K. W. Lehnert. Quantum state tomography of an itinerant squeezed microwave field. *Phys. Rev. Lett.* 106(2011): 220502.

Mück, M., J. Kycia, and J. Clarke. Superconducting quantum interference device as a near-quantum-limited amplifier at 0.5 GHz. *App. Phys. Lett.* 78(2001): 967.

Mück, M., M. André, J. Clarke, J. Gail, and C. Heiden. Radio-frequency amplifier based on a niobium dc superconducting quantum interference device with microstrip input coupling. *App. Phys. Lett.* 72(1998): 2885.

Peccei, R.D. and Quinn, H.R. CP conservation in the presence of pseudoparticles. *Phys. Rev. Lett.* 38(1977): 1440.

Shokair, T., J. Root, K.A. van Bibber, B. Brubaker, Y.V. Gurevich, S.B. Cahn, S.K. Lamoreaux, M.A. Anil, K. W. Lehnert, B.K. Mitchell, A. Reed, and G. Carosi. Future directions in the microwave cavity search for dark matter axions. *Int. J. of Mod. Phys A.* 29, 19(2014): 1443004.

Sikivie, P. Experimental tests of the 'invisible' axion. *Phys. Rev. Lett.* 51(1983): 1415–1417; erratum *Phys. Rev. Lett.* 54(1984): 695.

Weinberg, S. A new light boson? *Phys. Rev. Lett.* 40(1978): 223.

Wilczek, F. Problem of strong P and T invariance in the presence of instatons. *Phys. Rev. Lett.* 40(1978): 279.

Wuensch, W., S. De Panfilis-Wuensch, Y. K. Semertzidis, J. T. Rogers, A. C. Melissinos, H. J. Halama, B. E. Moskowitz, A. G. Prodell, W. B. Fowler, and F. A. Nezrick.Results of a laboratory search for cosmic axions and other weakly coupled light particles. *Phys. Rev. D* 40(1989):3153.

Xi, X., J. Hwang, C. Martin, D.B. Tanner, and G.L. Carr. Far-infrared conductivity measurements of pair breaking in superconducting $Nb_{0.5}Ti_{0.5}N$ thin films induced by an external magnetic field. *Phys. Rev. Lett.* 105(2010): 257006.

14

Low Temperature Materials and Mechanisms: Applications and Challenges

Ray Radebaugh and Yoseph Bar-Cohen

CONTENTS

14.1 Introduction

Civilizations have used temperatures near 0°C for thousands of years. The ability to reach significantly lower temperatures and into the cryogenic range has only become feasible in the last two centuries with the development of effective coolers as described in Chapters 1 and 6. Specifically, the use of cryogenic temperatures offers numerous benefits as listed:

1. Preservation of biological material and food
2. High fluid densities (liquefaction and separation of gases)

3. Macroscopic quantum phenomena (superconductivity and superfluidity)
4. Reduced thermal noise
5. Low vapor pressures (cryopumping)
6. Temporary or permanent property changes
7. Tissue destruction (cryoablation)

Applications of cryogenics make use of one or more of these benefits. In some cases, the benefits are so profound that the use of an ambient-temperature solution is completely impractical. One important example is the use of superconducting magnets for magnetic resonance imaging (MRI). Magnetic fields of 1.5 T are required for obtaining a reasonable resolution. The use of copper electromagnets to produce such fields at room temperature over the volume of a human body would require megawatts of power to overcome the resistive loss in the wire and a massive stream of water to provide the necessary cooling to remove the heat generated from Joule heating.

However, the production of cryogenic temperatures presents several challenges or disadvantages. Thermodynamic laws, which dictate an increased power input, cannot be overcome, but the mechanisms for producing the low temperatures can continually be improved through the use of innovation and motivation. For any application to be useful or marketable, the benefits must outweigh the disadvantages. Typical disadvantages are such things as inconvenience, low reliability, high input power, capital and operating costs, size and weight, vibration, noise, electromagnetic interference (EMI), and heat rejection. The relative importance of these disadvantages depends on the application. In general, the disadvantage becomes serious if it dominates the behavior of the complete system. For example, if the reliability of the entire system is limited by the cryocooler, that can be a serious problem and hamper the marketability of the system. If the cost of a cryocooler is low compared with the entire system, then the cost of the cryocooler no longer becomes a disadvantage. For space applications, the cost of a cryocooler may be rather high to ensure high reliability, but system costs are also usually rather high. For some space applications, mechanisms must be designed to operate at cryogenic temperatures even though there may not be an advantage to doing so. An example of this would be the exploration of distant bodies in the solar system that are extremely cold.

Another challenge to the use of cryogenics involves the proper selection of materials for systems or mechanisms operating at these temperatures or that often span temperatures between the low temperature and ambient temperature. Part of the difficulty is finding data on material properties at cryogenic temperatures. In some cases, data may exist over a different temperature range than what is needed, and some theoretical guidance is thus needed to extrapolate the data to the desired temperature range. This chapter gives examples for the applications of cryogenics that make use of each of the seven benefits discussed above. The chapter then goes on to discuss the challenges facing researchers in cryogenics to minimize the disadvantages of operating at cryogenic temperatures with the ultimate goal of making the cryogenic operation invisible to the end user.

14.2 Benefits of Cryogenics

Processes and material properties are strongly affected by temperature, probably more so than by any other physical parameter, including pressure, magnetic field, electric field, and so forth. Some properties, such as electrical conductivity and thermal conductivity, can change by several orders of magnitude when cooled from room temperature to 4 K or below. Most quantum effects occur only at cryogenic temperatures, whereas most metals liquefy only at high temperatures. The ability to harness and apply temperature effects is a unique feature of mankind, and it has contributed to great advances in our civilization. Mankind has discovered abundant benefits and uses for high temperatures, beginning in prehistoric times with the use of fire for warmth, light, and cooking. Later, but still more than 20 centuries ago, mankind learned to forge tools and make crude pottery using heat from fires. As civilization advanced and higher temperatures could be achieved, stronger metals, such as iron, could be forged into tools, and much stronger pottery and china could be produced by the higher-temperature firing (sintering) of clay. The industrial revolution ushered in the steam engine and the ability to generate tremendous power for efficient manufacturing and transportation. The enhancement of chemical reactions at higher temperatures has been exploited for the production of vast amounts of new and improved materials in the last century or so [Bar-Cohen, 2014].

High-temperature applications began rather early in the history of civilization due to the ease of producing increasingly hotter fires. In contrast, mankind's use of low temperatures has lagged behind that of high temperatures due to the increased difficulty in producing low temperatures. Low temperature applications were limited for many centuries to the use of naturally occurring ice. The practice of using natural ice to treat injuries and inflammation was carried out by Egyptians as early as 2500 BC [Freiman and Bouganim, 2005], and the Chinese began to use crushed ice in food around 2000 BC. Although ice was first created artificially in the laboratory in 1755, it was not until near the mid-1800s with the development of the steam engine and practical compressors in the Industrial Revolution that artificial ice could be produced in sufficient quantities to replace natural ice cut from lakes. Until then the sole use of low temperatures was with natural ice for food preservation and a few medical procedures.

Around the mid-1800s, the science of thermodynamics was being developed with an understanding of how to produce lower temperatures through the liquefaction of the permanent gases, such as oxygen, nitrogen, hydrogen, and helium. Researchers learned that these gases could be liquefied in a cascade scheme by compression and expansion of each gas. The initial impetus for achieving lower temperatures was, for the most part, purely scientific, but entrepreneurs soon recognized the benefits of cryogenic temperatures. The production of oxygen from the fractional distillation of liquid air around 1900 was the first significant application of cryogenics. The oxygen was needed for oxy-acetylene welding, and the fractional distillation process was much cheaper than the chemical process used prior to that time [Scurlock, 1992]. The seven general benefit areas of nearly all cryogenic applications were described in Section 14.1, and applications employing these benefits are described in the next section. Other applications were covered in the other chapters of this book.

14.3 Applications of Cryogenics

14.3.1 Long-Term Preservation of Biological Material and Food

14.3.1.1 *Preservation of Biological Material*

After geneticists had refined the process of artificial insemination in cattle and other farm animals in the 1940s, a major breakthrough came in England in 1949 regarding successful freezing of chicken sperm by including glycerol. Other cryoprotectors had been tried earlier without much success [Foote, 2001]. Storage of frozen cattle sperm treated with glycerol was initially carried out with dry ice at 194 K, but in the 1950s, liquid nitrogen became the preferred refrigerant because some biological changes were noted at 194 K but not at 77 K using liquid nitrogen.

The first successful human pregnancy using frozen sperm occurred in 1953 [Bunge and Sherman, 1953], and by 1972, the first commercial cryobanks were founded. In 1983, the first pregnancy resulting from the use of a frozen embryo took place in Australia [Trounson and Mohr, 1983], and twins, a boy and a girl, were born in Australia as a result of the use of frozen eggs in 1986 [Chen, 1986]. A successful birth using sperm frozen with liquid nitrogen for 20 years was carried out at the Tyler Medical Clinic in 1998 [www.tylermedicalclinic.com, 2005]. It is possible to store blood for over 20 years utilizing liquid nitrogen and cryopreservation agents, although the process is expensive. It is particularly useful for the storage of cord blood and stem cells. Further research is continuing in this field.

14.3.1.2 Food Freezing

The freezing of food for preservation has been practiced for centuries and does not require cryogenic temperatures. However, the use of liquid nitrogen allows for quick freezing of foods, which can prevent surface dehydration and improves taste and color. Fast freezing prevents large ice crystals from forming that do more damage to cell walls than do small crystals. It also reduces the tendency for individual pieces to stick together. Cryogenic quick freezing is three to four times faster than that obtained with conventional refrigeration systems and allows for higher throughput and lower costs. Typically the fast freezing is carried out with the food placed on a flat belt that moves into a tunnel with cold nitrogen vapor passing over the food before further cooling with the spray of liquid nitrogen. Flat-belt LN_2 tunnels are designed so that the LN_2 is sprayed on the food product at the exit end of the freezer, and the cold vapor forced back toward the entrance of the freezer. In this way, the available refrigeration in the vapor is used most efficiently. Figure 14.1 shows an example of food that was frozen using liquid nitrogen. Typical liquid nitrogen flow rates in one of these systems vary from about 200–2000 kg/h. The fast freezing techniques and equipment became especially popular when liquid nitrogen supplies were readily available beginning in the 1960s.

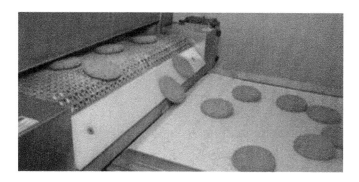

FIGURE 14.1
Liquid nitrogen fast food freezer. (Courtesy of Air Products.)

14.3.2 High-Density Fluids (Liquefaction and Separation)

14.3.2.1 Air Separation

The benefit of liquefaction to separate oxygen and nitrogen from air by distillation was the first application of cryogenics, which appeared around 1900 to provide large quantities of low-cost oxygen for the welding industry. The air separation industry had grown to a rather large size by 1950, and it has continued to grow rapidly in the last 50 years. The liquefaction of a gas increases its density by about 600–900 times compared to the gas at ambient temperature and pressure. Thus, much smaller volumes are needed for the same energy or mass of fluid. The shipping of large quantities of gas is much more economical if done in liquid form.

Figure 14.2 shows how oxygen production in the USA has grown by about a factor of 30 compared with that in 1954 [Royal, 2005]. Nitrogen has grown even faster, increasing from 0.53×10^6 t in 1960 to 21.3×10^6 t in 1985, a factor of 40 [Baker and Fisher, 1992; Bureau, 2005; Grenier and Petit, 1986]. Data after 2005 is now difficult to obtain because the US Census Bureau discontinued its report on industrial gases in 2005. The typical size of an air separation plant increased from about 100 t/d in 1954 to about 500 t/d in 2005, but the largest is now about 4000 t/d from a single train. A large eight-train facility in Qatar produces 30,000 t/d of oxygen. Cryogenic air separation units (ASUs) are used in medium- and large-scale plants and for high-purity (98–99.5%) gases. A cryogenic ASU is more economical than pressure swing absorption (PSA) or membrane processes for systems larger than about 60 t/d and for high purities. Figure 14.3 shows a modern cryogenic air separation plant for liquefied argon, nitrogen, oxygen, and carbon dioxide. Generally, liquid hydrogen production uses liquid nitrogen precooling, so the two facilities are often located together. Large deliveries of any of these gases are only economical if carried out in the liquid form, as shown by the insulated tank truck in this figure. Figure 14.4 compares the oxygen demand in Japan from various industries [Kato et al., 2003]. This figure shows that the steel-making industry with its blast furnaces (basic oxygen furnace) and electric furnace dominates the oxygen market. Figure 14.5 shows the production rate of oxygen and nitrogen in Japan between 1960 and 1985 [Oshima and Aiyama, 1992]. We see the oxygen demand follow closely to the production of crude steel. Figure 14.6 shows the world production of crude steel from 1950 to 2013 [World Steel, 2014]. The rapid increase since about 1995 is driven by

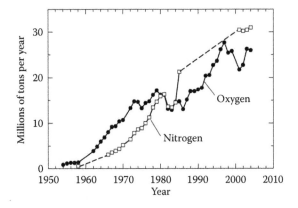

FIGURE 14.2
Annual US oxygen and nitrogen production.

FIGURE 14.3
Modern air separation plant for liquefied argon, nitrogen, oxygen, and carbon dioxide. (Courtesy of Praxair.)

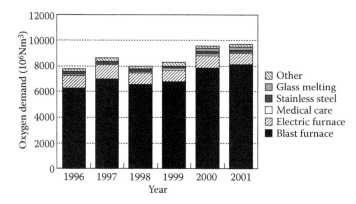

FIGURE 14.4
Japanese annual oxygen demand and usage. (Based on data published in Kato et al., *International Energy Workshop (IEW)*, 2003.)

much new construction in China, which now is the world's largest producer and consumer of steel [World Steel, 2014].

Liquid oxygen is used in the space program as an oxidizer in rockets but the total amount used is less than about 5% of the total USA oxygen production. In 1965, the USA space program consumed 0.40×10^6 t of liquid oxygen compared with the total USA production of 6.8×10^6 t. For this application, it is the much higher density of the liquid phase that enables the use of pure oxygen in rockets. For exploration of the solar system with return capabilities, such as manned missions to Mars, the production of liquid oxygen on Mars will be an important part of the mission. The oxygen would be generated chemically from the carbon dioxide atmosphere, liquefied, and then stored in empty tanks brought on board the lander. Very efficient and reliable oxygen liquefiers will be required for such a mission [Marquardt and Radebaugh, 2000]. Recent developments in space cryocoolers in the last 10 years can satisfy these difficult requirements.

The large increase in the use of nitrogen in the last 20–30 years has been driven by the food freezing industry (discussed earlier), as an inerting gas in the stainless steel and semiconductor industry and to pressurize oil wells to increase the extraction rate.

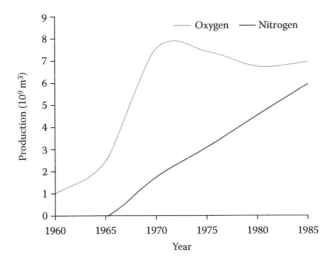

FIGURE 14.5
Oxygen and nitrogen production in Japan. (Based on data from Oshima and Aiyama, In: *History and Origins of Cryogenics*, ed. R.G. Scurlock, pp. 520–46. Oxford: Clarendon Press, 1992.)

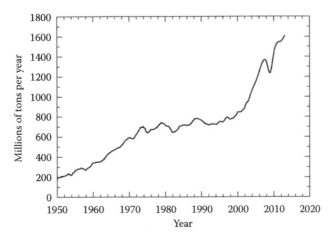

FIGURE 14.6
Annual world crude steel production.

In fact one of the world's largest air separation plants (3000 t/d) is for the purpose of nitrogen gas pressurization of oil wells off the coast of Mexico. Air separation plants in the United States required about $50 \geq 10^{15}$ J of electrical energy in 2004 [ORNL, 2005], about 0.35% of the total US electrical energy output at that time. The total value of oxygen, nitrogen, argon, and hydrogen produced in the United States in 2002 was about $4 billion [Bureau, 2004].

14.3.2.2 Liquid Hydrogen

Demands for liquid hydrogen began to increase rapidly after about 1958 to provide the rocket fuel needed for the space programs of many countries. The first space flight of Sputnik I by the Soviet Union in 1957 initiated the space age with many countries developing capability to send satellites into space over the next several years. The "Papa Bear" hydrogen liquefier established in Florida for the secret Air Force program on a hydrogen-fueled airplane was ready to supply liquid hydrogen for NASA's development of rocket engines after cancellation of the Air Force program [Sloop, 1978]. Additional hydrogen liquefiers were built soon after 1960, and by 1965, the USA hydrogen liquefaction capacity reached 200 t/d compared with 14 t/d in 1959 [Tishler, 1966].

For the manned mission to the moon, NASA developed the Saturn V rocket with three stages. The first-stage engine was fueled with kerosene and liquid oxygen and produced a thrust of 33.4 MN. The second and third stages were fueled with liquid hydrogen and liquid oxygen with thrusts of 5 and 1 MN, respectively. In July 1969, the Saturn V carried Apollo 11 to the moon for the first manned lunar landing. The launch is shown in Figure 14.7. About 50% of the liquid hydrogen used by NASA for flight operations is lost because of transient chilldown of warm cryogenic equipment, purging, and heat leaks into ground storage tanks, transportation vessels, flight vessels, and transfer lines. A new project under development at NASA/Kennedy, called Integrated Ground Operations Demonstration Unit for Liquid Hydrogen (GODU-LH2), aims to decrease the loss to less than 20%. It includes onsite refrigeration to provide zero boiloff of storage tanks and densification of the liquid by chilling to 16 K [Notardonato, 2014].

A 1990 NASA study indicated that the USA liquid hydrogen production rate was 140 t/d that year. By 2008, the rate had increased to about 210 t/d [EPA, 2008]. The largest hydrogen liquefiers today can produce about 20,000 L/h or about 34 t/d. The total US production rate of hydrogen was about 3000 t/d in 2001 and 4100 t/d in 2004, but most of the hydrogen is used onsite for oil refining or ammonia production [Bureau, 2005]. If the hydrogen economy were to progress, the production rate of hydrogen will increase dramatically, and so will the liquefaction rate. At this time, the studies show that because of its high density, liquid hydrogen–fueled automobiles provide the greatest range compared with other hydrogen storage methods, such as compressed gas or metal hydrides. New automotive hydrogen dewars with liquid air shielding show zero boiloff hold times of up to 2 weeks [Trill, 2002].

FIGURE 14.7
Launch of Apollo 11 to the moon with Saturn V rocket. (Courtesy of NASA.)

14.3.2.3 Liquefied Natural Gas

Liquefied natural gas (LNG) was first exported from Lake Charles to Canvey Island, United Kingdom, in 1959 via a converted crude-oil tanker renamed the *Methane Pioneer* with free-standing aluminum tanks and insulated with balsa wood and fiber glass [Haselden, 1992]. The total load was 5000 m^3.

In 2002, the United States produced 84% of its total natural gas needs, with most of the remainder coming via pipeline from Canada. In that year, about 1% of the total US supply was imported as LNG from overseas via ship. After about 2005, the production of natural gas in the United States increased dramatically because of the hydraulic fracturing of shale gas fields and the use of horizontal drilling. Figure 14.8 shows this rapid increase in US natural gas production. Previous to that time, the Energy Information Administration (EIA) of the Department of Energy (DoE) had predicted that by 2025 LNG imports would contribute about 21% of the US natural gas needs. As of 2012, EIA now predicts an export of LNG from the United States amounting to about 10% of the US production in 2015, as shown in Figure 14.8 [DOE/EIA, 2014].

As of 2005, there were only four LNG marine receiving terminals in the United States: (1) Everett, Massachusetts, built in 1971; (2) Cove Point, Maryland, built in 1978; (3) Elba Island, Georgia, built in 1978; and (4) Lake Charles, Louisiana, built in 1982. Figure 14.9 shows the

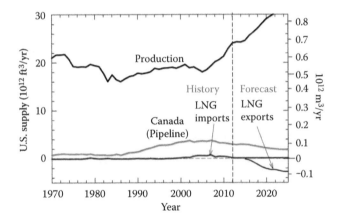

FIGURE 14.8
US annual natural gas supply. (Data from EIA/DoE.)

FIGURE 14.9
Elba Island (Savannah) LNG import terminal and regasification facility. (Courtesy of Kinder Morgan.)

Elba Island facility in 2004 before completion of the new unloading docks to the left in a channel off from the main Savannah River. The total storage capacity of all four of these facilities is currently 0.53 billion m^3 (18.8 billion ft^3) (gas equivalent), sufficient for only 8 h of the US total natural gas usage. In 2014, there were 11 import terminals in the United States, but because of the rapid increase in domestic natural gas production that started around 2005, many of these facilities are adding liquefaction plants and export capability [CLNG, 2015]. The oldest active marine terminal in the United States was constructed in Kenai, Alaska, in 1969 to export LNG to Japan, the creators of the world LNG market. The Kenai facility has a liquefaction plant, whereas the 11 receiving terminals currently have only vaporizers. In 2005, there were 57 storage facilities in the United States with lique-faction plants for use in peak shaving and another 56 smaller niche-market LNG facili-ties. Because of the rapid increase in US natural gas production, the demand for import LNG facilities rapidly decreased except for certain regions, such as the northeast United States. However, there are now nine sponsors of applications for LNG export facilities in the United States. One import facility that has been completed in 2015 is the Sabina Pass terminal in Louisiana. It went into production online in the beginning of 2016. This facility consists of four liquefaction trains with a total liquefaction capacity of 77.8 million m^3/d (2.76 Bcf/d). Two more liquefaction trains are scheduled to be added later. Each train has a liquefaction capacity of about 4.5 million tons per annum (mtpa).

On May 21, 2003, Alan Greenspan, Federal Reserve Chairman, said, "Our limited capac-ity to import liquefied natural gas effectively restricts our access to the world's abundant supplies of natural gas." Preparations to bring about an enormous increase in LNG supply to the United States began in the early to mid-2000s. As of July 2005, there were more than 50 proposals for new receiving terminals in the United States, but those proposals were quickly abandoned as US production rapidly increased. Currently, Japan has 23 receiving stations and obtains nearly 100% of its natural gas as LNG. Many other countries with insufficient natural gas reserves are now looking to follow in Japan's footsteps by import-ing LNG to at least supplement their domestic production. The LNG market then begins to mimic the oil market and will play a significant role in global energy supplies in the 21st century.

To meet the growing demands for global natural gas requires a large increase in the number of LNG ships and liquefaction facilities. At the end of 2011, there were 359 LNG tankers in operation worldwide, most of them with capacities of 120,000 m^3 (liquid) or more. There are 55 new ships under construction with 46 of them having a capacity of 138,000 m^3. These ships are among the most complex and expensive merchant ships ever built, but the construction costs have decreased from about \$270M in 1990 to about \$155M in 2005. The length of a typical LNG tanker is about 300 m (1000 ft). Figure 14.10 shows one type of LNG tanker. The other major type contains the liquid within the hull and uses a membrane. About 0.15%–0.25% of the LNG being transported boils off each day and is used to partially fuel the ship. Typically about 1%–5% of the cargo boils off during a shipment (depending on the distance traveled). The shipping cost is about 10%–30% of the LNG cargo value (depending on the distance traveled). An excellent review of the LNG shipping industry and market is given in a report by Alavi with DVB [DVB, 2003].

A new approach to the processing, liquefaction, and exporting of natural gas is the development of floating liquefied natural gas (FLNG) platforms. Construction on the first one (known as Prelude) began in 2012 and is expected to be operational 200 km off the northwest coast of Western Australia in 2017 [Schilling, 2014]. The construction cost is expected to be \$52 billion. It is 488 m long, which makes it the largest ship in the world. It will extract, process, liquefy, and offload LNG and other liquid products as it is moored

FIGURE 14.10
LNG tanker Chelles traveling on the Rouen, France.

above the Prelude and Concerto gas fields in the Browse Basin. It will produce 3.6 million tons per annum (mtpa) of LNG, 1.3 mtpa of condensate, and 0.4 mtpa of LPG. Plans for several other FLNG platforms are under study.

14.3.3 Macroscopic Quantum Phenomena (Superconductivity and Superfluidity)

Decreasing the temperature of any material has the effect of reducing the thermal excitations of the atoms. At sufficiently low temperatures, the thermal excitations can become less than some quantized first energy level. The atoms or pairs of atoms may then "condense" into the lowest energy in which their de Broglie wavelengths become very long and overlap with other atoms. The interaction causes a coherence of the atoms such that they can move together as one and take on entirely different properties than at higher temperatures. Such quantum effects in bulk systems are unique to low temperatures. The two common examples are superconductivity where there is no resistance to the flow of electricity and superfluidity where there is no resistance to the flow of the fluid and the thermal conductivity is extremely high. Superconductivity, in particular, has many very useful applications.

14.3.3.1 Bulk Superconductor Applications

Soon after the discovery of superconductivity by Onnes in 1911 at a temperature of 4.2 K in mercury, other metals were found that were also superconducting. Unfortunately, the superconducting state would revert back to the normal state with rather low values of applied magnetic fields. The critical fields of the pure elements are less than about 0.2 T at 0 K, so they are of little use in carrying large currents and generating high magnetic fields.

Metallic alloys and compounds were also studied between 1911 and 1950 and found to have higher critical temperatures and fields. In 1953, V_3Si was discovered to have a critical temperature of 17.5 K, the highest value at that time. The critical magnetic field in V_3Si near 0 K is about 2.3 T, too low for useful magnets. The materials were also difficult to make into wire form. In addition, only few laboratories at that time had the ability to achieve the temperature of 4.2 K that would be required for any superconducting magnet. Thus, in the 1950s, there were no practical applications of superconductivity. In fact, there were no theories available at that time to satisfactorily explain superconductivity.

Figure 14.11 gives a time line of the important discoveries and applications dealing with bulk superconductivity after 1950. The first was the discovery by Matthias in 1954 of Nb_3Sn with a transition temperature of 18 K, the highest at that time [Matthias et al., 1954]. The BCS theory [Bardeen et al., 1957] of superconductivity was then published in 1957, 46 years after superconductivity was discovered by Onnes. A very significant discovery was made in 1961 that quickly led to applications of superconductivity. The superconducting compound Nb_3Sn that was found 7 years earlier to have such a high transition temperature was found to have a critical field in excess of 8 T at 4.2 K, about an order of magnitude higher than anything found before [Kunzler et al., 1961]. Unfortunately, it was very brittle and difficult to fabricate. A year later the ductile alloy NbTi was found by scientists at Westinghouse to be superconducting at 9.5 K and to have an upper critical field of about 11 T at 4.2 K [Hulm and Blaugher, 1961]. Most importantly, it was a ductile alloy easily fabricated into wires. It became the first commercial superconducting wire.

By the mid-1960s, liquid helium had become readily available either through the use of the commercial Collins helium liquefier or through air shipments of liquid helium from large liquid helium suppliers, which afforded the first significant commercial application of superconducting magnets for MRI [Schwall, 1987]. In 1971, Raymond Damadian showed that the nuclear magnetic relaxation times of tissues and tumors differed [Damadian, 1971], thus motivating scientists to consider magnetic resonance for the detection of disease. MRI was first demonstrated on small test tube samples in 1973 by Paul Lauterbur [Lauterbur, 1973]. A whole-body image [Damadian, 1977] of the chest, shown in Figure 14.12, was first made in 1977 using the system patented by Damadian [Damadian, 1974] in 1974, which utilized a NbTi superconducting magnet consisting of a 1.3-m-diameter Helmholtz pair that reached a

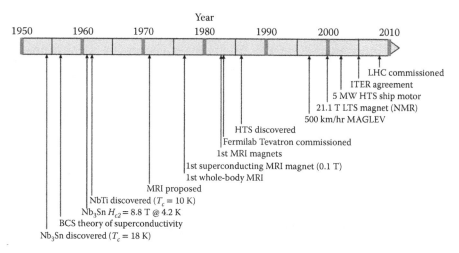

FIGURE 14.11
Time line for significant discoveries or inventions dealing with bulk superconductivity.

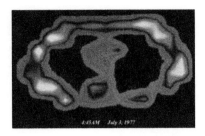

FIGURE 14.12
The First MR scan of a live human body (1977). (Courtesy of FONAR Corporation.)

maximum field of 0.10 T [Goldsmith et al., 1977]. Improved imaging techniques were introduced by others [Kumar et al., 1975; Mansfield, 1977; Mansfield and Maudsley, 1976] that are in common use today. A recent MRI image showing a brain tumor is given in Figure 14.13. Some of the early MRI systems used normal or permanent magnets, which could provide fields of 0.15 T, but greatly improved resolution was provided by going to 1.5 T, which could only be provided by NbTi superconducting magnets. Most systems today use 1.5 T superconducting magnets, but there are a few new systems with 3.0 T magnets. MRI has now become the largest commercial application of superconductivity with over 22,000 superconducting MRI systems in operation worldwide. Approximately 1000 systems per year are sold [Muller and Hopfel, 1998] at a cost of about $2 million each. About 100 t/yr of NbTi are required to produce the superconducting magnets [Scanlan et al., 2004]. Considering the magnetic fields required for good images, normal magnets simply cannot compete. Figure 14.14 shows a typical MRI system of today, which uses a field of 1.5 T. Two-stage Gifford McMahon cryocoolers capable of providing 1.5 W at 4.2 K are often used to eliminate any liquid helium boiloff. About 7000 such cryocoolers had been sold between 1995 and 2003 [Kuriyama, 2003]. Figure 14.15 shows the rapid growth of MRI systems from 1980 to 2004 and again since about 2012 [Andrews, 1988; Devred et al., 1998; Schwall, 1987; Magnetic Resonance, 2015; CCAS, 2015]. The field is expected to see continued steady growth, especially as newer applications develop in the identification of Alzheimer's disease and multiple sclerosis. The MRI and nuclear magnetic resonance (NMR) markets dominate the worldwide superconductivity market, which is projected to exceed about US$8.8 billion by 2020 [Global Industry Analyst, 2015].

NMR is used in MRI for imaging. It can also be used for spectroscopy to analyze biomacromolecules. Over the past 50 years, NMR has become the preeminent technique for

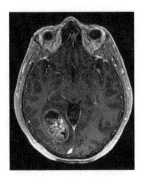

FIGURE 14.13
Recent MRI image of the brain showing a brain tumor (in white). (Courtesy of Mike Chen, City of Hope, CA.)

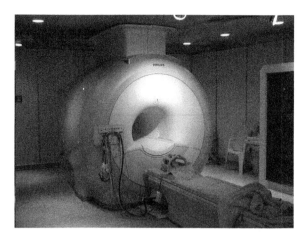

FIGURE 14.14
Modern MRI system with a superconducting magnet and cryocooler. (Courtesy of Kasuga Huang [http://commons.wikimedia.org/wiki/File:Modern_3T_MRI.JPG].)

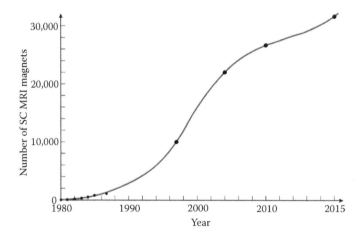

FIGURE 14.15
Cumulative number of MRI systems in use around the world.

determining the structure of organic compounds, especially proteins. There are two main ways to study the structure of proteins: NMR spectroscopy and x-ray crystallography. X-ray crystallography can be used only with proteins that can be made into single crystals, which include about 75% of the 50,000–100,000 proteins existing in the human body. The structure of the remaining proteins can be determined only by the use of NMR spectroscopy. This technique can be used for proteins in solution, that is, in an environment similar to that in a living cell. In fact the use of this technique for the study of protein structure and function led to two Nobel Prizes in Chemistry, one to Richard Ernst in 1991 and one to Kurt Wuthrich in 2002. Very high magnetic fields are required for NMR spectroscopy of large macromolecules such as proteins. A magnetic field of 21 T is required for resonant frequencies of 900 MHz, which allows the study of proteins with molecular weights up to about 30,000. Figure 14.16 shows an example of a 900 MHz NMR system. In order to study heavier proteins, higher resonance frequencies are required, which in turn requires higher magnetic fields. A system operating at 1 GHz frequency with a magnetic field of 23.5 T has been introduced in the last

FIGURE 14.16
900 MHz (21.1 T) NMR system. (Courtesy of NIGMS, NIH.)

couple of years. Obviously, such high magnetic fields are only possible with superconducting magnets. The magnets are made with an Nb_3Sn coil inside an NbTi coil, and the helium bath is vacuum pumped to a temperature of about 2.2 K. Many laboratories throughout the world are now using high-field NMR spectrometers for protein studies. One magnet supplier states that over 6000 of their superconducting magnets are being used for such studies. It would appear that many more such magnets will be needed for this rapidly growing field of research, especially for magnets with fields of 21 T or higher.

At about the same time, MRI systems with superconducting magnets were being developed. The Fermilab Energy Doubler or Tevatron was under construction using about 1000 NbTi superconducting magnets in a ring of 6.3 km circumference to bend and focus the proton beam [Edwards, 1985]. The dipole magnets for bending the beam are 6.1 m long and provide a magnetic field of 4.4 T [Gourlay et al., 2004]. To cool the magnets, the helium liquefier system (central plant plus satellites) delivers a peak refrigeration power of 24 kW at 4.5 K through a circuit of 27 km. The Tevatron was commissioned in 1983 and named in 1993 as an International Historic Mechanical Engineering Landmark by the American Society of Mechanical Engineers. Since then many other high-energy physics accelerators have used superconducting magnets. The largest, completed in 2008 in Switzerland at CERN, is the Large Hardon Collider (LHC). It is the most powerful accelerator ever built, in which two counter-rotating beams of protons at energies of 7 TeV each can be made to collide head on. Its ring circumference is 27 km, and each of the 1232 NbTi dipole magnets are 14.2 m long [Gourlay et al., 2004]. According to CERN, an energy of 1 TeV is equivalent to the energy of motion of a flying mosquito, but with the LHC, the 14 TeV is concentrated into a volume about 10^{12} times smaller than a mosquito. The high energy concentration splits the proton into elementary particles that help us understand the structure of the early universe. A particle matching the Higgs boson was already discovered in 2013 with the LHC running at about half its rated power. It is expected that several new particles will be discovered with the LHC when it is run at full power starting in 2015. Figure 14.17 shows the assembly of one LHC dipole section surrounded by its vacuum can. Each of

FIGURE 14.17
One dipole magnet being assembled into a cryostat for the LHC at CERN. The two tubes near the center are the two beam tubes. (Courtesy of CERN, Switzerland.)

the 392 quadrupole magnets is 3 m long. About 7000 km of superconducting wire was required for all the magnets. If normal magnets had been used, the ring circumference would have been 120 km, and the required input power would have been astronomical. The specified magnetic field is 8.36 T, which requires the magnets be cooled to about 1.9 K in superfluid helium in order to achieve the higher field. The total mass to be cooled is 37,000 t. The quantum nature of superfluid gives it enormous thermal conductivity (hundreds of times that of OFHC copper), extremely low viscosity, and a specific heat about 10^5 times that of the superconducting magnets per unit mass. The helium liquefier plant is split into eight liquefier stations spaced around the ring, each of which services about 3.3 km of the ring. Figure 14.18 shows a portion of the ring where the helium lines connect into the main ring from an auxiliary helium line. Both 4.5 K liquid and 1.9 K liquid at 0.13 MPa (1.3 bar) are produced at each station. The heat load on the 1.9 K superfluid helium is about 2.3 kW for each station. The total equivalent 4.5 K refrigeration capacity for the complete ring is 144 kW, which makes the LHC the world's most powerful helium refrigeration system. The total input power is about 40 MW [Delikaris and Tavian, 2014].

FIGURE 14.18
The LHC tunnel and dipole magnet sections at CERN. Superfluid helium is supplied to the tubes from the tower located behind the beam path. (Courtesy of CERN, Switzerland.)

Superconducting magnets are often used in particle detectors also because observing a particle's trajectory in a magnetic field is the easiest way to determine its momentum. The magnetic field is the easiest tool to determine the momentum of a particle by observing its trajectory in the field. In the LHC, the high-energy protons collide in four regions around the ring, housing four unique detector experiments. The two largest experiments, ATLAS and CMS, will use superconducting magnets. The ATLAS detector is the largest ever built for particle physics. The barrel toroid has an outer diameter of 20.1 m and a length of 25.3 m [Gourlay et al., 2004]. The peak field is about 3.9 T. The CMS detector uses the largest solenoid ever built. It has a central field of 4 T with a diameter of 6 m and a length of 12.5 m.

Superconducting magnets are also used in magnetic confinement systems for fusion studies and in some inertial confinement systems. International agreement was obtained in 2005 for the location of the International Thermonuclear Experimental Reactor (ITER), the largest of the tokamaks under design by an international collaboration. A tokamak is a device that uses a magnetic field to confine a plasma in the shape of a torus. It has a plasma major radius of 6.2 m and a minor radius of 2 m. The toroidal field at the major radius is 5.3 T, but fields of 12–13 T are required at other locations [Gourlay et al., 2004]. The higher fields are generated with Nb_3Sn magnets, whereas the lower fields are generated with NbTi magnets [Shimomura, 2001]. All coils are cooled with supercritical helium that is forced through the tubes containing the cables at a temperature of 4.4–4.7 K. The liquid helium is produced by four identical liquid helium modules similar to the modified CERN 18 kW helium refrigerators [Kalinine et al., 2004].

One of the most interesting applications of superconductivity is for magnetic levitation. Levitation can be accomplished with either an attractive system, called electromagnetic suspension (EMS), which requires feedback control, or a repulsive system, called electro-dynamic suspension (EDS), which is inherently stable and requires no feedback control. The attractive systems can use normal magnets, such as those used in the Maglev mono-rail train between Shanghai and the Shanghai Pudong International Airport, a distance of 30 km. The regular speed is 430 km/h, although it has achieved a top speed of 501 km/h. It is the fastest commercial railway system in the world. The repulsive (EDS) system has the advantage of simpler control systems and a levitated height of about 10 cm, thus relaxing the requirement on the straightness of the guideway. Japan began experimental work on a superconducting EDS system in 1972 that achieved the 10 cm levitation with a maximum speed of 60 km/h. The current test trains travel nominally at 500 km/h on the 43-km long Yamanashi Test Track, although a top speed of 581 km/h was achieved in 2003 with a manned three-car test train. Nonpaying passengers have been invited to ride the five-car test train since 2001, and in 2004, the cumulative traveled distance exceeded 400,000 km. Figure 14.19 shows the 2003 five-passenger test train, which the author was invited to ride in 2003 at a speed of 500 km/h. There are four NbTi magnets in each cryostat, and one cryostat is placed on each side of the coupling between two cars. Each cryostat requires 7 W of refrigeration at 4.2 K, which is provided by a helium JT cryocooler precooled with a Gifford-McMahon cryocooler. The "track" is a U-shaped guideway with two sets of aluminum rings on each wall. As the superconducting magnet passes by one set at a speed greater than about 50 km/h, it levitates the train. An electromagnetic wave travels along the other set of rings in the guideway to propel the train like a surfboard riding a wave [Railway, 2004]. In late 2005, one of the LTS magnets was replaced with a Bi2223 HTS magnet with the same field of 1 T, but operating at 20 K by direct conduction cooling with a Gifford-McMahon cryocooler. Performance of the train was not changed [Kuriyama, 2005]. In 2011, the Japanese government granted JR Central permission to operate their SCMaglev system between Tokyo and Nagoya by 2027 and to Osaka by 2045.

FIGURE 14.19
Japanese Maglev train that uses superconducting magnets for levitation and has a rated speed of 500 km/h. (Courtesy of Saruno Hirobano, Central Japan Railway Company [http://commons.wikimedia.org/wiki/File:Series_L0.JPG]).

In 1986, a breakthrough discovery was made in the field of superconductivity. Bednorz and Muller found that a brittle ceramic compound became superconducting at temperatures around 30 K [Bednorz and Muller, 1986]. A flurry of activity ensued, and similar materials with much higher critical temperatures T_c were found. One year later, the material $YBa_2Cu_3O_7$, known as YBCO, was found to be superconducting at 92 K, the first superconductor to have a transition temperature above liquid nitrogen temperature [Wu et al., 1987]. Subsequent measurements on this material showed it to have an upper critical field at 0 K of about 130 T. This material is now being manufactured as tape and is called second-generation HTS wire. It can be operated at temperatures of about 50-70 K, depending on the magnetic field strength. In 1988, the ceramic $Bi_2Sr_2Ca_2Cu_3O_{10+x}$, known as BSCCO-2223, was discovered to have a transition temperature of 110 K [Maeda et al., 1988]. This superconductor is marketed as first-generation HTS tape for power and magnet applications. It has a lower critical field than YBCO and must be operated at about 30–35 K for high-field applications. Figure 14.20 shows the comparison of the upper critical field of useful LTS and HTS materials [Scanlan et al., 2004]. The highest confirmed T_c is 138 K, discovered in 1995 in a thallium-doped mercuric cuprate [Dai et al., 1995]. Generally, applications of HTS materials are limited by a lower characteristic field, the so-called "irreversibility field," where the critical current and other hysteretic properties disappear, which is much less than the upper critical field [Malozemoff, 2015].

The high-temperature superconductors are now being developed into many large-scale power applications, such as motors, generators, transformers, synchronous condensers, fault current limiters, and transmission lines [Malozemoff et al., 2005]. Several demonstration AC transmission lines have been tested or are being developed in many countries. In early 2005, a 500-m 77-kV single-phase HTS cable was tested in Japan. In the United States, a consortium of companies built a 660-m, 138-kV HTS cable to connect to the Long Island Power Authority grid. It became operational in 2008. The primary advantage of HTS cables is their much higher rms current density of 10,000 A/cm^2 compared with 100 A/cm^2 for copper cables. After adding the dielectric, cryostat, and structure, the HTS cables can still carry about two to five times that of conventional cables [Malozemoff et al., 2005]. As a result, in urban areas where underground utilities are crowded, additional electric power can be added simply by replacing the conventional cable in a given conduit with a HTS cable. Currently, expensive first-generation BSCCO cables are being used, but soon sufficient lengths of lower-cost second-generation YBCO cables will be available. Typically cooling is accomplished with

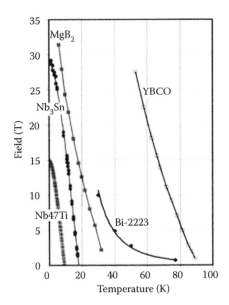

FIGURE 14.20
Upper critical magnetic field for several practical superconductors [Scanlan et al., 2004]. (Courtesy of Alex Malozemoff, AMSC.)

forced flow of liquid nitrogen subcooled to about 65 K to eliminate two-phase flow. The refrigeration power is typically about 10 kW at 65 K for these test transmission lines.

The use of HTS in rotating machinery, such as generators or motors, offers the advantage of reduced losses, weight, size, and cost. The need for a cryocooler with HTS systems restricts the usefulness of such systems to those larger than about 1 MW. In the case of large motors or generators, the losses may be reduced by a factor of two compared with conventional motors or generators, but even conventional machines are quite efficient. The major advantage is that HTS machinery can be about one-third the weight and volume of conventional machines of the same power [Malozemoff et al., 2005]. Such size reductions are particularly attractive for use in ships. A demonstration model 5 MW HTS ship motor with a speed of 230 rpm was constructed and tested in 2003 using BSCCO wire and operating at about 35 K. The higher magnetic field in motors and generators compared with transmission lines means that they must operate at a lower temperature compared with transmission lines (see Figure 14.20). The largest superconducting motor developed to date is a 36.5-MW, 120-rpm ship propulsion motor (about 50,000 hp). Such a motor is sufficiently large to use on passenger cruise ships as well as larger military vessels. Figure 14.21 compares the size of such a motor to that of a conventional motor of the same power. The much smaller size of HTS motors makes way for greater cargo or passenger capacity. Figure 14.22 shows the rapid increase in the size of developmental HTS motors since 1990. Industry analysts estimate the electric drive market for ship propulsion to grow to about $2.4 billion annually in the next 10 years. HTS motors are expected to capture about one-half of that market [Corporation, 2005]. A thorough review of both LTS and HTS motors and generators is given by Kalsi et al. [2004].

14.3.3.2 Electronic Applications of Superconductivity

Figure 14.23 shows the important discoveries or inventions in the last 50 years that have had a major influence on the development of electronic applications of superconductivity. A better understanding of superconductivity that came with the BCS theory

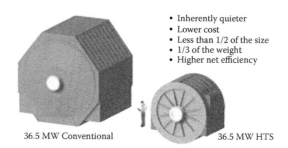

- Inherently quieter
- Lower cost
- Less than 1/2 of the size
- 1/3 of the weight
- Higher net efficiency

36.5 MW Conventional   36.5 MW HTS

FIGURE 14.21
Size comparison between a conventional ship motor and a superconducting ship motor. (Courtesy of the American Superconductor Corp. [AMSC].)

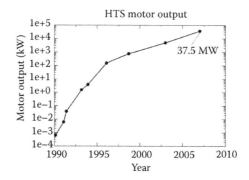

FIGURE 14.22
HTS motor output versus year.

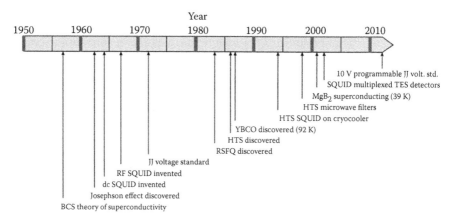

FIGURE 14.23
Time line for significant discoveries or inventions dealing with superconducting electronics.

[Bardeen et al., 1957] in 1957 was instrumental in leading Brian Josephson to predict in 1962 that a junction of two superconducting electrodes separated by a thin insulating barrier would allow Cooper pairs to tunnel through the barrier with a zero voltage difference as long as the current was less than some critical value [Josephson, 1962]. A voltage would appear for currents greater than the critical value. The long-range order associated with the quantum nature of supercoductivity could pass through the thin insulating junction and maintain coherence between the two superconductors. The prediction was verified experimentally in 1963 [Anderson and Rowell, 1963; Shapiro, 1963]. This Josephson tunneling behavior is the foundation for most electronic applications of superconductivity.

In 1964, the superconducting quantum interference device (SQUID) was invented. The first type invented was the dc SQUID, which consists of two Josephson junctions in parallel connected by a superconducting loop and operating in the voltage state with a current bias [Jaklevic et al., 1964]. An increase in the magnetic flux through the superconducting loop causes the voltage to oscillate with a period of the flux quantum $\Phi_0 = h/2e = 2.07 \times 10^{-15}$ Tm^2, where h is Planck's constant and e is the electron charge. A count of the voltage peaks determines the number of flux quanta and the resulting magnetic field. The small size of the flux quantum gives the dc SQUID extreme sensitivity to small magnetic fields. By detecting a small change in voltage within a single quanta, one typically can determine a change in magnetic flux as small as 10^{-6} Φ_0. The second type of SQUID is the RF SQUID, invented in 1967 [Silver and Zimmerman, 1967]. It consists of a superconducting loop with a single Josephson junction. The loop is then inductively coupled to the inductor of an LC-resonant circuit that is driven with an RF current in the range of a few tens of megahertz to several gigahertz [Kleiner et al., 2004]. The amplitude of the oscillating voltage is periodic in the applied flux, with a period of Φ_0. The RF SQUID typically can be used to detect flux changes as small as 10^{-5} Φ_0. Increased sensitivity to magnetic fields is usually achieved by using a superconducting flux transformer with a loop area much larger than that of the SQUID. The limiting magnetic field noise achieved with LTS SQUIDs is about 1 fT/Hz$^{1/2}$ [Cantor et al., 1996; Drung et al., 2001]. Most SQUIDs have been made with LTS materials, primarily niobium, but the advent of HTS led to SQUIDs being made with YBCO. However, the higher temperature leads to more Johnson thermal noise and less sensitivity with the HTS SQUIDs. In addition, junction quality in YBCO is inferior to that with Nb, partly due to material characteristics and partly due to the very small coherence lengths at temperatures around 77 K. The lowest magnetic noise achieved at about 1 kHz with HTS SQUIDs is about 10 fT/Hz$^{1/2}$ using a flux transformer, but noise at 1 Hz is much higher [Drung et al., 1996].

The extreme sensitivity of SQUIDs to small changes in magnetic flux has led to their being useful in many applications. One of the primary applications is as magnetometers for medical applications, biosensors, geological exploration, and high-frequency amplifiers. For medical applications, they can be used for magnetocardiography (MCG) to study the heart or for magnetoencephalography (MEG) to study brain activity. In these applications, the SQUIDs are paired to form gradiometers to cancel out most of the effect of Earth's dc field or magnetic noise from medium-range to far-range sources. Typically MCG can provide better information than electrocardiograms (ECG), but because ECG is usually satisfactory, there is little market for MCG with the added cost of cooling the SQUID even to 80 K for YBCO devices. An advantage of the MCG is that there is no need to attach electrodes to the patient as is the case with ECG. If the cost and electromagnetic interference (EMI) associated with a small 80 K cryocooler can be reduced, such a market could develop.

The very low magnetic signals from the brain (about 50 fT) make it difficult for conventional magnetometers to provide a useful signal-to-noise ratio for this application. Thus, SQUID gradiometers have not had much competition for this application and have found a good market in whole-head MEG systems. A typical modern helmet contains an array of about 151 × 275 SQUID sensors, including a number of reference sensors for noise cancellation, all cooled with liquid helium to 4.2 K. Figure 14.24 shows a commercial MEG system in use. Various stimuli such as sound, touch, and light can be used to produce magnetic signals in the brain and help locate the corresponding portion of the brain receiving that signal. MEG is

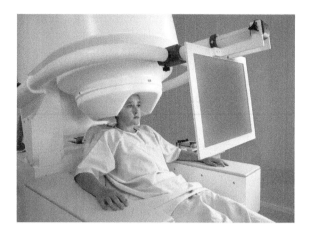

FIGURE 14.24
Superconducting magnetoencephalogram (MEG) system for brain function. (Courtesy of MEG International Services Ltd., Canada.)

most often used to map the function of the brain in the vicinity of a tumor to aid the surgeon in finding the least invasive path. Other MEG applications include the study of Alzheimer's disease, Parkinson's disease, schizophrenia, head trauma, epilepsy, and the effects of strokes.

The extreme sensitivity of SQUIDs to magnetic fields (also to voltage or current when they are coupled to a coil) makes them very useful to a wide variety of scientific measurements. In one example, the presence of antigens is detected when they are selectively labeled with superparamagnetic particles. Such particles are commercially available in sizes of about 20–100 nm, typically consisting of a cluster of γ-Fe_2O_3 subparticles each about 10 nm in diameter [Kleiner et al., 2004]. In an assay, the magnetic particles are attached to the antibody appropriate to the particular antigen being sought. By applying a magnetic field and then removing it, the decay rate of magnetism can be measured to determine if the particles have attached to an antigen. In one experiment, a different decay rate was detected when the antigen was switched from the bacteria *Listeria monocytogenes* to *Escherichia coli* [Grossman et al., 2004]. A SQUID microscope was used for the analysis. In the microscope, a sample at room temperature and pressure can be brought within about 100–200 μm of a high-T_c SQUID held at 77 K in a vacuum at the end of a sapphire rod that is attached to a bath of liquid nitrogen or to the tip of a mixed-gas JT cryocooler. A thin sapphire window separates the sample from the SQUID [Grossman et al., 2004; Kleiner et al., 2004].

Josephson had predicted that if the two superconductors separated by a thin insulating barrier were current biased with a frequency f, the junction would develop a region of constant voltage steps at the values $nhf/2e$, where n is an integer. Shapiro verified the prediction experimentally in 1963 [Shapiro, 1963]. If the ac current applied to the junction is within a certain range, it will cause the flux quanta passing through the junction to phase lock to the applied frequency. With this phase lock, the voltage across the junction is precisely $hf/2e$, and is known as the ac Josephson effect [Benz and Hamilton, 2004]. Phase lock can also occur at various harmonics of the Josephson effect. This precise voltage is independent of the junction quality and can be used for a voltage standard. In the early 1970s, many laboratories began assigning a value to the Josephson constant $K_J = 2e/h$ and using it as a voltage standard [Benz and Hamilton, 2004]. By international agreement in 1990, the constant K_{J-90} was assigned the value 483,597.9 GHz/V and adopted by all standards laboratories [Quinn, 1989]. Because frequencies can be measured with extremely high accuracy, the voltage defined by the Josephson constant improved the accuracy of the standard volt by orders of magnitude. Instead of relying on a chemical cell that had to be transported for voltage comparisons, laboratories anywhere in the world could fabricate a Josephson junction and use it to obtain the same voltage steps as anyone else. Typically the total uncertainty of a measurement using the Josephson effect is a few nanovolts. Many commercial instrument manufacturers now rely on a commercially available Josephson junction voltage standard in which liquid helium is used to maintain the Nb junctions in the superconducting state. Junction fabrication has advanced to the point where arrays of thousands of junctions have variations of critical current and resistance of only a few percent. Such arrays can then be used to provide a voltage standard at higher voltage values and to even program any desired precise output voltage up to 10 V from dc to 100 Hz [Benz and Hamilton, 2004].

Josephson junctions are being used as detectors and mixers of low-level, high-frequency signals from a variety of sources. They provide enhanced resolution for millimeter and sub-millimeter astrophysics signals [Zmuidzinas and Richards, 2004]. When a SQUID is used to detect small changes in resistance of a superconducting film at a temperature in the transition region between the normal and superconducting states, the detector is known as a transition-edge sensor (TES detector). Such detectors typically operate at a temperature near 100 mK where the material specific heat is extremely low. Thus, very small power inputs will cause relatively rapid and large temperature changes in the detector (known as a bolometer) and result in a resistance change detected by the SQUID. When these devices are used to detect integrated energies, they are known as microcalorimeters. They are very sensitive to low-energy x-rays and can be used for x-ray astronomy [White and Tananbaum, 1999] or for x-ray spectroscopic microanalysis of semiconductors [Wollman et al., 1999]. Large arrays of TES detectors amplified with multiplexed SQUID arrays are now used for a variety of astronomical studies. Figure 14.25a shows a 10,240-pixel TES array amplified with a 5120-pixel SQUID amplifier array (Figure 14.25b), both developed at NIST for use on the SCUBA-2 submillimeter telescope in Hawaii. The TES array is cooled to about 100 mK with an adiabatic demagnetization refrigerator, which in turn is precooled to 4 K with a pulse tube cryocooler.

A superconducting film can be made to switch from the superconducting state to the normal state by applying a magnetic field from a nearby control line. Such a switch, known as a cryotron, can be used for digital electronics and was first studied in 1956 by Buck [Buck, 1956]. The switching was quite slow because it relied on a thermal effect

(a)　　　　　　　　　　　　　　　　　　(b)

FIGURE 14.25
(a) Transition-edge sensor (TES) array for the SCUBA-2 submillimeter telescope. (b) SQUID current amplifier array with 10,240 pixels for amplifying signals from the TES array shown on left.

[Hayakawa et al., 2004]. After the development of the Josephson junction devices, various binary switching mechanisms were studied that made use of the Josephson junction and were much faster than the original cryotron. During the period of 1983–1987, various high-speed switching mechanisms using single flux quanta (SFQ) were investigated [Hayakawa et al., 2004]. Then, in 1987, a very high speed logic system emerged that was called rapid SFQ (RSFQ) [Mukhanov et al., 1987] that could be operated at frequencies up to 100 GHz or higher. Such superconducting digital electronic systems have been developed worldwide for logic systems [Hayakawa et al., 2004]. The use of SFQ logic for the following applications has been investigated: digital signal processing for digital filtering and multiuser detectors, digital sensor readout, switches for routers, and high-end computing. High-end computing studies are focusing on a goal of 1 exaflops (10^{18} floating point operations per second) by 2020. Weather modeling over a 2-week time span may require zettaflops (10^{21} flops). The K-computer in Japan with a speed of 10 petaflops (10^{16} flops) uses semiconductor circuits that require about 1 nJ per operation. A simple scaling to 1 exaflops indicates a power demand of 1 GW for each computer of that speed. With RSFQ logic, the switching energy is only about 10^{-19} J. Recent developments showed that the DC bias resistors with their relatively high power dissipation could be eliminated [Herr, 2011; Mukhanov, 2011].

In 2005, the National Security Agency (NSA) convened a panel of experts to investigate the use of RSFQ logic for petaflops-scale high-end computing at speeds not possible with semiconductor logic. The study recommended the development of a processor with approximately one million gates operating at a 50 GHz clock rate [National Security Agency, August 2016]. A computer or processor of this complexity will require about one million Josephson junctions per chip with about ten chips. In order to keep time of-flight short between various parts of the processor, the processor needs to be kept rather small, on the order of a cubic meter, and with very careful architecture to achieve that clock speed. Niobium technology would be used for the Josephson junctions.

A major application of superconducting electronics that does not involve Josephson junctions is the superconducting microwave filter systems for cellular telephone base stations. They use HTS microstrip designs fabricated on wafer substrates. Because no junctions are involved, HTS material provides very good performance with low noise. The most common substrate material to date has been MgO [Simon et al., 2004]. Both YBCO and TBCCO have been used for the HTS material. Figure 14.26 shows two common geometries used for the superconducting film in the filter. Figure 14.27 shows the comparison of the insertion

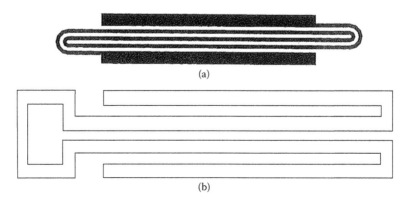

FIGURE 14.26
(a) Geometry of SISO resonator. (b) Geometry of clip resonator for HTS microwave filters.

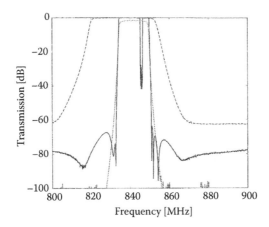

FIGURE 14.27
Insertion loss for three base station filters: dashed line, normal filter; solid line, HTS filter [Simon et al., 2004]. (Courtesy of Randy Simon.)

loss versus the frequency for the superconducting filter (solid line) compared with a conventional filter.

14.3.3.3 Applications of Superfluidity

When liquid ⁴He is cooled below 2.17 K at atmospheric pressure, it undergoes a phase transition to a superfluid state known as helium II. In contrast, normal liquid helium is referred to as helium I. The phase transition temperature is called the lambda point. Superfluid helium is a result of macroscopic quantum phenomena in which a portion of the fluid in the two-fluid model has "condensed" into the quantum mechanical ground state and makes no contribution to the entropy. In the superfluid state, the wave functions of each atom overlap and cause the atoms to act in unison as though they were one particle. Atoms acting in unison do not collide with each other and have no friction. The superfluid portion increases as the temperature is lowered below the lambda point. This portion of the fluid has zero viscosity and can flow through molecular-sized pores with no friction. If a vessel containing superfluid helium is kept below the lambda point, a thin superfluid film about 30 nm thick will creep up and over the wall and drip off the bottom of the container like a self-starting siphon. The zero viscosity of the superfluid portion also gives helium II an extremely high effective thermal conductivity brought about by the ease at which the superfluid can flow to the location of any heat input to absorb the heat. The thermal conductivity can approach infinity under ideal low-heat-flow conditions and is hundreds of times that of copper in practical conditions. The high thermal conductivity causes the helium to evaporate at the saturation point without boiling. The lack of boiling is sometimes useful where vibration from boiling can interfere with sensitive experiments. The third unique and useful feature of superfluid helium is its very high specific heat, especially near the lambda point where considerable heat is required to change the superfluid portion to normal fluid. The specific heat can be as high as 10^5 times that of superconductors.

The three unique properties of superfluid helium mentioned earlier (zero viscosity, high thermal conductivity, and high specific heat) make it very useful for cooling superconducting magnets. First, the lower temperature of helium II coolant causes the critical magnetic field of a superconductor to increase. Then, the zero viscosity of superfluid helium makes it easy to circulate it through the windings of a superconducting magnet to keep it cold. The high thermal conductivity of superfluid helium allows it to conduct heat over long distances in regions of stagnant flow. The high specific heat allows it to act as a good stabilizer to prevent quenching of superconducting magnets in which a local disturbance in the superconductor leads to a propagating hot spot that drives the superconductor into the normal state.

As discussed in Section 14.3.3.1, the 8.36 T magnetic field required of the dipole magnets in LHC at CERN was achieved by cooling the NbTi superconductor windings to 1.9 K with superfluid helium. The total refrigeration power at 1.9 K for all eight stations in the LHC is about 18 kW. Several earlier accelerators also used superfluid helium in the range of 1.8–1.9 K, but refrigeration powers were much lower. The acceleration of electrons in the Continuous Electron Beam Acceleration Facility (CEBAF) at Jefferson Lab makes use of 338 radio frequency (RF) cavities made with superconducting niobium [Delayen, 1998]. The recent upgrade from 6 to 12 GeV makes use of additional superconducting cavities. All are cooled to 2 K with superfluid helium for enhanced heat transfer and reduced RF losses in the niobium. The energy lost to heat in a niobium cavity is about 10^5 times smaller than in a copper cavity. Even so, 20 MW of input power is required to provide the 2 K cooling for the cavities and the 4.5 K cooling for superconducting bending magnets. A large variety of superconducting RF cavities are being used in particle accelerators because of the improved beam quality and high duty factor that they provide.

The zero entropy of the superfluid fraction in the two-fluid model of superfluid helium is a key to the operation of the ^{3}He-^{4}He dilution refrigerator, which can provide refrigeration down to about 2 mK. The superfluid fraction is close to one for temperatures below 1 K. When ^{3}He is added to superfluid ^{4}He, the thermodynamic properties of the mixture are dominated by that of the dissolved ^{3}He. The zero entropy of the ^{4}He component makes it act like a vacuum and simply separate the ^{3}He atoms as though they were a dilute gas. The enthalpy and entropy of the dilute ^{3}He are higher than those of pure ^{3}He. The maximum solubility of ^{3}He in ^{4}He is about 6.4 mol% at temperatures less than about 0.1 K [Radebaugh, 1967]. Any additional ^{3}He phase separates and floats on top of the dilute mixture as nearly pure ^{3}He. The dilute and pure ^{3}He are in equilibrium with each other, analogous to that of a gas and liquid. When ^{3}He is removed from the dilute phase, more ^{3}He from the concentrated phase enters the dilute phase to maintain the saturated ^{3}He concentration of 6.4%. In the dilution process, the increased enthalpy of the ^{3}He causes it to absorb heat. This process takes place in the mixing chamber of a dilution refrigerator. Separation of the ^{3}He from the dilute phase takes place in a still at a higher temperature of about 0.7 K where the vapor pressure of the dilute ^{3}He is high enough to be pumped away easily with a vacuum pump. The vapor pressure of ^{4}He at this temperature is much lower, so the gas phase has only 1%–2% of ^{4}He. The ^{4}He component is then stationary in the dilution refrigerator. The ^{3}He then diffuses through the superfluid ^{4}He in the tube connecting the still with the mixing chamber. In practice, the tube is one side of a heat exchanger with the incoming pure ^{3}He. A room-temperature vacuum pump circulates the nearly pure ^{3}He.

14.3.4 Reduced Thermal Noise

The temperature of any object determines the thermal fluctuations of the atoms and electrons. Such fluctuations will lead to Johnson electrical noise if the device is part of an electrical circuit. The thermal noise power can be given by $P = 4kT\Delta f$, where k is Boltzmann's constant and Δf is the bandwidth. The voltage fluctuations across a resistor R become $V_n = (4kTR\Delta f)^{1/2}$. This noise voltage is reduced by cooling the resistor. Thus, in low-noise applications it may be desirable to cool low-noise amplifiers (LNA). In fact, the microwave filter systems discussed in the previous section also make use of cryogenically cooled low-noise amplifiers to keep system noise as low as possible.

Another form of noise is thermally radiated noise. The power associated with such radiation is proportional to temperature to the fourth power. Thus, temperature reduction has a very pronounced effect on reducing radiated power. The temperature of an object also influences the intensity of the radiation. The wavelength at which the intensity is a maximum becomes greater with decreasing temperature. This wavelength is given by Wien's displacement formula,

$$\lambda_{max}T = 2.898 \times 10^{-3} \, \text{K m}$$

At 300 K, $\lambda_{max} = 9.66$ µm, whereas at 30 K the maximum occurs at 96.6 µm. These wavelengths are in the main part of the infrared spectrum, which spans the range from about 0.7 to 1000 µm. For the detection of thermal radiation, the detector must be much colder than the object to be detected for the best sensitivity. Night vision equipment relies on detecting the thermally radiated electromagnetic spectrum. For military applications where high detectivity is desired to obtain good images of objects with small temperature variations, the infrared sensors or focal plane arrays are usually cooled to about 80 K for observation of objects near 300 K. The Stirling cryocoolers shown in Chapter 6 are typical of what is used by the military for this application. Over 140,000 such coolers have been made since the early 1970s [Dunmire, 1998]. There have been some limited spinoffs of this cooled infrared sensor technology to other areas, such as police and rescue operations, security, and process monitoring. In process monitoring, small temperature changes in a product outside what is normal can be observed immediately, which allows the process to be modified or corrected in some way before a large inventory of unsatisfactory products accumulate.

Space applications of cooled infrared sensors are for either Earth observations or astronomy observations. In civilian Earth observations, the applications are for detection of small temperature changes in the atmosphere for aid in long-range weather forecasting, for studies of the greenhouse effect and global warming, and for studies of the ozone hole. For military applications, the need is for night vision of objects of interest. The military is also interested in observing objects that may be in space or heading out into space. If they are not being propelled by a rocket at that point in time, the sensors must be cold to observe the object, which could be at 300 K or below. For astronomy, most of the missions for the next 10–20 years will be for the study of the long-wavelength infrared spectrum. There is much interest in the infrared wavelengths because (1) much of the universe is cold (planets), (2) there are regions that obscure other wavelengths, (3) dust in the universe is cold and can be studied in the infrared, (4) chemical compositions can be determined from molecular spectral lines that are common in the infrared, and (5) primordial light from the early universe is highly red-shifted due to cosmic expansion; in fact, only a small portion of the mass in the universe can be observed in the visible wavelengths. The need for very efficient, lightweight, and reliable cryocoolers for this application has had a dramatic impact on improvements to cryocoolers over the past 20 years. Some of these improvements have found their way into the commercial marketplace, which have helped in other applications of cryogenics. Further details on these space-qualified cryocoolers are given by R. Ross elsewhere in this publication.

14.3.5 Low Vapor Pressures (Cryopumping)

Figure 14.28 shows that vapor pressure of fluids decreases very fast as the temperature is lowered. For a surface at about 150 K, the vapor pressure of water is reduced to 10^{-9} torr ($\approx 10^{-7}$ Pa). At 15 K, even air or oxygen has a vapor pressure too small to be shown in the graph of Figure 14.28. Gases such as hydrogen, which may be outgassing products from semiconductor processing, still have a high vapor pressure at 15 K. Its vapor pressure and that of helium can be greatly reduced at 15 K by the use of adsorption on a material with very high surface area, such as charcoal. The first atomic layer adsorbed on a surface has a pressure much the normal vapor pressure. The vapor pressure curves for helium and hydrogen in equilibrium with an adsorbed monolayer are similar to those shown in Figure 14.28, except that they are shifted to higher temperature by about an order of magnitude. The shift for other fluids decreases as the temperature is increased. For water, there is little difference compared with the vapor pressure curve shown in Figure 14.28.

Cryopumps typically have a quantity of charcoal cooled to 15 K with the second stage of a Gifford-McMahon cryocooler (see example in Figure 14.29), to pump hydrogen and helium, and baffles cooled to about 60 K to cryopump most of the other gases to keep them from overloading the charcoal. This technique to produce high vacuum results in the ultimate cleanliness of the vacuum space, which is highly desirable for the semiconductor processing industry, especially as line spacing becomes less and less and any contaminant could short-circuit the lines. The semiconductor processing industry has been the major user of such cryopumps. During the peak periods in semiconductor processing, about 20,000 cryopumps/year were being sold. Within the last 10 years, another cryopumping procedure has become of some interest, wherein a dry turbo vacuum pump is used to pump most gases except water. The water is then cryopumped at very high speed with baffles cooled to about 110–120 K to assist the turbopump.

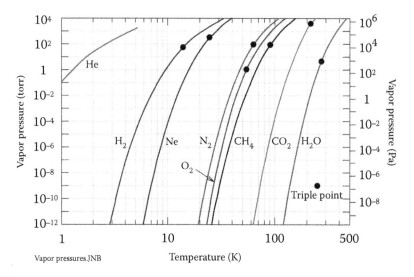

FIGURE 14.28
Vapor pressures of various gases at low temperatures. The graphs was created using data published in IUVSTA [2015].

FIGURE 14.29
Two-stage coldheads of Gifford-McMahon cryocooler. (Courtesy of Oxford Cryosystems Ltd. [http://www.oxcryo.com/gm-cryocoolers/two-stage-coldheads/].)

14.3.6 Property Changes (Temporary or Permanent)

Reducing the temperature of any material will always change its properties to some extent. Some materials undergo large property changes that can be useful for some applications. For example, the electrical resistivity of pure metals, such as copper or aluminum, decreases by several orders of magnitude when cooled from 300 to 4 K. For commercially pure copper, such as grade 101, the change is about a factor of 100. Such a reduction in resistance can be useful in creating a high magnetic field without using superconducting wires.

This change in resistivity of a metal is a temporary change, and the resistivity will revert back to the original value after the material is warmed back to the starting temperature.

A significant commercial application of cryogenics relies on the transition to a brittle glassy state of certain soft materials. The two primary examples are rubber and spices. At room temperature, these soft materials are difficult to grind and require large amounts of energy. The recycling of rubber tires is accomplished by grinding the tire into small particles about 10 mm in diameter or smaller. Often a fine powder is desired for some applications, and with cryogenic grinding powder, sizes down to 50–300 μm are possible, since the energy required is greatly reduced. There is also less dust generated in the cryogrinding process compared with room-temperature grinding. For safety reasons, a new tire contains less than 10%–15% of recycled rubber. Thus, recycled tires are used mostly in other applications, such as various surfacing materials. Currently, about 13% of scrap tires in the United States are recycled with about 1% being done cryogenically (Shatten, 2004). One interesting approach now being investigated is the exploitation of the "free" cold available at LNG regasification terminals to reduce the total energy required for cryogenic grinding [Shatten, 2004; Shatten et al., 2003].

Another cryogrinding application is that for spices and herbal medicines. India is the world's largest grower and consumer of spices with an annual output of more than two million tons [Jacob et al., 2000]. In conventional grinding of spices, about 99% of the applied energy gets dissipated in the form of heat; temperatures of 363 K have even been reported in some conventional grinding at high speeds [Pesek et al., 1985]. This heating of the spices causes them to lose much of their highly volatile components, such as those contributing to aroma, and as a result, their value suffers. Conversely, cryogrinding prevents overheating, which prevents a loss of oils and moisture. Throughput in a cryogenic grinding pilot plant was increased by a factor of 2.25 compared with ambient grinding [Jacob et al., 2000]. In the cryogrinding of Chinese herbal medicines, one study found that about 1–5 kg of liquid nitrogen was used per kilogram of ground material [Li et al., 1991].

A search on the web under cryogenics brings up many topics and company websites dealing with cryogenic treatment of metals and other materials. This appears to have become a rather large business, impacting much of the general public with examples such as cryotreated razor blades, automobile parts, and machine tools, affording them all extended wear life. Scientific studies have shown that proper cryogenic treatment (slow cooling to liquid nitrogen temperature over a period of a day or more followed by a soak period of a day or so and slow warming) can impart increased wear life to certain tool steels. A recent study in India [Lal et al., 2001] and an earlier study by Barron [Barron, 1982] describe the process and the effect on improved wear life in tool steels. Increased wear life of two to six times have been reported. These articles point out the importance of proper heat treatment prior to the cryotreatment. In essence, the lower temperature brings about a more complete transformation of the austenitic microstructure to the harder martensitic phase. Although hardness of the metal does not change much, its wear life is significantly increased. This effect is being recognized by more manufacturers and adopted in a wide variety of applications. This is a permanent effect and requires only one treatment. Further research in this area is needed to better understand the effects and determine the limitations.

14.3.7 Tissue Ablation (Cryosurgery)

Catheters with diameters of 3 mm or smaller are capable of accessing many internal organs through arteries and veins. They provide the medical community with tools to operate on internal organs without the need to cut through the body to gain access. As a result, the use of catheters greatly speeds recovery times, reduces costs, and often reduces risks. The removal of unwanted tissue, such as cancer tumors or malfunctioning tissue, can be carried out through cryoablation with cryogenic catheters. The effectiveness of these catheters depends very much on enhanced heat transfer at the very small tip. The recent development of small-diameter and flexible cryogenic catheters has opened up the possibility of treating many abnormalities, including cancer, in internal organs without major surgery to gain access to the organ or remove the organ. Instead of a several-day hospital stay and about a 6-week recovery period, the cryosurgical procedure can usually be performed as an outpatient service with local anesthesia that has a one-day recovery period.

Cryosurgery had its beginnings in 1851 when James Arnott used iced saline solutions to treat carcinomas of the breast and cervix [Arnott, 1851; Fraser, 1979]. The minimum temperature of –21°C achievable with this technique is just at the upper limit for cell destruction [Dobak, 1998], so Arnott had little success in curing the cancers, but achieved some beneficial effects, such as pain relief and reduction of bleeding. The use of solid carbon dioxide and liquid nitrogen around 1900 for cryosurgery [Le Pivert, 1984] permitted much lower temperatures and more complete tissue destruction. These lower temperatures were commonly used in dermatological applications after 1900, but were not used much for deep-seated malignancies. One exception was the work of the neurosurgeon Temple Fay between 1936 and 1940 [Fay, 1959]. He used implanted metal capsules connected to an external refrigeration system to treat brain tumors. He also used refrigerated liquids to treat large inoperable cancers of the cervix and breast. Few advances in cryosurgery followed until in 1961 the New York neurosurgeon Dr. Irving Cooper designed and used a cryosurgical probe cooled with liquid nitrogen to treat Parkinson's disease [Cooper and Lee, 1961]. This event is often taken as the beginning of modern cryosurgery, although subsequent developments in medication have continued to replace the cryosurgical approach for this disease.

In the mid-1960s, Gonder and colleagues developed a modified liquid nitrogen cryosurgical system for prostate cryosurgery [Gonder et al., 1964]. Extensive animal experiments were carried out that led to a broader clinical use of cryosurgery. Up to that point, liquid nitrogen was the most common cryogen used in cryosurgery, but it had certain disadvantages. It could not be stored for long periods of time and required bulky supporting equipment. Thus, the equipment could not be easily moved from room to room, and was not easy to use or control. The freezing process could not be stopped quickly because of the time for the liquid to drain from the probe. Vacuum-insulated probes were required to prevent freezing along the length of the probe. Many advances in cryosurgical probes have occurred since 1961, and a much better understanding of tissue destruction at cold temperatures has developed. Advances in cryobiology have shown that cell temperatures of –40°C or lower result in complete destruction of cancerous tissue, although somewhat higher temperatures up to about –20°C may lead to destruction of some healthy tissue [Gage, 1979, 1992]. It has also been found that the lethal cell temperature can be increased by repeated freeze/thaw cycles and by rapid cooling and slow warming [Dobak, 1998]. In the last 40 years, several cooling methods have been developed and used for cryogenic catheters and probes. It was during this period that Joule–Thomson (JT) systems with high-pressure argon and nitrous oxide (N_2O) were introduced. Expansion of these fluids from high to low pressure results in cooling and liquefaction of the fluid. Liquid argon (–186°C), liquid nitrous oxide (N_2O) (–88°C), and various hydroflourocarbons (HFCs) and fluorocarbons (FCs) (–30°C to –80°C) are now used as refrigerants in addition to liquid nitrogen (–196°C). Mixtures of pure fluids have recently been used [Marquardt et al., 1998; Missimer, 1994; Radebaugh, 1996] that allow for a wide range of boiling temperatures between –30°C and –196°C.

The 1980s saw the development of a competing technology, that of electrosurgical or radio frequency (RF) catheters that destroy tissue by heating them to temperatures above about 42°C. These RF catheters are simpler and easy to control. However, they are limited to the amount of tissue they can destroy at one location because the temperature difference between the 42°C destruction temperature and the catheter tip is limited by the 100°C boiling point of the water in the tissue. The RF catheters have been widely accepted by the medical community for the past three decades. The later development of cryogenic catheters has hindered their use in place of the RF catheters.

Beginning in the 1990s, the development of improved cryosurgical probes along with advances in ultrasound and MRI imaging to locate the ice front have resulted in considerable interest within the medical community in the use of cryosurgery for a variety of applications [Rubinsky, 2000]. Early applications were mostly for the treatment of abnormalities near the body surface. However, the recent development of small-diameter and flexible cryogenic catheters has opened up the possibility of treating many abnormalities, including cancer, in internal organs without major surgery to gain access to the organ or remove the organ. Instead of a several-day hospital stay and about a 6-week recovery period, the cryosurgical procedure can usually be performed as an outpatient service with local anesthesia that has a one-day recovery period. For example, a new cryogenic catheter [Dobak et al., 2000] received approval from the US Food and Drug Administration (FDA) in 2001 to treat women with abnormal menstrual bleeding by freezing the uterine lining instead of surgically removing the uterus in a hysterectomy. Another new cryogenic catheter has been developed for the treatment of cardiac arrhythmias, or irregular heart beating [Ryba, 2001]. Over 5 million people in the world's developed countries suffer from some form of heart arrhythmia. Current treatments involve medication or the use of RF catheters. Heart arrhythmia occurs when a portion of the heart distorts its electrical signals. Ablating the tissue that distorts these electrical signals with either RF or cryogenic catheters eliminates the arrhythmia in almost all cases. Though newer than RF catheters, the cryogenic cardiac catheters offer the following advantages: (1) cryomapping capability, (2) not susceptible to forming blood clots, and (3) will stick to the tissue once freezing begins and not be moved out of place by heart movement. The cryogenic cardiac catheters are one of the most challenging cryogenic catheters in terms of refrigeration and heat transfer issues. The catheter is about 3 mm in diameter and about 1 m long. It is inserted through a small incision into a large vein in the leg in the groin area. For treatment of atrial fibrillation, it enters the right atrium of the heart via this vein and continues into the left atrium through a hole punctured previously in the wall between the two chambers. As shown in Figure 14.30, the catheter tip is placed at the junction to the pulmonary vein coming from the lungs, at which time a small balloon is inflated at the tip to temporarily block the blood flow. Finally, refrigerant flow into the balloon is initiated to cause freezing of the necessary tissue around the perimeter of the junction. Recent reviews of the procedure show it to be as successful as RF catheters with some advantages and improved safety [Ozcan et al., 2011].

Further improvements in cryosurgical probes could open up many more medical applications, including the treatment of cancers of the liver, breast, and lung. Successful application of cryosurgical probes is very much dependent on providing a well-controlled cooling rate at the tip with little or no cooling along the length of the probe. To reach further into the body, catheters need to be made smaller and smaller. Large amounts of heat (often tens of watts) need to be removed at the catheter tip in order to freeze a large-enough region at a sufficiently fast rate to provide well-defined regions of tissue destruction. Cryosurgery applications have the potential to expand considerably and become a major medical procedure if the following developments occur: (1) the medical device community has the information

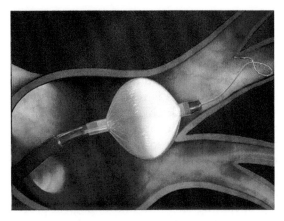

FIGURE 14.30
Cryoballoon catheter for the treatment of atrial fibrillation shown in place at the junction to the pulmonary vein. N_2O refrigerant is then flowed into the balloon for about 5 min to perform the ablation. (Courtesy of Medtronic, Inc.)

needed to help them select the optimum cooling method in a cryosurgical probe for a particular application; (2) a given heat removal rate can be achieved in smaller and smaller cryogenic catheters; and (3) compact, reliable, and easy-to-use catheter systems are readily available that can be applied to various procedures. Recent reviews of cryosurgery have been given by Dobak, [1998], Rubinsky [2000], Theodorescu [2004], and Gage et al. [2009].

14.4 Challenges to the Use of Cryogenic Temperatures

Cryogenic temperatures can provide many benefits, but any particular application will not be commercially successful if the challenges of reaching such temperatures become too great. The use of liquid cryogens has been a common practice in many applications, but the need to periodically replenish the cryogen can become a challenge and ultimately hinder the use of cryogenic temperatures. This is especially true with the case of liquid helium as the price of helium is increasing rapidly and its availability is limited. The use of liquid cryogens is being replaced in many applications with the use of closed-cycle refrigeration systems or cryocoolers [Chapter 6]. However, cryocoolers have their own disadvantages that often present challenges in adapting them for a particular application. These disadvantages were listed at the beginning of this chapter. Research on cryocoolers in the last 50 years has made great strides in minimizing these disadvantages. Space applications in particular have been a strong driver for increasing cryocooler lifetimes from a few hundred hours to as much as 10 years. Space applications have also led to significantly increased efficiencies and reduced sizes of cryocoolers, and commercial applications have been a strong driver to reduce costs. The ultimate goal for all cryogenic applications is to hide the fact that cryogenic temperatures are needed to produce the benefit that it provides. In most cases, the end user cares only about the benefits and does not want the cryogenics to be obvious in terms of problems, such as reliability, power, size, noise, cost, and so forth. Significant reductions of these disadvantages will inevitably lead to more applications. Many advances in cryocoolers have been made in the last 50 years that have had a major impact, and led to many applications of cryogenics that would not have been possible 50 years ago.

14.5 Summary/Conclusions

Operating at cold temperatures is essential to many fields of science and engineering including space exploration, electronics, and medicine. One of the areas that require low-temperature materials and mechanisms is related to planetary exploration of bodies in the solar system that are extremely cold. This includes potential NASA *in situ* exploration missions to Europa and Titan where the ambient temperature is in the range of –200°C. Generally, material properties change at low temperatures and these changes affect the strength, thermal conductivity, ductility, and electrical resistance.

With the increased availability of effective cooling mechanisms and the reduction in their cost, there is a growing interest in technologies that are applicable at low temperatures. As a method of slowing or halting chemical and biological processes, cooling is widely used as a means of preserving food, chemicals, as well as biological tissues and organs. Further, as a method of reducing electrical conductivity, cooling to produce superconductivity enables unique capabilities including levitation, highly efficient electromagnets, and others. The subject of low temperature materials and mechanisms is multidisciplinary and involves various related science and engineering disciplines including chemistry, material science, electrical engineering, mechanical engineering, metallurgy, and physics.

Various methods are applied to reach cryogenic temperatures including the use of boiling cryogens, compression and expansion of gases, adiabatic demagnetization, and fluid mixing. The tools that allow reaching cryogenic temperatures are called cryocoolers and they are mostly based on heat exchange. Applications of cryogenic temperatures include such fields as physics, chemistry, materials science, and biology.

There are many challenges to producing and applying low temperatures and they significantly increase as the temperature drops. The challenges are continually being addressed as the number of applications grows.

Acknowledgments

Some of the research reported in this chapter was conducted at the Jet Propulsion Laboratory (JPL), California Institute of Technology (Caltech), under a contract with the National Aeronautics and Space Administration (NASA). The authors would like to thank Adam M. Swanger, Cryogenics Test Laboratory, NASA Kennedy Space Center, FL; and Ron Ross, Jet Propulsion Lab/Caltech, Pasadena, CA, for reviewing this chapter and for providing valuable technical comments and suggestions.

References

Anderson P.W., Rowell J.M. 1963. Probable observation of the Josephson superconducting tunnel effect. *Phys. Rev. Lett.* 10: 230.

Andrews D.E. 1988. Magnetic Resonance Imaging in 1987. In *Proceedings of the 1987 International Cryogenic Materials Conference (ICMC)*, A. F. Clark and R. P. Reed (Eds.), St. Charles, Illinois, ISBN 0-306-42802-4.

Arnott J. 1851. *On the Treatment of Cancer through the Regulated Application of an Anaesthetic Temperature.* London, England: Churchill.

Baker C.R., Fisher T.F. 1992. Industrial cryogenic engineering in the USA. In: *History and Origins of Cryogenics*, ed. R.G. Scurlock, pp. 217–54. London, NY: Clarendon Press.

Bar-Cohen Y. (ed.). *High Temperature Materials and Mechanisms*, pp. 1–551. ISBN 10: 1466566450, ISBN 13: 9781466566453, Boca Raton, FL: CRC Press, Taylor & Francis.

Bardeen J., Cooper L.N., Schrieffer J.R. 1957. Theory of superconductivity. *Phys. Rev.* 108: 1175.

Barron R.F. 1982. Cryogenic treatment of metals to improve wear resistance. *Cryogenics* 22: 409–13.

Bednorz J.G, Muller K.A. 1986. Possible high Tc superconductivity in the Ba-La-Cu-O system. *Z. Phys. B Condens. Matter.* 64: 189–93.

Benz S.P., Hamilton C. 2004. Application of the Josephson effect to voltage metrology. *Proc. IEEE, Special Issue on Applications of Superconductivity* 92: 1617–29.

Buck D.A. 1956. The cryotron—A superconducting computer component. *Proc. IRE* 44: 482.

Bunge R.G., Sherman J.K. 1953. Fertilizing capacity of frozen spermatoza. *Nature* 172: 767.

Bureau USC. 2004. *Industrial Gas Manufacturing: 2002EC02-311-325120.* Washington, DC: U.S. Census Bureau.

Bureau USC. 2005. *Industrial Gas Production.* Washington, DC: U.S. Census Bureau.

Cantor R., V. Vinetskiy, and A. Matlashov. 1996. A low-noise, integrated dc SQUID magnetometer for applications in biomagnetism. In *Biomag 96: Proceedings of the 10th International Conf. on Biomagnetism*, C. J. Aine, Y. Okada, G. Stroink, S. J. Swithenby, and C. C. Wood (Eds.), Springer Verlag, New York, NY, pp. 15–19.

CCAS 2015. Coalition for the Commercial Application of Superconductivity, http://www.ccas-web.org/superconductivity/medicalimaging/.

Chen C. 1986. Pregnancy after human oocyte cryopreservation. *Lancet* 1: 884.

CLNG, *Center for Liquefied Natural Gas*, http://www.lngfacts.org - website was visited on Feb. 15, 2016

Cooper I.S., Lee A.S. 1961. Cryostatic congelation: A system for producing a limited controlled region of cooling or freezing of biological tissues. *J. Nerv. Ment. Dis.* 133: 259–69.

Dai P., Chakoumakos B.C., Sun G.F., Wong K.W., Xin Y., Lu D.F. 1995. Synthesis and neutron powder diffraction study of the superconductor HgBa2Ca2Cu3O8+ by Tl substitution. *Physica C* 243: 201–6.

Damadian R. 1977. NMR in cancer: XVI. Fonar image of the live human body. *Physiol. Chem. Phys.* 9: 97–100.

Damadian R.V. 1971. Tumor detection by nuclear magnetic resonance. *Science* 171: 1151–3.

Damadian R.V. 1974. US Patent No. 3,789,832.

Delayen JR. 1998. The Jefferson Lab Superconducting Accelerator, in *Adv. Cryogenic Engineering*, U. B. Balachandran, K. T. Hartwig, W. H. Warnes, D. G. Gubser, R. P. Reed, and V. A. Bardos (Eds.) vol. 43, pp. 37–42, New York: Plenum Press.

Delikaris D., Tavian L. 2014. The LHC cryogenic system and operational experience from the first three years run. *J. Cryo. Super. Soc. Jpn.* 49: 590–600.

Devred A., Desportes H., Kircher F., Lesmond C., Meuris C., et al. 1998. Superconducting magnet technology. In: *Handbook of Cryogenic Engineering*, J.G. Weisend II (Ed.), pp. 321–64. Philadelphia, PA: Taylor & Francis.

Dobak J. 1998. A review of cryobiology and cryosurgery. In: *Advances in Cryogenic Engineering*, Volume 43 of the series Advances in Cryogenic Engineering, Q.-S. Shu (Ed), Springer, New York, NY, pp 889–896.

Dobak J., Ryba E., Kovalcheck S. 2000. A new closed-loop cryosurgical device for endometrial ablation. *J. Am. Assoc. Gynecol. Laparosc.* 7: 245–9.

DOE/EIA. 2014. *Annual Energy Outlook 2014, DOE/EIA-0383 (2014).* Washington, DC: Department of Energy/Energy Information Administration.

Drung D., Bechstein S., Franke K-P., Schreiner M., Schurig T. 2001. Improved direct-coupled DC SQUID read-out electronics with automatic bias voltage tuning. *IEEE Trans. Appl. Superconductivity* 11: 880–3.

Drung D., Ludwig F., Muller W., Steinhoff U., Trahms L., Koch H., Shen Y.Q., Jensen M.B., Vase P., Holst T., Freltoft T., Curio G. 1996. Integrated YBCO magnetometer for biomagnetic measurements. *Appl. Phys. Lett.* 68: 1421–3.

Dunmire H. 1998. *U.S. Army cryocooler status update.* Presented at Second Workshop on Military and Commercial Applications for Low Cost Cryocoolers (MCALCII), San Diego, CA.

DVB SA. 2003. LNG Tanker Market Report, DVB Research & Strategic Planning, Rotterdam 101: 129–34.

Edwards H.T. 1985. The tevatron energy doubler: A superconducting accelerator. *Annu. Rev. Nucl. Part. Sci.* 35: 605–60.

EPA 2008. *Technical Support Document for Hydrogen Production: Proposed Rule for Mandatory Reporting of Greenhouse Gases.* Washington, DC: Office of Air and Radiation, U.S. Environmental Protection Agency.

Fay T. 1959. Early experiences with local and generalized refrigeration of the human brain. *J. Neurosurg.* 16: 239–59.

Foote R.H. 2001. The history of artificial insemination: Selected notes and notables. *Am. Soc. Anim. Sci.* 2002: 1–10.

Fraser J. 1979. Cryogenic techniques in surgery. *Cryogenics* 19: 375–81.

Freiman A., Bouganim N. 2005. History of cryotherapy. *Dermatol. Online J.* 11, Article #9.

Gage A.A. 1979. What temperature is lethal for cells. *J. Derm. Surg. Oncol.* 464: 459–60.

Gage A.A. 1992. Cryosurgery in the treatment of cancer. *Surg. Gynecol. Obstet.* 174: 73–92.

Gage A.A., Baust J.M., Baust J.G. 2009. Experimental cryosurgery investigations in vivo. *Cryobiology* 59: 229–43.

Global Industy Analyst. 2015, http://www.prweb.com/releases/superconductors/super_conductors/prweb12176943.htm.

Goldsmith M., Damadian R., Stanford M., Lipkowitz M. 1977. NMR in cancer: XVIII. A superconductive NMR magnet for a human sample. *Physiol. Chem. Phys.* 9: 105–7.

Gonder M., Soannes W., Smith V. 1964. Experimental prostate cryosurgery. *Invest. Urol.* 1: 610–9.

Gourlay S.A., Sabbi G., Kircher F., Martovetsky N., Ketchen D. 2004. Superconducting magnets and their applications. *Proc. IEEE on Special Issue on Applications of Superconductivity* 92: 1675–87.

Grenier M., Petit P. 1986. Cryogenic air separation: The last twenty years. In: *Advances in Cryogenic Engineering*, pp. 1063–70. New York, NY: Plenum Press.

Grossman H.L., Myers W.R., Vreeland V.J., Bruehl R., Alper M.D., Bertozzi C.R., Clarke J. 2004. Detection of bacteria in suspension using a superconducting quantum interference device. *Proc. Nat. Acad. Sci.* 101: 129–34.

Haselden G.G. 1992. The history of liquefied natural gas (LNG). In: *History and Origins of Cryogenics*, ed. R.G. Scurlock, pp. 599–619. Clarendon Press. Oxford, UK.

Hayakawa H., Yoshikawa N., Yorozu S., Fujimaki A. 2004. Superconducting digital electronics. *Proc. IEEE on Special Issue on Applications of Superconductivity* 92: 1549–63.

Herr Q.P. 2011. Ultra-low power superconducting logic. *J. Appl. Phys.* 109: 103903.

Hulm J. K., and R. D. Blaugher. 1961. Superconducting solid solution alloys of the transition elements. *Phys. Rev. Lett.* 123: 1569–80.

IUVSTA. 2015. International Union for Vacuum Science, Technique, and Applications, http://www.iuvsta.org/iuvsta2/index.php?id=643.

Jacob S., Kasthurirengan S., Karunanithi R., Behera U. 2000. Development of pilot plant for cryogrinding of spices: A method for quality improvement. In: *Advances in Cryogenic Engineering*, pp. 1731–8. New York, NY: Plenum Publishers.

Jaklevic R.C., Lambe J., Silver A.H., Mercereau J.E. 1964. Quantum interference effects in Josephson tunneling. *Phys. Rev. Lett.* 12: 159–60.

Josephson B.D. 1962. Possible new effects in superconducting tunneling. *Phys. Lett.* 1: 251–3.

Kalinine V., Haange R., Shatil N., Millet F., Jager B., et al. 2004. Design and operating features of the ITER 4.5 K cryoplant. In: *Advances in Cryogenic Engineering*, pp. 176–83. College Park, MD: American Institute of Physics.

Kalsi S.S., Weeber K., Takesue H., Lewis C., Neumueller H.W., Blaugher R.D. 2004. Development status of rotating machines employing superconducting field windings. *Proc. IEEE on Special Issue on Applications of Superconductivity* 92: 1688–704.

Kato T., Kubota M., Kobayashi N., Suzuoki Y. 2003. Effective utilization of by-product oxygen of electrolysis hydrogen production. Presented at *International Energy Workshop (IEW)*, Laxenburg, Austria.

Kleiner R., Koelle D., Ludwid F., Clarke J. 2004. Superconducting quantum interference devices: State of the art and applications. *Proc. IEEE on Special Issue on Applications of Superconductivity* 92: 1534–48.

Kumar A., Welti D., Ernst R.R. 1975. NMR Fourier zeugmatography. *J. Magn. Res.* 18: 69–83.

Kunzler J.E., Buehler E., Hsu F.S.L., Wernick J.H. 1961. *Phys. Rev. Lett.* 6: 89.

Kuriyama T. 2003. Private communication, ed. R Radebaugh.

Kuriyama T. 2005. Private communication, ed. R Radebaugh.

Lal D.M., Renganarayanan S., Kalanidhi A. 2001. Cryogenic treatment to augment wear resistance of tool and die steels. *Cryogenics* 41: 149–55.

Lauterbur P.C. 1973. Image formation by induced local interactions: examples employing nuclear magnetic resonance. *Nature* 242: 190–1.

Le Pivert P. 1984. Cryosurgery: Current issues and future trends, *10th International Cryogenic Engineerig Conf.*, Butterworth, Surrey, 551–7.

Li S., Ge S., Huang Z., Wang Q., Zhao H., Pan H. 1991. Cryogenic grinding technology for traditional Chinese herbal medicine. *Cryogenics* 31: 136–7.

Maeda H., Tanaka T., Fukutomi M., Asano T. 1988. A new superconductor without a rare earth element. *J. Appl. Phys.* 27: L209.

Magnetic Resonance. 2015. Chapter 21–02, www.magnetic-resonance.org.

Malozemoff A.P. 2015. AMSC, personal e-mail communication with the coauthor, Y. Bar-Cohen.

Malozemoff A.P., Mannhart J., Scalapino D. 2005. High-temperature cuprate superconductors get to work. *Phys. Today:* 41–7.

Mansfield P. 1977. Multi-planar image formation using NMR spin-echos. *J. Phys. C: Solid State Phys.* 10: L55–L58.

Mansfield P., Maudsley A.A. 1976. Planar spin imaging by NMR. *J. Phys. C: Solid State Phys.* 9: L409–11.

Marquardt E.D., Radebaugh R. 2000. Pulse tube oxygen liquefier. In: *Advances in Cryogenic Engineering*, pp. 457–64. Plenum Press, New York, NY.

Marquardt E.D., Radebaugh R., Dobak J.A. 1998. Cryogenic catheter for treating heart arrhythmia. In: *Advances in Cryogenic Engineering*, pp. 903–10.

Matthias B.T., Geballe T.H., Geller S., Corenzwit E. 1954. Superconductivity of Nb3Sn. *Phys. Rev.* 95: 1435.

Missimer D.J. 1994. *Tenth Intersociety Cryogenic Symposium (AIChE Spring National Meeting)*, Houston, Texas.

Mukhanov O. 2011. Energy-efficient single flux quantum technology, *IEEE Trans. Appl. Superconductivity* 21: 760–9.

Mukhanov O., Semenov V., Likharev K. 1987. Ultimate performance of the RSFQ logic circuits. *IEEE Trans. Magn.* vol. MAG-23: 759–62.

Muller WH-G., Hopfel D. 1998. Magnetic resonance imaging and spectroscopy (medical applications), In: *Handbook of Applied Superconductivity*, pp. 1213–48. Institute of Physics Publishing, Bristol, GB.

National Security Agency. https://www.nsa.gov/public_info/prepub/ website last visited on Feb. 15, 2016.

Notardonato W.U. 2014. Development of a ground operations demonstration unit for liquid hydrogen at Kennedy Space Center, *25th International Cryogenic Engineering Conference and the International Cryogenic Materials Conference*. Amsterdam, the Netherlands: Elsevier B.V.

ORNL. 2005. *Materials for Separation Technologies: Energy and Emission Reduction Opportunities.* Washington, DC: Department of Energy.

Oshima K., Aiyama Y. 1992. The development of cryogenics in Japan. In: *History and Origins of Cryogenics*, ed. R.G. Scurlock, pp. 520–46. Clarendon Press, Oxford, GB.

Ozcan C., Ruskin J., Mansour M. 2011. Cryoballoon catheter ablation in atrial fibrillation. *Cardiol. Res. Proact.* 2011: Article ID 256347, 6 pages http://dx.doi.org/10.4061/2011/256347

Pesek C.A., Wilson L.A., Hammond E.G. 1985. Spice quality: Effect of cryogenic and ambient grinding on volitiles. *J. Food Sci.* 50: 559.

Quinn T.J. 1989. News from the BIPM. *Metrologia* 26: 69–74.

Radebaugh R. 1967. Thermodynamic properties of ^{3}He-^{4}He mixtures with applications to the ^{3}He-^{4}He Dilution Refrigerator, NBS Tech. Note 362.

Radebaugh R. 1996. *Proc. 19th International Congress of Refrigeration*, 973–89. The Hague, the Netherlands.

Railway J. 2004. Overview of Maglev R&D. Railway Technical Research Institute, Japan Railway, http://www.rtri.or.jp/eng/rd/seika/2004/index_E.html.

Royal J. 2005. Private communication with the lead author, R. Radebaugh.

Rubinsky B. 2000. Cryosurgery. *Annu. Rev. Biomed. Eng.* 2: 157–87.

Ryba E. 2001. Private communication, R. Radebaugh.

Scanlan R.M., Malozemoff A.P., Larbalestier D.C. 2004. Superconducting materials for large scale applications. *Proc. IEEE on Special Issue on Applications of Superconductivity* 92: 1639–54.

Schilling D.R. 2014. World's Largest Ship Ever Built and First Floating Liquefied Natural Gas (FLNG) Platform to Begin Drilling in 2017, Industry Tap, October 16, 2014.

Schwall R.E. 1987. MRI—Superconductivity in the marketplace. *IEEE Trans. on Superconductivity*, MAG-23: 1287–93.

Scurlock, R.G. 1992. *History and Origins of Cryogenics*, pp. 28–9. Clarendon Press. Oxford, UK.

Shapiro S. 1963. Josephson currents in superconducting tunnelling: The effect of microwaves and other observations. *Phys. Rev. Lett.* 11: 80–2.

Shatten, R.A. 2004. US patent issued for Super Cool's innovative energy conservation and cryogenic recycling method, In: *Cold Facts (Cryogenic Society of America)*, Winter 20: 34.

Shatten R.A., Carrier J., Jackson J.D. 2003. System and method for cryogenic cooling using liquefied natural gas.0 US Patent No. 6,668,562. USA.

Shimomura Y., R. Aymar, V.A. Chuyanov, M. Huguet, H. Matsumoto, T. Mizoguchi, Y. Murakami, A.R. Polevoi, and M. Shimada. 2001. ITER-FEAT operation. *Nucl Fusion*, 41(3), 309–316.

Silver A.H., Zimmerman J.E. 1967. Quantum states and transitions in weakly connected superconducting rings. *Phys. Rev.* 157: 317–41.

Simon R.W., Hammond R.B., Berkowitz S.J., Willemsen B.A. 2004. Superconducting microwave filter systems for cellular telephone base stations. *Proc. IEEE on Special Issue on Applications of Superconductivity* 92: 1585–96.

Sloop J.L. 1978. Liquid Hydrogen as a Propulsion Fuel, 1945–1959, NASA SP 4404. The NASA History Series, NASA, Washington, D.C.

Theodorescu D. 2004. Cancer cryotherapy: Evolution and biology. *Rev. Urol.* 6, suppl. 4: S9–S19.

Tishler A.O. 1966. The impact of the space age on cryogenics. In: *Advances in Cryogenic Engineering*, pp. 1–10. New York, NY: Plenum Press.

Trill R. 2002. Hydrogen as alternative fuel. Presented at *BMW Clean Energy Seminar*. Sacramento, CA.

Trounson A., Mohr L. 1983. Human pregnancy following cryopreservation, thawing, and transfer of an eight cell embryo. *Nature* 305: 707.

White N., Tananbaum H. 1999. The constellation X-ray mission. *Astrophys. Lett. Comm.* 39: 933–6.

Wollman D.A., Hilton G.C., Irwin K.D., Bergren N.F., Rudman D.A., Newbury D.E., Martinis J.M. 1999. Cryogenic microcalorimeters for X-ray microanalysis. In: *Proc. 1999 NCSL Workshop and Symp. (Nat. Conf. Standards Laboratories)*, pp. 811–9.

World Steel 2014. World Steel in Figures, 2014 Edition, World Steel Association, https://www.worldsteel.org/dms/internetDocumentList/bookshop/World-Steel-in-Figures-2014/document/World%20Steel%20in%20Figures%202014%20Final.pdf last time visited Feb. 15, 2016.

Wu M.K., Ashburn J.R., Torng C.J., Hor P.H., Meng R.L., et al. 1987. Superconductivity at 93 K in a new mixed-phase Y-Ba-Cu-O compound system at ambient pressure. *Phys. Rev. Lett.* 58: 908.

Zmuidzinas J., Richards P.L. 2004. Superconducting detectors and mixers for millimeter and submillimeter astrophysics. *Proc. IEEE on Special Issue on Applications of Superconductivity* 92: 1597–616.

Internet References for further reading

NIST. Cryogenic Technologies Group—http://cryogenics.nist.gov.

NIST Cryogenic Material Properties Database—http://cryogenics.nist.gov/MPropsMAY/material-properties.htm.

Cryogenic Society of America, Inc., - http://www.cryogenicsociety.org/

Cryocooler—http://en.wikipedia.org/wiki/Cryocooler.

Superconductivity—http://www.superconductors.org/Uses.htm; http://www.superconductors.org/index.htm.

www.eia.doe.gov.

www.tylermedicalclinic.com. 2005.

15

Safety Considerations When Working with Cryogenic Liquids

Mimi Ton

CONTENTS

15.1 Introduction

Working with cryogenic fluids presents health and safety hazards that have to be addressed. The hazards include exposure to ultra-cold temperatures, flammability, oxygen deficiency/asphyxiation, and high-pressure gas resulting in over-pressurization of containers. To work safely with cryogenic liquids, a number of factors need to be considered. This chapter provides a general discussion of the properties of cryogenic liquids, storage containers, potential hazards, and possible safety controls that should be considered when working with cryogenic materials in a research laboratory-type environment.

Requirements for specific operations and types of cryogenic liquids and emergency procedures may be found in the Compressed Gas Association's Pamphet-12 titled "Safe Handling of Cryogenic Liquids." In addition, in developing appropriate safety measures for a planned cryogenic liquid operation, one may need to consult with his/her workplace safety professionals, pressure vessel engineers, and local regulatory agencies for assistance.

15.2 Properties of Cryogenic Liquids

15.2.1 Extremely Low Temperature

As described in Chapters 1 and 6, cryogenic liquids are gases that are liquid at extremely cold temperatures. Their normal boiling point is below −130°F (−90°C) (Compressed Gas Association Pamphlet-12, 2000).

15.2.2 High Liquid-to-Gas Expansion Ratios

All cryogenic liquids produce large volumes of gas when vaporized. For example, 1 vol of liquid nitrogen at its boiling temperature at 1 atm vaporizes to approximately 700 vol of nitrogen gas when warmed to room temperature at 1 atm. Other examples of cryogenic liquid properties compared to water are given in Table 15.1.

15.2.3 Low Critical Temperature

The critical temperature is the highest temperature at which a material changes from a gas to a liquid regardless of the pressure. Cryogenic liquids, namely nitrogen and helium, have very low critical temperatures, 126.3 K (−232°F) for nitrogen and 5.2 K (−450°F) for helium. Due to the extremely low critical temperatures, when cryogenic liquids are heated (i.e., exposed to room temperature), they turn very rapidly into gas. If confined inside a container, the pressure from vaporization of the liquid can be quite high.

TABLE 15.1

Cryogenic Liquid Properties Compared with Water

Substance	Boiling Point (°C)	Critical Temperature (°C)	Volume Expansion Ratio	Gas Type
Water	100	374	1673:1	
Oxygen	−183.0	−119	860:1	Oxidizer
Argon	−185.7	−122	847:1	Inert
Fluorine	−187.0	−129	888:1	Reactive
Nitrogen	−195.8	−147	696:1	Inert
Hydrogen	−252.7	−240	851:1	Flammable
Helium	−269.0	−268	757:1	Inert

15.3 Storage Containers

15.3.1 Dewars

Unpressurized, vacuum-jacketed containers designed to hold cryogens are called dewars. A diagram of a typical dewar can be seen in Figure 15.1. They have a loose-fitting dust cap over the outlet of the neck tube that prevents atmospheric moisture from plugging the neck and allows gas produced from the vaporized liquid to escape (*Safetygram-27: Cryogenic Liquid Containers*, Air Products and Chemicals, Inc., Allentown, PA, 2000). Dewars are typically used for relatively small quantity transfer of cryogens from a cryogenic liquid cylinder or storage tank to a piece of equipment, or is used to store materials submerged in a cryogenic liquid.

15.3.2 Cryogenic Liquid Cylinders

Cryogenic liquid cylinders are typically used for storing and transporting of cryogenic liquids. Often incorrectly referred to as dewars, cryogenic liquid cylinders are insulated, vacuum-jacketed, pressure vessels that are equipped with safety relief valves and rupture disks to protect the cylinder from over-pressurization. These containers can operate at pressures up to 350 psig and have typical capacities of about 180 L. Figure 15.2 shows the design of a typical cryogenic liquid cylinder.

The product that is contained in the cryogenic liquid cylinders may be withdrawn as a gas by passing liquid through an internal vaporizer or as a liquid under its own vapor pressure. For more details on the construction and operation of cryogenic liquid cylinders, one can consult Air Products' *Safetygram-27: Cryogenic Liquid Containers*.

15.3.3 Cryogenic Storage Tanks

Cryogenic storage tanks are typically installed at a site requiring large quantities of cryogenic liquid. They may typically hold from 500 to 420,000 gallons. Such cryogenics storage systems usually consist of a tank, vaporizer, and pressure control manifold.

(a)

(b)

FIGURE 15.1
Typical dewar diagram (a) and photo (b). (Diagram is courtesy of Air Products.)

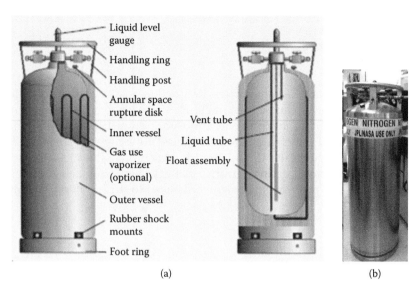

(a)   (b)

FIGURE 15.2
Cryogenic liquid cylinder diagram (a) and photo (b). (Diagram is courtesy of Air Products.)

15.4 Potential Hazards of Cryogens

15.4.1 Temperature

Cryogenic liquids and their vapors can rapidly freeze human tissue and cause many common materials such as carbon steel, rubber, and plastics to become brittle or even break under stress. Cryogenic liquids in containers and piping having temperatures at or below the boiling point of liquefied air (–194°C [–318°F]) can actually condense the surrounding air and cause a localized oxygen-enriched atmosphere or form ice.

Ice plugs may form if the top cap, exhaust port, or gas valve of a cryogenic vessel is left open where moisture can be drawn into the neck of the vessel. When the moisture turns into ice, a potentially explosive situation occurs as the cryogenic liquid in the vessel warms and builds up pressure.

Similar to heat burns, exposure of the skin to cryogens can result in local freezing, tearing, or removal of skin. Due to the decrease in feeling, unattended burns can lead to frost bite. Prolonged exposure of cold vapor or gas can damage lungs and the eyes. Due to a cryogen's low viscous nature, it will penetrate woven and other porous clothing faster than water, so special protection is required at all times when transferring or working with cryogenic liquids (refer to Section 15.5.3).

Personnel training is critical to personnel safety. An organized, well-thought-out training program will significantly reduce personnel injuries in this area.

15.4.2 Pressure

A cryogenic liquid cannot be indefinitely maintained as a liquid even in well-insulated containers. If these liquids are vaporized in a sealed container, they can produce enormous pressures that could rupture the container. Injuries or damages may occur as a result of the pressure force or the projectiles from the damaged container or the propulsion of nearby items. For this reason, pressurized cryogenic containers are normally protected with multiple devices for over-pressure prevention.

Portions of an operation where over-pressurization can occur include

- Pressurized liquid cylinders
- Tubing and hoses used to transfer the cryogen
- Experimental volume, even if the cryogen is only in contact with the exterior
- Bath space surrounding experimental volume
- Vacuum spaces in contact with the cryogen
- Lines between two valves

15.4.3 Oxygen Deficiency/Asphyxiation

The high liquid to gas expansion ratio has the ability to displace oxygen in the atmosphere. Normal environment consists of about 21% oxygen. Oxygen levels less than 19.5% are considered oxygen deficient. Most cryogens are considered simple asphyxiants with no good warning properties. Most cryogens are odorless and colorless making it difficult to identify an oxygen-deficient environment. The effects of oxygen-deficient exposure are listed in Table 15.2.

15.4.4 Physical

Cryogenic liquid cylinders and dewars are large, heavy, and awkward in shape. Depending on the cryogenic liquid, typical cryogenic liquid cylinders weigh approximately 118 kg (260 lb.) when empty and at least 254 kg (560 lb.) when full. When having to move these containers, it is necessary to use mechanical lifting devices, and the container always needs to be pushed, rather than be pulled [CGA P-12, 2009]. Illustrations of how to safely handle cryogenic containers are shown in Figure 15.3.

TABLE 15.2

Effects of Oxygen-Deficient Exposure

Oxygen Concentration (% vol)	Health Effects of Persons at Rest
19	Some adverse physiological effects occur, but they may not be noticeable.
15–19	Impaired thinking and attention. Increased pulse and breathing rate. Reduced coordination. Decreased ability to work strenuously. Reduced physical and intellectual performance without awareness.
12–15	Poor judgment. Faulty coordination. Abnormal fatigue upon exertion. Emotional upset.
10–12	Very poor judgment and coordination. Impaired respiration that may cause permanent heart damage. Possibility of fainting within a few minutes without warning. Nausea and vomiting.
<10	Inability to move. Fainting almost immediate. Loss of consciousness. Convulsions. Death.

Source: Courtesy of Air Products. Reference: *Safetygram-17: Dangers of oxygen-deficient atmospheres*, Air Products, 2014.

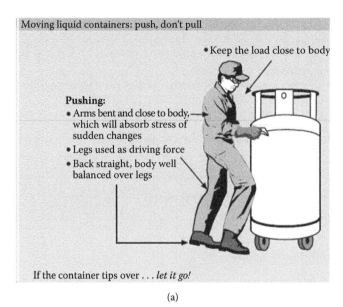

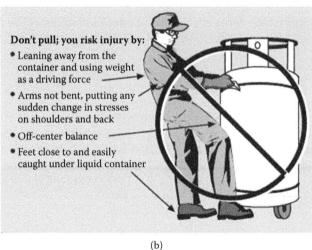

FIGURE 15.3
(a) and (b) Safe handling of cryogenic containers. (Courtesy of Air Products.)

Special four-wheeled carts specially designed for the movement of cryogenic dewars should be used at all times when wheeled dewars are not available.

15.4.5 Chemical

It is important to consider the unique chemical properties of individual cryogenic liquids. Hydrogen is highly flammable and can cause hydrogen embrittlement. Oxygen is an oxidizer, and fluorine is highly reactive. Additional safety measures may need to be implemented beyond the usual cryogenic liquid hazards and they are described in CGA P-12 [2009].

15.5 Safety Control Measures

15.5.1 Engineering Controls

Engineering controls involve the design of a work environment or the job itself to eliminate hazards or reduce potential sources of exposure to cryogens. This is the first and best strategy, since it controls the hazard at its source. Table 15.3 provides a list of possible engineering controls that could be used in a cryogenic liquid operation.

15.5.2 Administrative Controls/Safe Work Practices

Operation/specific rules or established safe work practices are examples of administrative controls. These are typically used in conjunction with other controls. Table 15.4 provides a list of some administrative controls that could be implemented for cryogenic liquid operations.

15.5.3 Personal Protective Equipment

Generally, cryogenic exposure is attributable to the wearing of inadequate personal protective equipment (PPE). Use of PPE is needed when exposure to hazards cannot be engineered completely out of normal operations or maintenance work or administrative controls cannot provide sufficient additional protection. Examples of PPE that could be used for cryogenic liquid operations are listed in Table 15.5.

TABLE 15.3

Sample Engineering Controls

Engineering Control	Hazard Type	Comments
General ventilation	Oxygen deficiency	Needs to ensure the presence of sufficient air flow, including the use of additional fans/open doors. Enclosed spaces such as cold rooms or enclosed chambers should be avoided.
Local exhaust/chemical fume hood	Oxygen deficiency Temperature Pressure	Isolates and exhausts cryogenic liquid vapors.
Cryogen and chemical compatible materials/equipment (including dewars and cryogenic liquid cylinders)	Temperature Pressure	Prevents leaks and failure due to embrittlement.
Pressure relief devices/burst disks	Pressure	Prevents cryogenic liquid containing equipment or cryogenic liquid cooled equipment from over-pressurizing.
Grounding/bonding of equipment	Chemical	Recommended when working with flammable cryogens.
Mechanical lifting devices (i.e., cylinder dolly, crane)	Physical	Minimizes bodily injury due to use of excessive force to move a heavy object.
ASME-certified pressure manifold system	Pressure	Ensures that all parts of a pressurized system are appropriately designed to operate safely.
Vacuum jacketed piping/hoses	Pressure Temperature	Prevents cold contact burns and potential of condensation of moisture at low temperatures.

TABLE 15.4

Sample Administrative Controls

Administrative Controls	Hazard Type	Comments
Review of Safety Data Sheet (SDS)	Oxygen deficiency Chemical Physical	Document provided by the vendor/manufacturer that includes an overview of the specific product safety hazards.
Oxygen deficiency sensors	Oxygen deficiency	Depending on the quantity, location, and planned use, a local or an area sensor may be needed. Sensors should be regularly calibrated. Cryogen vapor characteristics should be considered when installing sensors.
Warning signage and labeling	Oxygen deficiency Chemical	Signage and labeling should clearly identify what cryogens are in use and if there are any special precautions.
Routine inspections	Temperature Pressure Oxygen deficiency	Should be done pre and post operation and at a regular frequency to ensure that equipment/operations are within acceptable parameters.
Standard operating procedure (SOP)	Temperature Pressure Oxygen deficiency Physical Chemical	Cryogenic liquid operations should have detailed SOPs that incorporate safety measures and be communicated to those involved in the operation.
Training	Temperature Pressure Oxygen deficiency Physical Chemical	Personnel new to the operation should attend appropriate safety training and receive on-the-job training with experienced individuals before commencing work.

TABLE 15.5

Sample Personal Protective Equipment

Affected Body Part	PPE	Comments
Hand	Cryogenic insulated gloves	Gloves should be highly insulated, liquid penetration resistant, and cryogen temperature embrittlement resistant, and should be able to be tossed off readily if they become soaked with cryogens. When handling liquid oxygen, gloves must be flame resistant, show resistance to static charge, and have a high limiting oxygen index (LOI) value.
Eye/Face	Safety glasses/ goggles/full face shield	Full face shields together with goggles shall be used: • When a cryogen is poured. • For open transfer of cryogens. • If fluid in an open container is likely to bubble.
Body	Long sleeves/ apron, or lab coat Noncuffed long pants	Preferred ensemble is long-sleeved clothing made of nonabsorbent material with trousers worn outside of boots. An insulated apron should be used when handling large quantities of cryogens. If handling liquid oxygen, clothing must be flame resistant, static charge resistant, and have a high LOI value.
Feet	Closed-toe shoes	Closed-toe shoes covering the top of the foot or boots (extend trousers over the boot).
Ear	Ear plugs or ear muffs	Ear plugs or ear muffs need to be used when excessive noise levels may occur near filling and venting operations.

15.6 Safe Handling Precautions/Prework Planning

Before using cryogenic liquids, a thorough risk assessment has to be performed. Factors that should be considered during this evaluation may include

- Quantity
- Intended use
- System/equipment compatibility
- Location of dewars/cylinders and piping
- Location of oxygen monitor sensors versus source/cryogen
- Location of other personnel who may be affected by cryogen work
- Potential hazards from oxygen deficiency, skin contact, embrittlement, pressure, and flammability
- Consequences of spills and system leaks

In addition, industry standards and local, state, and federal regulations may also dictate additional requirements that would need to be incorporated. Examples of relevant industry and regulatory agencies may include the federal Occupational Safety and Health Administration (OSHA), American Society of Mechanical Engineers (ASME), Compressed Gas Association (CGA), and National Fire Protection Association (NFPA), to name a few.

Acknowledgments

Some of the research reported in this chapter was conducted at the Jet Propulsion Laboratory (JPL), California Institute of Technology, under a contract with the National Aeronautics and Space Administration (NASA).

The author would like to express her appreciation for Michael C. Kumpf, University of California, Berkeley, CA; Philip F. Simon, University of California, Berkeley, CA; Ron Welch, JPL/Caltech, Pasadena, CA; and Robert Boyle, NASA—Goddard Space Flight Center (GSFC), Greenbelt, MD, for reviewing this chapter and for providing valuable technical comments and suggestions.

References

American Society of Heating. *Refrigerating and Air Conditioning (ASHRAE)*, ASHRAE Handbook: Refrigeration. SI Edition, 2014.

American Society of Mechanical Engineers (ASME). *Boiler and Pressure Vessel Code: Section VIII: Division 1: Rules for Construction of Pressure Vessels*, ASME BPVC, 2013.

Compressed Gas Association (CGA). *Safe Handling of Cryogenic Liquids*. 5th Edition, CGA, p. 12, 2009.

Edeskuty, F.J., and W.F. Stewart. *Safety in the Handling of Cryogenic Fluids*. New York, NY: Plenum Press, 1996.

National Fire Protection Association (NFPA). *Compressed Gases and Cryogenic Fluids Code*. NFPA 55, 2013.

NFPA. Chapter 63: Compressed Gases and Cryogenic Fluids. *Fire Code*. NFPA 1, 2015.

Safetygram-16. *Safe Handling of Cryogenic Liquids*. Allentown, PA: Air Products and Chemicals, Inc., 2010.

Safetygram-17. *Dangers of Oxygen-Deficient Atmospheres*. Allentown, PA: Air Products and Chemicals, Inc., 2014.

Safetygram-27. *Cryogenic Liquid Containers*. Allentown, PA: Air Products and Chemicals, Inc., 2000.

Index

Printed and bound by CPI Group (UK) Ltd, Croydon, CR0 4YY

01/11/2024

01782603-0010